# Real and Complex Singularities

CONTEMPORARY MATHEMATICS

675

# Real and Complex Singularities

XIII International Workshop
Real and Complex Singularities
July 27–August 8, 2014
Universidade de São Paulo, Campus São Carlos,
São Carlos, Brazil

Ana Claudia Nabarro
Juan J. Nuño-Ballesteros
Raúl Oset Sinha
Maria Aparecida Soares Ruas
Editors

American Mathematical Society
Real Sociedad Matemática Española

American Mathematical Society
Providence, Rhode Island

2010 *Mathematics Subject Classification*. Primary 58Kxx, 32Sxx, 53Axx, 57Rxx, 14Pxx, 37Dxx.

**Library of Congress Cataloging-in-Publication Data**

Names: International Workshop on Real and Complex Singularities (13th : 2014 : Universidade de São Paulo) — Nabarro, Ana Claudia, 1972- editor. — Real Sociedad Matemática Española.
Title: Real and complex singularities : XIII International Workshop on Real and Complex Singularities, July 27-August 8, 2014, Universidade de São Paulo, São Carlos, SP, Brazil / Ana Claudia Nabarro [and four others], editors.
Description: Providence, Rhode Island : American Mathematical Society, 2016. — Series: Contemporary mathematics ; volume 675 — "Real Sociedad Matemática Española." — Includes bibliographical references.
Identifiers: LCCN 2016010844 — ISBN 9781470422059 (alk. paper)
Subjects: LCSH: Singularities (Mathematics)–Congresses. — AMS: Global analysis, analysis on manifolds – Theory of singularities and catastrophe theory – Theory of singularities and catastrophe theory. msc — Several complex variables and analytic spaces – Singularities – Singularities. msc — Differential geometry – Classical differential geometry – Classical differential geometry. msc — Manifolds and cell complexes – Differential topology – Differential topology. msc — Algebraic geometry – Real algebraic and real analytic geometry – Real algebraic and real analytic geometry. msc — Dynamical systems and ergodic theory – Dynamical systems with hyperbolic behavior – Dynamical systems with hyperbolic behavior. msc
Classification: LCC QA614.58 .I527 2014 — DDC 514/.746–dc23
LC record available at http://lccn.loc.gov/2016010844
Contemporary Mathematics ISSN: 0271-4132 (print); ISSN: 1098-3627 (online)
DOI: http://dx.doi.org/10.1090/conm/675



∞ The paper used in this book is acid-free and falls within the guidelines
established to ensure permanence and durability.
Visit the AMS home page at http://www.ams.org/

10 9 8 7 6 5 4 3 2 1    21 20 19 18 17 16

*This volume is dedicated to Carmen Romero Fuster*
*on the occasion of her 60th birthday*

# Contents

# Preface

This volume contains the proceedings of the XIII International Workshop on Real and Complex Singularities, held in honour of María del Carmen Romero Fuster in celebration of her sixtieth birthday. The meeting was part of a highly successful series of biennial conferences organized by the Singularity Theory group at São Carlos, Universidade de São Paulo, Brazil. Its main theme was the singularity theory of spaces and maps, and the applications of singularity theory to differential geometry.

The 2014 workshop was held in two weeks. In the first week, 17 plenary and 47 parallel talks were presented by specialists in their fields and 27 posters were presented by young researchers. In addition, one mini-course on a current topic of research was also delivered by Jean-Paul Brasselet. The program of the second week consisted of 4 mini-courses, 2 public lectures, 10 short talks and 5 posters. The mini-courses were taught by Marcelo Hernandes, Shyuichi Izumiya, David Massey, and José Luis Cisneros and José Seade, the public lectures were delivered by Stanislaw Janeczko and Marcio Gomes Soares.

The meeting was attended by 183 participants from 25 countries (Algeria, Brazil, Bulgaria, China, Colombia, Denmark, England, EUA, France, Germany, Greece, Hungary, Iceland, Iran, Iraq, Japan, Mexico, Nigeria, Peru, Poland, Russia, Spain, Sweden, Turkey, Vietnam).

The high level of the book reflects the atmosphere of the conference, where the most recent and most important results in singularity theory and its applications were presented. The volume contains the notes of the mini-courses taught by J.-P. Brasselet on intersection homology and by D. Massey on non-isolated hypersurface singularities and Lê cycles. The remaining contributions are research articles which cover topics from the foundations of singularity theory (including classification theory and invariants) to topology of singular spaces (links of singularities and semi-algebraic sets), as well as applications to topology (cobordism and Lefschetz fibrations), dynamical systems (Morse-Bott functions) and differential geometry (affine geometry, Gauss-maps, caustics, frontals and non-Euclidean geometries). All the manuscripts have been carefully peer-reviewed. We thank the authors for their valuable contributions, and the referees for their careful and diligent work. Our thanks also go to the editor of the RSME-AMS series in Contemporary Mathematics Pedro J. Paúl Escolano as well as to the editorial staff of the American Mathematical Society for their patience and efficient help with the publication process of the volume.

We could not have organized the workshop without the help of many people and institutions. We start by thanking the members of the Organizing Committee: João Carlos Ferreira Costa, Thaís Maria Dalbelo, Nivaldo Grulha, Ton Marar, Luciana Martins, Aurélio Menegon Neto, Bruna Oréfice Okamoto, Miriam da Silva Pereira, Raimundo Nonato Araújo dos Santos and João Nivaldo Tomazella. We also thank the members of the Scientific Committee: Carolina Araújo, James Damon, Alexandre Gurgel Fernandes, Javier Fernández de Bobadilla, Shyuichi Izumiya, Stanislaw Janeczko, Isabel Labouriau, Anne Pichon, Federico Sánchez Bringas, Márcio Soares for their support. We are also very grateful to the admistrative staff of the ICMC-USP. The workshop was funded by the Brazilian agencies and institutions FAPESP, CNPq, CAPES, PROEX-USP, INCTMAT, ICMC-USP, SBM, the Spanish DGCYT and the Japanese funding agency JSPS, whose support we gratefully acknowledge.

Ana Claudia Nabarro
Juan José Nuño-Ballesteros
Raúl Oset Sinha
Maria Aparecida Soares Ruas

## Carmen Romero Fuster

The meeting in 2014 was organized by the São Carlos and the València groups of singularities, and had the great pleasure to celebrate the *60th* birthday of Carmen Romero Fuster, professor at the Universitat de València, Spain. Carmen is one of the leading experts in applications of singularity theory to geometry, widely known by her pioneer work on generic geometry of submanifolds in Euclidean and hyperbolic spaces. Carmen and the São Carlos group started their scientific collaboration in 1985, when she was an assistant professor at the Universidade Federal de Viçosa, Brazil. Her fruitful interaction with the Brazilian singularists includes about thirty joint papers and six former students spread across four Brazilian states, São Paulo, Minas Gerais, Rio Grande do Sul and Bahia. Furthermore, some of her Spanish students have spent long periods of time in São Carlos as postdocs or visiting professors, it was therefore natural to celebrate her *60th* birthday at the ICMC-USP, São Carlos.

Carmen completed her undergraduate degree in the Universitat de València in 1976. Soon after that, in 1977, she started her Master's degree at the University of Warwick with E. C. Zeeman. Followingly, she went to Southampton to begin her PhD under the supervision of S. A. Robertson. Her PhD thesis, finished in 1981, was entitled *Sphere Stratifications and the Gauss Map* and was about global convexity properties of hypersurfaces in the Euclidean space. The results of the thesis were cited in the books of Arnold, Gussein-Zade and Varchencko, "Singularités des applications differéntiables" (Ed. Mir, 1986) and Banchoff, Gaffney and MacCrory, "Cusps of Gauss Mappings" (Res. Not. Math. Pitman, 1982), two of the main references in Singularity Theory at that moment.

Soon enough, Carmen started applying singularity theory techniques to differential geometry also in higher codimension submanifolds. Her generalization of the four vertex theorem for generic space curves lead to fruitful collaborations with V. D. Sedykh or to the thesis of her first student J. J. Nuño Ballesteros. In 1985 Carmen met M. A. S. 'Cidinha' Ruas and after some joint publications they co-supervised the thesis of D. K. H. Mochida, which layed the foundations of generic geometry of surfaces in $\mathbb{R}^4$ in a series of papers which are still amongst Carmen's most cited papers up to date. Other PhD students in the context of generic geometry of submanifolds in Euclidean spaces were E. Sanabria Codesal and R. A. Gonçalves in Spain and S. M. Moraes and R. R. Binotto in Brazil. In the 2000's, Carmen started a collaboration with S. Izumiya on geometry of submanifolds in the Minkowski space. Her interaction with the Japanese singularity group has been one of the most prolific and counts up to 12 joint publications.

In the same period, Carmen started to be interested in Vassiliev type invariants of spaces of mappings. This lead to the thesis of C. Mendes de Jesus on Arnold's invariants, which was continued by defining certain weighted graphs dual to the singular set of a map from a surface to the plane as a global topological invariant which complements the Vassiliev invariants. The study of topological invariants produced two more students, R. Oset Sinha, who studied Vassiliev type invariants of stable maps from 3-manifolds to $\mathbb{R}^3$ in his thesis, and C. Casonatto, who studied the case of 3-manifolds to $\mathbb{R}^4$.

The conference was a great success and most of Carmen's students attended the meeting. There was a plenary lecture prepared by J. J. Nuño Ballesteros, R. Oset Sinha and M. A. S. Ruas on Carmen's work, and at the conference dinner a video produced by E. Sanabria Codesal showed the non-research aspects of Carmen's academic life, a life in which she has made many good and lasting friendships around the world and in which she has shared her passion for mathematics with anyone who has been lucky enough to have their paths cross with hers.

Ana Claudia Nabarro
Juan José Nuño-Ballesteros
Raúl Oset Sinha
Maria Aparecida Soares Ruas

Contemporary Mathematics
Volume **675**, 2016
http://dx.doi.org/10.1090/conm/675/13581

# Hurwitz equivalence for Lefschetz fibrations and their multisections

R. İnanç Baykur and Kenta Hayano

ABSTRACT. In this article, we characterize isomorphism classes of Lefschetz fibrations with multisections via their monodromy factorizations. We prove that two Lefschetz fibrations with multisections are isomorphic if and only if their monodromy factorizations in the relevant mapping class groups are related to each other by a finite collection of modifications, which extend the well-known Hurwitz equivalence. This in particular characterizes isomorphism classes of Lefschetz pencils. We then show that, from simple relations in the mapping class groups, one can derive new (and old) examples of Lefschetz fibrations which cannot be written as fiber sums of blown-up pencils.

## 1. Introduction

Lefschetz fibrations became a central tool in differential geometry and topology following Donaldson's insight in the late 1990s [**5**] that one can effectively explore the topological aspects of manifolds by studying smooth maps on them which have locally holomorphic character. Since then they have gained a prominent role in symplectic topology, where a striking balance between flexibility and rigidity makes it possible to carry out topological constructions and geometric obstructions at the same time. In dimension 4, a firmer grasp on the quickly developing theory of Lefschetz fibrations is pursued via factorizations in mapping class groups of surfaces [**11**, **12**, **14**].

Our goal is to better understand how surfaces in symplectic 4–manifolds arise in this setting. Surfaces in 4–manifolds play a crucial role in our exploration of the topology of smooth and symplectic 4–manifolds; they help determining the homeomorphism type, and distinguishing the diffeomorphism type. As observed by Donaldson and Smith [**6**], any symplectic surface can be seen as a *multisection* or an *n–section* of a Lefschetz fibration, which is a branched surface intersecting all the fibers positively $n$ times. In [**4**], we initiated an extensive study of symplectic surfaces via factorizations in more elaborate mapping class groups of surfaces. The current article aims to add to this effort by refining and exploring the correspondence between symplectic surfaces in symplectic 4–manifolds, multisections of Lefschetz fibrations, and positive factorizations in surface mapping class groups.

2010 *Mathematics Subject Classification.* Primary 57N13, 57R22.

*Key words and phrases.* Lefschetz fibration, Hurwitz equivalence, multisectin, mapping class group, positive factorization.

An isomorphism between two Lefschetz fibrations $(X_i, f_i)$ with $n$–sections $S_i$, $i = 1, 2$, is given by a pair of orientation-preserving diffeomorphisms between the total spaces $X_i$ and the base 2–spheres, which commute with $f_i$ and match $S_i$. The mapping class group $\mathrm{Mod}(\Sigma_g; \{s_1, \ldots, s_n\})$, which consists of orientation-preserving self-diffeomorphisms of the genus-$g$ surface $\Sigma_g$ preserving the set of marked points $\{s_1, \ldots, s_n\}$, is the host to the lift of any monodromy factorization of $(X_i, f_i)$ prescribed by an $n$–section $S_i$. There are simple modifications of monodromy factorizations that naturally arise from a few choices made in the process of extracting these factorizations from a given fibration —which we describe and study in detail in the later sections. Two monodromy factorizations in $\mathrm{Mod}(\Sigma_g; \{s_1, \ldots, s_n\})$ will be called Hurwitz equivalent if one can be obtained from the other by applying a sequence of this finite set of simple modifications.

Our main theorem in this article is the following:

THEOREM 1.1. *For $g, n \geq 1$ there exists a one-to-one correspondence*

$$\left\{ \begin{array}{c} \textit{Genus–}g\textit{ Lefschetz fibrations} \\ \textit{with } n\textit{–sections,} \\ \textit{up to isomorphism} \end{array} \right\} \longleftrightarrow \left\{ \begin{array}{c} \textit{Monodromy factorizations} \\ \textit{in } \mathrm{Mod}(\Sigma_g; \{s_1, \ldots, s_n\}), \\ \textit{up to Hurwitz equivalence} \end{array} \right\}$$

When $g \geq 2$ and $n = 0$, i.e. when the multisections are omitted, the above result is classical, due to Kas [**12**] and Matsumoto [**14**].

Theorem 1.1 is proved in Section 3, after a review of background results in Section 2. We also provide an extension of this correspondence to one between *framed* multisections of Lefschetz fibrations and monodromy factorizations in framed mapping class groups introduced in [**4**]. In turn, through their monodromy factorizations, we obtain a full characterization of Lefschetz *pencils*, up to isomorphisms that can permute base points; see Corollary 3.10.

In the last section, we turn to an intriguing question regarding the diversity of Lefschetz fibrations versus that of pencils, which allows us to demonstrate how geometric, topological, and algebraic aspects of the theory of Lefschetz fibrations can be nicely brought together. In [**17**], Stipsicz asked whether every Lefschetz fibration can be obtained as fiber sums of blown-up pencils; that is to say, whether Lefschetz pencils are the building blocks of all Lefschetz fibrations via blow-ups and fiber sums. In Section 4, we will illustrate a way to produce counter-examples, using the well-known 5–chain relation in the mapping class group of a genus–2 surface, along with monodromy modifications involving multisections, and a variety of geometric and topological results packaged in a recipe from [**4**] we will be following here. We moreover show –by appealing to above Hurwitz equivalences– that the only other counter-example we know of, a genus–2 Lefschetz fibration of Auroux (shown to be a counter-example by Sato in [**15**]), can also be derived from the same scheme.

*Acknowledgements.* The results of this article were partially presented by the second author at the 13th International Workshop on Real and Complex Singularities held in August 2014 at Sao Carlos, Brazil. He would like to thank the organizers of the workshop for the opportunity and for their invitation to prepare this manuscript. The authors would also like to thank the anonymous referee for helpful comments. The first author was partially supported by the Simons Foundation Grant 317732 and the NSF Grant DMS-1510395. The second author was supported by JSPS and

CAPES under the Japan-Brazil research cooperative program and JSPS KAKENHI (26800027).

## 2. Multisections of Lefschetz fibrations via positive factorizations

In this section we will briefly review the basic definitions and properties of Lefschetz fibrations and their multisections, focusing on how they can be captured and studied as certain factorizations in mapping class groups of surfaces. For a more detailed exposition, the reader can turn to [**11**] and [**4**].

Throughout the paper, all manifolds we work with are assumed to be closed, connected and oriented, unless otherwise noted.

**2.1. Lefschetz pencils, fibrations, and multisections.** A *Lefschetz fibration* $(X, f)$ is a smooth map $f : X \to S^2$, from a 4–manifold $X$ onto the 2–sphere, which only has *nodal singularities*, that is, for any $x$ in the *critical locus* $\mathrm{Crit}(f)$, there exist orientation-preserving complex coordinate neighborhoods $(U, \varphi)$ at $x \in X$ and $(V, \psi)$ at $f(x) \in S^2 \cong \mathbb{CP}^1$, such that

$$\psi \circ f \circ \varphi^{-1}(z, w) = z^2 + w^2 .$$

So $f$ is a submersion at all but finitely many points, where we have the local model of a complex nodal singularity. For $g$ the genus of a regular fiber, we call $(X, f)$ a *genus–$g$ Lefschetz fibration.*

A *Lefschetz pencil* $(X, f)$ is a Lefschetz fibration $f : X \setminus B \to S^2$, where $B$ is a *non-empty* discrete set in $X$, called the *base locus*, such that there exist an orientation-preserving complex coordinate neighborhood $(U, \varphi)$ around each *base point* $x \in B$ and a diffeomorphism $\psi : S^2 \to \mathbb{CP}^1$, which together satisfy

$$\psi \circ f \circ \varphi^{-1}(z, w) = [z : w].$$

We say $(X, f)$ is a *genus–$g$ Lefschetz pencil with $n$ base points* for $g$ the genus of the regular fiber (compactified by adding the base points), and $n = |B|$. Given a genus–$g$ Lefschetz pencil $(X, f)$ with $n$ base points, we can obtain a genus–$g$ Lefschetz *fibration* $f' : X' = X \# n\overline{\mathbb{CP}}^2 \to S^2$ with $n$ disjoint sections $S_j$ of self-intersection $-1$, each arising as an exceptional sphere of the blow-up at the base point $x_j$. The correspondence is canonical, as one can blow-down all the $S_j$ in the fibration $(X', f')$ to recover the pencil $(X, f)$ (e.g. [**11**, §.8.1]).

Recall that a *symplectic structure* is a closed non-degenerate 2–form $\omega$ on a smooth manifold, such as the Kähler form on a complex algebraic variety. A *symplectic 4–manifold* is then a pair $(X, \omega)$. The prominent role of Lefschetz fibrations in differential geometry and topology is mostly due to Donaldson's amazing result from the late 1990s, who showed that an analogue of the classical Lefschetz hyperplane theorem for complex algebraic surfaces holds in this more flexible setting: every *symplectic* 4–manifold admits a Lefschetz pencil [**5**]. Conversely, generalizing an idea of Thurston, Gompf observed that every 4–manifold admitting a Lefschetz pencil or a non-trivial (i.e. with non-empty critical locus) Lefschetz fibration is symplectic [**11**]. Furthermore, one can strike a compatibility condition between the pairs $(X, \omega)$ and $(X, f)$, which asks for the fibers of $f$ to be symplectic surfaces with respect to $\omega$.

THEOREM 2.1 (Donaldson, Gompf). *Every symplectic 4–manifold $(X, \omega)$ admits a compatible Lefschetz pencil, and every Lefschetz pencil / non-trivial fibration $(X, f)$ can be equipped with a compatible symplectic form.*

The main companion of a Lefschetz fibration $(X,f)$ in this paper will be an embedded surface $S$ which sits in $X$ in a rather special way with respect to $f$.

DEFINITION 2.2 ([**4**]). A (possibly disconnected) closed oriented surface $S \subset X$ is called a *multisection*, or an $n$*–section*, of a Lefschetz fibration $(X,f)$ if it satisfies the conditions:

(1) The restriction $f|_S$ is an $n$–fold simple branched covering,
(2) The restriction of the differential $df_x : N_xS \to T_{f(x)}S^2$ preserves the orientation for any branched point $x \in S$ of $f|_S$, where $N_xS \subset T_xX$ is a normal space of $T_xS$, which has the canonical orientation derived from that of $X$,
(3) For any branched point $x \in S \cap \mathrm{Crit}(f)$ of $f|_S$, there exist complex coordinate neighborhoods $(U,\varphi)$ and $(V,\psi)$ which make the following diagram commute:

$$\begin{array}{ccc} (U, U\cap S) & \xrightarrow{\ \varphi\ } & (\mathbb{C}^2, \mathbb{C}\times\{0\}) \\ {\scriptstyle f}\big\downarrow & & \big\downarrow{\scriptstyle (z,w)\mapsto z^2+w^2} \\ V & \xrightarrow{\ \psi\ } & \mathbb{C}. \end{array}$$

A triple $(X,f,S)$ will denote a Lefschetz fibration $(X,f)$ and its multisection $S$.

Just like how a Lefschetz fibration locally behaves like a holomorphic map, a multisection behaves like a holomorphic curve, intersecting the fibers all positively, and so that the restriction of the fibration map to it is a holomorphic branched covering onto $\mathbb{CP}^1$. Multisections are found in abundance, as observed by Donaldson and Smith (who referred to them as *standard surfaces*): for any symplectic surface $S$ in a symplectic 4–manifold $(X,\omega)$, there exists a compatible Lefschetz pencil $(X,f)$, such that $S$ is a multisection of $f|_{X\setminus B}$, and conversely, for any triple $(X,f,S)$, there exists a compatible symplectic form $\omega$ making the fibers and $S$ symplectic [**6**].

**2.2. Local and global monodromies, positive factorizations.** We will make a few additional assumptions on $(X,f,S)$, merely to simplify our upcoming discussions on how to describe Lefschetz fibrations and their multisections in terms of certain factorizations in surface mapping class groups. First, we will assume that $f$ is injective on $\mathrm{Crit}(f)$, and also that each branched point of a multisection, if not contained in $\mathrm{Crit}(f)$, does not lie on a singular fiber, i.e. not contained in $f^{-1}(\mathrm{Crit}(f))$. Both of these can be always achieved after a small perturbation. These assumptions will allow us to get standard local models for $(X,f,S)$ for the fibration over $S^2 \setminus f(\mathrm{Crit}(f)\cup \mathrm{Crit}(f|_S))$ –and not only *around* the isolated points in $\mathrm{Crit}(f)\cup\mathrm{Crit}(f|_S)$.

It is also customary to assume that $f$ is *relatively minimal*, that is, no fiber contains a sphere with self-intersection $-1$, which otherwise could be blown-down without destroying the rest of the fibration. A Lefschetz *pencil* $(X,f)$ is said to be *relatively minimal*, if no fiber component is a self-intersection $-1+k$ sphere containing $k$ points of $B$, to ensure that the associated Lefschetz fibration $(X',f')$ is relatively minimal. As we will see shortly, this assumption is needed to guarantee that no information on the local topology is lost when we look at the monodromy of the fibration.

For a surface $\Sigma$ and points $s_1,\ldots,s_n \in \Sigma$, let $\mathrm{Diff}(\Sigma;\{s_1,\ldots,s_n\})$ be the group of orientation-preserving diffeomorphisms of $\Sigma$ which preserve the set $\{s_1,\ldots,s_n\}$.

We call $\mathrm{Mod}(\Sigma; \{s_1, \dots, s_n\}) = \pi_0(\mathrm{Diff}(\Sigma; \{s_1 \dots, s_n\})$ the *mapping class group of* $\Sigma$ *with marked points* $\{s_1, \dots, s_n\}$. It consists of elements of $\mathrm{Diff}(\Sigma; \{s_1, \dots, s_n\})$, modulo isotopies fixing the set $\{s_1, \dots, s_n\}$, where the group structure is induced by compositions of maps, that is, $[\varphi_1]\cdot[\varphi_2] = [\varphi_1 \circ \varphi_2]$ for $\varphi_1, \varphi_2 \in \mathrm{Diff}(\Sigma_g; \{s_1, \dots, s_n\})$.

Let $\Sigma_g^n$ denote a genus–$g$ surface with $n$ boundary components, and take points $u_1, \dots, u_n \in \partial\Sigma_g^n$ which cover the elements of $\pi_0(\partial\Sigma_g^n)$. The *framed mapping class group* $\mathrm{Mod}(\Sigma_g^n; \{u_1, \dots, u_n\})$ consists of orientation-preserving diffeomorphisms of $\Sigma_g^n$ which preserve set of marked points $\{u_1, \dots, u_n\}$, modulo isotopies fixing the same data [**4**]. Clearly, one can pass to a closed surface $\Sigma_g = \Sigma_g^0$ by capping the boundaries by disks, the centers of which we label as $s_1, \dots, s_n \in \Sigma_g$. This boundary capping map induces a surjective homomorphism

$$\mathrm{Mod}(\Sigma_g^n; \{u_1, \dots, u_n\}) \twoheadrightarrow \mathrm{Mod}(\Sigma_g; \{s_1, \dots, s_n\}) .$$

Now let $(X, f)$ be a Lefschetz fibration with $l$ critical points, $S$ its $n$–section and $\mathrm{Crit}(f|_S) \setminus \mathrm{Crit}(f) = \{b_1, \dots, b_k\} \subset X$ the set of branched points of $f|_S$ away from $\mathrm{Crit}(f)$. Set $f(\mathrm{Crit}(f) \cup \mathrm{Crit}(f|_S)) = \{a_1, \dots, a_{k+l}\}$, and take paths $\alpha_1, \dots, \alpha_{k+l} \subset S^2$ with a common initial point $p_0 \in S^2 \setminus f(\mathrm{Crit}(f) \cup \mathrm{Crit}(f|_S))$ such that

- $\alpha_1, \dots, \alpha_{k+l}$ are mutually disjoint except at $p_0$,
- $\alpha_i$ connects $p_0$ with $a_i$,
- $\alpha_1, \dots, \alpha_{k+l}$ are ordered counterclockwise around $p_0$, i.e. there exists a small loop around $p_0$ oriented counterclockwise, hitting each $\alpha_i$ only once in the given order.

We take a loop $\widetilde{\alpha}_i$ with the base point $p_0$ by connecting $\alpha_i$ with a small counterclockwise circle with center $a_i$. Let $\mathcal{H}$ be a horizontal distribution of $f|_{X\setminus(\mathrm{Crit}(f)\cup\mathrm{Crit}(f|_S))}$, that is, $\mathcal{H} = \{\mathcal{H}_x\}_{x\in X\setminus(\mathrm{Crit}(f)\cup\mathrm{Crit}(f|_S))}$ is a plane field such that $\mathrm{Ker}(df_x) \oplus \mathcal{H}_x = T_xX$ for any $x \in X \setminus (\mathrm{Crit}(f) \cup \mathrm{Crit}(f|_S))$. We assume that $\mathcal{H}_x = T_xS$ for any $x \in S \setminus \mathrm{Crit}(f|_S)$. Using $\mathcal{H}$, we can take a lift of the direction vector field of $\widetilde{\alpha}_i$ and a flow of this lift gives rise to a self-diffeomorphism of $f^{-1}(p_0)$. We call this diffeomorphism a *parallel transport* of $\widetilde{\alpha}_i$ and its isotopy class a *local monodromy* around $a_i$. Note that a local monodromy does not depend on the choice of $\mathcal{H}$. Indeed, for any horizontal distribution $\mathcal{H}$ we can find a Riemannian metric $g$ such that $\mathcal{H}_x$ is equal to $(\mathrm{Ker}(df_x))^\perp$, in particular for any two horizontal distribution there exists a one-parameter family of horizontal distributions connecting the given two distributions.

Under an identification of the pair $(f^{-1}(p_0), f^{-1}(p_0)\cap S)$ with $(\Sigma_g, \{s_1, \dots, s_n\})$, we can regard a parallel transport as a diffeomorphism in $\mathrm{Diff}(\Sigma_g; \{s_1, \dots, s_n\})$, and thus, a local monodromy as a mapping class in $\mathrm{Mod}(\Sigma_g; \{s_1, \dots, s_n\})$. We denote this mapping class by $\xi_i$. Since the concatenation $\widetilde{\alpha}_1 \cdots \widetilde{\alpha}_{k+l}$ is null-homotopic in $S^2 \setminus f(\mathrm{Crit}(f) \cup \mathrm{Crit}(f|_S))$, the composition $\xi_{k+l} \cdots \xi_1$ is the unit element of the group $\mathrm{Mod}(\Sigma_g; \{s_1, \dots, s_n\})$. The factorization

$$\xi_{k+l} \cdots \xi_1 = 1 \quad \text{in } \mathrm{Mod}(\Sigma_g; \{s_1, \dots, s_n\})$$

is called a *monodromy factorization of the triple* $(X, f, S)$, which we will denote in short by $W_{X,f,S}$ (as a *word* in $\xi_i \in \mathrm{Mod}(\Sigma_g; \{s_1, \dots, s_n\})$.

Analyzing the local models around $\mathrm{Crit}(f)$ and $\mathrm{Crit}(f|_S)$ (which, remember, might contain in common the type (3) branched points in Definition 2.2), we identify three standard elements in the mapping class group $\mathrm{Mod}(\Sigma_g; \{s_1, \dots, s_n\})$ that appear as a factor $\xi_i$ above [**4**]: If the fiber $f^{-1}(a_i)$ contains a Lefschetz critical point which is not a branched point of $f|_S$, then $\xi_i$ is a right-handed Dehn twist along some simple closed curve $c \subset \Sigma_g \setminus \{s_1, \dots, s_n\}$, which is called a *vanishing cycle* of a Lefschetz critical point in $f^{-1}(a_i)$ [**12**]. (Relative minimality of $(X, f)$ now guarantees that $c$ is not null-homotopic, so we do not have a "hidden" Dehn twist factor.) If $f^{-1}(a_i)$ contains a branched point of $S$ away from $\mathrm{Crit}(f)$, then $\xi_i$ is a half twist $\tau_\gamma$ along some path $\gamma \subset \Sigma_g$ between some $s_j$ and $s_{j'}$. Lastly, if $a_i$ is the image of a point in $\mathrm{Crit}(f) \cap \mathrm{Crit}(f|_S)$, we get a mapping class $\xi_i$ which is represented by a Dehn twist $\widetilde{t_c}$ shown in Figure 1 for some simple closed curve $c \subset \Sigma_g$ going through $s_j$ and $s_{j'}$.

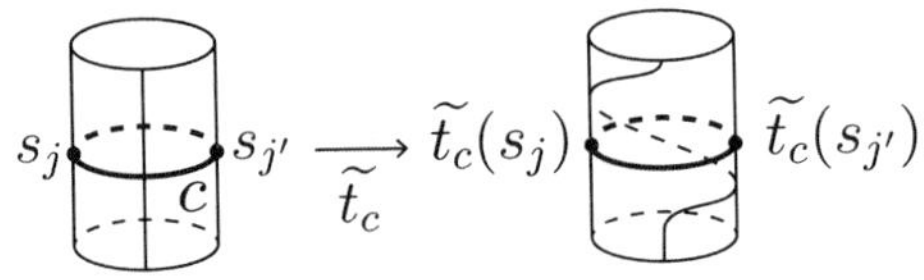


FIGURE 1. A lift of a Dehn twist.

Observe that under the forgetful homomorphism, only $\xi_i$ that are Dehn twists (possibly going through $\mathrm{Crit}(f|_S)$) survive, yielding the standard monodromy factorization $W_{X,f}$ of $(X, f)$ of the form $t_{c_l} \cdots t_{c_1} = 1$ in $\mathrm{Mod}(\Sigma_g)$. In other words, the factorization $W_{X,f,S}$ is a *lift* of the factorization $W_{X,f}$.

It is worth noting that each standard element $\xi_i$ discussed above, let it be a Dehn twist or an arc twist, comes with a preferred orientation, corresponding to positive (right-handed) Dehn twists and arc/braid twists. Any factorization $\xi_{k+l} \cdots \xi_1 = 1$ in $\mathrm{Mod}(\Sigma_g; \{s_1, \dots, s_n\})$, which consists of only these three types of elements is called a *positive factorization* (of the identity) in $\mathrm{Mod}(\Sigma_g; \{s_1, \dots, s_n\})$, and it conversely gives rise to a triple $(X, f, S)$.

We can summarize these as follows (which is a direct corollary of [**4**, Theorem 1.1] obtained by applying the boundary capping homomorphism to the framed mapping class group):

THEOREM 2.3 ([**4**]). *Let $(X, f, S)$ be a genus-$g$ Lefschetz fibration with $l$ critical points, where $S$ is a connected $n$-section which has $k$ branched points away from $\mathrm{Crit}(f)$ and $r$ branched points on $\mathrm{Crit}(f)$. Then $(X, f, S)$ has a monodromy factorization $W_{X,f,S}$ of the form $\xi_{k+l} \cdots \xi_1 = 1$ in $\mathrm{Mod}(\Sigma_g; \{s_1, \dots, s_n\})$, where among $\xi_i$ $k$ many are half-twists $\tau_{\gamma_i}$, $r$ many are Dehn twists $\widetilde{t_{c_i}}$ through two marked points in $\{s_1, \dots, s_n\}$, and the rest are Dehn twists along curves missing the marked points. Moreover, $g(S) = \frac{1}{2}(k + r) - n + 1$ and the monodromy group $\langle \xi_1, \dots, \xi_{k+l} \rangle \subset \mathrm{Mod}(\Sigma_g; \{s_1, \dots, s_n\})$ acts on the set $\{s_1, \dots, s_n\}$ transitively.*

*Conversely, from any such positive factorization of 1 in $\mathrm{Mod}(\Sigma_g; \{s_1, \dots, s_n\})$, subject to conditions listed above, one can construct a genus-$g$ Lefschetz fibration $(X, f)$ with $l$ vanishing cycles $c_1, \dots, c_l$, and a connected $n$-section $S$ of genus $g(S) = \frac{1}{2}(k + r) - n + 1$.*

REMARK 2.4. We can modify Theorem 2.3 so that it also holds for a *disconnected* multisection: in this case the group $\langle \xi_1, \dots, \xi_{k+l} \rangle$ acts intransitively on $\{s_1, \dots, s_n\}$ (the number of orbits of the action coincides with that of components of $S$), and the Euler characteristic $\chi(S)$ is equal to $2n - (k + r)$.

REMARK 2.5. As we have shown in [4] the positive factorization $W_{X,f,S}$ of the identity element in $\mathrm{Mod}(\Sigma_g; \{s_1, \ldots, s_n\})$ lifts to another positive factorization of a product of boundary parallel Dehn twists in $\mathrm{Mod}(\Sigma_g^n; \{u_1, \ldots, u_n\})$. This geometrically corresponds to removing a framed tubular neighborhood of $S$. The latter positive factorization consists of standard factors $t_{c_i}$, which are Dehn twists in the interior of $\Sigma_g^n$, and the lifts of $\tau_\gamma$ and $\widetilde{t_c}$ as shown in Figure 2, both of which interchange the two boundary components in prescribed ways.

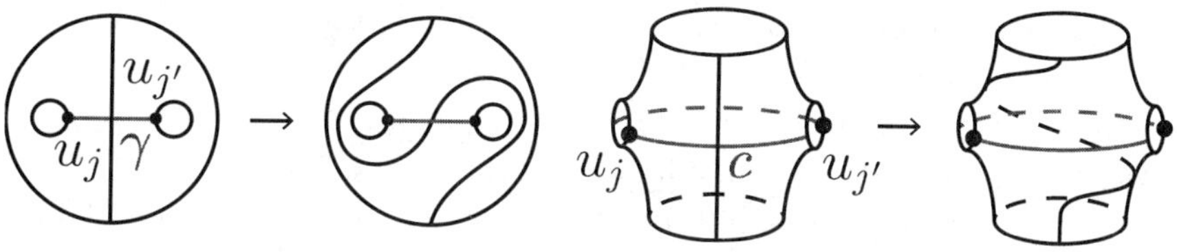


FIGURE 2. Lifts of $\tau_\gamma$ and $\widetilde{t_c}$.

For simplicity, we denote these lifts by $\tau_\gamma$ and $\widetilde{t_c}$ as well. This monodromy factorization in the framed mapping class group allows us to capture the self-intersection number of $S$.

### 3. Equivalence of Lefschetz fibrations with multisections

The goal of this section is to establish a one-to-one correspondence between triples $(X, f, S)$, where $S$ is an $n$–section of a genus–$g$ Lefschetz fibration $(X, f)$, and positive factorizations $W_{X,f,S}$ of the form $\xi_{k+l} \cdots \xi_1 = 1$ in $\mathrm{Mod}(\Sigma_g; \{s_1, \ldots, s_n\})$, modulo natural equivalence relations on both sides, which we will spell out right away.

The triples $(X_i, f_i, S_i)$, $i = 1, 2$, are said to be *equivalent* (or *isomorphic*) if there exist orientation-preserving diffeomorphisms $\Phi : X_1 \to X_2$ and $\phi : S^2 \to S^2$ such that $\Phi(S_1) = S_2$ and $f_2 \circ \Phi = \phi \circ f_1$. Clearly, a necessary condition for $(X_i, f_i, S_i)$ to be equivalent is that both fibrations $f_i$ should have the same genus $g$, and the multisections $S_i$ should have the same covering degree $n$.

As we noted in the previous section, a monodromy factorization $(X, f, S)$ does not depend on the choice of a horizontal distribution $\mathcal{H}$. It does however depend on the choice of paths $\alpha_1, \ldots, \alpha_{k+l}$ and that of an identification of $(f^{-1}(p_0), f^{-1}(p_0) \cap S)$ with $(\Sigma_g, \{s_1, \ldots, s_n\})$. Identical to the well-known case of a monodromy factorization of a *pair* $(X, f)$ one can (see e.g. [11]) easily verify that two monodromy factorizations for a triple $(X, f, S)$, derived from different choices of paths and identifications can be related by successive applications of the following two types of modifications:

(1) *Elementary transformation*, which changes a factorization as follows:

$$\xi_{k+l} \cdots \xi_{i+1}\xi_i \cdots \xi_1 \longleftrightarrow \xi_{k+l} \cdots (\xi_{i+1}\xi_i\xi_{i+1}^{-1})\xi_{i+1} \cdots \xi_1.$$

Note that $(\xi_{i+1}\xi_i\xi_{i+1}^{-1})$ in the right hand side represents a single factor of the factorization.

(2) *Global conjugation*, which changes each member of a factorization by the conjugation of some mapping class $\psi \in \mathrm{Mod}(\Sigma_g; \{s_1, \dots, s_n\})$:

$$\xi_{k+l} \cdots \xi_1 \longleftrightarrow (\psi \xi_{k+l} \psi^{-1}) \cdots (\psi \xi_1 \psi^{-1}).$$

We will thus call two factorizations of the unit element $1 \in \mathrm{Mod}(\Sigma_g; \{s_1, \dots, s_n\})$ *Hurwitz equivalent* if one can be obtained from the other after a sequence of these two types of modifications.

**3.1. Equivalence of monodromy factorizations.** This subsection will be devoted to the proof of the following theorem, which, together with Theorem 2.3, implies the main result of our paper, Theorem 1.1.

THEOREM 3.1. *Let $(X_i, f_i, S_i)$, $i = 1, 2$ be a genus–$g$ Lefschetz fibration with an $n$–section $S_i$. Suppose that $2-2g-n$ is negative, that is, $f_i^{-1}(p_0) \setminus (f_i^{-1}(p_0) \cap S_i)$ is a hyperbolic surface for a regular value $p_0$. The triples $(X_1, f_1, S_1)$ and $(X_2, f_2, S_2)$ are equivalent if and only if their monodromy factorizations $W_{X_1,f_1,S_1}$ and $W_{X_2,f_2,S_2}$ are Hurwitz equivalent.*

To prove Theorem 3.1, we will need a few preliminary results on mapping classes in $\mathrm{Mod}(\Sigma_g; \{s_1, \dots, s_n\})$.

LEMMA 3.2. *Let $\gamma_1, \gamma_2 \subset \Sigma_g$ be simple paths between distinct marked points $s_i$ and $s_j$. Then $\tau_{\gamma_1} = \tau_{\gamma_2}$ in $\mathrm{Mod}(\Sigma_g; \{s_1, \dots, s_n\})$ if and only if $\gamma_1$ and $\gamma_2$ are isotopic relative to the points $s_1, \dots, s_n$.*

PROOF. The "if" part is obvious. To prove the "only if" part we assume that $\gamma_1$ and $\gamma_2$ are not isotopic and show that $\tau_{\gamma_1}$ and $\tau_{\gamma_2}$ are not equal. For simple curves $d_1, d_2$, we denote the geometric intersection number of $d_1$ and $d_2$ by $i(d_1, d_2)$, that is, $i(d_1, d_2)$ is the minimum number of intersections between two curves isotopic (relative to $s_1, \dots, s_n$) to $d_1$ and $d_2$.

Without loss of generality we may assume that $\gamma_1$ and $\gamma_2$ are in minimal position. Let $d$ be the boundary of a regular neighborhood of $\gamma_1$. The curve $d$ does not intersect $\gamma_1$. On the other hand, $i(d, \gamma_2)$ is not equal to 0. To see this, we will check that there is no bigon between sub-paths of $d$ and $\gamma_2$ (see the bigon criterion in [**10**]). As shown in Figure 3 there are three types of regions which are candidates of such bigons. The two shaded regions in Figure 3(a) cannot be bigons since $\gamma_1$

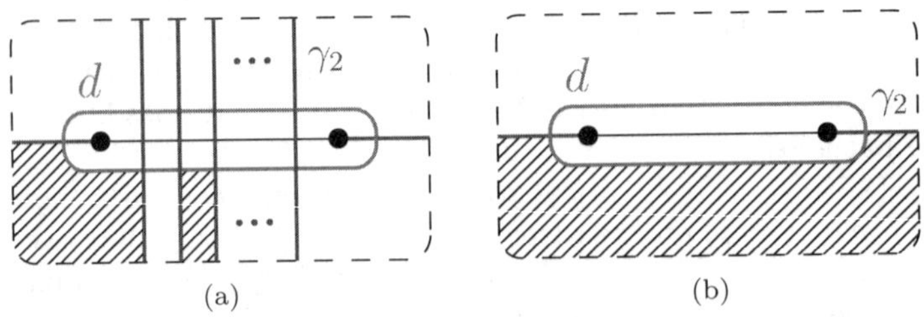


FIGURE 3. Candidates of bigons around $d$.

and $\gamma_2$ are in minimal position. Since $\gamma_1$ and $\gamma_2$ are not isotopic, the shaded region in Figure 3(b) is not a bigon.

Since $i(d, \tau_{\gamma_1}(d))$ is equal to 0, it is sufficient to prove that $i(d, \tau_{\gamma_2}(d))$ is not equal to 0. Let $d'$ be the simple closed curve obtained by changing a parallel copy of

$d$ around $\gamma_2$ as shown in Figure 4. The curve $d'$ represents the isotopy class $\tau_{\gamma_2}(d)$. It is easy to see that the number of the intersections between $d$ and $d'$ is equal to

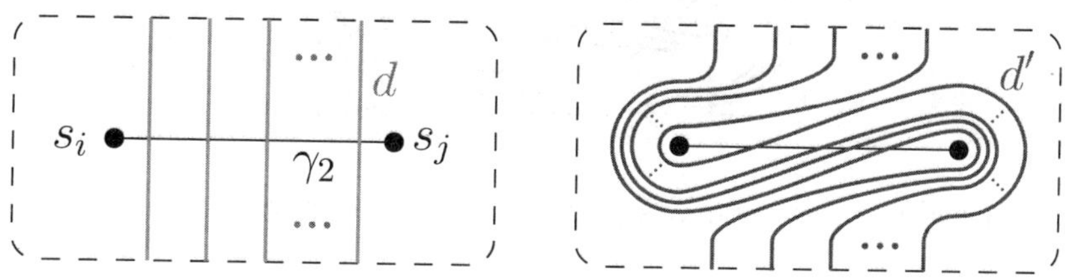


FIGURE 4. The curves $d$ and $d'$ around $\gamma_2$.

$2i(d, \gamma_2)^2$. In what follows, we will prove that $d$ and $d'$ are in minimal position using the bigon criterion. As shown in Figure 5 there are six types of regions which are candidates of innermost bigons. If the far right region in Figure 5(a) were a

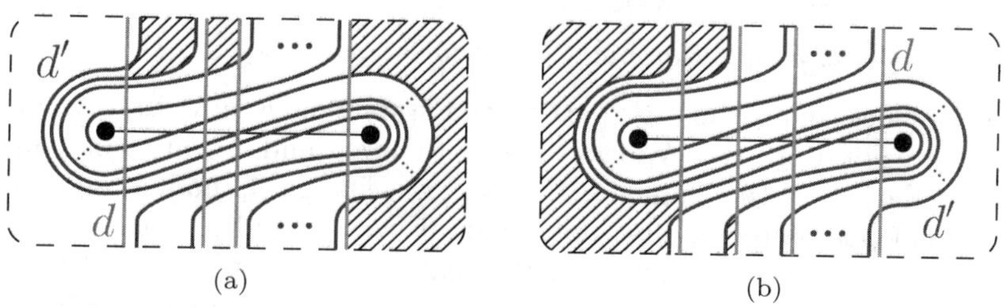


FIGURE 5. Shaded regions are candidates of bigons.

bigon, $d$ would be isotopic to a small circle with center $s_j$, but it is not the case since $d$ does not intersect $\gamma_1$. Similarly, we can also verify that the far left region in Figure 5(b) is not a bigon. If either of the rest of two regions in Figure 5(a) were a bigon, then $d$ and $\gamma_2$ would form a bigon, which contradicts the assumption that $d$ and $\gamma_2$ are in minimal position. As for the rest of two regions in Figure 5(b), either the boundary of them contain at least two sub-paths of $d$, or they contain the point $s_i$ or $s_j$. In either case, these regions cannot be bigons. We can eventually conclude that $d'$ and $d$ are in minimal position, and thus, $i(d, \tau_{\gamma_2}(d)) = 2i(d, \gamma_2)^2 \neq 0$. □

LEMMA 3.3. *Let $c \subset \Sigma_g$ be a simple closed curve going through $s_i$ and $s_j$ which is not null-homotopic (as a curve in $\Sigma_g$), and $\gamma \subset \Sigma_g$ the closure of a component of $c \setminus \{s_i, s_j\}$. For any $N \in \mathbb{Z} \setminus \{0\}$ the mapping classes $\widetilde{t_c}$ and $\tau_\gamma^N$ do not commute, that is, $[\widetilde{t_c}, \tau_\gamma^N] \neq 1$ in $\mathrm{Mod}(\Sigma_g; \{s_1, \ldots, s_n\})$.*

PROOF. Since $\widetilde{t_c}$ commutes with $\tau_\gamma^N$ if and only if $\widetilde{t_{\tau_\gamma^N(c)}}$ commutes with $\tau_\gamma^{-N}$, we may assume $N > 0$ without loss of generality. Let $\gamma'$ be the closure of the complement $c \setminus \gamma$ and $d$ and $d'$ the boundaries of regular neighborhoods of $\gamma$ and $\gamma'$, respectively. The curve $\widetilde{t_c}(d)$ is isotopic to $d'$, while $\tau_\gamma^N \widetilde{t_c} \tau_\gamma^{-N}(d)$ is isotopic to $\tau_\gamma^N(d')$. We will prove that $d'$ and $\tau_\gamma^N(d')$ are not isotopic by showing $i(\tau_\gamma^N(d'), c) \neq i(d', c) = 2$.

The curve $\tau_\gamma^N(d')$ is described in Figure 6. It intersects with $c$ at $4N+2$ points. We easily see that none of the regions made by $\tau_\gamma^N(d')$ and $c$ in Figure 6, except for

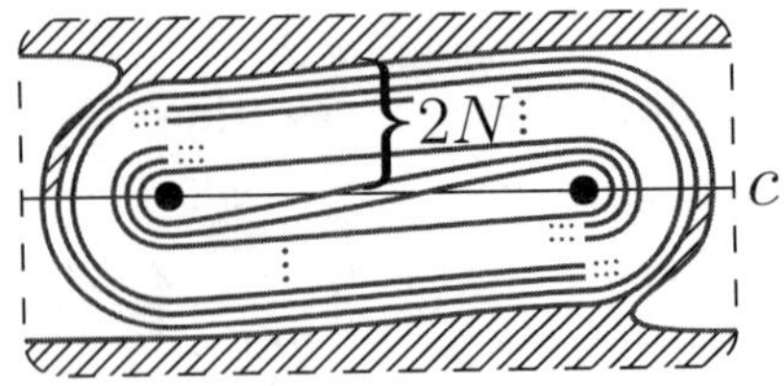


FIGURE 6. The curve $\tau_\gamma^N(d')$. The central bold curve is $\gamma$.

the shaded ones, can be bigons. Furthermore, neither of the shaded regions can be a bigon since $c$ is not null-homotopic in $\Sigma_g$. Thus the curve $\tau_\gamma^N(d')$ is in minimal position with $c$, and $i(c,\tau_\gamma^N(d')) = 4N+2 \neq 2$ for $N \neq 0$, as claimed. □

LEMMA 3.4. *Let $c_1, c_2 \subset \Sigma_g$ be simple closed curves going through $s_i$ and $s_j$ which are not null-homotopic (as curves in $\Sigma_g$). Then $\widetilde{t_{c_1}} = \widetilde{t_{c_2}}$ in*

$$\mathrm{Mod}(\Sigma_g; \{s_1, \ldots, s_n\})$$

*if and only if $c_1$ and $c_2$ are isotopic relative to the points $s_1, \ldots, s_n$.*

PROOF. Once again the "if" part of the statement is obvious. Assume that $c_1$ and $c_2$ are not isotopic. We will prove that $\widetilde{t_{c_1}}$ and $\widetilde{t_{c_2}}$ are not equal. By an isotopy relative to the points $s_1, \ldots, s_n$, we change $c_1$ and $c_2$ so that these are in minimal position. We first note that, if we can find a simple closed curve $c_3 \subset \Sigma_g \setminus \{s_1, \ldots, s_n\}$ away from $c_1$ such that $i(c_2, c_3)$ is not equal to 0, we can prove that $\widetilde{t_{c_1}}$ and $\widetilde{t_{c_2}}$ are different mapping classes in the same way as in the proof of [**10**, Fact 3.6].

**Case 1** : Suppose that both of the components of $c_1 \setminus \{s_i, s_j\}$ intersect $c_2$. We may assume that, at each of the points $s_i$ and $s_j$, either $c_1$ and $c_2$ intersect transversely or these are tangent to each other. Let $\nu c_1$ be a tubular neighborhood of $c_1$ and $U_i, U_j$ small neighborhoods of $s_i, s_j$. If $c_1$ and $c_2$ are tangent to each other at both of the points, then one of the following holds:

- the intersections $U_i \cap c_2$ and $U_j \cap c_2$ are contained in the same component of $\nu c_1 \setminus c_1$ (see Figure 7(c)),
- the component of $\nu c_1 \setminus c_1$ containing $U_i \cap c_2$ is different from that containing $U_j \cap c_2$ (see Figure 7(d)).

Altogether we have to consider the four cases described in Figures 7(a) – 7(d). For each case we take a parallel copy $c_1'$ of $c_1$ as shown in the figures. We can easily verify that no region made by $c_1'$ and $c_2$ in Figures 7(a), 7(b) and 7(c) can be a bigon. (Recall that $c_1$ and $c_2$ are assumed to be in minimal position.) In these cases, we put $c_3 = c_1'$, which satisfies the desired conditions (i.e. away from $c_1$ and $i(c_2, c_3) \neq 0$). In particular we can deduce that $\widetilde{t_{c_1}}$ and $\widetilde{t_{c_2}}$ are different mapping classes. As for the last case, the shaded region in Figure 7(d) can be a bigon. If this region is a bigon, we move $c_1'$ to $c_1''$ so that it avoids all the bigons nested around $s_j$ (see Figure 7(e)). It is easy to see that $c_1'' \cap c_2$ is not empty and no region made by $c_1''$ and $c_2$ in Figure 7(e) can be a bigon. Thus $c_3 = c_1''$ satisfies the desired conditions.

**Case 2** : Suppose that one of the components of $c_1 \setminus \{s_i, s_j\}$ intersects $c_2$ but the other one does not. As in Case 1, we have to consider four cases according to the

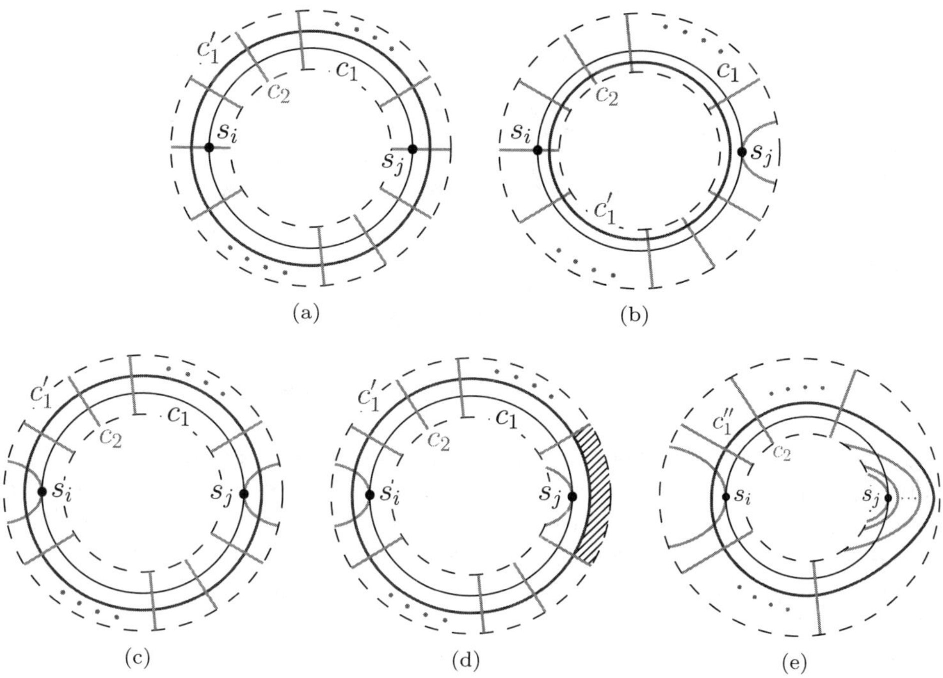


FIGURE 7. The curves $c_1'$ and $c_1''$.

configuration of $c_2$ around $s_i$ and $s_j$. In each case we take parallel copies of $c_1$ as shown in Figures 8, 9 and 10. It is easy to verify that either of the shaded regions

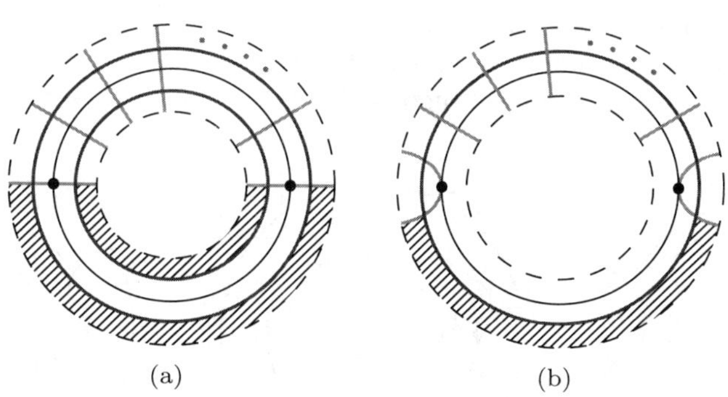


FIGURE 8. The curves parallel to $c_1$ and candidates of bigons.

in Figure 8(a) is not a bigon. Thus, either of the parallel curves of $c_1$ in the figure is in minimal position with $c_2$. If the shaded region in Figure 8(b) is not a bigon, then the parallel curve in the figure is in minimal position with $c_2$. We can move $c_2$ by an isotopy so that it intersects $c_1$ transversely on both $s_i$ and $s_j$ if the shaded region in Figure 8(b) is a bigon. In this case, we can take a curve $c_3 \subset \Sigma_g \setminus \{s_1, \dots, s_n\}$ so that it is away from $c_1$ and $i(c_2, c_3) \neq 0$.

None of the regions made by $c_2$ and the parallel copy of $c_1$ in Figure 9(a), except for the shaded one, can be bigons. If the shaded region in Figure 9(a) is

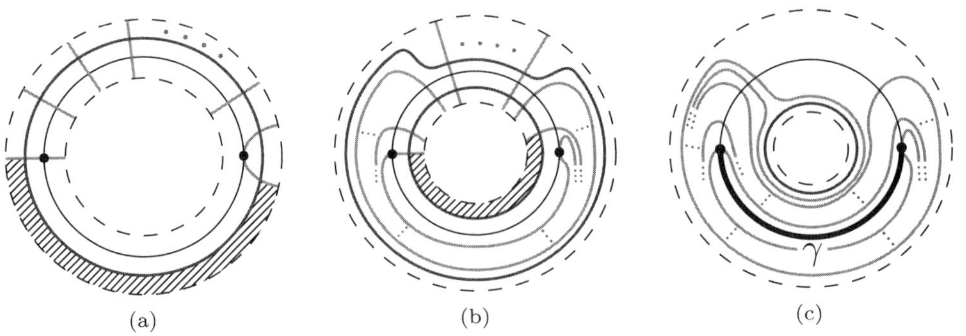


FIGURE 9. The curves parallel to $c_1$ and candidates of bigons.

not a bigon, then the parallel copy of $c_1$ in the figure is in minimal position with $c_2$. If the shaded region in Figure 9(a) is a bigon, we move the parallel copy of $c_1$ so that it avoids all the bigons nested around the shaded region (see Figure 9(b)). It is easily verified (using bigon criterion) that the resulting curve $d$ is in minimal position with $c_2$. If $d$ has non-empty intersection with $c_2$, we can deduce that $\widetilde{t_{c_1}}$ and $\widetilde{t_{c_2}}$ are different mapping classes. If $d$ is away from $c_2$, we take another parallel copy of $c_1$ "inside" $c_1$ as shown in Figure 9(b). This copy is in minimal position provided that the shaded region in Figure 9(b) is not a bigon. If it is a bigon, we again move the copy so that it avoids all the bigons nested around the shaded region. It is easily verify that the resulting curve $d'$ is in minimal position with $c_2$. We can deduce $\widetilde{t_{c_1}} \neq \widetilde{t_{c_2}}$ provided that $d'$ intersects with $c_2$. If $d'$ is away from $c_2$, then $c_2$ is as shown in Figure 9(c), in particular it is isotopic to $\tau_\gamma^N(c_1)$ for some $N > 0$, where $\gamma$ is a path between $s_i$ and $s_j$ in Figure 9(c). We can thus deduce from Lemma 3.3 that $\widetilde{t_{c_1}}$ and $\widetilde{t_{c_2}} = \tau_\gamma^{-N}\widetilde{t_{c_1}}\tau_\gamma^N$ are different mapping classes.

None of the regions made by $c_2$ and the parallel copy of $c_1$ in Figure 10(a), except for the shaded one, can be bigons. Thus we can deduce $\widetilde{t_{c_1}} \neq \widetilde{t_{c_2}}$ provided

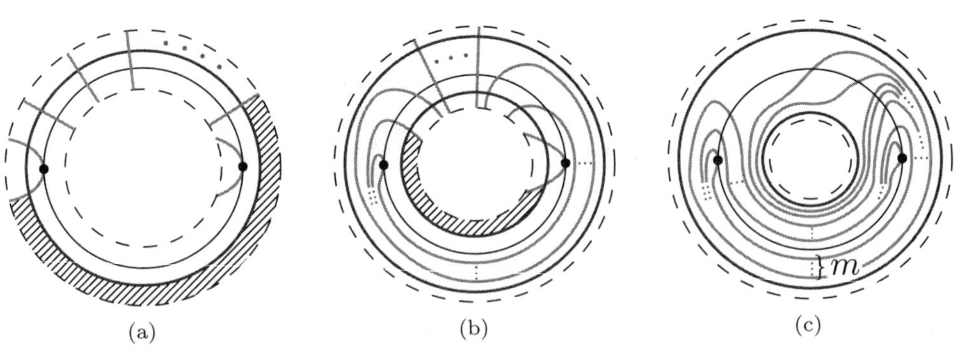


FIGURE 10. The curves parallel to $c_1$ and candidates of bigons.

that the shaded region in Figure 10(a) is not a bigon. If the shaded region is a bigon, we move the parallel copy by an isotopy so that it avoids all the bigons nested around the shaded one in Figure 10(a) (see Figure 10(b)). It is easily verified that the resulting curve $\tilde{d}$ is in minimal position with $c_2$, so we can deduce $\widetilde{t_{c_1}} \neq \widetilde{t_{c_2}}$

provided that $\tilde{d}$ intersects with $c_2$. If $\tilde{d}$ is away from $c_2$, we take another parallel copy of $c_1$ "inside" $c_1$ as shown in Figure 10(b). This copy is in minimal position provided that the shaded region in Figure 10(b) is not a bigon. If it is a bigon, we again move the copy so that it avoids all the bigons nested around the shaded region. It is easy to check that the resulting curve $\tilde{d}'$ is in minimal position with $c_2$. The curve $\tilde{d}'$ must intersect with $c_2$. For, if $\tilde{d}'$ were away from $c_2$, $\tilde{d}'$ is as shown in Figure 9(c), in particular it would be disconnected if the number $m$ of strands of paths is odd, or null-homotopic (as a curve in $\Sigma_g$) if $m$ is even, but both of the consequences contradict our initial assumptions. We conclude that $\widetilde{t_{c_1}}$ and $\widetilde{t_{c_2}}$ are different mapping classes.

**Case 3** : Suppose that neither of the components of $c_1 \setminus \{s_i, s_j\}$ intersect $c_2$. As before, we consider four cases according to the configuration of $c_2$ around $s_i$ and $s_j$. In each case we take parallel copies of $c_1$ as shown in Figure 11. If both of

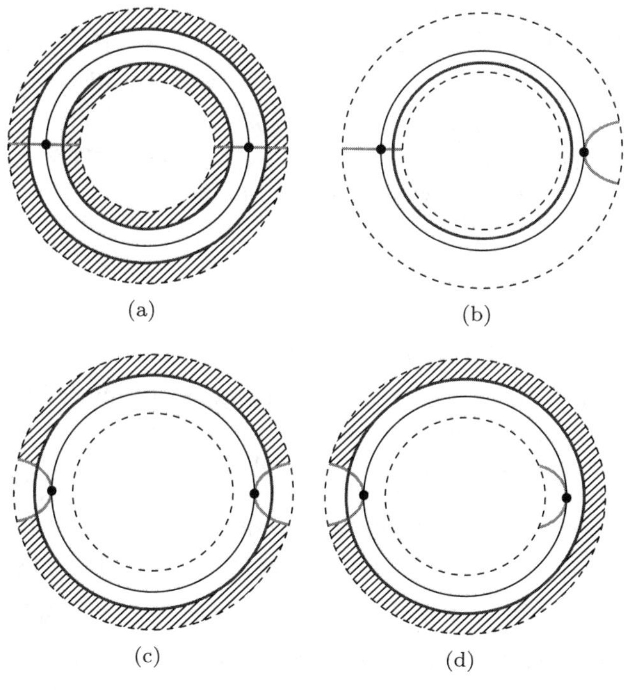

FIGURE 11. The curve $c_1'$ and candidates of bigons.

the parallel copies of $c_1$ in Figure 11(a) formed bigons with $c_2$, either $c_2$ would be homotopic to $c_1$ or $c_2$ is null-homotopic (as a simple closed curve in $\Sigma_g$), which contradicts the assumptions. Thus either one of the copies in Figure 11(a) is in minimal position with $c_2$, and we can deduce $\widetilde{t_{c_1}} \neq \widetilde{t_{c_2}}$. The parallel copy of $c_1$ in Figure 11(b) is in minimal position with $c_2$ since these intersect at a single point. If both of the shaded regions in Figure 11(c) are not bigons, $c_2$ and the copy in the figure are in minimal position. If either of the shaded regions in Figure 11(c) is a bigon, we can move $c_1$ by a isotopy so that $c_1$ and $c_2$ intersect transversely at $s_i$ and $s_j$. In both cases, we can deduce $\widetilde{t_{c_1}} \neq \widetilde{t_{c_2}}$. If the shaded region in Figure 11(d) were a bigon, $c_2$ would be homotopic to $c_1$, which contradicts the assumptions. Thus, $c_2$ and the copy in Figure 11(d) are in minimal position. $\square$

REMARK 3.5. We should point out that one cannot state Lemmas 3.2 and 3.4 for the lifts of the elements featured in them to the framed mapping class group $\mathrm{Mod}(\Sigma_g^n; \{u_1, \dots, u_n\})$ instead. For example, for the paths $\gamma$ and $\gamma' = t_{\delta_i} t_{\delta_j}(\gamma)$ between $s_i$ and $s_j$, which are not isotopic (relative to the boundary of $\Sigma_g^n$) in general, we can see that $\tau_\gamma$ is equal to $\tau_{\gamma'}$ in $\mathrm{Mod}(\Sigma_g^n; \{u_1, \dots, u_n\})$. There are in fact infinitely many such lifts of arc twists in $\mathrm{Mod}(\Sigma_g; \{s_1, \dots, s_n\})$, which is the underlying cause for this ambiguity.

Next is a variation of a classical result of Earle and Schatz [**8**]:

LEMMA 3.6. *If* $2 - 2g - n$ *is negative, then* $\pi_1(\mathrm{Diff}(\Sigma_g; \{s_1, \dots, s_n\}), \mathrm{id})$ *is trivial.*

PROOF. As in the proof of [**10**, Theorem 4.6], we can obtain the following exact sequence (note that we omit the base points for simplicity):

$$\begin{aligned} \pi_2(F_{0,n}(\Sigma_g)) \to \pi_1(\mathrm{Diff}(\Sigma_g; \{s_1, \dots, s_n\})) \to \pi_1(\mathrm{Diff}(\Sigma_g)) \\ \to \pi_1(F_{0,n}(\Sigma_g)), \end{aligned} \tag{1}$$

where $F_{0,n}(\Sigma_g)$ is the configuration space defined in [**9**], which is aspherical if $g \geq 1$. Since the group $\pi_1(\mathrm{Diff}(\Sigma_g))$ is trivial for $g \geq 2$ ([**7**]), so is $\pi_1(\mathrm{Diff}(\Sigma_g; \{s_1, \dots, s_n\}))$ if $g \geq 2$. Furthermore, the following diagram commutes:

$$\begin{array}{ccc} & & \pi_1(T^2) \\ & \swarrow^{\cong} & \uparrow \pi_* \\ \pi_1(\mathrm{Diff}(T^2)) & \xrightarrow{\mathrm{ev}_*} & \pi_1(F_{0,n}(T^2)), \end{array}$$

where $\pi_*$ is induced by the natural projection and $\mathrm{ev}_*$ is induced by the evaluation map, which is the same map as that in (1). In particular, $\mathrm{ev}_*$ is injective. We can thus deduce from the exact sequence (1) that $\pi_1(\mathrm{Diff}(T^2; \{s_1, \dots, s_n\}))$ is trivial for $n > 0$.

We can also obtain the following exact sequence:

$$\begin{aligned} \pi_2(F_{n-1,1}(S^2)) \to \pi_1(\mathrm{Diff}(S^2; \{s_1, \dots, s_n\})) \to \pi_1(\mathrm{Diff}(S^2; \{s_1, \dots, s_{n-1}\})) \\ \to \pi_1(F_{n-1,1}(S^2)). \end{aligned} \tag{2}$$

The configuration space $F_{n-1,1}(S^2)$ is aspherical for $n \geq 2$ ([**9**]). Thus, if the fundamental group $\pi_1(\mathrm{Diff}(\Sigma_g; \{s_1, \dots, s_{n-1}\}))$ is trivial, so is $\pi_1(\mathrm{Diff}(\Sigma_g; \{s_1, \dots, s_n\}))$. Using (2) we can verify that $\pi_1(\mathrm{Diff}(S^2; \{s_1\}))$ is an infinite cyclic group generated by the loop $\theta \to \phi_{2\pi\theta} \in \mathrm{Diff}(S^2; \{s_1\})$, where $\phi_\theta$ is the $\theta$–degree rotation of $S^2$ fixing $s_1$. Since $F_{1,1}(S^2) = \mathbb{R}^2$, especially $\pi_1(F_{1,1}(S^2)) = 1$, $\pi_1(\mathrm{Diff}(\Sigma_g; \{s_1, s_2\}))$ is also an infinite cyclic group generated by $\phi_{2\pi\theta}$. It is easy to see that $[\phi_{2\pi\theta}] \in \pi_1(\mathrm{Diff}(\Sigma_g; \{s_1, s_2\}))$ is sent to the generator of $\pi_1(F_{2,1}(S^2)) \cong \mathbb{Z}$. We can eventually conclude that $\pi_1(\mathrm{Diff}(\Sigma_g; \{s_1, \dots, s_n\}))$ is trivial for any $n \geq 3$. $\square$

REMARK 3.7. The above lemma can be possibly derived as a corollary of the contractibility of the identity component of $\mathrm{Diff}(\Sigma_g^n)$, as shown in [**8**], provided $\mathrm{Diff}(\Sigma_g^n)$ is seen to be homotopy equivalent to $\mathrm{Diff}(\Sigma_g; \{s_1, \dots, s_n\})$. For our purposes however, it is sufficient to calculate the fundamental group of $\mathrm{Diff}(\Sigma_g; \{s_1, \dots, s_n\})$. So we have given a direct proof of Lemma 3.6.

We are now ready to prove the theorem.

PROOF OF THEOREM 3.1. We first assume that $(X_1, f_1, S_1)$ and $(X_2, f_2, S_2)$ are equivalent. We can take diffeomorphisms $\Phi : X_1 \to X_2$ and $\phi : S^2 \to S^2$ such that $\phi \circ f_1 = f_2 \circ \Phi$ and $\Phi(S_1) = S_2$. Let $\alpha_1, \ldots, \alpha_{k+l} \subset S^2$ be reference paths for $f_1$ with the common initial point $p_0$ which give rise to the factorization $W_{X_1,f_1,S_1}$ under an identification $\Theta : (f_1^{-1}(p_0), f_1^{-1}(p_0) \cap S_1) \xrightarrow{\cong} (\Sigma_g, \{s_1, \ldots, s_n\})$. Let $p_0' = \phi(p_0)$, $\alpha_i' = \phi(\alpha_i)$ and $\Theta' = \Theta \circ \Phi^{-1}$, which is an identification of $(f_2^{-1}(p_0'), f_2^{-1}(p_0') \cap S_2)$ with $(\Sigma_g, \{s_1, \ldots, s_n\})$. It is easily verify that the monodromy factorization of $f_2$ obtained from $\alpha_1', \ldots, \alpha_{k+l}'$ and $\Theta'$ coincides with $W_{X_1,f_1,S_1}$. Since any two factorizations of $(X_2, f_2, S_2)$ are Hurwitz equivalent, so are $W_{X_1,f_1,S_1}$ and $W_{X_2,f_2,S_2}$.

In what follows we assume that $W_{X_1,f_1,S_1}$ and $W_{X_2,f_2,S_2}$ are Hurwitz equivalent. We first consider the case that $f_i$ has no critical points and $f_i|_{S_i}$ has no branched points. In this case, $f_i$ can be obtained by pasting two trivial surface bundles over the disk so that the marked points corresponding $n$–sections match. By Lemma 3.6 such a pasting map is unique up to isotopy preserving fibration structures. Thus $f_1$ and $f_2$ are equivalent.

Assume that $f_i$ has critical points or $f_i|_{S_i}$ has branched points. We can take reference paths $\alpha_1^i, \ldots, \alpha_{k+l}^i \subset S^2$ for $f_i$ so that the local monodromies associated with $\alpha_j^1$ and $\alpha_j^2$ coincide.

By composing a self-diffeomorphism of $S^2$ to $f_2$, we assume that $f_1(\mathrm{Crit}(f_1) \cup \mathrm{Crit}(f_1|_{S_1}))$ and $f_2(\mathrm{Crit}(f_2) \cup \mathrm{Crit}(f_2|_{S_2}))$ coincide. Let $f_1(\mathrm{Crit}(f_1) \cup \mathrm{Crit}(f_1|_{S_1})) = \{a_1, \ldots, a_{k+l}\}$ and $D_j \subset S^2$ a sufficiently small disk neighborhood of $a_j$. Since all the local monodromies of $f_1$ and $f_2$ coincide, we can take a diffeomorphism $H : f_1^{-1}(S^2 \setminus \amalg_j D_j) \to f_2^{-1}(S^2 \setminus \amalg_j D_j)$ sending the intersection $f_1^{-1}(S^2 \setminus \amalg_j D_j) \cap S_1$ to $f_2^{-1}(S^2 \setminus \amalg_j D_j) \cap S_2$ such that the following diagram commutes:

$$\begin{array}{ccc} f_1^{-1}(S^2 \setminus \amalg_j D_j) & \xrightarrow{H} & f_2^{-1}(S^2 \setminus \amalg_j D_j) \\ \Big\downarrow f_1 & \swarrow f_2 & \\ S^2 \setminus \amalg_j D_j. & & \end{array}$$

In what follows we will extend $H$ to a diffeomorphism with the source containing the preimages $f_1^{-1}(D_1), \ldots, f_1^{-1}(D_{k+l})$.

If $f_i^{-1}(a_j)$ contains a Lefschetz critical point which is not a branched point of $f_i|_{S_i}$, then in the same manner as in the proof of [**14**, Theorem 2.4] we can extend $H$ to a diffeomorphism with the source containing $f_1^{-1}(D_j)$ (in this procedure we need Lemma 3.6 instead of the contractibility of $\mathrm{Diff}(\Sigma_g)$ used in [**14**, p.133]).

Assume that $f_i^{-1}(a_j)$ contains a Lefschetz critical point $x_{ij} \in X_i$ which is also a branched point of $f_i|_{S_i}$. There exist complex coordinate neighborhoods $(U^i, \varphi^i)$ and $(V^i, \psi^i)$ at $x_{ij} \in X_i$ and $f_i(x_{ij}) \in S^2$, respectively, which satisfy the following properties:

(1) $\psi^i \circ f_i \circ (\varphi^i)^{-1}(z, w) = z^2 + w^2$,
(2) $\varphi^i(U^i \cap S_i) = \mathbb{C} \times \{0\}$.

Using the disk theorem as in the proof of [**14**, Theorem 2.4], we may assume that $\psi^1$ and $\psi^2$ coincide without loss of generality. By Lemma 3.4 the vanishing cycles associated with $\alpha_j^1$ and $\alpha_j^2$ coincide up to isotopy relative to the points $s_1, \ldots, s_n$. Thus, in the same way as that in the proof of [**14**, Lemma 2.5], we can change $H$

by a vertical isotopy (in the sense of [**14**]) sending $S_1$ to $S_2$ at all times so that $\varphi^2 \circ H = \varphi^1$ on a neighborhood (in $f_1^{-1}(\partial D_j)$) of the vanishing cycle of $f_1$ associated with $\alpha_j^1$. The arguments following the proof of [**14**, Lemma 2.5] can be applied to our situation, and we can eventually extend $H$ to a diffeomorphism with the source containing $f_1^{-1}(D_j)$.

Lastly, if $f_i^{-1}(a_j)$ contains a branched point of $f_i|_{S_i}$ away from $\mathrm{Crit}(f_i)$, then we can extend $H$ to a diffeomorphism with the source containing $f_1^{-1}(D_j)$ in a manner quite similar to that in the previous paragraph, where we invoke Lemma 3.2 instead of Lemma 3.4 this time. □

**3.2. Monodromy factorizations in the framed mapping class group.** As discussed in the previous section (cf. Remark 2.5) we can take a lift of the factorization $W_{X,f,S}$ to that of a product of Dehn twists along boundary components in the framed mapping class group $\mathrm{Mod}(\Sigma_g^n; \{u_1, \ldots, u_n\})$. Such a lift is needed to fully capture the local topology of the multisection $S$.

Two such lifts $\widetilde{W}_{X_i,f_i,S_i}$ of $W_{X_i,f_i,S_i}$, for $i = 1, 2$, are not necessarily related by elementary transformations and simultaneous conjugations even if $(X_1, f_1, S_1)$ and $(X_2, f_2, S_2)$ are equivalent. There is indeed no canonical way to choose lifts of $\tau_\gamma$ and $\widetilde{t_c}$. For instance, each one of the paths $\gamma_i$, $i = 1, 2$ in Figure 12 is a lift of $\gamma$, and in turn, $\tau_{\gamma_i} \in \mathrm{Mod}(\Sigma_g^n; \{u_1, \ldots, u_n\})$ is a lift of $\tau_\gamma \in \mathrm{Mod}(\Sigma_g; \{s_1, \ldots, s_n\})$.

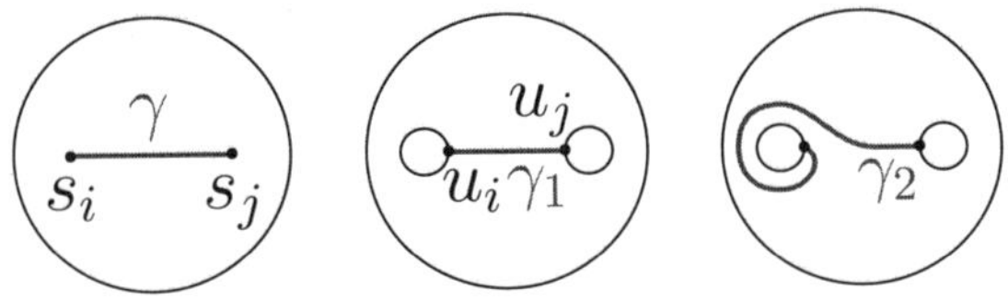


FIGURE 12. Examples of paths between $s_i$ and $s_j$.

It is easy to see that $\tau_{\gamma_2}$ is equal to $t_{\delta_i} \tau_{\gamma_1} t_{\delta_i}^{-1} = \tau_{\gamma_1} t_{\delta_j} t_{\delta_i}^{-1}$, where $\delta_k$ is a simple closed curve along the boundary component containing $u_k$. Thus, different choices of lifts of $\tau_\gamma$ and $\widetilde{t_c}$ in $W_{X_i,f_i,S_i}$ might yield a factorization $\widetilde{W}_{X_i,f_i,S_i}$ with distinct products, in particular the Hurwitz equivalence class of $\widetilde{W}_{X_i,f_i,S_i}$ depends on the choices of lifts of members in $W_{X_i,f_i,S_i}$.

For lifts of $\tau_\gamma$ and $\widetilde{t_c}$ are uniquely determined up to conjugations by Dehn twists along boundary components however, $\widetilde{W}_{X_i,f_i,S_i}$ is uniquely determined up to elementary transformations, simultaneous conjugations, plus a third modification:

(3) *Framing conjugation*, which changes a factorization as follows:

$$\xi_{k+l} \cdots \xi_{i+1} \xi_i \xi_{i-1} \cdots \xi_1 \longleftrightarrow \xi_{k+l} \cdots \xi_{i+1} (t_\delta \xi_i t_\delta^{-1}) \xi_{i-1} \cdots \xi_1,$$

where $\delta$ is a simple closed curve along a boundary component of $\Sigma_g^n$.

A framing conjugation does not affect usual Dehn twists among the $\xi_i$ factors since any simple closed curve can be isotoped away from the boundary (but does affect lifts of Dehn twists). In particular, this move is not needed to relate monodromy factorizations in $\mathrm{Mod}(\Sigma_g^n; \{u_1, \ldots, u_n\})$ associated to *pure multisections*, i.e. disjoint union of sections.

We say that two positive factorizations of products of Dehn twists along boundary components in $\mathrm{Mod}(\Sigma_g^n;\{u_1,\dots,u_n\})$ are *Hurwitz equivalent*, if one can be obtained from the other by a sequence of elementary transformations, simultaneous conjugations and framing conjugations.

For $i=1,2$, let $(X_i,f_i,S_i)$ be a genus–$g$ Lefschetz fibration with an $n$–section $S_i$, with monodromy factorization $W_{X_i,f_i,S_i}$ in $\mathrm{Mod}(\Sigma_g;\{s_1,\dots,s_n\})$. Let $\widetilde{W}_{X_i,f_i,S_i}$ be a lift of $W_{X_i,f_i,S_i}$, a positive factorization of the form

$$\widetilde{W}_{X_i,f_i,S_i}:\widetilde{\xi}_{k+l}\cdots\widetilde{\xi}_1=t_{\delta_1}^{a_1}\cdots t_{\delta_n}^{a_n}$$

in $\mathrm{Mod}(\Sigma_g^n;\{u_1\dots,u_n\})$. Then Theorem 3.1, together with the above observation, thus yields to a one-to-one correspondence in this setting as well:

COROLLARY 3.8. *Suppose that $2-2g-n$ is negative. The triples $(X_1,f_1,S_1)$ and $(X_2,f_2,S_2)$ are equivalent if and only if $\widetilde{W}_{X_1,f_1,S_1}$ and $\widetilde{W}_{X_2,f_2,S_2}$ are Hurwitz equivalent.*

REMARK 3.9. It is worth noting that a framing conjugation can also affect the right hand side of a positive factorization

$$\xi_{k+l}\cdots\xi_1=t_{\delta_1}^{a_1}\cdots t_{\delta_n}^{a_n}$$

in the framed mapping class group, simultaneously increasing and decreasing the powers of the involved boundary twists. In order to make the effect of a framing conjugation clear, let us define the following surjective homomorphism

$$\Lambda:\mathrm{Mod}(\Sigma_g^n;\{u_1,\dots,u_n\})\twoheadrightarrow\mathfrak{S}_n$$

defined by the action of a mapping class on the set $\{u_1,\dots,u_n\}$, where $\mathfrak{S}_n$ is the symmetric group of order $n$. A Dehn twist $t_c$ is contained in the kernel of $\Lambda$, while a half twist $\tau_\gamma$ and a lift $\widetilde{t}_c$ of a Dehn twist are sent to transpositions by $\Lambda$. Clearly $\xi\, t_{\delta_i}$ is equal to $t_{\delta_{\Lambda(\xi)(i)}}\xi$ for any $\xi\in\mathrm{Mod}(\Sigma_g^n;\{u_1,\dots,u_n\})$, so a framing conjugation changes the right hand side of the factorization as follows:

$$\xi_{k+l}\cdots\xi_{i+1}(t_{\delta_j}\xi_i t_{\delta_j}^{-1})\xi_{i-1}\cdots\xi_1=t_{\delta_{\Lambda(\xi_{k+l}\cdots\xi_{i+1})(j)}}t^{-1}_{\delta_{\Lambda(\xi_{k+l}\cdots\xi_i)(j)}}t_{\delta_1}^{a_1}\cdots t_{\delta_n}^{a_n}.$$

**3.3. Equivalence of Lefschetz pencils.** Given a genus–$g$ Lefschetz pencil $(X,f)$ with base locus $B=\{x_1,\dots,x_n\}$, recall that we can pass to a genus–$g$ Lefschetz *fibration* $(X',f')$, with a distinguished *pure* $n$–section $S$ that consists of $n$ disjoint sections $S_j$ of self-intersection $-1$, each arising as an exceptional sphere in $X'=X\#n\overline{\mathbb{CP}}^2$ of the blow-up at the base point $x_j\in X$. Since one can blow-down all $S_j$ to recover the pencil $(X,f)$, we can work with the well-known monodromy factorization $\widetilde{W}_{X',f',S}$ of $(X',f',S)$ of the form

$$t_{c_l}\cdots t_{c_1}=t_{\delta_1}\cdots t_{\delta_n}$$

in the framed mapping class group $\mathrm{Mod}(\Sigma_g^n;\{u_1,\dots,u_n\})$. Note that this factorization is in fact contained in the subgroup $\mathrm{Mod}_{\partial\Sigma_g^n}(\Sigma_g^n)$ (whose elements restrict to identity along $\partial\Sigma_g^n$) which only captures *pure* $n$–sections, but we need the larger group in order to factor in pencil automorphisms which swap base points, which we will discuss shortly. This associated factorization is what we will call monodromy factorization of the *pencil* $(X,f)$, and with the above correspondence in mind, we will denote it simply by $W_{X,f}$.

Lefschetz pencils $f_i : X_i \setminus B_i \to S^2$, $i = 1, 2$, are said to be *equivalent* if there exist orientation-preserving diffeomorphism $\Phi : X_1 \to X_2$ and $\phi : S^2 \to S^2$ such that $\Phi(B_1) = B_2$ and $\phi \circ f_1 = f_2 \circ \Phi$. Clearly, for $(X_i, f_i)$ to be equivalent pencils, they should both have the same fiber genus and the same number of base points $|B_1| = |B_2|$. It now follows from Theorem 3.1 that:

COROLLARY 3.10. *Two Lefschetz pencils $(X_i, f_i)$ of genus $g \geq 1$ with $n$ base points are equivalent if and only they have Hurwitz equivalent monodromy factorizations $W_{X_i,f_i}$ in* $\mathrm{Mod}(\Sigma_g^n; \{u_1, \ldots, u_n\})$.

PROOF. Since $g \geq 1$ and $n > 0$, we have $2-2g-n < 0$. If $f_1$ and $f_2$ are equivalent, then the corresponding pairs of Lefschetz fibrations and these sections are also equivalent. Thus, $W_{X_1,f_1}$ and $W_{X_2,f_2}$ are Hurwitz equivalent by Corollary 3.8. Suppose that $W_{X_1,f_1}$ and $W_{X_2,f_2}$ are Hurwitz equivalent. Let $f_i' : X_i' = X_i \# n\overline{\mathbb{CP}}^2 \to S^2$ be the associated Lefschetz fibration. We deduce from Theorem 3.1, and the assumption that there exist diffeomorphisms $\Phi' : X_1' \to X_2'$ and $\phi : S^2 \to S^2$ such that $\Phi'$ sends the union of exceptional spheres of $X_1'$ to that of $X_2'$ and $\phi \circ f_1' = f_2' \circ \Phi'$. So $\Phi'$ induces a diffeomorphism $\Phi : X_1 \to X_2$ which satisfies $\Phi(B_1) = B_2$ and $\phi \circ f_1 = f_2 \circ \Phi$, providing an equivalence between $(X_1, f_1)$ and $(X_2, f_2)$. □

REMARK 3.11. The Hurwitz equivalence in the statement of Corollary 3.10 is not the classical one for fibrations, it is our (extended) Hurwitz equivalence for monodromy factorizations in the framed mapping class group, allowing the exceptional sections / base points to be interchanged. Although we believe the above corollary to be known to experts, we are not aware of any proof of it in the literature.

## 4. Lefschetz fibrations which do not arise from pencils

Although every Lefschetz pencil gives rise to a Lefschetz fibration on a blow-up of its total space, the converse is known to be false. As shown by Stipsicz [**17**], and independently by Smith [**16**], if $(X, f)$ is a fiber sum of two nontrivial Lefschetz fibrations,[1] it cannot have any exceptional sections, i.e. sections of self-intersection $-1$, and thus it is not a blow-up of a pencil. Motivated by this, Stipsicz conjectured in the same article that every Lefschetz fibration $(X, f)$, if not a blow-up of a pencil, is a fiber sum of such, which amounts to having Lefschetz pencils as building blocks of any Lefschetz fibration via fiber sums.

In [**15**], Sato proved that an interesting genus–2 Lefschetz fibration constructed by Auroux in [**2**], which could not be a fiber sum of nontrivial Lefschetz fibrations, did not have any exceptional sections either. This remained as the only known counter-example until recently, where in [**4**], we obtained several other genus–2 and 3 counter-examples. The purpose of this section is to demonstrate the recipe of [**4**] to generate such examples. We will do this while producing quick examples from a well-known relation in the mapping class group. We will then show that Auorux's example can also be derived in this very scheme. For various background results that goes into this recipe, we advise the reader to turn to [**4**].

[1]That is, the monodromy factorization $W_{X,f}$, up to Hurwitz equivalence, can be expressed as a product of two nontrivial positive factorizations of 1.

**4.1. Examples derived by monodromy substitutions.** Let $c_1, \ldots, c_5$ be the simple closed curves on $\Sigma_2^2$ as shown in Figure 14, and $\delta_1, \delta_2$ denote the two boundary components with marked points $u_1, u_2$. The chain relation of length 5

$$(t_{c_1} t_{c_2} t_{c_3} t_{c_4} t_{c_5})^6 = t_{\delta_1} t_{\delta_2}$$

in $\mathrm{Mod}(\Sigma_2; \{u_1, u_2\})$ (see [**10**, Proposition 4.12]), prescribes a triple $(X_0, f_0, S_0)$, which is a genus–2 Lefschetz fibration $(X_0, f_0)$ with a pure 2–section $S$ that consists of two exceptional sections $S_1, S_2$. It is well-known that $X_0$ is the K3 surface blown-up twice, the symplectic canonical class of which is represented by $[S] = [S_1] + [S_2]$ in $H_2(X_0; \mathbb{Z})$.

We will need the following *braiding lantern relation*, which is a generalization of the lantern relation in the framed mapping class group:

LEMMA 4.1 ([**4**]). *Let the curves $a, b, c, d, x, \delta_1, \delta_2$, pairs of arcs $y, z$ and points $u_1, u_2$ in $\Sigma_0^6$ be as shown in Figure 13, where $a, b, c, d, \delta_1, \delta_2$ are parallel to boundary components. Denote the boundary components parallel to $\delta_i$ by $S_i$. Then the relation $\widetilde{t_z} t_x \widetilde{t_y} = t_a t_b t_c t_d t_{\delta_2}$ holds in $\mathrm{Mod}_{\partial\Sigma_0^6 \setminus (S_1 \sqcup S_2)}(\Sigma_0^6; \{u_1, u_2\})$ (whose elements restrict to identity along the four boundary components without marked points).*

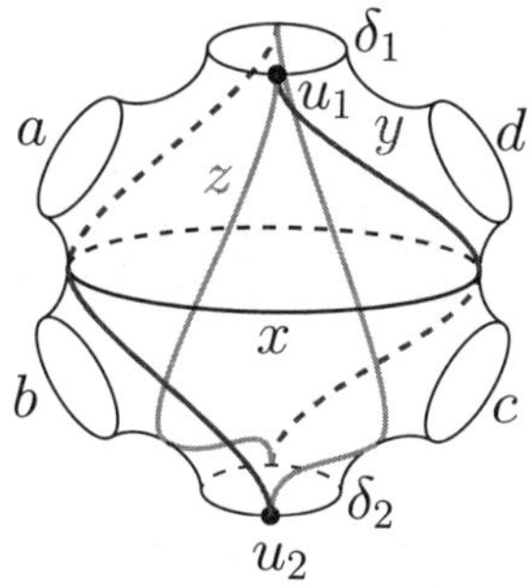


FIGURE 13. Curves in $\Sigma_0^6$.

Substituting the subword on the right hand side of the above relation in the framed mapping class group (of a surface that contains the above subsurface with marked boundary components) with the subword on the left is then called a *braiding lantern substitution.* Its importance and relevance to our current discussion is due to our observation in [**4**] that whenever the two marked boundary components correspond to two exceptional sections $S_1, S_2$, we get a new exceptional 2–section $S_{12}$ after the substitution, which we view as a result of *braiding* $S_1$ and $S_2$ together.

Forgetting the two marked boundary components, one gets the usual lantern relation. In this case, the subword $t_a t_b t_c t_d$ indicates that by clustering the corresponding Lefschetz critical points on the same singular fiber, we can obtain a fiber *component* $V$, which is a sphere of self-intersection $-4$. This $V$, which we will call a *lantern sphere*, can be assumed to be symplectic with respect to a Gompf-Thurston form. Remembering the two marked boundary components, we conclude that the exceptional classes $S_1$ and $S_2$ each intersect $V$ positively at one point.

Now, one can easily find a lantern sphere $V$ in the monodromy factorization of $(X_0, f_0)$ hit once by each exceptional section $S_i$, $i = 1, 2$. Remarkably, this holds for *any* lantern sphere (and there are many of them; at least six disjoint ones [**1**]). This is because for $V$ symplectic, the adjunction equality implies that its intersection with the canonical class $[S_1]+[S_2]$ is 2, whereas each $S_i$ intersects the fibers positively.

Applying the braiding lantern substitution results in a new triple $(X_2, f_2, S_{12})$, a genus–2 Lefschetz fibration with an exceptional 2–section $S_{12}$. As discussed in [**4**], an observation of Gompf shows that $X_2$ is diffeomorphic to an ordinary blow-down of $X_0$, so $X_2 = \mathrm{K3}\#\overline{\mathbb{CP}}^2$, which has only one exceptional class, already represented by $S_{12}$. (See [**13**, Corollary 3]. Note that $\mathrm{K3}\#\overline{\mathbb{CP}}^2$ is not rational nor ruled.) It follows that $(X_2, f_2)$ does not have any other exceptional *sections*. On the other hand, it was shown by Usher in [**18**] that a nonminimal symplectic 4–manifold cannot be a nontrivial fiber sum (also see [**3**] for a simpler proof for Lefschetz fibrations), so the presence of $S_{12}$ also implies that $(X_2, f_2)$ cannot be a fiber sum of any two nontrivial Lefschetz fibrations. Hence, we have obtained another example of a Lefschetz fibration which cannot arise from Lefschetz pencils via blow-ups and fiber sums.

**4.2. Auroux's genus–2 fibration with an exceptional 2–section.** Let $\delta_1$ and $\delta_2$ be simple closed curves in $\Sigma_2^2$ parallel to the boundary components containing $u_1$ and $u_2$, respectively. We take non-separating simple closed curves $c_1, c_2, c_3, c_4, c_5 \subset \Sigma_2^2$, a path $\gamma \subset \Sigma_2^2$ between $u_1$ and $u_2$ and a pair $\sigma \subset \Sigma_2^2$ of paths connecting the two boundary components as shown in Figure 14. We will denote

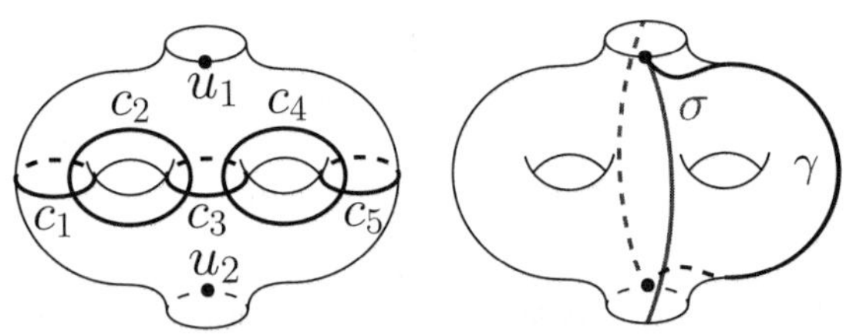


FIGURE 14. Simple closed curves and paths in $\Sigma_2^2$.

the right-handed Dehn twist along $c_i$ by $t_i \in \mathrm{Mod}(\Sigma_2^2)$.

PROPOSITION 4.2. *The following relation holds in* $\mathrm{Mod}(\Sigma_2^2; \{u_1, u_2\})$*:*

$$\tau_\gamma t_5 t_4 t_3 t_2 t_1 t_1 t_2 t_3 t_4 t_5 (t_3 t_2 t_1 t_4 t_3 t_2 t_5 t_4 t_3)^2 \widetilde{t_\sigma} = t_{\delta_1}^2 t_{\delta_2}^2.$$

PROOF. Let $L = t_5 t_4 t_3 t_2 t_1 t_1 t_2 t_3 t_4 t_5$ and $T = t_3 t_2 t_1 t_4 t_3 t_2 t_5 t_4 t_3$. It is easy to see (by the Alexander method (see [**10**, §. 2.3]), for example) that $\widetilde{t_\sigma}$ is equal to $(t_2 t_1)^3 (t_5 t_4)^3 [\iota]$, where $[\iota]$ is the mapping class of the hyperelliptic involution $\iota$ given in Figure 15.

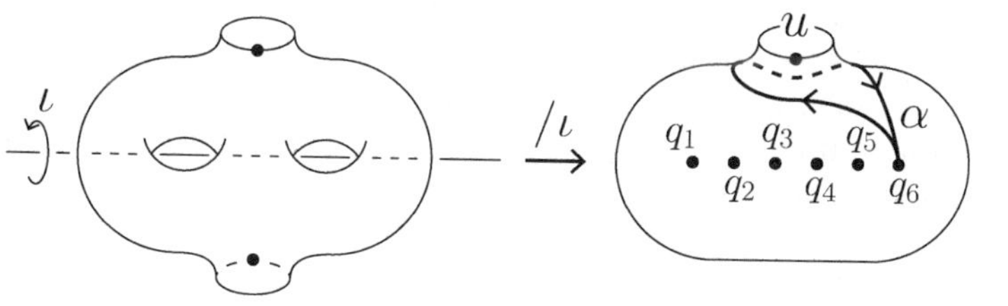


FIGURE 15. Left : the hyperelliptic involution $\iota$. Right : the quotient surface $\Sigma_2^2/\iota$.

Thus, the product $T^2\widetilde{t_\sigma}$ is calculated as follows:

$$\begin{aligned}
T^2\widetilde{t_\sigma} =& Tt_3t_2t_1t_4t_3t_2t_5t_4t_3(t_2t_1)^3(t_5t_4)^3[\iota]\\
=& Tt_3t_2t_1t_4t_3t_2(t_5t_4t_3t_2t_1)(t_2t_1)^2(t_5t_4)^3[\iota]\\
=& Tt_3t_2t_1(t_5t_4t_3t_2t_1)t_5t_4t_3(t_2t_1)^2(t_5t_4)^3[\iota]\\
=& Tt_3t_2t_1(t_5t_4t_3t_2t_1)^2t_2t_1(t_5t_4)^3[\iota]\\
=& T(t_5t_4t_3t_2t_1)^3(t_5t_4)^3[\iota]\\
=& T(t_2t_1)^3(t_5t_4t_3t_2t_1)^3[\iota]\\
=& (t_5t_4t_3t_2t_1)^6[\iota]\\
=& t_{\delta_1}t_{\delta_2}[\iota],
\end{aligned}$$

where the last equality follows from the chain relation of length 5. We can naturally regard the mapping class $L$ as an element in $\pi_0(C_\emptyset(\Sigma_2^2,\{u_1,u_2\};\iota))$, where $C_\emptyset(\Sigma_2^2,\{u_1,u_2\};\iota)$ is defined in [**4**, §.3.1]. By [**4**, Lemma 3.1], the kernel of the homomorphism

$$\iota_* : \pi_0(C_\emptyset(\Sigma_2^2,\{u_1,u_2\};\iota)) \to \mathrm{Mod}(\Sigma_2^2/\iota; u, \{q_1,q_2,q_3,q_4,q_5,q_6\})$$

induced by the quotient map $\Sigma_2^2 \to \Sigma_2^2/\iota$ is generated by the class $[\iota]$, where $u$ denotes the point $[u_1]=[u_2]\in\Sigma_2^2/\iota$ and $\mathrm{Mod}(\Sigma_2^2/\iota; u, \{q_1,q_2,q_3,q_4,q_5,q_6\})$ consists of orientation preserving diffeomorphisms fixing $u$ (resp. $\{q_1,q_2,q_3,q_4,q_5,q_6\}$) pointwise (resp. setwise) modulo isotopies fixing the same data. It is easy to see that the image $\iota_*(L)$ is equal to the pushing map $\mathrm{Push}(\alpha)$ along the loop $\alpha$ in Figure 15. The mapping class $\tau_\gamma^{-1}t_{\delta_1}t_{\delta_2}$ can also be regarded as an element in $\pi_0(C_\emptyset(\Sigma_2^2,\{u_1,u_2\};\iota))$, and it is sent to $\mathrm{Push}(\alpha)$ by $\iota_*$. Thus, $L^{-1}\tau_\gamma^{-1}t_{\delta_1}t_{\delta_2}$ is contained in the kernel of $\iota_*$. Since $L^{-1}\tau_\gamma^{-1}t_{\delta_1}t_{\delta_2}$ interchanges the points $u_1$ and $u_2$, this is equal to $[\iota]$. In particular the following relation holds in $\mathrm{Mod}(\Sigma_2^2;\{u_1,u_2\})$:

$$L = \tau_\gamma^{-1}t_{\delta_1}t_{\delta_2}[\iota].$$

Thus, the product $\tau_\gamma LT^2\widetilde{t_\sigma}$ is calculated as follows:

$$\begin{aligned}
\tau_\gamma LT^2\widetilde{t_\sigma} =& \tau_\gamma\tau_\gamma^{-1}t_{\delta_1}t_{\delta_2}[\iota]t_{\delta_1}t_{\delta_2}[\iota]\\
=& t_{\delta_1}t_{\delta_2}t_{\delta_2}t_{\delta_1}[\iota][\iota]\\
=& t_{\delta_1}^2t_{\delta_2}^2.
\end{aligned}$$

This completes the proof. □

As we explained in Section 2, we can regard the surface $\Sigma_2^2$ as a subsurface of $\Sigma_2$ by capping $\partial\Sigma_2^2$ by two disks with the centers $s_1, s_2 \in \Sigma_2$. The pair of paths $\sigma$ and the path $\gamma$ respectively give rise to a simple closed curve in $\Sigma_2$ going through $s_1$ and $s_2$ and a simple path between $s_1$ and $s_2$. To simplify the notation, we use the same symbols $\sigma$ and $\gamma$ to represent these curves. We also denote the Dehn twist along $c_i \subset \Sigma_2$ by $t_i \in \mathrm{Mod}(\Sigma_2; \{s_1, s_2\})$.

Now by Prop. 4.2, we obtain the following factorization in $\mathrm{Mod}(\Sigma_2; \{s_1, s_2\})$:

$$\tau_\gamma t_5 t_4 t_3 t_2 t_1 t_1 t_2 t_3 t_4 t_5 (t_3 t_2 t_1 t_4 t_3 t_2 t_5 t_4 t_3)^2 \widetilde{t_\sigma} = 1, \tag{3}$$

which prescribes a triple $(X_1, f_1, S_1)$ where $f_1 : X_1 \to S^2$ is a genus–2 Lefschetz fibration with a sphere 2–section $S_1$ by Theorem 2.3. Under the forgetful homomorphism $\mathrm{Mod}(\Sigma_2; \{s_1, s_2\}) \to \mathrm{Mod}(\Sigma_2)$ this positive factorization maps to the monodromy factorization of Auroux's aforementioned genus–2 Lefschetz fibration given in [**2**]. On the other hand, we can calculate the self-intersection number of $S_1$ using the positive factorization in the framed mapping class group we had in Proposition 4.2 (which of course is a lift of the monodromy factorization $W_{X_1, f_1, S_1}$!) and Theorem 1.1 of [**4**]. Hence $(X_1, f_1, S_1)$ is Auroux's genus–2 Lefschetz fibration, where $S_1$ is the exceptional 2–section.

We are now ready to show that $(X_1, f_1, S_1)$ can be reproduced using our recipe discussed in the previous subsection. Let $(X_0, f_0, S_0)$ denote the genus–2 Lefschetz fibration with a pure 2–section $S_0$, which is a disjoint union of two exceptional sections, as prescribed by the 5–chain relation $(t_1 t_2 t_3 t_4 t_5)^6 = 1$ in $\mathrm{Mod}(\Sigma_2; \{s_1, s_2\})$.

PROPOSITION 4.3. *The triple $(X_1, f_1, S_1)$, where $(X_1, f_1)$ is Auroux's genus–2 fibration with the exceptional 2–section $S_1$, is equivalent to a triple obtained from $(X_0, f_0, S_0)$ by a single braiding lantern substitution, followed by a perturbation of the 2–section.*

PROOF. We first prove that the factorization $(t_1 t_2 t_3 t_4 t_5)^6 = 1$ in

$$\mathrm{Mod}(\Sigma_2; \{s_1, s_2\})$$

is Hurwitz equivalent to the following factorization:

$$t_3 t_2 t_1 t_4 t_3 t_2 t_5 t_4 t_3 t_5 t_4 t_3 t_2 t_1 t_2 t_3 t_4 t_5 t_3 t_2 t_1 t_4 t_3 t_2 t_5 t_4 \underline{t_1 t_1 t_5 t_5} = 1. \tag{4}$$

The factorization (4) can be changed by elementary transformations as follows (in each line elementary transformations are applied to the underlined part to obtain the next line):

$$\begin{aligned}
&t_3 t_2 t_1 t_4 t_3 t_2 t_5 t_4 t_3 t_5 t_4 t_3 t_2 t_1 t_2 t_3 t_4 t_5 t_3 t_2 t_1 t_4 t_3 t_2 \underline{t_5 t_4 t_1 t_1 t_5} t_5 \\
\sim &t_3 t_2 t_1 t_4 t_3 t_2 t_5 t_4 t_3 t_5 t_4 t_3 t_2 t_1 t_2 t_3 t_4 t_5 t_3 t_2 t_1 \underline{t_4 t_3 t_2 t_4} t_5 t_4 t_1 t_1 t_5 \\
\sim &t_3 t_2 t_1 t_4 t_3 t_2 t_5 t_4 t_3 t_5 t_4 t_3 t_2 t_1 t_2 t_3 t_4 t_5 \underline{t_3 t_2 t_1 t_3} t_4 t_3 t_2 t_5 t_4 t_1 t_1 t_5 \\
\sim &t_3 t_2 t_1 t_4 t_3 t_2 t_5 t_4 t_3 t_5 t_4 t_3 \underline{t_2 t_1 t_2 t_3 t_4 t_5} t_2 t_3 \underline{t_2 t_1 t_4 t_3 t_2 t_5} t_4 t_1 t_1 t_5 \\
\sim &t_3 t_2 t_1 t_4 t_3 t_2 t_5 t_4 t_3 t_5 t_4 \underline{t_3 t_1 t_2 t_3 t_4 t_5 t_1 t_2 t_3 t_4 t_5} t_2 \underline{t_1 t_3 t_2 t_4 t_1} t_1 t_5 \\
\sim &t_3 t_2 t_1 t_4 t_3 t_2 t_5 t_4 t_3 t_5 t_4 (t_1 t_2 t_3 t_4 t_5)^2 t_1 t_2 t_3 \underline{t_2 t_1 t_2 t_4 t_1 t_5} \\
\sim &t_3 t_2 t_1 t_4 t_3 t_2 t_5 t_4 t_3 t_5 t_4 \underline{(t_1 t_2 t_3 t_4 t_5)^2 t_1 t_2 t_3 t_4 t_5 t_2 t_1 t_2 t_1} \\
\sim &\underline{t_3 t_2 t_1 t_4 t_3 t_2 t_5 t_4 t_3 t_5 t_4 t_5 t_4 t_5 t_4} (t_1 t_2 t_3 t_4 t_5)^3 \\
\sim &\underline{t_2 t_1 t_2} t_1 t_2 \underline{t_1 t_3 t_2 t_1} t_4 t_3 t_2 t_5 t_4 t_3 (t_1 t_2 t_3 t_4 t_5)^3
\end{aligned}$$

$$
\begin{aligned}
&\sim t_1t_2t_1t_1\underline{t_2t_3t_2}t_1\underline{t_2t_4t_3t_2}t_5t_4t_3(t_1t_2t_3t_4t_5)^3\\
&\sim t_1t_2\underline{t_1t_1t_3}t_2\underline{t_3t_1t_4t_3}t_2\underline{t_3t_5t_4t_3}(t_1t_2t_3t_4t_5)^3\\
&\sim t_1t_2t_3\underline{t_1t_1t_2t_1t_4}t_3\underline{t_4t_2t_5t_4}t_3t_4(t_1t_2t_3t_4t_5)^3\\
&\sim t_1t_2t_3t_4\underline{t_1t_1t_2t_1t_3t_2t_5}t_4t_5t_3t_4(t_1t_2t_3t_4t_5)^3\\
&\sim t_1t_2t_3t_4t_5t_1\underline{t_1t_2t_1}t_3t_2t_4t_5t_3t_4(t_1t_2t_3t_4t_5)^3\\
&\sim t_1t_2t_3t_4t_5t_1t_2t_1\underline{t_2t_3t_2}t_4t_5t_3t_4(t_1t_2t_3t_4t_5)^3\\
&\sim t_1t_2t_3t_4t_5t_1t_2\underline{t_1t_3}t_2t_3t_4t_5t_3t_4(t_1t_2t_3t_4t_5)^3\\
&\sim t_1t_2t_3t_4t_5t_1t_2t_3\underline{t_1t_2t_3t_4t_5}t_3t_4(t_1t_2t_3t_4t_5)^3\\
&\sim t_1t_2t_3t_4t_5t_1t_2t_3t_4t_5t_1t_2t_3t_4t_5(t_1t_2t_3t_4t_5)^3\\
&=(t_1t_2t_3t_4t_5)^6.
\end{aligned}
$$

We take pairs of paths $\xi,\zeta$ and a simple closed curve $a$ as shown in Figure 16. By Lemma 4.1, $t_1t_1t_5t_5$ is equal to $\widetilde{t_\xi}t_3\widetilde{t_\sigma}$. Applying the substitution by this relation

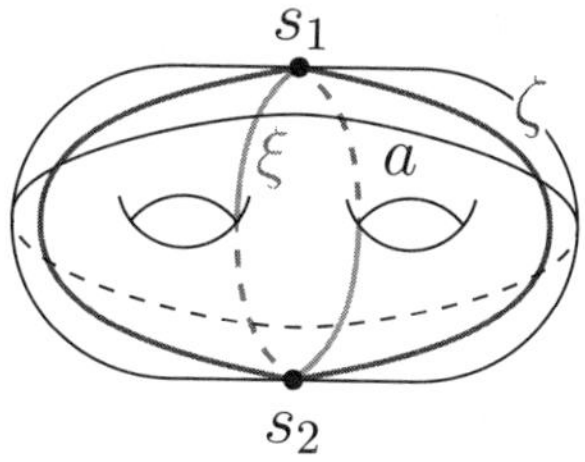



FIGURE 16. Pair of paths and a simple closed curve in $\Sigma_2$.

to the underlined part in (4), we obtain the following factorization:

$$
\begin{aligned}
&t_3t_2t_1t_4t_3t_2t_5t_4t_3t_5t_4t_3t_2t_1t_2t_3t_4t_5\underline{t_3t_2t_1t_4t_3t_2t_5t_4\widetilde{t_\xi}}t_3\widetilde{t_\sigma}\\
&\sim t_3t_2t_1t_4t_3t_2t_5t_4t_3t_5t_4t_3t_2t_1t_2t_3t_4t_5\widetilde{t_\zeta}t_3t_2t_1t_4t_3t_2t_5t_4t_3\widetilde{t_\sigma},
\end{aligned}
\tag{5}
$$

where the equivalence above holds since $t_3t_2t_1t_4t_3t_2t_5t_4t_3(\xi)$ is equal to $\zeta$.

According to [**4**, Remark 3.9], we can perturb the 2–section of the fibration corresponding (5) so that it is away from the Lefschetz critical point corresponding $\widetilde{t_\zeta}$, and the resulting fibration with a 2–section has the following factorization:

$$
\begin{aligned}
&t_3t_2t_1t_4t_3t_2t_5t_4t_3t_5t_4t_3t_2t_1\underline{t_2t_3t_4t_5t_a}\tau_\gamma t_3t_2t_1t_4t_3t_2t_5t_4t_3\widetilde{t_\sigma}\\
&\sim \underline{t_3t_2t_1t_4t_3t_2t_5t_4t_3t_5t_4t_3t_2t_1t_1t_2t_3t_4t_5\tau_\gamma} t_3t_2t_1t_4t_3t_2t_5t_4t_3\widetilde{t_\sigma}\\
&\sim \underline{t_5t_4t_3t_2t_1t_1t_2t_3t_4t_5\tau_\gamma}(t_3t_2t_1t_4t_3t_2t_5t_4t_3)^2\widetilde{t_\sigma}\\
&\sim \tau_\gamma t_5t_4t_3t_2t_1t_1t_2t_3t_4t_5(t_3t_2t_1t_4t_3t_2t_5t_4t_3)^2\widetilde{t_\sigma}.
\end{aligned}
$$

The last factorization coincides with the monodromy factorization of $(X_1, f_1, S_1)$ in the Equation (3) above.

We have thus derived $(X_1, f_1, S_1)$ from $(X_0, f_0, S_0)$, after a braiding lantern substitution, followed by a perturbation of the resulting 2–section to move one of its branched points off the Lefschetz critical locus —while applying Hurwitz moves for Lefschetz fibrations with multisections in various steps of the proof. □

## References


[1] Anar Akhmedov and Jun-Yong Park, *Lantern substitution and new symplectic 4-manifolds with* ${b_2}^+ = 3$, Math. Res. Lett. **21** (2014), no. 1, 1–17, DOI 10.4310/MRL.2014.v21.n1.a1. MR3247035

[2] Denis Auroux, *Fiber sums of genus 2 Lefschetz fibrations*, Turkish J. Math. **27** (2003), no. 1, 1–10. MR1975329

[3] R. İnanç Baykur, *Minimality and fiber sum decompositions of Lefschetz fibrations*, Proc. Amer. Math. Soc. **144** (2016), no. 5, 2275–2284, DOI 10.1090/proc/12835. MR3460185

[4] R. I. Baykur and K. Hayano *Multisections of Lefschetz fibrations and topology of symplectic 4–manifolds*, Geom. & Topol. (to appear).

[5] S. K. Donaldson, *Lefschetz pencils on symplectic manifolds*, J. Differential Geom. **53** (1999), no. 2, 205–236. MR1802722

[6] Simon Donaldson and Ivan Smith, *Lefschetz pencils and the canonical class for symplectic four-manifolds*, Topology **42** (2003), no. 4, 743–785, DOI 10.1016/S0040-9383(02)00024-1. MR1958528

[7] Clifford J. Earle and James Eells, *A fibre bundle description of Teichmüller theory*, J. Differential Geometry **3** (1969), 19–43. MR0276999

[8] C. J. Earle and A. Schatz, *Teichmüller theory for surfaces with boundary*, J. Differential Geometry **4** (1970), 169–185. MR0277000

[9] Edward Fadell and Lee Neuwirth, *Configuration spaces*, Math. Scand. **10** (1962), 111-118. MR0141126

[10] Benson Farb and Dan Margalit, *A primer on mapping class groups*, Princeton Mathematical Series, vol. 49, Princeton University Press, Princeton, NJ, 2012. MR2850125

[11] Ian Hambleton and Matthias Kreck, *Cancellation, elliptic surfaces and the topology of certain four-manifolds*, J. Reine Angew. Math. **444** (1993), 79–100. MR1241794

[12] A. Kas, *On the handlebody decomposition associated to a Lefschetz fibration*, Pacific J. Math. **89** (1980), no. 1, 89–104. MR596919

[13] Tian-Jun Li, *Smoothly embedded spheres in symplectic* 4*-manifolds*, Proc. Amer. Math. Soc. **127** (1999), no. 2, 609–613, DOI 10.1090/S0002-9939-99-04457-3. MR1459135

[14] Yukio Matsumoto, *Lefschetz fibrations of genus two—a topological approach*, Topology and Teichmüller spaces (Katinkulta, 1995), World Sci. Publ., River Edge, NJ, 1996, pp. 123–148. MR1659687

[15] Yoshihisa Sato, *2-spheres of square* $-1$ *and the geography of genus-2 Lefschetz fibrations*, J. Math. Sci. Univ. Tokyo **15** (2008), no. 4, 461–491 (2009). MR2546906

[16] Ivan Smith, *Geometric monodromy and the hyperbolic disc*, Q. J. Math. **52** (2001), no. 2, 217–228, DOI 10.1093/qjmath/52.2.217. MR1838364

[17] András I. Stipsicz, *Indecomposability of certain Lefschetz fibrations*, Proc. Amer. Math. Soc. **129** (2001), no. 5, 1499–1502, DOI 10.1090/S0002-9939-00-05681-1. MR1712877

[18] Michael Usher, *Minimality and symplectic sums*, Int. Math. Res. Not., posted on 2006, Art. ID 49857, 17, DOI 10.1155/IMRN/2006/49857. MR2250015



Department of Mathematics and Statistics, University of Massachusetts, Amherst, Massachusetts 01003-9305

*E-mail address*: `baykur@math.umass.edu`

Department of Mathematics, Faculty of Science and Technology, Keio University, Yokohama, Kanagawa, 223-8522, Japan

*E-mail address*: `k-hayano@math.keio.ac.jp`

Contemporary Mathematics
Volume **675**, 2016
http://dx.doi.org/10.1090/conm/675/13582

# The curvature Veronese of a 3-manifold immersed in Euclidean space

Rosane R. Binotto, Sueli I. R. Costa, and M. Carmen Romero Fuster

ABSTRACT. We study the behaviour of the curvature locus of a 3-manifold immersed in $\mathbb{R}^n, n \geq 5$, describing their possible topological types at the different points of a 3-manifold immersed in $\mathbb{R}^6$. We analyze the connections between the structure of the curvature locus with the behaviour of the principal configurations associated to normal fields on the 3-manifold, as well as with the local convexity of the 3-manifold.

## 1. Introduction

The concept of curvature ellipse at a point of a surface immersed in 4-space has been known since a long time ago [**6**] and it has proven to be a useful tool in the study of the geometrical properties from both, the local and global viewpoint [**3**, **7**, **8**]. Its natural generalization to higher dimensional manifolds is given by the image of a convenient linear projection of a Veronese submanifold of order 2 in the normal space of the manifold at each point [**7**]. We call it the *curvature locus* or *curvature veronese*. The behaviour of the curvature locus of codimension two submanifolds in $\mathbb{R}^n$ was investigated in [**12**], where a especial emphasis was made in the study of submanifolds with flat normal bundle, for which it was shown that the curvature locus becomes a polihedron at each point.

An important feature of the curvature locus for submanifolds of dimension $n \geq 3$ immersed in $\mathbb{R}^{n+k}, k > 2$, is the fact that, for certain pairs $(n, k)$, it may present several topological types. Another interesting aspect is the possible interpretation of its singularities in terms of the behaviour of the principal directions of the normal fields at a given point and hence the possible description of geometrical properties relative to the behaviour of the family of principal configurations on the submanifold in terms of these singularities. Such singularities become especially interesting for some values $(n, k)$. We consider here the case of 3-manifolds immersed in $\mathbb{R}^n, n \geq 5$ and provide the analytical description of their curvature locus in terms of the coefficients of their fundamental form at each point. We focalize our attention mainly in the study of the pair $(3, 3)$ which represents the richest possibility from the viewpoint of the analysis of shapes of the curvature locus of a 3-manifold. In this case, we show that *all the possible curvature locus models are the Steiner surfaces of types 1,3,5,6, the ellipsoid, a truncated cone, a convex planar region*

2010 *Mathematics Subject Classification.* Primary 53A05, 58K20, 58K05.
*Key words and phrases.* 3-manifolds, curvature locus, principal configurations, convexity.

*(that may adopt different geometrical shapes), a segment and a point.* We show that the planar region may only occur at isolated points of generic immersions. We also study the connection of the singularities of the curvature locus with the behaviour of the set of principal directions and other geometrical properties, such as convexity and existence of quasiumbilic and umbilic directions at a given point.

A relevant question in connection with this study is the investigation of closed 3-manifolds immersed in $\mathbb{R}^6$ whose curvature locus has constant type. A trivial example of this is given by a round 3-sphere, whose curvature locus at each point coincides with the center of the sphere. Other examples may be furnished by convenient products of lower dimensional round spheres. An interesting open question is the existence of a generic embedding of a closed 3-manifold in $\mathbb{R}^6$ with constant shape of its curvature locus.

## 2. Second fundamental form and shape operators

Let $M$ be a compact 3-manifold in $\mathbb{R}^{3+k}, k \geq 1$, locally given as the image of an embedding $f : U \to \mathbb{R}^n$. Denote by $(x, y, z)$ the local coordinates on $M$. Then for each $p = f(x, y, z)$ we consider the basis of $T_pM$ given by $B^f = \{f_x, f_y, f_z\}_p$, where $f_x = \frac{\partial f}{\partial x}$, etc. Let $\{e_1, e_2, \ldots, e_k\}$ be an orthonormal normal frame in $U$, such that the orientation of the frame $\{f_x, f_y, f_z, e_1, \ldots, e_{n-3}\}$ coincides with the orientation of the space $\mathbb{R}^n$.

Given a normal field $\nu$ on $M$, consider the linear map

$$d\nu : T_pM \to T_p\mathbb{R}^{3+k} = T_pM \oplus N_pM$$

and let $\pi^T : T_pM \oplus N_pM \to T_pM$ be the tangential projection. The map

$$S_p^\nu = -\pi^T \cdot d\nu : T_pM \to T_pM$$

is known as the *$\nu$-shape operator* at $p$.

The *second fundamental form of $M$ along a normal vector field $\nu$* is the bilinear map $II_p^\nu : T_pM \times T_pM \to \mathbb{R}$, defined as $II_p^\nu(\mathbf{v}, \mathbf{w}) = \langle \nu, d^2f(\mathbf{v}, \mathbf{w})\rangle$. Observe that for each $p = f(u_0) \in M$ and $\mathbf{v} \in N_pM$, the coefficients of $II_p^{\mathbf{v}}$ do not depend on the choice of the vector field $\nu$, $\nu(0) = \mathbf{v}$.

The second fundamental form $II_p : T_pM \times T_pM \to N_pM$ is the associated bilinear map given by $II_p(\mathbf{v}, \mathbf{w}) = \pi^\perp d^2f(\mathbf{v}, \mathbf{w})$ that projects $d^2f(\mathbf{v}, \mathbf{w})$ onto the normal space at $p$ and induces a linear map

$$\begin{array}{rcccl} A_p & : & N_pM & \longrightarrow & \mathbb{Q}_2 \\ & & v & \longmapsto & II_p^v, \end{array}$$

where $\mathbb{Q}_2$ is the 6-dimensional space of quadratic forms on $T_pM$. We say that a point $p \in M$ *is of type $M_i$* provided rank $A_p = i, i = 1, \cdots, 6$. It is well known that $S_p^\nu$ is a self-adjoint operator and satisfies

$$\langle S_p^\nu(v), w\rangle = II_p^\nu(v, w), \forall v, w \in T_pM.$$

The eigenvectors of $S_p^\nu$ are called *$\nu$-principal directions* and its corresponding eigenvalues are the *$\nu$-principal curvatures.* The points at which two of the $\nu$-principal curvatures coincide are called *$\nu$-quasiumbilic points.* This means that there exists a tangent plane made of $\nu$-principal directions. The points at which all the tangent directions are $\nu$-principal directions are said to be *$\nu$-umbilic points.* The $\nu$-quasiumbilic and $\nu$-umbilic points form the critical set of the $\nu$-principal configurations foliations on $M$. A 3-manifold $M$ is said to be *totally $\nu$-umbilic* it all its points are $\nu$-umbilic.

## 3. Curvature locus

Given a point $p \in M \subset \mathbb{R}^n, n \geq 5$ and a unit vector $v \in S^2 \subset T_pM$, denote by $\gamma_v$ the *normal section* of $M$ in the direction $v$, that is, $\gamma_v = M \cap \mathbb{H}_v$, where $\mathbb{H}_v = \{\lambda v\} \oplus N_pM$ is a hyperplane through $p$ in $\mathbb{R}^n$. The normal curvature vector $\eta(v)$ of $\gamma_v$ at $p$ belongs to $N_pM$ and varying $v$ in $S^2 \subset T_pM$, we obtain a surface in the normal space $N_pM$. Since the normal curvature vector $\eta(v)$ is the same at antipodal points of $S^2 \subset T_pM$, we can view the image of $\eta$ as the image of a projective plane $\mathbb{P}^2$ in $N_pM \equiv \mathbb{R}^k$. We call this surface the *curvature locus* of $M$ at $p$. A straightforward calculation shows that the curvature locus can also be seen as the image of the unit tangent sphere at $p \in M$ via $II(v, v)$.

LEMMA 3.1. *By taking spherical coordinates in the unit tangent sphere,*

$$\begin{array}{rcccl} \eta & : & S^2 \subset T_pM & \longrightarrow & N_pM \\ & & (\theta, \phi) & \longmapsto & \eta(\theta, \phi), \end{array}$$

*we obtain the following expression for the curvature locus,*

$$\begin{aligned} \eta(\theta, \phi) &= H + \big(1 + 3\cos(2\phi)\big)B_1 + \cos(2\theta)\big(\sin(\phi)\big)^2 B_2 \\ &+ \sin(2\theta)\big(\sin(\phi)\big)^2 B_3 + \cos(\theta)\sin(2\phi)B_4 + \sin(\theta)\sin(2\phi)B_5, \end{aligned}$$

*where*

$$H = \tfrac{1}{3}\big(f_{xx} + f_{yy} + f_{zz}\big),\ B_1 = \tfrac{1}{12}\Big(-f_{xx} - f_{yy} + 2f_{zz}\Big),$$
$$B_2 = \tfrac{1}{2}\Big(f_{xx} - f_{yy}\Big),\ B_3 = f_{xy},\ B_4 = f_{xz} \text{ and } B_5 = f_{yz}.$$

**Proof:** Suppose that $M$ is locally given at $p = \psi(0,0,0) \in M$ by an immersion in the Monge form $f(x,y,z) = xf_x + yf_y + zf_z + f_1(q)e_1 + \cdots + f_k(q)e_k$, where $\{e_1, e_2, \ldots, e_k\}$ is an orthonormal frame in a small enough neighbourhood $V$ of $p \in M$, as above. The matrix of the second fundamental form with respect to this frame is given by

$$[\alpha_f(p)] = \begin{pmatrix} a_1 & b_1 & c_1 & d_1 & r_1 & s_1 \\ a_2 & b_2 & c_2 & d_2 & r_2 & s_2 \\ & & & \vdots & & \\ a_{n-3} & b_{n-3} & c_{n-3} & d_{n-3} & r_{n-3} & s_{n-3} \end{pmatrix},$$

where

$$a_i = f_{xx} \cdot e_{i+3} = \frac{\partial^2 f_i}{\partial x^2}, \quad b_i = f_{xy} \cdot e_{i+3} = \frac{\partial^2 f_i}{\partial x \partial y}, \quad c_i = f_{xz} \cdot e_{i+3} = \frac{\partial^2 f_i}{\partial x \partial z}$$

$$d_i = f_{yy} \cdot e_{i+3} = \frac{\partial^2 f_i}{\partial y^2}, \quad r_i = f_{yz} \cdot e_{i+3} = \frac{\partial^2 f_i}{\partial y \partial z}, \quad s_i = f_{zz} \cdot e_{i+3} = \frac{\partial^2 f_i}{\partial z^2},$$

the different derivatives being evaluated at $p$. Now, if we denote

$$H = \tfrac{1}{3}\sum_{i=1}^{n-3}(a_i + d_i + s_i)e_{i+3} = \frac{1}{3}(f_{xx} + f_{yy} + f_{zz}),$$

$$B_1 = \tfrac{1}{12}\sum_{i=1}^{n-3}\big(-a_i - d_i + 2s_i\big)e_{i+3} = \frac{1}{12}\Big(-f_{xx} - f_{yy} + 2f_{zz}\Big),$$

$$B_2 = \tfrac{1}{2}\sum_{i=1}^{n-3}\big(a_i - d_i\big)e_{i+3} = \frac{1}{2}\Big(f_{xx} - f_{yy}\Big), \quad B_3 = \sum_{i=1}^{n-3} b_i\, e_{i+3} = f_{xy},$$

$$B_4 = \sum_{i=1}^{n-3} c_i\, e_{i+3} = f_{xz} \quad \text{and} \quad B_5 = \sum_{i=1}^{n-3} r_i\, e_{i+3} = f_{yz}$$

we obtain the required expression for the curvature locus. □

REMARK 3.2. *The value of the normal field $H$ at $p \in M$, is the mean curvature vector of $M$ at $p$.*

The *first normal space* of $M$ a $p$ is given by

$$N_p^1 M = \big\langle f_{xx},\ f_{xy},\ f_{xz},\ f_{yy},\ f_{yz},\ f_{zz}\big\rangle_p.$$

This clearly coincides with the subspace $\big\langle H,\ B_1,\ B_2,\ B_3,\ B_4,\ B_5\big\rangle_{(p)}$. Denote by $Aff_p$ the affine hull of the curvature locus in $N_pM$ and by $E_p$ the linear subspace of $N_p^1M$ which is parallel to $Aff_p$. Clearly $E_p = \big\langle B_1,\ B_2,\ B_3,\ B_4,\ B_5\big\rangle_{(p)}$. Also observe that $p \in Aff_p$ if and only if $H \in E_p$. It follows that the curvature locus is contained in an affine subspace of $N_pM$ with dimension $\leq 5$.

REMARK 3.3. *An alternative expression for the curvature locus is given by*

$$\eta(u, v, w) = H + \big(4 - 6u^2 - 6v^2\big)B_1 + \big(u^2 - v^2\big)B_2 + 2uvB_3 + 2uwB_4 + 2vwB_5,$$

*where* $u^2 + v^2 + w^2 = 1$.

**3.1. The curvature locus as a projection of the Veronese surface.** We shall interpret now the curvature locus as the image of the classical Veronese surface of order 2 through a convenient linear map. We recall the expression of the classical Veronese surface of order 2, which is given by the image of the 2-sphere in $\mathbb{R}^3$ through the map $\xi : \mathbb{R}^3 \to \mathbb{R}^6$, given by

$$\xi(u, v, w) = \Big(u^2, v^2, w^2, \sqrt{2}uv, \sqrt{2}uw, \sqrt{2}vw\Big).$$

The subset $\xi(S^2)$ lies in a hyperplane (more precisely, in a 4-sphere given by the intersection of a hyperplane with a hypersphere) of $\mathbb{R}^6$. An expression for the Veronese surface in $\mathbb{R}^5$ is given, up to rotation, by

$$\xi(u, v, w) = \Big(\tfrac{-1}{\sqrt{6}}, \tfrac{1}{\sqrt{2}}, 0, 0, 0\Big) + \Big(\tfrac{3}{\sqrt{6}}v^2,\ \tfrac{1}{\sqrt{2}}(-2u^2 - v^2),\ \sqrt{2}uv, \sqrt{2}uw, \sqrt{2}vw\Big)$$

$$= H_1 + \tilde{\xi}(u, v, w).$$

Consider the following matrices in the respective canonical basis of $\mathbb{R}^5$ and $\mathbb{R}$:

$$[\tilde{\xi}] = \begin{pmatrix} 0 & \frac{3}{\sqrt{6}} & 0 & 0 & 0 \\ \frac{-2}{\sqrt{2}} & \frac{-1}{\sqrt{2}} & 0 & 0 & 0 \\ 0 & 0 & \sqrt{2} & 0 & 0 \\ 0 & 0 & 0 & \sqrt{2} & 0 \\ 0 & 0 & 0 & 0 & \sqrt{2} \end{pmatrix} \quad \text{and} \quad [W] = \begin{pmatrix} u^2 \\ v^2 \\ uv \\ uw \\ vw \end{pmatrix},$$

We can write

$$\xi(u,v,w) \;=\; H_1 \;+\; [\tilde{\xi}][W],\ \forall (u,v,w) \in S^2.$$

Where we observe that the expression $[\tilde{\xi}][W]$ represents a rotation of the Veronese surface. We take now the expression of the curvature locus at a point $p \in M$ as in Remark 3.3,

$$\eta(u,v,w) \;=\; H_2 \;+\; 6(-u^2-v^2)B_1 \;+\; (u^2-v^2)B_2 \;+\; 2uvB_3 \;+\; 2uwB_4 \;+\; 2vwB_5,$$

where $H_2 = H + 4B_1$. Consider the linear transformation $T : \mathbb{R}^5 \to N_pM \equiv \mathbb{R}^{n-3}, n \geq 5$, determined by requiring $T(e_1) = B_1, T(e_2) = B_2, T(e_3) = B_3, T(e_4) = B_4$ and $T(e_5) = B_5$, where $\{e_1,\ e_2,\ e_3,\ e_4,\ e_5\}$ is the canonical basis of $\mathbb{R}^5$. Now, in the respective canonical basis $\mathbb{R}^5$ and $\mathbb{R}^{n-3}$, we can take the matrices

$$[\tilde{\eta}] = \begin{pmatrix} -6 & -6 & 0 & 0 & 0 \\ 1 & -1 & 0 & 0 & 0 \\ 0 & 0 & 2 & 0 & 0 \\ 0 & 0 & 0 & 2 & 0 \\ 0 & 0 & 0 & 0 & 2 \end{pmatrix},$$

and

$$[T] = \begin{pmatrix} \frac{1}{12}\big(-a_1-d_1+2s_1\big) & \frac{1}{2}\big(a_1-d_1\big) & b_1 & c_1 & r_1 \\ \frac{1}{12}\big(-a_2-d_2+2s_2\big) & \frac{1}{2}\big(a_2-d_2\big) & b_2 & c_2 & r_2 \\ & & \vdots & & \\ \frac{1}{12}\big(-a_{n-3}-d_{n-3}+2s_{n-3}\big) & \frac{1}{2}\big(a_{n-3}-d_{n-3}\big) & b_{n-3} & c_{n-3} & r_{n-3} \end{pmatrix}.$$

From this we arrive to the following expression for the curvature locus at $p$,

$$\eta(u,v,w) \;=\; H_2 \;+\; [T][\tilde{\eta}][W],$$

for all $(u,v,w) \in S^2$. Observe that we can find an invertible matrix $[G]$ of order $5 \times 5$, such that

$$[\tilde{\eta}] = [G][\tilde{\xi}].$$

Then, if $G$ is the diffeomorphism associated to the matrix $[G]$ we get,

$$\eta(S^2) \;=\; H_2 + T \circ \tilde{\eta}(S^2) \;=\; H_2 + T \circ G \circ \tilde{\xi}(S^2).$$

Here, we notice that

$$\eta(S^2) \;=\; H_2 + T \circ (G \circ \tilde{\xi})(S^2) \subset Aff_p \subset N_pM \equiv \mathbb{R}^{n-3}.$$

As a consequence, we can state the following:

PROPOSITION 3.4. *Given a 3-manifold $M$ in $\mathbb{R}^{3+k}, k \geq 1$ and $p \in M$, the curvature locus at $p$ is isomorphic (and thus diffeomorphic) to the classical Veronese surface of order 2 composed with an affine map from $\mathbb{R}^5$ to $N_pM \equiv \mathbb{R}^k$, which is determined by the second fundamental form at each point. The rank of the associated linear transformation coincides with the dimension of the subspace $Aff_p \subseteq N_pM$.*

EXAMPLE 3.5. *The following examples illustrate different possibilities for the curvature locus on a smooth 3-manifolds immersed in* $\mathbb{R}^n, n \geq 4$.

(1) ***Classical Veronese surface:*** *The immersion,* $f : (\mathbb{R}^3, 0) \to (\mathbb{R}^9, 0)$ *defined as* $f(x, y, z) = \left(x, y, z, \frac{x^2}{2}, \frac{y^2}{2}, \frac{z^2}{2}, \sqrt{2}xy, \frac{\sqrt{2}xz}{2}, \frac{\sqrt{2}yz}{2}\right)$ *has the following curvature locus at the origin,*

$$\eta(u, v, w) = \left(u^2, v^2, w^2, \sqrt{2}uv, \sqrt{2}uw, \sqrt{2}vw\right),$$

*with* $u^2 + v^2 + w^2 = 1$. *This is a classical Veronese surface of order 2 substantially embedded in an affine 5-subspace of* $\mathbb{R}^6$ *and is transversal to the mean curvature vector* $H = \frac{1}{4}(1, 1, 2, 0, 0, 0)$.

(2) ***Embedded projective plane:*** *The immersion* $f : (\mathbb{R}^3, 0) \to (\mathbb{R}^7, 0)$ *given by* $f(x, y, z) = \left(x, y, z, \frac{\sqrt{2}}{2}xz, \frac{\sqrt{2}}{2}yz, \frac{\sqrt{2}}{4}(x^2 - y^2), \sqrt{2}xy\right)$, *has the following curvature locus at the origin*

$$\eta(u, v, w) = \left(\sqrt{2}uw, \sqrt{2}vw, \frac{\sqrt{2}}{2}(u^2 - v^2), \sqrt{2}uv\right),$$

*with* $u^2+v^2+w^2 = 1$. *This surface is an orthogonal projection of the classical Veronese surface of order 2 and defines an embedding of the projective plane* $\mathbb{P}^2$ *into an affine 4-dimensional subspace.*

(3) ***Roman Steiner surface:*** *The immersion,* $f : (\mathbb{R}^3, 0) \to (\mathbb{R}^6, 0)$ *defined as* $f(x, y, z) = \left(x, y, z, \sqrt{2}xy, \frac{\sqrt{2}}{2}xz, \frac{\sqrt{2}}{2}yz\right)$ *has the following curvature locus at the origin,*

$$\eta(\theta, \phi) = \frac{\sqrt{2}}{2}\left((\sin(\phi))^2 \sin(2\theta),\ \sin(2\phi)) \cos(\theta),\ \sin(2\phi) \sin(\theta)\right),$$

*with* $0 \leq \theta, \psi \leq \pi$, *which is a Roman Steiner surface contained in an affine 3-dimensional subspace.*

(4) ***Eliptical region:*** *The immersion,* $f : (\mathbb{R}^3, 0) \to (\mathbb{R}^5, 0)$ *defined as* $f(x, y, z) = \left(x, y, z, \frac{\sqrt{2}}{2}xz, \frac{\sqrt{2}}{2}yz,\right)$ *has the following curvature locus at the origin,*

$$\eta(\theta, \phi) = \frac{\sqrt{2}}{2}\left(\sin(2\phi) \cos(\theta),\ \sin(2\phi) \sin(\theta)\right),$$

*com* $0 \leq \theta, \phi \leq \pi$. *Its image is a circular region with radius* $\frac{\sqrt{2}}{2}$, *centered at the origin.*

(5) ***Segment:*** *The curvature locus of the immersion* $f : (\mathbb{R}^3, 0) \to (\mathbb{R}^4, 0)$, *defined as* $f(x, y, z) = \left(x, y, z, z^2\right)$, *at the point* $p \equiv (0, 0, 0)$ *is a non radial segment given by*

$$\eta(\theta, \phi) = \left(1 + \cos(2\phi)\right),$$

*with* $0 \leq \theta, \phi \leq \pi$.

(6) ***Point:*** *The curvature locus at the origin of the immersion*

$$f \; : \; (\mathbb{R}^3,0) \longrightarrow (\mathbb{R}^4,0)$$
$$(x,y,z) \longmapsto \Big(x,y,z,x^2+y^2+z^2\Big).$$

*reduces to the point* $\eta(\theta,\phi) = \Big(2,0,0,0\Big), \quad 0 \leq \theta,\phi \leq \pi.$

REMARK 3.6. ***1)*** *The curvature locus is invariant through isometries of* $\mathbb{R}^m$ *into* $\mathbb{R}^m$ *and is preserved through the canonical embeddings of* $\mathbb{R}^m$ *into* $\mathbb{R}^{m+r}, r>0$.

Given a 3-manifold $M$ immersed in $\mathbb{R}^{3+k}, k \geq 1$, we say that a point $p \in M$ *is of type* $M_i, i=0,1,\cdots,6$ if and only if dim $N^1_pM = i$, where $N^1_pM$ denotes the first normal space of $M$ at $p$, i.e. the orthogonal complement of the kernel of the second fundamental form of $M$ at $p$. Clearly, $N^1_pM = \langle H, B_1, B_2, B_3, B_4, B_5\rangle$.

REMARK 3.7. *According to the rank of the second fundamental form at the point p, we have the following possibilities:*

(1) $p \in M_6$ *if and only if the curvature locus at* $p$ *is difeomorphic to the Veronese surface which is substancial in the 5-space* $Aff_p$ *and* $H(p) \notin E_p$.
(2) $p \in M_k$, *for* $k=4,5$, *if and only if the curvature locus at* $p$ *is either substancial in the* $(k-1)$*-space* $Aff_p$ *and* $H(p) \notin E_p$, *or it is a substancial surface in the k-space, given by a projection of the Veronese surface to* $E_p = Aff_p$.
(3) $p \in M_3$ *if and only if the curvature locus at* $p$ *is either a closed region contained in the plane* $Aff_p$, *with* $H$ *transversal to* $E_p$, *or it is a substancial compact surface given by a projection of the Veronese surface to the 3-space* $E_p = Aff_p$.
(4) $p \in M_2$ *if and only if the curvature locus at* $p$ *is either a non radial segment non parallel to* $H$, *or a closed region given by a projection of the Veronese surface to the plane* $E_p = Aff_p$.
(5) $p \in M_1$ *if and only if the curvature locus at* $p$ *is either a radial segment, or a point distinct from* $p$.
(6) $p \in M_0$ *if and only if the curvature locus at* $p$ *coincides with* $p$.

From now on, we shall refer to the curvature locus of a 3-manifold $M \subset \mathbb{R}^{3+k}, k \geq 3$ at a point $p$ as the *curvature veronese* of $M$ at $p$.

THEOREM 3.8. *Given a compact 3-manifold* $M$, *there is a residual subset of immersions of* $M$ *into* $\mathbb{R}^{3+k}$ *satisfying the following:*

(1) *If* $k>6$, *every point of* $M$ *is of type* $M_6$.
(2) *If* $k=6$, *there exists a codimension* 1 *subset* $S$ *(possibly empty) in* $M$ *such that the points in* $M-S$ *are of type* $M_6$ *and those of* $S$ *are of type* $M_5$.
(3) *For* $k=5,4,3,2,1$, *every point of* $M$ *is of type* $M_k$.

**Proof:** We have that a point $p$ is of type $M_k$ if and only if the second fundamental form has rank $k$ at $p$, which means that the subspace spanned by the six vectors $\{H, B_1, B_2, B_3, B_4, B_5\}$ has dimension $k$ in $N_pM$. Clearly, these vectors just depend on the second derivatives of the immersion, so the condition for a point to be of type $M_i, i \leq 6$ can be given in terms of the intersection of the image of the 2-jet map

$$j^2 f : M \to J^2(M, \mathbb{R}^6)$$

with a convenient algebraic variety of $J^2(M, \mathbb{R}^{3+k})$. Then by analyzing the codimension of the algebraic variety corresponding to each case, we can prove each assertion as an immediate consequence of Thom's Transversality Theorem [**5**]. In fact,

1) For $k > 6$, we have that a point $p \in M$ is not of type $M_6$ if and only if the vectors $\{H, B_1, B_2, B_3, B_4, B_5\}$ are linearly dependent in $N_pM$, where dim $N_pM > 6$. But this is equivalent to asking the vanishing of all the 6 minors in the $k \times 6$ matrix of the second fundamental form at $p$. These equations determine an algebraic submanifold $V$ of codimension higher than 3 in $J^2(M, \mathbb{R}^{3+k})$. We now get from Thom's Transversality Theorem that for a generic immersion $f : M \to \mathbb{R}^{3+k}$, the image of $j^2 f$ does not hit $V$ which implies the required result.
2) If $k = 6$, the condition for a point to be of type $M_k, k < 6$ is equivalent to asking that the squared $6 \times 6$-matrix $[Q]$ given by the coeficients of the second fundamental form has vanishing determinant. This gives rise to a codimension 1 algebraic submanifold $V$ in $J^2(M, \mathbb{R}^{3+k})$. According to Thom's transversality theorem, for a residual subset of immersions, the pull-back of $V$ by $j^2 f$ may be either empty, or a codimension 1 algebraic submanifold $M'$ in $M$. Given a point $p \in M'$, we have that $p \in M_4$ if and only if all the 5-dimensional minors of $[Q]$ vanish. This defines an algebraic subset of codimension higher than 3 and from this we get the required result.
3) For $k = 5, 4, 3, 2, 1$, the second fundamental form has a $k \times 6$-matrix. Since $k > 6$, not having maximal rank implies the vanishing of the 6 minors $k \times k$. By using similar arguments to the above ones, this defines an algebraic variety $V$ of codimension 6 in $J^2(M, \mathbb{R}^{3+k})$, and again, as a consequence of Thom's Transversality Theorem, we conclude that a generic immersion of $M$ into $\mathbb{R}^{3+k}$ will not have points of type $M_r$ with $r < k$. $\square$

## 4. Topological types for the curvature locus

We explore in this section the topological types that the curvature veronese of a 3-manifold immersed in $\mathbb{R}^n$ may adopt at different points. We observe first that, in order to analyze the shape of the curvature veronese, we can discard the translation associated to the vector $H_2$ above and study the behaviour of the following expression in terms of the properties of the matrix $[T]$.

$$\eta(S^2) \cong \Big(T \circ (G \circ \tilde{\xi})\Big)(S^2).$$

Clearly, these depend on the matrix of $T$ associated to the second fundamental form at the considered point. We analyze the different possibilities through the following strategy: Elementary operations on the columns (or the files) of the matrix $[T]$ preserve the properties that depend on its rank. Now, triangulating $[T]$ through elementary operations with lines is equivalent to composing the linear map associated to $[T]$ with an isomorphism of $N_pM$ and this preserves the topological type of the curvature veronese. On the other hand, we can also compose $[T]$, on the right hand side, with a diagonal $(5 \times 5)$-matrix with non vanishing determinant without changing the topological type of the associated curvature veronese. This procedure allows us to determine representatives of all the possible cases and we analyze in each case the corresponding topological type.

For 3-manifolds immersed in $\mathbb{R}^n, n \geq 8$ for which $rank(T) = 5$ it is easy to see that the curvature veronese is diffeomorphic to the classical Veronese surface of order 2. For $rank(T) = 4$, it can be shown that it is either an embedding of the projective plane into $\mathbb{R}^4$, or a map with isolated corank 1 singularities. We shall describe with detail the more interesting cases concerning $rank(T) = 3, 2$. For this analysis we can consider without lost of generality that the 3-manifold is immersed in $\mathbb{R}^6$.

(1) For $\mathrm{rank}(T) = 3$, considering all the possible combinations of the vectors $B_1, B_2, B_3, B_4$ and $B_5$ such that 3 of them are linearly independent leads to the following representatives of the matrix $[T]$ in the triangular form,

$$[\tilde{T_1}] = \begin{pmatrix} 0 & 0 & 1 & 0 & 0 \\ 0 & 0 & 0 & 1 & 0 \\ 0 & 0 & 0 & 0 & 1 \end{pmatrix}, \quad [\tilde{T_2}] = \begin{pmatrix} 0 & 1 & a_{13} & 0 & 0 \\ 0 & 0 & 0 & 1 & 0 \\ 0 & 0 & 0 & 0 & 1 \end{pmatrix},$$

$$[\tilde{T_3}] = \begin{pmatrix} 1 & a_{12} & a_{13} & 0 & 0 \\ 0 & 0 & 0 & 1 & 0 \\ 0 & 0 & 0 & 0 & 1 \end{pmatrix}, \quad [\tilde{T_4}] = \begin{pmatrix} 0 & 1 & 0 & a_{14} & 0 \\ 0 & 0 & 1 & a_{24} & 0 \\ 0 & 0 & 0 & 0 & 1 \end{pmatrix},$$

$$[\tilde{T_5}] = \begin{pmatrix} 1 & a_{12} & 0 & a_{14} & 0 \\ 0 & 0 & 1 & a_{24} & 0 \\ 0 & 0 & 0 & 0 & 1 \end{pmatrix}, \quad [\tilde{T_6}] = \begin{pmatrix} 0 & 1 & 0 & 0 & a_{15} \\ 0 & 0 & 1 & 0 & a_{25} \\ 0 & 0 & 0 & 1 & a_{35} \end{pmatrix},$$

$$[\tilde{T_7}] = \begin{pmatrix} 1 & a_{12} & 0 & 0 & a_{15} \\ 0 & 0 & 1 & 0 & a_{25} \\ 0 & 0 & 0 & 1 & a_{35} \end{pmatrix}, \quad [\tilde{T_8}] = \begin{pmatrix} 1 & 0 & a_{13} & a_{14} & 0 \\ 0 & 1 & a_{23} & a_{24} & 0 \\ 0 & 0 & 0 & 0 & 1 \end{pmatrix},$$

$$[\tilde{T_9}] = \begin{pmatrix} 1 & 0 & a_{13} & 0 & a_{15} \\ 0 & 1 & a_{23} & 0 & a_{25} \\ 0 & 0 & 0 & 1 & a_{35} \end{pmatrix}, \quad [\tilde{T_{10}}] = \begin{pmatrix} 1 & 0 & 0 & a_{14} & a_{15} \\ 0 & 1 & 0 & a_{24} & a_{25} \\ 0 & 0 & 1 & a_{34} & a_{35} \end{pmatrix}.$$

(2) For $\mathrm{rank}(T) = 2$ we get,

$$[\tilde{T_1}] = \begin{pmatrix} 0 & 0 & 0 & 1 & 0 \\ 0 & 0 & 0 & 0 & 1 \end{pmatrix}, \quad [\tilde{T_2}] = \begin{pmatrix} 0 & 0 & 1 & a_{14} & 0 \\ 0 & 0 & 0 & 0 & 1 \end{pmatrix},$$

$$[\tilde{T_3}] = \begin{pmatrix} 0 & 1 & a_{13} & a_{14} & 0 \\ 0 & 0 & 0 & 0 & 1 \end{pmatrix}, \quad [\tilde{T_4}] = \begin{pmatrix} 0 & 0 & 1 & 0 & a_{15} \\ 0 & 0 & 0 & 1 & a_{25} \end{pmatrix},$$

$$[\tilde{T_5}] = \begin{pmatrix} 1 & a_{12} & a_{13} & a_{14} & 0 \\ 0 & 0 & 0 & 0 & 1 \end{pmatrix}, \quad [\tilde{T_6}] = \begin{pmatrix} 0 & 1 & a_{13} & 0 & a_{15} \\ 0 & 0 & 0 & 1 & a_{25} \end{pmatrix},$$

$$[\tilde{T_7}] = \begin{pmatrix} 1 & a_{12} & a_{13} & 0 & a_{15} \\ 0 & 0 & 0 & 1 & a_{25} \end{pmatrix}, \quad [\tilde{T_8}] = \begin{pmatrix} 0 & 1 & 0 & a_{14} & a_{15} \\ 0 & 0 & 1 & a_{24} & a_{25} \end{pmatrix},$$

$$[\tilde{T_9}] = \begin{pmatrix} 1 & a_{12} & 0 & a_{14} & a_{15} \\ 0 & 0 & 1 & a_{24} & a_{25} \end{pmatrix}, \quad [\tilde{T_{10}}] = \begin{pmatrix} 1 & 0 & a_{13} & a_{14} & a_{15} \\ 0 & 1 & a_{23} & a_{24} & a_{25} \end{pmatrix}.$$

As we have seen in Theorem 3.8, a generic immersion of a 3-manifold $M$ into $\mathbb{R}^6$ is exclusively made of $M_3$ points. We describe next all the possible topological types for the curvature locus at the points of these manifolds.

THEOREM 4.1. *The curvature locus at a point of type $M_3$ of a 3-manifold $M$ immersed in $\mathbb{R}^6$ may have one of the following shapes:*

1) *A Roman Steiner surface, a Cross-Cap surface, a Steiner surface of type 5, a Cross-Cup surface, an ellipsoid, or a (compact) cone, provided the mean curvature vector $H(p)$ lies in $E_p$.*
2) *In the case that the mean curvature vector $H(p)$ does not lie in $E_p$, the curvature veronese is a convex planar region linearly equivalent to one of the following: an elliptic region, a triangle, a (compact) planar cone or a planar projection (of type 1, 2 or 3) of the Veronese surface.*

**Proof:** The curvature veronese is a projection of the classical Veronese surface of order 2 into a linear 3-space and all the possible projections are determined by the above matrices. Moreover, it is a quadratically parametrizable compact surface, so we can use the information contained in [**2**], where an exhaustive description of the quadratically parametrized surfaces of order greater that 2 in $\mathbb{R}^3$ is given. These surfaces have an implicit form in terms of polynomials of degree at most 4. In the case we are interested here, we must also add those of order 2, which may either be an ellipsoid, or a truncated cone. On the other hand, we must discard those cases obtained in [**2**] that are non compact, in particular those corresponding to ruled surfaces, whose implicit expression is given in terms of polynomials of degree 3 and cannot be the image of a projective plane in $\mathbb{R}^3$. This leads to the models listed above. We shall show now, by means of specific examples, that all these cases appear as the curvature veronese at a given point of some immersion. To do this, we identify sufficient conditions in the entries of the matrix that warrantee each particular model. Once such coefficients are determined, it is easy to find examples of immersions leading to them. We observe that such choices are not unique, but we do not include here an exhaustive study of all the possibilities, because it is not necessary for our purpose in this proof.

We first observe that Theorem 3.8 shows that a generic 3-manifold $M$ into $\mathbb{R}^6$ can only have $M_3$ points. This means that dim $N_p^1 M = 3$ at every point, which may happen if and only if either $N_p^1 M = E_p$, or dim $E_p = 2$ and the mean curvature vector $H$ does not belong to the subspace $E_p$.

We start with the case of dim $E_p = 3$.

(1) **Quadratic surfaces:**
- **Elipsoid**: This corresponds to the matrix $[T_1]$. The curvature veronese at $p \equiv (0,0,0)$ is given by

$$\eta(\theta,\phi) = \Big(1+3\cos(2\phi),\ \cos(\theta)\sin(2\phi),\ \sin(\theta)\sin(2\phi)\Big).$$

- **Truncated cone**: By taking $a_{14} = a_{15} = a_{24} = a_{25} = a_{34} = a_{35} = 0$ in matrix $[T_{10}]$ we obtain a curvature veronese at the origin which is the cone with circular base $2Y^2 + Z^2 = 4$, $X = -2$ and vertex at $(4,0,0)$ given by

$$\eta(\theta,\phi) = \Big(1+3\cos(2\phi),\ \cos(2\theta)\big(\sin(\phi)\big)^2,\ \sin(2\theta)\big(\sin(\phi)\big)^2\Big).$$

(2) **Steiner Surfaces:** (Figure 1)

- **Roman Steiner surface**: By taking $a_{13} = 0$ in the matrix $[T_2]$, the curvature veronese at the origin is given by
$\eta(\theta, \phi) = \Big(\cos(2\theta)\big(\sin(\phi)\big)^2,\ \cos(\theta)\sin(2\phi),\ \sin(\theta)\sin(2\phi)\Big).$
- **Cross-Cap surface**: For the choices $a_{14} = a_{24} = 0$ in $[T_4]$, we get a curvature veronese at the origin given by
$\eta(\theta, \phi) = \Big(\cos(2\theta)\big(\sin(\phi)\big)^2,\ \sin(2\theta)\big(\sin(\phi)\big)^2,\ \sin(\theta)\sin(2\phi)\Big).$
- **Steiner surface of type 5**: If we take $a_{12} = a_{14} = a_{24} = 0$ in the matrix $[T_5]$, we obtain a curvature veronese at the origin given by
$\eta(\theta, \phi) = \Big(\ 1 + 3\cos(2\phi),\ \sin(2\theta)\big(\sin(\phi)\big)^2,\ \sin(\theta)\sin(2\phi)\Big).$
- **Cross-Cup surface**: Taking $a_{24} = 1$ e $a_{13} = a_{14} = a_{23} = 0$ in the matrix $[T_8]$ leads to a curvature locus at the origin given by
$\eta(\theta, \phi) = \Big(1 + 3\cos(2\phi),\ \cos(2\theta)\big(\sin(\phi)\big)^2 + \cos(\theta)\sin(2\phi),$
$\sin(\theta)\sin(2\phi)\Big).$

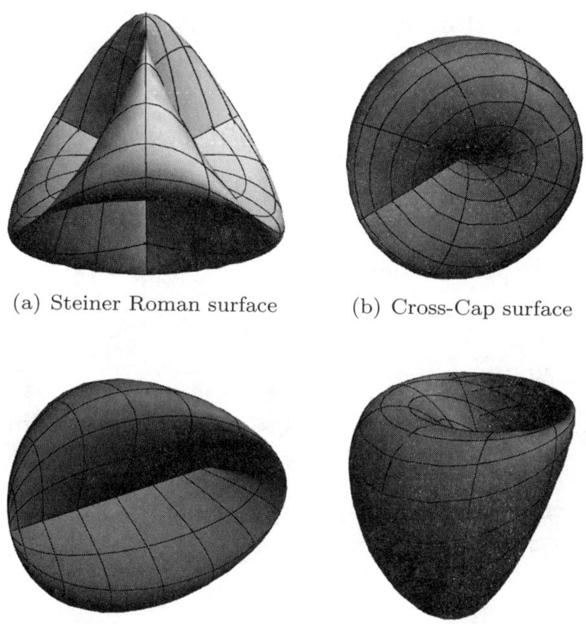

(a) Steiner Roman surface (b) Cross-Cap surface

(c) Steiner surface of type 5 (d) Cross-Cup surface

FIGURE 1. Curvature loci in 3-dimensional space

We consider now the case with dim $E_p = 2$. This means that the curvature locus at $p$ is a projection of the classical Veronese surface of order 2 corresponding to one of the matrices of rank 2 listed above. Notice that the curvature veronese in this case can also be seen as a linear projection of one of the above models in 3-dimensional space onto the plane and any one of such projections is convex.

We show now the existence of the different shapes listed above by choosing different values of the free entries in these matrices (Figure 2).

- **Eliptical region**: In matrix $[T_8]$ we can take $a_{14} = a_{15} = a_{24} = a_{25} = 0$ and the curvature veronese at $p \equiv (0,0,0)$ is given by

$$\eta(\theta,\phi) = \Big(\cos(2\theta)\big(\sin(\phi)\big)^2,\ \sin(2\theta)\big(\sin(\phi)\big)^2\Big).$$

- **Triangular region**: In matrix $[T_{10}]$, we take $a_{1j} = a_{2j} = 0,\ j = 1,2,3$ and then the curvature veronese at $p \equiv (0,0,0)$ is a triangular region given by

$$\eta(\theta,\phi) = \Big(1 + 3\cos(2\phi),\ \cos(2\theta)\big(\sin(\phi)\big)^2\Big).$$

- **Planar cone**: If we take $a_{13} = a_{15} = a_{25} = 0$ in the matrix $[T_6]$, the curvature veronese at $p \equiv (0,0,0)$ is given by

$$\eta(\theta,\phi) = \Big(\cos(2\theta)\big(\sin(\phi)\big)^2,\ \cos(\theta)\sin(2\phi)\Big).$$

- **Planar projection of type 1**: If we take $a_{25} = 1$ and $a_{12} = a_{14} = a_{15} = a_{24} = 0$ in matrix $[T_9]$, we obtain a curvature veronese at $p \equiv (0,0,0)$ is given by

$$\eta(\theta,\phi) = \Big(1 + 3\cos(2\phi),\ \sin(2\theta)\big(\sin(\phi)\big)^2 + \sin(\theta)\sin(2\phi)\Big).$$

- **Planar projection of type 2**: Take $a_{13} = 1$ and $a_{15} = a_{25} = 0$ in $[T_6]$. In such case, we obtain that the curvature locus at $p \equiv (0,0,0)$ is given by

$$\eta(\theta,\phi) = \Big(\cos(2\theta)\big(\sin(\phi)\big)^2 + \sin(2\theta)\big(\sin(\phi)\big)^2,\ \cos(\theta)\sin(2\phi)\Big).$$

- **Planar projection of type 3**: Putting $a_{14} = 0$ in $[T_2]$ gives rise to a curvature locus at the origin given by

$$\eta(\theta,\phi) = \Big(\sin(2\theta)\big(\sin(\phi)\big)^2,\ \sin(\theta)\sin(2\phi)\Big).$$

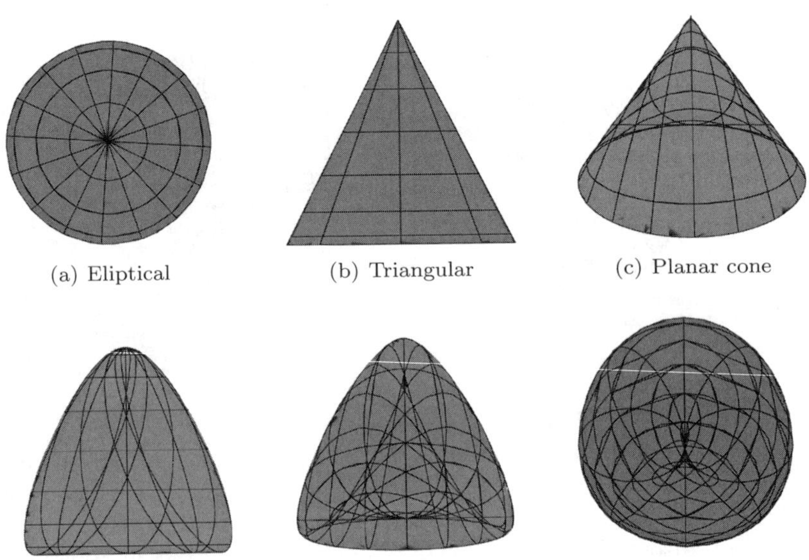

(a) Eliptical (b) Triangular (c) Planar cone

(d) Type 1 (e) Type 2 (f) Type 3

FIGURE 2. Planar curvature loci

A direct inspection of the geometry of the above described types of projections of the Veronese surface in 3-space is sufficient to show that the only possibilities for their planar projections are exhausted by this list. □

REMARK 4.2. *The above choices for the entries of matrix $T$ are meant to illustrate that all the possibilities in our list that can occur as a curvature locus at a point of type $M_3$ of a 3-manifold. Straightforward calculations show that alternative choices of these coefficients also lead to one of the above models. It is worth to notice that the Roman Steiner surface and the Cross-Cup surface appear with higher degree of freedom than all the other cases. An interesting problem, that shall not be considered here, consists in the investigation of the behaviour of the stratified subset that separates different shapes for the curvature locus at the points of a generic 3-manifold. Another interesting open problem is the existence of immersions of closed 3-manifolds in $\mathbb{R}^6$ having the same shape of the curvature locus at each point, as well as the determination of possible obstructions to this property.*

REMARK 4.3. *The Steiner surface of type 1 (or Roman Steiner surface) is given in implicit form by*

$$x^2y^2 + x^2z^2 + y^2z^2 - xyz = 0.$$

*It has 6 pinch points lying at the ends of 3 double curves with mutual intersection at a triple point.*

*The Steiner surface of type 3 (Cross-Cap surface) is given in implicit form by*

$$4x^2(x^2 + y^2 + z^2 + z) + y^2(y^2 + z^2 - 1) = 0.$$

*It has one real double line and two pinch points.*

*The Steiner surface of type 5 is given in implicit form by*

$$x^2(z-1)^2 + y^2(y^2 + z^2 - 1) = 0,$$

*It has two pinch points, one of them being a triple point too, joined by a double line.*

*The Steiner surface of type 6 (Cross-Cup surface) has one triple point and no singular points. Its expression in the implicit form is:*

$$-\frac{5}{4}x^4 + 3x^3y - \frac{5}{2}x^2y^2 + xy^3 - \frac{1}{4}y^4 - 3x^2z^2 + 4xyz^2 - y^2z^2 - z^4 + \frac{7}{2}x^3 -$$
$$\frac{11}{2}x^2y + \frac{5}{2}xy^2 - \frac{1}{2}y^3 + 5xz^2 - 3yz^2 - \frac{13}{4}x^4 + \frac{5}{2}xy - \frac{1}{4}y^2 - 2z^2 + x = 0.$$

*The Boy surface, which provides an immersion of projective plane in $\mathbb{R}^3$, is an algebraic variety of degree 6 and cannot be realized as a curvature locus.*

*The analysis made above shows the existence of a map from the projective space into $\mathbb{R}^3$ whose image is an ellipsoid. We must note that this is a very special map which has a closed curve made of corank one singularities. As a consequence of a well known result in Singularity Theory, we get that it has infinite codimension in the space of smooth maps from $\mathbb{R}P^2$ to $\mathbb{R}^3$. Examples explored by using the numerical program "ImmersionR3toR6", due A. Montesinos Amilibia* [**11**]*, illustrate that this shape may appear at isolated points of 3-manifolds, but it represents an unstable situation from the geometrical viewpoint.*

EXAMPLE 4.4. *Consider the 3-manifold $M$ in $\mathbb{R}^6$, parametrized by $f(x, y, z) = \Big(x, y, z, x^2 + y^2 - z^2, x^2 - y^2 + z^2, xyz\Big)$. The curvature veronese at a point $(x, y, z)$*

*is given by:*

$$\eta(\theta,\phi) = \left(\frac{2}{3},\frac{2}{3},0\right) + \Big(-\frac{2}{3}(1+3\cos(2\phi)),\ \frac{1}{3}(1+3\cos(2\phi)) + 2\cos(2\theta)(\sin(\phi))^2,$$

$$\big(\sin(2\theta)(\sin(\phi))^2\big)z + \big(\cos(\theta)\sin(2\phi)\big)y + \big(\sin(\theta)\sin(2\phi)\big)x\Big).$$

- *At the origin* $p=(0,0,0)$, *the curvature veronese is a triangular region.*
- *At the points* $p=(x,0,0)$, $p=(0,y,0)$ *and* $p=(0,0,z)$, *with* $x\neq 0$, $y\neq 0$, *and* $z\neq 0$, *it is a cone.*
- *At* $p=(x,y,0)$, $p=(x,0,z)$ *or* $p=(0,y,z)$, *with* $x\neq 0$, $y\neq 0$, $z\neq 0$, *it is a surface of type 5.*
- *Finally, the curvature veronese at the points* $p=(x,y,z)\neq(0,0,0)$ *is a Roman Steiner surface.*

## 5. Curvature veronese and principal configurations

We analyze now how certain relevant properties of the family of principal configurations on a 3-manifold can be deduced from the geometric behaviour of the curvature veronese.

LEMMA 5.1. [**13**] *Let* $(V,\langle,\rangle)$ *be a finite dimensional vector space with inner product and* $L:V\to V$ *a self-adjoint operator on* $V$. *Denote* $S=\{v\in V:\langle v,v\rangle = 1\}$ *and consider the function*

$$\begin{array}{rcccl} h_L & : & S & \longrightarrow & \mathbb{R} \\ & & v & \longmapsto & \langle L(v),v\rangle. \end{array}$$

*Then* $v_0$ *is a critical point of* $h_L$ *if and only if* $v_0$ *is an eigenvector of* $L$ *with eigenvalue* $h_L(v_0)v_0$.

An immediate consequence of this lemma is the following proposition characterizing the principal directions and curvatures of a normal field $\nu$ at a point $p$ of $M$ in terms of the projections of the curvature veronese at $p$ in the normal direction $\nu$.

PROPOSITION 5.2. *Given a normal field* $\nu$ *on a 3-manifold* $M$ *immersed into* $\mathbb{R}^{3+k}, k\geq 1$, *the* $\nu$*-principal directions at a point* $p\in M$ *are the critical points of the function*

$$\begin{array}{rcccl} h_\nu & : & S_p=\{v\in T_pM:\langle v,v\rangle = 1\} & \longrightarrow & \mathbb{R} \\ & & v & \longmapsto & \langle S_p^\nu(v),\nu\rangle. \end{array}$$

*Moreover, if* $\{w_i^\nu\}_{i=1}^3$ *denote the* $\nu$*-principal directions at* $p$, *then the corresponding* $\nu$*-principal curvatures are* $\kappa_i^\nu = h_\nu(w_i^\nu) = \langle S_p^\nu(w_i^\nu),\nu\rangle, i=1,2,3$.

As a consequence of Proposition 5.2, we can also detect the umbilic points of the normal fields on the 3-manifold in terms of the curvature locus:

PROPOSITION 5.3. *Given* $p\in M$ *and a normal field* $\nu$ *on* $M$, $p$ *is a* $\nu$*-umbilic point if and only if* $\nu(p)\perp L(p)$.

**Proof:** We have $\nu(p) \perp L(p) \Leftrightarrow h_\nu(v) = h_\nu(w), \forall v, w \in S_p \Leftrightarrow h_\nu$ is a constant function $\Leftrightarrow v$ is a critical point of $h_\nu, \forall v \in S_p \Leftrightarrow$ all tangent directions are principal directions $\Leftrightarrow p$ is $\nu$-umbilic. $\square$

Different normal fields may have the same principal direction at some points of the manifold. The following result reflects this fact in terms of the singularities of the curvature veronese.

PROPOSITION 5.4. *Given a 3-manifold $M$ immersed in $\mathbb{R}^6$, the corank $r = 1, 2$ singularities of the curvature locus map $\eta : S_p^{n-1} \to N_pM$ at $p \in M$ are principal directions of exactly $r+1$ linearly independent normal vectors on $M$ at $p$.*

**Proof:** Let $v \in S_p^{n-1}$ be a corank $r$ singularity of $\eta : S_p^2 \to N_pM$. Then the linear subspace $d\eta(p)(T_vS_p^2)$ is a $(3-r)$-dimensional subspace of $N_pM$. So we can find $r+1$ linearly independent normal directions at $p$, $\nu_i, i = 1, \cdots, r+1$, such that the orthogonal projections $P_{\nu_i} : N_pM \to \ell_{\nu_i}$ take the point $\eta(v)$ to a critical point of its image in $\ell_{\nu_i}$. But it follows from proposition 5.2 that in such case $v$ must be a $\nu_i$-principal direction, $1 = 1, \cdots, r+1$. $\square$

COROLLARY 5.5. *Let $p$ be a point of a 3-manifold $M$ immersed into $\mathbb{R}^6$.*

a) *If the curvature locus at $p$ is a cone, then all the normal fields at $p$ share one of the principal directions. This direction corresponds to the apex of the cone through the curvature locus map. Moreover, the boundary curve of the cone is made of corank 1 singularities of the curvature locus map. Therefore, each one of its points is the image, through the curvature locus map, of a common principal direction of two linearly independent normal fields at $p$.*
b) *If the curvature locus at $p$ is a triangle, then the 3 vertices are the images of the 3 principal directions, which are shared by all the normal fields at $p$, i. e., there is a unique basis of principal directions at $p$. Conversely, if $M$ admits a unique a basis of principal directions at $p$ then the curvature locus at $p$ is a triangle.*
c) *The curvature locus cannot be a polygon with more that 3 vertices.*

**Proof:** Assertions a) and b) are an immediate consequence of Proposition 5.4. For assertion c) observe that the existence of 4 or more vertices (i.e. singularities of corank 2 for the curvature locus map) would mean the existence of more that 3 different principal directions for each normal field, which cannot occur unless we have that each normal field is an umbilic field. But this would imply that the curvature locus degenerates to a point. $\square$

An alternative proof of assertion b) can be found in [**12**], proven in a wider context that relies on the concept of normal curvature of the normal bundle.

THEOREM 5.6. [**12**] *Given a 3-manifold in $\mathbb{R}^5$, we have that the curvature locus of $M$ at $p$ is a triangle, given by the convex hull of the points $\eta(X_i)$ in $N_pM$, if and only if $M$ admits a unique principal configuration (i.e., $M$ has vanishing normal curvature) and $\{X_i\}_{i=1}^3$ are the (univocally defined) principal directions at $p$.*

**5.1. Curvature Veronese and quasiumbilic fields.** We say that $p \in M$ is a *quasiumbilic point* for a normal field $\nu$ if two of the $\nu$-principal curvatures, $\kappa_1, \kappa_2$, of $M$ coincide at $p$. A field $\nu$ on $M$ is said to be *quasiumbilic* if it has a principal curvature $\kappa$ with multiplicity 2 at each point of $M$. This clearly implies the existence of a circle of $\nu$-principal directions associated to the value $\kappa_1 = \kappa_2$ of the $\nu$-curvature. As a consequence of the characterization of $\nu$-principal directions as singularities of the height function $h_\nu$ on the curvature veronese, we get that *a quasiumbilic direction $\nu(p)$ must be orthogonal to a plane which is tangent to the curvature veronese along a curve* (which may either be a parabolic conic or a torsal line). This curve is the image by $\eta : S_p^2 \to N_pM$ of a circle of principal directions of the shape operator $S_\nu$. By analyzing the geometry of the different possible curvature veronese types [**2**], we get:

PROPOSITION 5.7. (1) *The curvature veronese at a point $p$ is a cone if and only if there is a 1-paramenter family (circle) of quasiumbilic directions together with a linearly independent quasiumbilic direction at $p$.*

(2) *The curvature veronese at a point $p$ is a Roman Steiner surface if and only if there are exactly three quasiumbilic directions and no umbilic direction at $p$.*
(3) *If the curvature veronese is a Steiner surfaces of type 5, then $M$ has exactly two quasiumbilic directions and no umbilic directions at $p$.*
(4) *The curvature veronese at $p$ is a convex planar region if and only if there exists exactly one umbilic direction at $p$. Moreover, the cases of a planar region of type 1, a planar cone and a triangular region are respectively characterized by the existence of one, two and three curves of quasiumbilic directions. The remaining planar cases do not admit quasiumbilic fields.*

**Proof:** As a consequence of Proposition 5.2 we just need to analyze the behaviour of height functions on the different cases of curvature locus and the umbilic and quasiumbilic direction corresponds respectively to singularities of a corank 2 and 1 of these functions.

(1) If the curvature veronese at $p$ is a cone, then the normal to the cone is constant along each generatrix line. This provides a 1-paramenter family (circle) of quasiumbilic directions at $p$. Moreover, the normal to the basis of the cone is a quasiumbilic direction too.
(2) In the case of a Roman Steiner surface the perpendiculars to the three planes tangent along the three parabolic conics of the surface are the only three quasiumbilic directions.
(3) A Steiner surfaces of type 5 gives rise to two quasiumbilic directions corresponding to the normals along its two parabolic conics.
(4) We first observe that if the curvature locus is a planar region, the orthogonal direction $\nu$ to the plane determined by the curvature veronese is (the unique) umbilic direction. Conversely, the existence of a unique umbilic direction at $p$ implies that the curvature locus has dimension 2 and lies in a plane.

    Moreover, if the curvature locus is a planar region, it must be necessarily convex, because otherwise we would be able to find some normal direction $\nu'$, contained in the plane determined by the curvature locus, leading to a height function that has at least 4 critical points on the curvature locus. But as a consequence of Proposition 5.2 this would imply the existence of more than 3 principal directions at the given point, which in the case of a 3-manifold means that $\nu'$ needs to be an umbilici field too. In such case, we get that the curvature locus is contained in a subspace which is orthogonal to both directions $\nu$ and $\nu'$. This means that it is contained in a line and therefore cannot be a planar region.

Finally, in the cases of a planar region of type 1, a planar cone and a triangular region, the orthogonal direction to (respectively one, two and three segments) lying on its boundary determine quasiumbilic directions at the point. The linear combinations of these with the umbilic direction at $p$ lead respectively to one, two and three curves of quasiumbilic directions.

□

## 6. Curvature veronese and convexity

Consider the height function on the 3-manifold $M = f(\mathbb{R}^3) \subset \mathbb{R}^6$ in a direction $v \in \mathbb{R}^{3+k}$,

$$\begin{aligned} h_v : \quad & M \to \mathbb{R} \\ & p = f(u) \longmapsto \, < f(u), v > \end{aligned}$$

A point $p$ is singular for $h_v$ if and only if $v \in N_pM$. A normal direction $n$ at $p = f(u)$, such that $h_n$ has a degenerate (i.e., non Morse) singularity at $u$ is said to be a *binormal direction* at $p$. In such case the kernel of Hess$(h_n)$ has non trivial vectors and any tangent vector $v$ lying in the kernel of Hess$(h_n)$ is said to be an *asymptotic vector* of $M$ at $p$. The following result provides a characterization of the asymptotic directions on in terms of the curvature locus. This generalizes an analogous well known result for case of surfaces immersed in 4-space ([**7**], [**9**]).

LEMMA 6.1. *Given a 3-manifold $M$ immersed in $\mathbb{R}^{3+k}, k > 1$, a tangent direction $v \in T_pM$ such that the normal direction $\eta(v)$ is tangent to the curvature veronese $\eta(S_p^2)$ is an asymptotic direction at $p$.*

**Proof:** Given a unit normal vector $\nu \in N_pM$, let us parameterize $M$ in a neighborhood of $p$ with a Monge coordinate chart. For this, we can consider an orthonormal basis $\{e_1, e_2, e_3, \ldots, e_{3+k}\}$ of $\mathbb{R}^{3+k}$ such that the tangent plane $T_pM$ is the vector subspace of $\mathbb{R}^{n+k}$ generated by $e_1, e_2, e_3$ and the normal vector $\nu$ coincides with $e_{3+k}$. In this coordinate chart the Hessian matrix of the height function $h_\nu$ at $p$ coincides with the matrix of the shape operator $S_\nu$.

Now, given $w \in S^1 \subset T_pM$ such that $\eta(w)$ is tangent to the curvature veronese $\eta(S_p^2)$, any direction $\nu \in N_pM$ which is orthogonal to $\eta(w)$ will satisfy that the segment given by the projection of the curvature locus onto the line spanned by $\nu$ has one of its end points at the origin of $N_pM$, which implies that $S_p^\nu$ has a zero eigenvalue. But since Hess$(h_\nu)$ and $S^\nu$ have equivalent matrices, it follows that $\nu$ must be a binormal direction with asymptotic direction $w$ at $p$. □

REMARK 6.2. *Dreibelbis [**4**] has proven the following stronger result for $n$-manifolds immersed in $\mathbb{R}^{2n}$: A tangent direction $v$ is an asymptotic direction if and only if $v$ is a singular point of the curvature locus map $\eta : S_p^2 \to N_pM$, or the normal direction $\eta(v)$ is tangent to the curvature veronese $\eta(S_p^2)$.*

An easy consequence of this is the following characterization of the set of binormal directions at a point of a 3-manifold in $\mathbb{R}^6$.

PROPOSITION 6.3. *The cone of binormal directions at a point $p$ of a 3-manifold $M$ immersed in $\mathbb{R}^6, k > 2$, is made of all the orthogonal directions to the cone subtended by the curvature veronese from the origin $p$ of $N_pM$ together with the* *orthogonal directions to those determined by the singular points of the curvature veronese in* $N_pM$.

A tangent hyperplane $\Pi$ is said to be a *locally support hyperplane* for the submanifold $M$ at the point $p$ if $M$ is locally contained at $p$ in one of the two closed half-spaces determined by $\Pi$ in $\mathbb{R}^6$. We say that the submanifold $M$ is *locally convex* at $p \in M$ if there is a locally support hyperplane $\Pi$ of $\mathbb{R}^6$ at $p$. If this plane has a non-degenerate contact (i.e. of Morse type) with $M$, then we say that $M$ is *strictly locally convex* at $p$.

LEMMA 6.4. *A 3-manifold $M$ immersed in $\mathbb{R}^{3+k}, k \geq 2$ is strictly locally convex at $p$ if and only if there exists $v \in N_pM$ such that all the $v$-principal curvatures are positive.*

**Proof:** This follows easily from the fact that the shape operator $S_v$ associated to the normal vector $\nu$ and the Hessian of the height function $h_v$ have equivalent matrices. □

This leads to the following geometrical characterization of local convexity property in terms of the curvature veronese.

PROPOSITION 6.5. *A 3-manifold $M \subset \mathbb{R}^{3+k}, k \geq 2$ is strictly locally convex at $p$ if and only if the origin of the normal plane (identified with $p \in \mathbb{R}^{3+k}$) is not contained in the convex hull of the curvature veronese of $M$ at $p$.*

**Proof:** We have from Proposition 5.2 that $M$ is strictly convex at $p$ if and only if we can find a normal direction $v$ such that the projection of the curvature locus through the normal line in the direction $v$ does not contain the origin $p$. But this implies in turn that $p$ is not contained in the the convex hull of the curvature locus. □

COROLLARY 6.6. *The local convexity at a point $p \in M$ implies the existence of binormal (and hence of asymptotic) directions at $p$.*

In the particular case of a 3-manifold $M$ in $\mathbb{R}^5$, it was shown in [**9**] there is at least one binormal direction at each point of $M$. On the other hand, it is shown in [**12**] that for 3 manifolds in $\mathbb{R}^5$, local convexity implies the existence of exactly 3 binormal directions (counting their multiplicities).

## 7. Curvature veronese and distance squared functions

Given an immersion $f : M \to \mathbb{R}^6$ and a point $a \in \mathbb{R}^6$, consider the family of distance squared functions on $M$ associated to $f$,

$$\begin{aligned} D: \quad & M \times \mathbb{R}^6 \to \mathbb{R} \\ & (u, a) \longmapsto d_a(u) =< f(u) - a, f(u) - a > . \end{aligned}$$

A simple calculation shows that a point $p = f(u) \in M$ is a singular point for a distance squared function if and only if the vector $p - a$ is a normal vector to $M$ at $p$. The *focal set* $\mathcal{F}_pM$ at $p$ is the subset of points $a \in \mathbb{R}^6$ such that the function $d_a$ has a degenerate singularity at $p$. We say that a point $a \in \mathcal{F}_pM$ is a *semiumbilical center* of $M$ if the function $d_a$ has a corank 2 singularity at $p$. Those points $a \in \mathcal{F}_pM$ having a singulatity of corank 3 at $p$ are said to be *umbilical centers*. It follows from Montaldi's Genericity Theorem [**10**], that there is a residual subset in the set of immersions $\mathcal{I}(M, \mathbb{R}^6)$ with the Whitney $C^\infty$ topology, such that the distance squared family of functions associated to any immersion in $\mathcal{I}(M, \mathbb{R}^6)$ is a generic family of functions. This implies, in particular, that the functions $d_a$ may only have singularities up to $\mathcal{A}$-codimension 6.

PROPOSITION 7.1. *Given a 3-manifold $M$ immersed in $\mathbb{R}^6$ and points $p \in M$ and $a \in N_pM$,*

(1) *The point $a$ is a semiumbilic center at $p \Leftrightarrow p - a$ is a quasiumbilic direction at $p$.*
(2) *The point $a$ is an umbilic center at $p \Leftrightarrow p - a$ is an umbilic direction at $p$.*

**Proof:** Let $p = f(u) \in M$ and suppose that $a = f(u) + \lambda v \in N_pM$, with $v$ a unit vector in $N_pM$. A straightforward calculation shows that $\text{Hess}(d_a)(u) = \frac{1}{\lambda}\,\text{Id} + \text{Hess}\,(h_v)(u)$, where $h_v$ is the height function in the direction $v$. Now, in appropriate coordinates, we have that the matrices of Hess $(h_v)(u)$ and the shape operator $S_p^v$ coincide. It then follows that $u$ is a corank $k$ singularity of $d_a$ if and only if $\frac{1}{\lambda}$ is a principal curvature of multiplicity $k$. Since quasiumbilic and umbilic directions respectively correspond to the existence of a principal curvature of multiplicity 2 and 3 at the point $p$ we have the required results. □

As a consequence of this we have that the corank 3 singularities of distance squared functions are $M_3$ points at which the curvature veronese is of planar type. On the other hand, it follows from V. I. Arnold's classification of function germs [**1**] that the focal set of a generically immersed 3-manifold $M$ is a stratified subset of codimension one in $\mathbb{R}^6$ in which the semiumbilic centers form codimension 3 strata, corresponding simple singularities of types $D_k^{\pm}, k = 4, 5, 6$ and $E_6$, with $D_4^{\pm}$ along surfaces in $M$, $D_5$ along curves and $D_6^{\pm}$ and $E_6$ at isolated points. The umbilic centers must also be isolated points in $M$. So we can state the following,

COROLLARY 7.2. *Given a 3-manifold $M$ generically immersed in $\mathbb{R}^6$, the curvature veronese of $M$ may be of planar type just at isolated points.*

## References


[1] V. I. Arnol′d, S. M. Guseĭn-Zade, and A. N. Varchenko, *Singularities of differentiable maps. Vol. I*, Monographs in Mathematics, vol. 82, Birkhäuser Boston, Inc., Boston, MA, 1985. The classification of critical points, caustics and wave fronts; Translated from the Russian by Ian Porteous and Mark Reynolds. MR777682

[2] Adam Coffman, Art J. Schwartz, and Charles Stanton, *The algebra and geometry of Steiner and other quadratically parametrizable surfaces*, Comput. Aided Geom. Design **13** (1996), no. 3, 257–286, DOI 10.1016/0167-8396(95)00026-7. MR1390390

[3] S. I. R. Costa, S. M. Moraes, and M. C. Romero-Fuster, *Geometric contacts of surfaces immersed in $\mathbb{R}^n$, $n \geq 5$*, Differential Geom. Appl. **27** (2009), no. 3, 442–454, DOI 10.1016/j.difgeo.2008.10.020. MR2521904

[4] Daniel Dreibelbis, *Self-conjugate vectors of immersed 3-manifolds in $\mathbb{R}^6$*, Topology Appl. **159** (2012), no. 2, 450–456, DOI 10.1016/j.topol.2011.09.019. MR2868904

[5] M. Golubitsky and V. Guillemin, *Stable mappings and their singularities*, Springer-Verlag, New York-Heidelberg, 1973. Graduate Texts in Mathematics, Vol. 14. MR0341518

[6] Karl Kommerell, *Riemannsche Flächen im ebenen Raum von vier Dimensionen* (German), Math. Ann. **60** (1905), no. 4, 548–596, DOI 10.1007/BF01561096. MR1511325

[7] John A. Little, *On singularities of submanifolds of higher dimensional Euclidean spaces*, Ann. Mat. Pura Appl. (4) **83** (1969), 261–335. MR0271970

[8] Dirce Kiyomi Hayashida Mochida, Maria Del Carmen Romero Fuster, and Maria Aparecida Soares Ruas, *The geometry of surfaces in 4-space from a contact viewpoint*, Geom. Dedicata **54** (1995), no. 3, 323–332, DOI 10.1007/BF01265348. MR1326737

[9] D. K. H. Mochida, M. C. Romero-Fuster, and M. A. S. Ruas, *Osculating hyperplanes and asymptotic directions of codimension two submanifolds of Euclidean spaces*, Geom. Dedicata **77** (1999), no. 3, 305–315, DOI 10.1023/A:1005145104603. MR1716680

[10] James A. Montaldi, *On contact between submanifolds*, Michigan Math. J. **33** (1986), no. 2, 195–199, DOI 10.1307/mmj/1029003348. MR837577

[11] A. Montesinos Amilibia, *ImmersionR3toR6*. Program available by anonymous ftp at http://www.uv.es/montesin/.

[12] J. J. Nuño-Ballesteros and M. C. Romero-Fuster, *Contact properties of codimension 2 submanifolds with flat normal bundle*, Rev. Mat. Iberoam. **26** (2010), no. 3, 799–824, DOI 10.4171/RMI/617. MR2789366

[13] John A. Thorpe, *Elementary topics in differential geometry*, Springer-Verlag, New York-Heidelberg, 1979. Undergraduate Texts in Mathematics. MR528129

Curso de Matemática, Universidade Federal da Fronteira Sul, Campus Chapecó - UFFS 89802-112, Chapecó, Santa Catarina, Brasil
*E-mail address*: rosane.binotto@uffs.edu.br

Instituto de Matemática, Estatística e Computação Científica, IMECC Universidade Estadual de Campinas - UNICAMP 13083-859, Campinas, Sao Pãolo, Brasil
*E-mail address*: sueli@ime.unicamp.br

Departament de Geometria i Topologia, Universitat de València, UV 46100, Burjassot, Valéncia, Spain
*E-mail address*: carmen.romero@uv.es

Contemporary Mathematics
Volume **675**, 2016
http://dx.doi.org/10.1090/conm/675/13583

# Introduction to intersection homology with and without sheaves

Jean-Paul Brasselet

*São Carlos, July 2014*

ABSTRACT. These notes correspond to the mini-course "Introduction to intersection homology" given during the 13th International Workshop on Real and Complex Singularities, São Carlos, Brasil, 27th of July to 1st of August, 2014.

## Introduction

These notes correspond to the mini-course given during the 13th International Workshop on Real and Complex Singularities, São Carlos, Brasil. The purpose of the lectures is to bring people in an exploratory journey, the most pleasant and the most exciting possible, to the discovery of intersection homology, on the way to the perverse world. The notes do not pretend to any originality. They pretend only to show how the notion of intersection sheaf comes from very elementary calculus and also to provide numerous examples in order to facilitate the understanding of the introduced notions. The notes use and abuse of previous papers on the subject. A first version of the course has been delivered in IMPA, Rio de Janeiro, 2009. Complementary references are numerous and given in the text, in particular [**McP2**, **Ma**, **Kle**, **Kli**, **Rie**].

## 1. Poincaré - Lefschetz isomorphism

### 1.1. Homology with compact and closed supports.

DEFINITION 1.1. One says that a (locally compact) topological space $X$ is endowed with a *piecewise linear structure* (*PL structure*) if there is a class of locally finite triangulations of $X$ (say admissible triangulations) satisfying:

(1) every (barycentric or linear) subdivision of an admissible triangulation is admissible
(2) two admissible triangulations admit a common subdivision.

An example of triangulation of a non-compact space is given by the figure 1 of a triangulation of the (open) upper half-plane, with triangles of height $1/2^n$ based on the horizontal lines at levels $1/2^n$ for all $n \in \mathbb{N}$. The axis of coordinate $AB$

2010 *Mathematics Subject Classification.* Primary 55N33, 14F43, 32S60.

The author thanks the referree for valuable comments. While writing the course, the author was granted by CNPq, Bolsa Pesquisador Visitante Especial, Processo 400580/2012-8.

(at level 0) is not included in the half-plane. The triangulation is not finite but is *locally finite*. That means that each point $x$ in the open half-plane admits a neighborhood $U_x$ which meets a finite number of simplices.

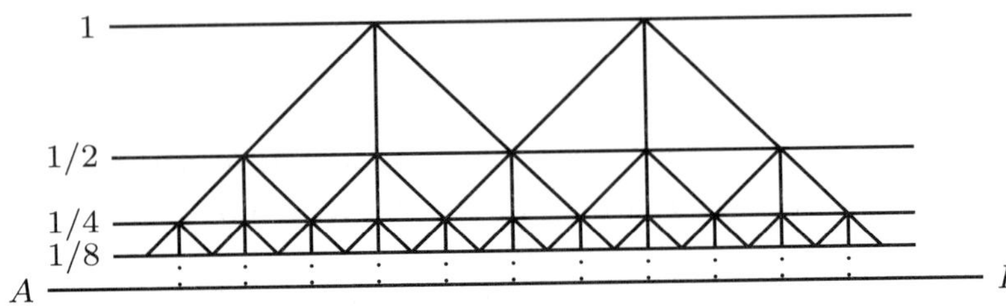


FIGURE 1. Locally finite triangulations of the half plane

Let $K$ be a triangulation of a PL space $X$ and consider $\xi = \sum \xi_\sigma \sigma$ a $K$-chain with coefficients in $\mathbb{Z}$. The support $|\xi|$ of $\xi$ is the union of (supports of) simplices $\sigma$ for which the coefficient $\xi_\sigma$ is non-zero. The support is closed, it is compact if and only if all $\xi_\sigma$ are zero but a finite number.

DEFINITION 1.2. Let $K$ be a triangulation of a PL space $X$, one denotes by $C_i(K)$ the group of *$i$-chains with compact supports*, *i.e.*, the linear combinations $\xi = \sum \xi_\sigma \sigma$ for which all coefficients $\xi_\sigma \in \mathbb{Z}$ are zero but a finite number.

DEFINITION 1.3. In the same way, one denotes by $C_i^{\mathrm{cl}}(K)$ the group of *$i$-chains with closed supports*, *i.e.*, the locally finite linear combinations with integer coefficients $\xi = \sum \xi_\sigma \sigma$, where $\sigma$ are $i$-simplices and $\xi_\sigma \in \mathbb{Z}$.

In other words, each point $x \in X$ admits an open neighbourhood $U_x$ wich meet a finite number of simplices of the support of $\xi$.

Let $K'$ be a subdivision of $K$, such that all simplices of $K$ and $K'$ are oriented. Let us consider two oriented simplices $\sigma \in K$ and $\tau \in K'$, one defines the incidence number:

$$\mathscr{O}(\tau : \sigma) = \begin{cases} 0 & \text{if } \tau \not\subset \sigma \\ +1 & \text{if } \tau \subset \sigma \text{ with the same orientation} \\ -1 & \text{if } \tau \subset \sigma \text{ with opposite orientations} \end{cases}$$

and one defines the map $\phi_K \colon C_i^{\mathrm{cl}}(K) \to C_i^{\mathrm{cl}}(K')$ by:

$$\phi_K\Big(\sum_{\sigma \in K} \xi_\sigma \, \sigma\Big) = \sum_{\sigma \in K} \sum_{\tau \in K'} \Big(\mathscr{O}(\tau : \sigma)\, \xi_\sigma\Big)\tau. \tag{1.4}$$

The group of PL $i$-chains of $X$ with closed support, denoted by $C_i^{\mathrm{cl}}(X)$ is the direct limit of the groups $C_i^{\mathrm{cl}}(K)$ for all triangulations of $X$, that is the union of groups $C_i^{\mathrm{cl}}(K)$ for all admissible triangulations of $X$, modulo the following equivalence relation: Let $\xi_1 \in C_i^{\mathrm{cl}}(K_1)$ and $\xi_2 \in C_i^{\mathrm{cl}}(K_2)$, one has $\xi_1 \sim \xi_2$ if and only if there is a common subdivision $K'$ of $K_1$ and $K_2$ such that $\phi_{K_1}(\xi_1) = \phi_{K_2}(\xi_2)$. Then one has $|\xi_1| = |\xi_2|$.

DEFINITION 1.5. The homology groups *with closed supports* (or Borel Moore homology groups) are the homology groups of the complex $C_*^{\mathrm{cl}}(X)$, with usual boundary $\partial_*$. We denote these homology groups by $H_i^{\mathrm{cl}}(X)$.

In the same way, the group of PL $i$-chains with compact support of $X$, denoted by $C_i(X)$, is defined as the direct limit of the groups $C_i(K)$ for all triangulations of $X$. Homology groups with compact supports are homology groups of the complex $C_*(K)$, with usual boundary $\partial_*$.

LEMMA 1.6. *Let $U \subset V$ be two open subsets in $X$, the inclusion of $U$ in $V$ induces morphisms*

$$\rho_{UV} : C_i^{\mathrm{cl}}(V) \to C_i^{\mathrm{cl}}(U) \qquad \textit{and} \qquad C_i(U) \to C_i(V).$$

In other words, the correspondence $U \mapsto C_i^{\mathrm{cl}}(U)$ is contravariant and the correspondence $U \mapsto C_i(U)$ is covariant functor from the category of topological spaces to the category of free abelian groups.

PROOF. *a) Definition of $\rho_{UV} : C_i^{\mathrm{cl}}(V) \to C_i^{\mathrm{cl}}(U)$:*

Let us consider $\xi \in C_i^{\mathrm{cl}}(V)$, there is a locally finite triangulation $K_V$ of $V$ such that $\xi$ can be written $\sum_{\sigma \in K} \xi_\sigma \sigma$. Let us consider a triangulation of $U$, it admits a subdivision $K_U$ such that every simplex $\tau$ in $K_U$ is contained in a simplex $\sigma(\tau)$ of a subdivision of $K_V$ and such that $\dim \tau = \dim \sigma(\tau)$. The chain $\rho_{UV}(\xi) \in C_i^{\mathrm{cl}}(U)$ is defined by

$$\rho_{UV}(\xi) = \sum_{\tau \in K_U} \mathscr{O}(\tau : \sigma(\tau))\, \xi_{\sigma(\tau)} \tau$$

and $|\rho_{UV}(\xi)| = |\xi| \cap U$.

*b) Definition of $C_i(U) \to C_i(V)$:*

A chain $\eta$ with compact support in $C_i(U)$ can be written $\eta = \sum_{\tau \in K_U} \xi_\tau \tau$ for a triangulation $K_U$ of $U$, where the coefficients $\xi_\tau$ are all zero but a finite number. Let $K_V$ be a triangulation of $V$. Let us consider a subdivision $K'_V$ of $K_V$ and a subdivision $K'_U$ of $K_U$ such that their restrictions to the support $|\eta|$ of $\eta$ coincide. Then $\phi(\eta) \in C_i(K'_U)$ (see 1.4) is a $K'_V$-chain with compact support and defines a chain in $C_i(V)$. □

We will see (Section 3) that the correspondence $U \mapsto C_i^{\mathrm{cl}}(U)$ defines a presheaf which is in fact a sheaf. That is not true for the chains with compact supports.

The two morphisms defined in Lemma 1.6 commute with boundaries. They induce morphisms in homology:

$$H_i^{\mathrm{cl}}(V) \to H_i^{\mathrm{cl}}(U) \qquad \text{and} \qquad H_i(U) \to H_i(V).$$

REMARK 1.7. Let us remark that (in general) there is no map $C_i(X) \to C_i(U)$ or $C_i(V) \to C_i(U)$ for $U \subset V$. A counter-example is provided by taking the chain $\xi = [-1, +1]$ in $X = \mathbb{R}$ and $U = ]0, +\infty[$. Then $\xi|_U = ]0, +\infty[$ is not a chain with compact support.

EXAMPLE 1.8. The cylinder: Let us consider the cylinder $X = \mathbb{R} \times \mathbb{S}^1$, one has two 1-dimensional cycles candidate to generate the 1-dimensional homology: the circle $\gamma = \{0\} \times \mathbb{S}^1$ and the line $\ell = \mathbb{R} \times \{p\}$ where $p$ is a fixed point in $\mathbb{S}^1$. Considering compact supports, $\gamma$ is a cycle which is not a boundary and $\ell$ is not a chain with compact support. Considering closed supports, $\gamma$ is the boundary of the 2-dimensional chain $[0, +\infty[ \times \mathbb{S}^1$ and $\ell$ is a cycle which is not a boundary. Homology groups of the cylinder are given in table 1. Let us remark that the fundamental class $[X]$ of the cylinder lies in $H_2^{\mathrm{cl}}(X)$, not in $H_2(X)$.

| $i$ | $H_i^{\text{cl}}(X)$ | $H_i(X)$ |
|---|---|---|
| 0 | 0 | $\mathbb{Z}_{pt}$ |
| 1 | $\mathbb{Z}_\ell$ | $\mathbb{Z}_\gamma$ |
| 2 | $\mathbb{Z}_{[X]}$ | 0 |

TABLE 1. Homology of the cylinder

### 1.2. Poincaré - Lefschetz isomorphism.

Let us recall that a regular point $x$ in a space $X$ is a point which admits a neighbourhood homeomorphic to an open ball in some $\mathbb{R}^n$. One says that $n$ is the dimension of $X$ at $x$. A singular point is a point which is not regular. If all regular points have the same dimension one says that the space is purely dimensional. The dimension of the (connected) purely dimensional space $X$ is its dimension in its regular points.

DEFINITION 1.9. An *$n$-manifold* (or $n$-topological manifold) is a (non-empty, Hausdorff) topological space $X$ such that each point admits a neighbourhood homeomorphic with an open subset of the Euclidean space $\mathbb{R}^n$.

An $n$-manifold is an $n$-homology manifold, that is for each point $\{x\}$ in $X$, the local homology group satisfies

$$H_i(X, X \setminus \{x\}; \mathbb{Z}) = \begin{cases} 0 & \text{if} \quad i \neq n \\ \mathbb{Z} & \text{if} \quad i = n. \end{cases} \tag{1.10}$$

DEFINITION 1.11. One says that two cycles $a$ and $b$ in a (topological) manifold $X$ are *dimensionally transverse* if either they do not meet or their dimensions are related by the formula:

$$\text{codim}\ (|a| \cap |b|) = \text{codim}\ |a| + \text{codim}\ |b|.$$

THEOREM 1.12 (Lefschetz). *In a topological manifold, the intersection of two dimensionally transverse cycles is a cycle.*

In a compact oriented topological manifold, if two dimensionally transverse cycles $a$ and $b$ have complementary dimensions, then the intersection $a \cap b$ is a finite number of points $\{x_i\}$. The cycles being oriented, in each of the points $x_i$ one defines the intersection index $I(a, b; x_i)$ in the following way: an oriented cycle is linear combination of oriented simplices, so it suffices to define the index for two elementary dimensionally transverse simplices (in $K = L'$).

DEFINITION 1.13. The intersection index at the intersection point $x_i$ of two elementary simplices is $+1$ if the orientation of $a$ followed by the orientation of $b$ is the orientation of the manifold and is $-1$ otherwise. One extends the definition by linearity.

In the case of smooth geometric cycles $a$ and $b$ in a differentiable manifold $X$, the intersection number can also be defined in the following way: On the one hand, let us denote by $m_a$ and $m_b$ the multiplicities of the cycles $a$ and $b$ respectively at the intersection point $x_i$; on the other hand, let us define $\varepsilon = +1$ if the orientation of the tangent space to $X$ at $x_i$, denoted by $T_{x_i}X$ is the one of $T_{x_i}(|a|) \oplus T_{x_i}(|b|)$ and $-1$ otherwise; then

$$I(a, b; x_i) = \varepsilon\, m_a \cdot m_b.$$

DEFINITION 1.14. Let $X$ be a compact oriented topological manifold. The *intersection index* of two dimensionally transverse and complementary dimensional cycles $a$ and $b$, denoted by $I(a, b)$ is the sum of the intersection indices at the intersection points of $a$ and $b$:

$$I(a, b) = \sum_{x_i \in a \cap b} I(a, b; x_i).$$

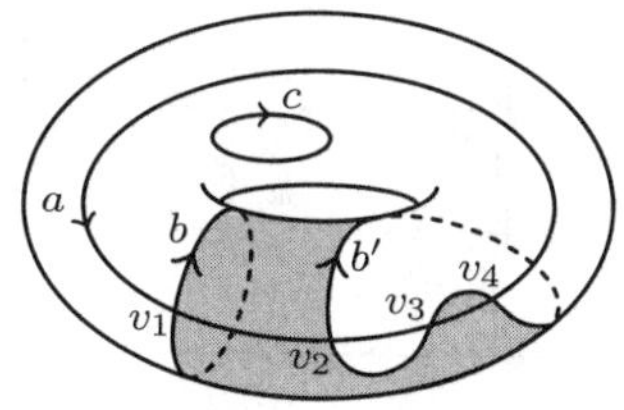


Orientation of the visible face: ↻

FIGURE 2. Transverse cycles in the torus

The intersection index defines an intersection product

$$\begin{array}{ccc} C_{n-p}(K) \times C_p(K) & \longrightarrow & \mathbb{Z} \\ a \quad , \quad b & \leadsto & I(a, b) \end{array} \tag{1.15}$$

which associates to each pair of dimensionally transverse and complementary dimensional cycles $(a, b)$ the intersection index $I(a, b)$.

EXAMPLE 1.16. In the figure 2, in the torus, the cycles $a$ and $c$ do not intersect, one has $I(a, c) = 0$. The intersection index of the cycles $a$ and $b$ at the point $v_1$ is $+1$ and $I(a, b) = +1$. The intersection index of the cycles $a$ and $b'$ is $+1$ at $v_2$ and $v_4$ and is $-1$ at $v_3$, then $I(a, b') = +1$. In this example, the cycles $b$ and $b'$ are homologous and one has $I(a, b) = I(a, b') = +1$. This property is a general property for topological manifolds:

THEOREM 1.17. *Let $X$ be a compact and oriented topological manifold, the intersection index $I(a, b)$ of two dimensionally transverse and complementary dimensional cycles $a$ and $b$ does not depend on the representative of the homology classes of the cycles $a$ and $b$.*

The following Theorem has been proved by Poincaré and Lefschetz (see [**Po1**, **Po2**, **Po3**, **Lef**]).

THEOREM 1.18 (Poincaré-Lefschetz duality). *When considering rational homology groups, the map 1.15 defines a non-degenerate bilinear map*

$$H^{\mathrm{cl}}_{n-p}(X; \mathbb{Q}) \times H_p(X; \mathbb{Q}) \longrightarrow H_0(X; \mathbb{Q}) \xrightarrow{\varepsilon} \mathbb{Q}$$

*where $\varepsilon$ is the evaluation map $\varepsilon : \sum n_i\{x_i\} \mapsto \sum n_i$.*

EXAMPLE 1.19. Let us consider the example (1.8) of the cylinder:

$$H_1^{\mathrm{cl}}(X;\mathbb{Q}) = \mathbb{Q} \quad \text{is generated by the cycle } \ell$$
$$H_1(X;\mathbb{Q}) = \mathbb{Q} \quad \text{is generated by the cycle } \gamma.$$

One has $I(\ell,\gamma) = \pm 1$ according to the orientations.

Let $X$ be a singular variety, then the Poincaré-Lefschetz duality is no longer true. Let us provide two simple (counter)-examples:

EXAMPLE 1.20. [The (double) cone] Let us consider the cone $X$ obtained by pinching the cylinder at level $\{0\}$. The pinched point is a singular point. One can easily compute the homology groups with closed and compact supports (see table 2). Poincaré duality fails for the cone $X$.

| $i$ | $H_i^{\mathrm{cl}}(X)$ | $H_i(X)$ |
|---|---|---|
| 0 | 0 | $\mathbb{Q}$ |
| 1 | $\mathbb{Q}$ | 0 |
| 2 | $\mathbb{Q}\oplus\mathbb{Q}$ | 0 |

TABLE 2. Homology of the (double) cone

EXAMPLE 1.21. [The pinched torus] Let us consider the pinched torus, *i.e.*, the pseudovariety obtained by pinching a meridian of the torus at a single point.

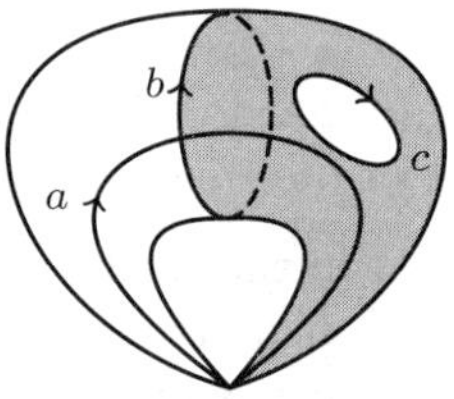


Orientation of the visible face: ↻

FIGURE 3. Transverse cycles in the pinched torus

The shaded area $A$ in Figure 3 is a chain whose boundary is $c - b$. The cycles $b$ and $c$ are homologous, however, one has $I(a,b) = +1$ and $I(a,c) = 0$. Theorem 1.17 is not true and one cannot define an intersection product as the level of homology classes. Rational homology groups of the pinched torus are:

$$H_0(X) = \mathbb{Q}_{[pt]} \qquad H_1(X) = \mathbb{Q}_{[a]} \qquad H_2(X) = \mathbb{Q}_{[X]}.$$

## 2. Intersection homology

The idea of intersection homology, due to Mark Goresky and Robert MacPherson, is to restrict ourselves to cycles which meet the singular part of the variety with a "controlled" dimension. That takes sense if we endow the variety with a suitable stratification. The singular varieties we will consider are pseudomanifolds.

**2.1. Pseudomanifolds.** Let $|K| = X$ be an $n$-manifold, then every $(n-1)$-simplex is a face of exactly two $n$-simplices. This property is one of the conditions for a space being a pseudomanifold:

DEFINITION 2.1. (Pseudomanifold - Combinatorial definition) One says that the polyhedron $|K|$ is an *$n$-pseudomanifold* if the simplicial complex $K$ satisfies the following properties:

(1) $|K|$ is the union of the $n$-simplices,
(2) Each $n-1$-simplex is face of exactly two $n$-simplexes.
(3) For every pair of $n$-simplices $\sigma$ and $\sigma'$ in $K$, there is a sequence of $n$-simplices $\sigma = \sigma_0, \sigma_1, \ldots, \sigma_p = \sigma'$ such that the intersection $\sigma_i \cap \sigma_{i+1}$ is an $(n-1)$-simplex for all $i$.

The topological definition of pseudomanifolds, which is equivalent to the combinatorial one in the case of triangulable topological space, goes as follows:

DEFINITION 2.2. (Pseudomanifold - Topological definition) One says that the (non-empty, paracompact, Hausdorff) topological space $X$ is an *$n$-pseudomanifold* if there is a subset $\Sigma \subset X$ such that:

(1) $X \setminus \Sigma$ is a $n$-topological manifold dense in $X$.
(2) $\dim \Sigma \leq n-2$.
(3) The set $X \setminus \Sigma$ is connected.

The subspace $\Sigma$ of the pseudomanifold $X$ contains the subset of singular points of $X$.

The pinched torus (example 1.21) and the suspension of the torus (see example 2.21) are examples of pseudomanifolds.

**2.2. Stratifications.** Let $X$ be a differentiable $\mathcal{C}^\infty$ manifold, one can define at each point a tangent space. That is no longer true for a singular variety. The points for which one cannot define a tangent space are precisely the singular points.

In the following, a stratification $\mathcal{S}$ of the pseudomanifold $X$ will be the data of a filtration $\mathcal{X}$

$$X = X_n \supset X_{n-1} = X_{n-2} \supset \cdots \supset X_1 \supset X_0 \supset X_{-1} = \emptyset \tag{2.3}$$

such that every stratum $S_i = X_i - X_{i-1}$ is either empty or a finite union of $i$-dimensional smooth submanifolds of $X$.

2.2.1. *Whitney stratifications.*

DEFINITION 2.4. One says that *Whitney conditions* are satisfied for the stratification $\mathcal{S}$ if for each pair of strata $(S_i, S_j)$ such that $S_i$ is in the closure of $S_j$, then one has:

a) Let $\{x_n\}$ be a sequence of points in $S_j$ with limit $y \in S_i$, let us assume that the sequence of tangent spaces $T_{x_n}(S_j)$ admits a limit $T$ for $n$ tending to $+\infty$, then $T_y(S_i) \subset T$.

b) Let $\{x_n\}$ be a sequence of points in $S_j$ with limit $y \in S_i$ and let $y_n$ be a sequence of points in $S_i$ tending to $y$. Let us assume that the sequence of tangent spaces $T_{x_n}(S_j)$ admits a limit $T$ for $n$ tending to $+\infty$ and that the sequence of lines $x_n y_n$ converges to a line $\lambda$ for $n$ tending to $+\infty$, then $\lambda \in T$.

A *Whitney stratification* is a stratification satisfying the Whitney conditions.

Before stating the next result, let us give two definitions:

- Let $\mathbb{B}^i$ be an open ball in $\mathbb{R}^i$ and $X$ a locally compact stratified pseudomanifold, whose strata are denoted by $S_k$. Then the product $\mathbb{B}^i \times X$ admits a "product stratification" whose strata are $\mathbb{B}^i \times S_k$.
- Let $L$ be a stratified variety whose strata are denoted by $L_k$, the open cone over $L$, denoted by $\mathring{c}(L) = L \times [0,1[/(x,0) \sim (x',0)$, admits strata

$$(\mathring{c}(L))_k = \mathring{c}(L_{k-1}) \quad \text{if } k > 0 \quad \text{and } (\mathring{c}(L))_0 = \{0\}.$$

THEOREM 2.5 (A' Campo [**Bo**, IV]). *Every Whitney stratification satisfies the following topological triviality property.*

DEFINITION 2.6. One says that the stratification $\mathcal{S}$ of the space $X$, given by the filtration (2.3) satisfies *local topological triviality* if each point $x$ in $S_i$ admits a *distinguished neighbourhood* $U_x \subset X$ together with a homeomorphism $\phi_x : U_x \to \mathbb{B}^i \times \mathring{c}(L)$ where:

- $\mathbb{B}^i$ is an open ball in $\mathbb{R}^i$,
- the "link" $L$ is a compact $(n-i-1)$-dimensional pseudomanifold independent (up to homeomorphism) of the point $x$ in the stratum $X_i - X_{i-1}$ and filtered in:

$$L = L_{n-i-1} \supset L_{n-i-3} \supset \cdots \supset L_0 \supset L_{-1} = \emptyset,$$

- $\mathring{c}(L)$ is the open cone over $L$. By definition, one has $\mathring{c}(\emptyset) = \{point\}$.

Moreover, the homeomorphism $\phi_x$ preserves the stratifications of $U_x$ and $\mathbb{B}^i \times \mathring{c}(L)$ respectively, that is there are restriction homeomorphisms $\phi_x|_{X_j} : U_x \cap X_j \to \mathbb{B}^i \times \mathring{c}(L_{j-i-1})$, for $j \geq i$.

**2.3. Intersection homology- Definition.** The idea of intersection homology, due to Mark Goresky and Robert MacPherson, is to restrict ourselves to cycles which meet the singular part of the variety with a "controlled" dimension. That makes sense if we endow the variety with a stratification satisfying the local triviality condition. The allowed chains and cycles will be those which meet each stratum of the singular part with a controlled and fixed transversality defect. This defect is called the *perversity* (in French: *Perversité*, in Portuguese: *Perversidade*, in German: *Toleranz*).

DEFINITION 2.7. Let $X$ be a singular variety, a *perversity* is an integer value function

$$\bar{p} : [0, \dim X] \cap \mathbb{Z} \to \mathbb{N}, \qquad p_\alpha := \bar{p}(\alpha)$$

such that $p_0 = p_1 = p_2 = 0$ and $p_\alpha \leq p_{\alpha+1} \leq p_\alpha + 1$ for $\alpha \geq 2$.

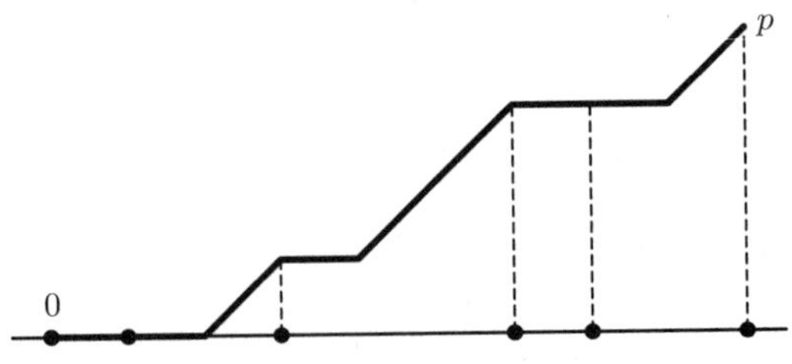


FIGURE 4. The perversity $\bar{p} = (0,0,0,1,1,2,3,3,3,4)$

EXAMPLE 2.8. Examples of perversities are

- the zero perversity $\bar{0} = (0, 0, \dots, 0)$,
- the maximal (or total) perversity $\bar{t} = (0, 0, 0, 1, 2, \dots, n-2)$,
- for $n$ even, $n \geq 4$, the upper middle $\bar{n} = (0, 0, 0, 1, 1, 2, 2, \dots, \frac{n}{2} - 1, \frac{n}{2} - 1)$ and the lower middle perversities $\bar{m} = (0, 0, 0, 0, 1, 1, \dots, \frac{n}{2} - 2, \frac{n}{2} - 1)$.

Figure 4 provides an example of perversity.

Let $\bar{p} = (p_0, p_1, p_2, \dots, p_n)$ be a perversity, the complementary perversity $\bar{q} = (q_0, q_1, q_2, \dots, q_n)$ is defined by $p_\alpha + q_\alpha = t_\alpha$.

Let us consider a singular variety $X$ endowed with a stratification (see 2.3):

$$X = X_n \supset X_{n-1} = X_{n-2} \supset \cdots \supset X_1 \supset X_0 \supset X_{-1} = \emptyset.$$

DEFINITION 2.9. An $i$-chain $\xi \in C_i(X)$ is said $\bar{p}$-*allowable*, or a $(\bar{p}, i)$-*chain*, if:

$$\dim(|\xi| \cap X_{n-\alpha}) \leq i - \alpha + p_\alpha \qquad \forall \alpha \tag{2.10}$$

REMARK 2.11. In the case of transversality $\xi \pitchfork X_{n-\alpha}$, one has

$$\dim(|\xi| \cap X_{n-\alpha}) = i - \alpha.$$

The condition (2.10) means that the perversity is the maximum admissible defect of transversality.

The boundary of a $\bar{p}$-allowable chain is not necessarily $\bar{p}$-allowable (easy examples). If one wants to define a complex of chains, one has to set:

DEFINITION 2.12. One denotes by $IC_i^{\bar{p}}(X)$ the subset of $C_i(X)$ consisting of chains $\xi$ such that $\xi$ and $\partial\xi$ are $\bar{p}$-allowable, that is

$$IC_i^{\bar{p}}(X) = \left\{ \xi \in C_i(X) \;\middle|\; \begin{array}{c} \dim(|\xi| \cap X_{n-\alpha}) \leq i - \alpha + p_\alpha \\ \\ \dim(|\partial\xi| \cap X_{n-\alpha}) \leq (i-1) - \alpha + p_\alpha \end{array} \quad \forall \alpha \geq 2 \right\}$$

Using usual boundary, one obtains a chain complex denoted by $(IC_*^{\bar{p}}(X), \partial_*)$.

DEFINITION 2.13. The *intersection homology* groups $IH_*^{\bar{p}}(X)$ are the homology groups of the complex $(IC_*^{\bar{p}}(X), \partial_*)$.

In the non-compact case, if one substitutes, in the previous definitions, chains with compact supports by chains with closed supports (that is locally finite linear combinations) then one obtains the intersection homology groups with closed supports, denoted by $IH_*^{\bar{p},\mathrm{cl}}(X)$. .

EXAMPLE 2.14. [The pinched torus]. The singular set is a point: the pinched point $\{0\}$. We consider the stratification given by the filtration

$$X \supset \Sigma = \{0\} \supset \emptyset$$

The only possible perversity is the perversity $\bar{0}$. The 1-dimensional intersection homology of the pinched torus is zero, while its 1-dimensional homology does not vanish.

$$IH_0^{\bar{0}}(X) = \mathbb{Q}_{[pt]} \qquad IH_1^{\bar{0}}(X) = 0 \qquad IH_2^{\bar{0}}(X) = \mathbb{Q}_{[X]}.$$

(compare with example 1.21).

| $i$ | $IH_i^{cl}(X)$ | $IH_i(X)$ |
|---|---|---|
| 0 | 0 | $\mathbb{Q}\oplus\mathbb{Q}$ |
| 1 | 0 | 0 |
| 2 | $\mathbb{Q}\oplus\mathbb{Q}$ | 0 |

TABLE 3. Intersection homology of the (double) cone

EXAMPLE 2.15. [The double cone]. While it is not a pseudomanifold, let us consider the double cone with vertex $\{0\}$ (example 1.20). The only possible perversity is the perversity $\bar{0}$. One can easily compute the intersection homology groups (table 3).

Compare with Table 2, page 50. One remarks that two points $a_1$ and $a_2$ contained in different connected components of $X \setminus \{0\}$ are not homologous (in intersection homology) as any 1-chain linking these two points contain the vertex $\{0\}$ and is not permitted. We notice that, in this example, we recover Poincaré duality.

More examples of computation of intersection homology groups can be found for instance in [**BG**, **DM**, **FK**].

**2.4. Intersection homology - Properties.** In general, results and proofs of this section come from Goresky and MacPherson: see [**GM1**, **GM2**]. In this case, references are omitted.

2.4.1. *Topological invariance.*

PROPOSITION 2.16. *Let $X$ be a locally compact stratified pseudomanifold and $\bar{p}$ a perversity, then the intersection homology groups $IH_*^{\bar{p}}(X)$ and $IH_*^{\bar{p},\mathrm{cl}}(X)$ do not depend on the Whitney stratification of $X$.*

REMARK 2.17. The intersection homology groups are not homotopy invariant. We will see that the intersection homology groups of a cone do not (all) vanish, while the cone is homotopic to a point, whose non-zero homology groups vanish.

2.4.2. *Intersection product.*

PROPOSITION 2.18 ([**GM1**]). *Let $\bar{p}$, $\bar{q}$ and $\bar{r}$ perversities such that $\bar{p}+\bar{q}\leq\bar{r}$, one has bilinear pairings*

$$IH_i^{\bar{p}}(X)\times IH_j^{\bar{q}}(X)\to IH_{i+j-n}^{\bar{r}}(X),$$

*if $X$ is compact and*

$$IH_i^{\bar{p},\mathrm{cl}}(X)\times IH_j^{\bar{q}}(X)\to IH_{i+j-n}^{\bar{r}}(X),$$

*otherwise.*

IDEA OF THE PROOF: The proof (see [**GM1**]) uses a result from Clint Mc Crory [**McC**] which shows that if $\xi$ and $\eta$ are two cycles in $IC_i^{\bar{p}}(X)$ and $IC_j^{\bar{q}}(X)$ respectively, then there is a cycle $\xi_1$ in $IC_i^{\bar{p}}(X)$ and a chain $\theta$ in $IC_{i+1}^{\bar{p}}(X)$ such that $\partial\theta = \xi - \xi_1$ and $|\xi_1| \cap |\eta|$ satisfies the perversity condition (with $i \mapsto i + j - n$ and $\bar{p} \mapsto \bar{r}$). In that case, the intersection product is defined by

$$[\xi] \cap [\eta] = [\xi_1 \cap \eta].$$

□

Assume $X$ is compact, then one deduces the generalised Poincaré duality:

PROPOSITION 2.19. *Let $X$ be a compact, stratified pseudomanifold and let $\bar{p}$ and $\bar{q}$ be two complementary pervesities, then the pairing*

$$IH_i^{\bar{p}}(X) \times IH_{n-i}^{\bar{q}}(X) \to IH_0^{\bar{t}}(X) \xrightarrow{\varepsilon} \mathbb{Z}$$

*followed by the evaluation map $\varepsilon$ (which counts points with their multiplicity order) is non-degenerate, when tensorised by $\mathbb{Q}$.*

In the non-compact case, one has in the same way, a pairing:

$$IH_i^{\bar{p},\mathrm{cl}}(X) \times IH_{n-i}^{\bar{q}}(X) \to IH_0^{\bar{t}}(X) \xrightarrow{\varepsilon} \mathbb{Z}.$$

PROPOSITION 2.20 ([**GM1**]). *Let $\bar{p}$ and $\bar{q}$ be complementary perversities, then one has an isomorphism*

$$IH_i^{\bar{p}}(X;\mathbb{Q}) \cong \mathrm{Hom}(IH_{n-i}^{\bar{q}}(X;\mathbb{Q}),\mathbb{Q})$$

*in the compact case, and*

$$IH_i^{\bar{p},\mathrm{cl}}(X;\mathbb{Q}) \cong \mathrm{Hom}(IH_{n-i}^{\bar{q}}(X;\mathbb{Q}),\mathbb{Q})$$

*otherwise.*

EXAMPLE 2.21. [Suspension of the torus].
That is the original Poincaré example for which Poincaré duality fails for usual homology. Let us consider a $k$-simplex $\sigma = [v_0, \dots, v_k]$ and a $p$-simplex $\tau = [w_0, \dots, w_p]$ in the Euclidean space $\mathbb{R}^n$ such that the points $v_0, \dots, v_k, w_0, \dots, w_p$ are linearly independent. The *join* of $\sigma$ and $\tau$ is defined as the $(k+p+1)$-simplex $[v_0, \dots, v_k, w_0, \dots, w_p]$ generated by these points, that is their convex hull. Given two simplical complexes $K$ and $L$, if the collection of all simplices $\sigma * \tau$ for $\sigma \in K$ and $\tau \in L$ is a simplicial complex, then this complex is called the *join* of $K$ and $L$ and denoted by $K * L$.

If $K$ is reduced to a vertex $\{v\}$, the join of $K$ and $L$ is the cone over $L$, that we denoted by $c(L)$. If $K$ consists in two vertices, the join $K * L$ is the suspension of $L$, denoted by $\Sigma(L)$. The suspension of the circle $\mathbb{S}^1$ is homeomorphic to $\mathbb{S}^2$ (figure 5, left) and in general, the suspension of $\mathbb{S}^n$ is homeomorphic to $\mathbb{S}^{n+1}$. The suspension of the torus (figure 5, right) is a 3-dimensional singular variety with two singular points $A$ and $B$: the link of $A$ (or $B$) is a torus, not a sphere. See the alternative nice picture of the suspension of the torus in [**GM1**]).

Another way to define the suspension of $X$ is to take the product (cyclinder) $X \times [-1, +1]$ and to "pinch" the cyclinder at the levels $\{-1\}$ and $\{+1\}$. That is

$$\Sigma(X) = X \times [-1, +1]/(X \times \{-1\}) \cup (X \times \{+1\}).$$

In the suspension of the torus, the points $A$ and $B$ are the pinched points.

The natural stratification of the suspension of the torus is

$$X \supset X_0 = \{A, B\} \supset \emptyset.$$

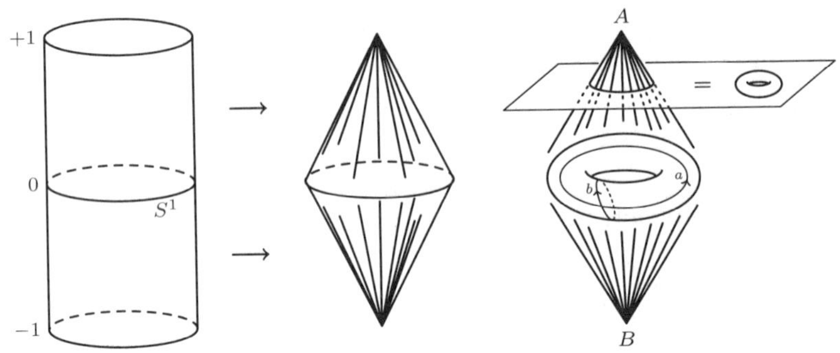


FIGURE 5. Suspension of the circle $\mathbb{S}^1$ and of the torus

There are two possible perversities:

$$\bar{p} = \bar{0} = (0,0,0,0) \qquad \text{and} \qquad \bar{p} = \bar{t} = (0,0,0,1)$$

An $i$-dimensional chain $c$ containing one (or two) of the singular points $A$ and $B$ is allowable if

$$0 = \dim(|c| \cap X_0) \leq i - 3 + p_3,$$

that means, if $\bar{p} = \bar{0}$, then $i \geq 3$ and if $\bar{p} = \bar{t}$, then $i \geq 2$.

Let us compute the intersectiojn homology groups $IH_i^{\bar{p}}(X;\mathbb{Q})$ corresponding to the two perversities:

For $i = 0$, a point $x \in X \setminus X_0$ is a 0-dimensional allowable cycle for each of the two perversities, then $IH_0^{\bar{p}}(X;\mathbb{Q}) = \mathbb{Q}_{\{x\}}$ in the two cases.

For $i = 1$ and $\bar{p} = \bar{0}$ or $\bar{t}$, a cycle containing $A$ or $B$ is not allowable. the candidates for generators of the 1-dimensional intersection homology groups are the cycle $a$ and $b$, generators of the homoloy of the torus. Now, one has

$$a = \partial(c(a)) \qquad \text{and} \qquad b = \partial(c(b))$$

where $c(a)$ and $c(b)$ are the cones on $a$ and $b$ respectively, with vertex $A$. They are 2-dimensional chains. In the case $\bar{p} = \bar{0}$, the chain $c(a)$ is not allowable. The allowable cycle $a$ is not an allowable boundary, so its homology class is non zero. One has the same for $b$ and then:

$$IH_1^{\bar{0}}(X;\mathbb{Q}) = \mathbb{Q}_a \oplus \mathbb{Q}_b.$$

In the case $\bar{p} = \bar{t}$, the chain $c(a)$ is allowable. The allowable cycle $a$ is an allowable boundary, so its homology class is zero. One has the same for $b$ and one has:

$$IH_1^{\bar{t}}(X;\mathbb{Q}) = 0.$$

For $i = 2$, the candidates to be generators of the 2-dimensional intersection homology groups are the suspensions $\Sigma(a)$ and $\Sigma(b)$ of the cycles $a$ and $b$ and the torus $T$ itself. In the case $\bar{p} = \bar{0}$, then the 2-dimensional cycles $\Sigma(a)$ and $\Sigma(b)$ contain the singular points and are not allowable. The cycle $T$ is allowable (it does not contain $A$ or $B$), but it is the boundary of the allowable chain $c(T)$, cone over $T$ with vertex $A$, One concludes that

$$IH_2^{\bar{0}}(X;\mathbb{Q}) = 0.$$

In the case $\bar{p} = \bar{t}$, the 2-dimensional cycles $\Sigma(a)$ and $\Sigma(b)$ are allowable, and they are not boundary. The cycle $T$ is still the boundary of the allowable chain $c(T)$. One concludes that

$$IH_2^{\bar{t}}(X;\mathbb{Q}) = \mathbb{Q}_{\Sigma(a)} \oplus \mathbb{Q}_{\Sigma(b)}.$$

For $i = 3$, the fundamental class of $X = \Sigma(T)$ is an allowable cycle, for each of the two perverstites and one has $IH_2^{\bar{p}}(X;\mathbb{Q}) = \mathbb{Q}_{[X]}$ in the two cases.

These computations are resumed in the table 4:

| Perversities | $\bar{p} = \bar{0}$ | $\bar{p} = \bar{t}$ |
|---|---|---|
| $i$ | $IH_i^{\bar{0}}(X)$ | $IH_i^{\bar{t}}(X)$ |
| 0 | $\mathbb{Q}_{\{x\}}$ | $\mathbb{Q}_{\{x\}}$ |
| 1 | $\mathbb{Q}_a \oplus \mathbb{Q}_b$ | 0 |
| 2 | 0 | $\mathbb{Q}_{\Sigma(a)} \oplus \mathbb{Q}_{\Sigma(b)}$ |
| 3 | $\mathbb{Q}_{[X]}$ | $\mathbb{Q}_{[X]}$ |

TABLE 4. Intersection homology of the suspension of the torus

Let us compute the intersection matrix and verify that, for $\bar{p} = \bar{0}$, $\bar{q} = \bar{t}$ and $i = 1$, the intersection product

$$IH_i^{\bar{p}}(X;\mathbb{Q}) \times IH_{n-i}^{\bar{q}}(X:\mathbb{Q}) \to \mathbb{Q}$$

is non-degenerate.

On the one hand, one remarks that $a$ and $\Sigma(a)$ are not dimensionally transverse , but the cycle $a$ can be moved as a cycle $a'$ in the same intersection homology class so that $a' \cap \Sigma(a) = \emptyset$ and $I(a, \Sigma(a)) = 0$. On the other hand, the intersection $b \cap \Sigma(a)$ is a single point with intersection number $I(b, \Sigma(a)) = \pm 1$ depending on the choices of orientations. The intersection matrix is:

$$\begin{array}{c} \\ a \\ b \end{array} \begin{array}{c} \begin{array}{cc} \Sigma(a) & \Sigma(b) \end{array} \\ \begin{pmatrix} 0 & \mp 1 \\ \pm 1 & 0 \end{pmatrix} \end{array}$$

whose determinant is $\pm 1$.

2.4.3. *Restriction morphism.* Let $U \subset V$ be two open subsets in $X$. In the same way than in (1.6), for every perversity $\bar{p}$, one defines natural maps

$$\rho_{UV} : IC_i^{\bar{p},\mathrm{cl}}(V) \to IC_i^{\bar{p},\mathrm{cl}}(U). \tag{2.22}$$

If $\xi \in IC_i^{\bar{p},\mathrm{cl}}(V)$ then the chain $\rho_{UV}(\xi)$ defined in (1.6) satisfies the perversity conditions for the perversity $\bar{p}$, *i.e.*, $\rho_{UV}(\xi) \in IC_i^{\bar{p},\mathrm{cl}}(U)$.

We will see (section 3) that the correspondence $U \mapsto IC_i^{\bar{p},\mathrm{cl}}(U)$ defines a presheaf which is a sheaf.

2.4.4. *Relative homology.* Let $X$ be a stratified pseudomanifold, and $U$ an open subset in $X$, then $U$ inherits a structure of stratified pseudomanifold induced by the one of $X$. For every perversity $\bar{p}$, the complex of intersection chains of $U$ with compact supports $IC_*^{\bar{p}}(U)$, is a sub-complex of $IC_*^{\bar{p}}(X)$.

Let us define $IC_*^{\bar{p}}(X,U) = IC_*^{\bar{p}}(X)/IC_*^{\bar{p}}(U)$, one obtains a relative complex and one has a long exact sequence:

$$\cdots \to IH_i^{\bar{p}}(U) \to IH_i^{\bar{p}}(X) \to IH_i^{\bar{p}}(X,U) \to IH_{i-1}^{\bar{p}}(U) \to \cdots$$

2.4.5. *Excision.*

LEMMA 2.23. *Let $X$ be a locally compact stratified pseudomanifold and $U$ and $V$ two open subsets in $X$, then the inclusion $(U, U \cap V) \hookrightarrow (U \cup V, V)$ induces an isomorphism of intersection homology groups with compact supports*

$$IH_i^{\bar{p}}(U, U \cap V) \cong IH_i^{\bar{p}}(U \cup V, V).$$

PROPOSITION 2.24. *Let $X$ be a locally compact stratified pseudomanifold, $U$ an open subset in $X$ and $B$ a closed subset in $U$. Let $\bar{p}$ be any perversity, then the inclusion $(X - B, U - B) \hookrightarrow (X, U)$ induces an isomorphism of intersection homology groups with compact supports*

$$IH_i^{\bar{p}}(X,U) \cong IH_i^{\bar{p}}(X - B, U - B).$$

2.4.6. *Künneth formulae.* In general, Künneth formula in no longer true for intersection homology. However, one has Künneth formulae when one of the element of the product is a smooth manifold (see [**CGJ**]).

PROPOSITION 2.25. *Let us denote by $\mathbb{B}^k$ the $k$-dimensional ball. Let $X$ be a locally compact stratified pseudomanifold, $U$ an open subset in $X$ and $\bar{p}$ any perversity, one has*

$$IH_i^{\bar{p}}(X) \cong IH_i^{\bar{p}}(X \times \mathbb{B}^k) \tag{2.28a}$$

$$IH_i^{\bar{p},\mathrm{cl}}(X) \cong IH_{i+k}^{\bar{p},\mathrm{cl}}(X \times \mathbb{B}^k) \tag{2.28b}$$

$$IH_i^{\bar{p}}(X,U) \cong IH_i^{\bar{p}}(X \times \mathbb{B}^k, U \times \mathbb{B}^k) \tag{2.28c}$$

$$IH_i^{\bar{p}}(X,U) \cong IH_{i-k}^{\bar{p}}(X \times \mathbb{B}^k, (U \times \mathbb{B}^k) \cup (X \times \partial\mathbb{B}^k \times \mathbb{B}^1)) \tag{2.28d}$$

PROPOSITION 2.26. *Let $X$ be a locally compact stratified pseudomanifold and $M$ a manifold. Let $\bar{p}$ be a perversity, one has a split exact sequence:*

$$0 \to (H_*(M) \otimes IH_*^{\bar{p}}(X))_i \to IH_i^{\bar{p}}(X \times M) \to (IH_*^{\bar{p}}(X) * H_*(M))_{i-1} \to 0$$

**2.5. Local calculus.** The local calculus, and precisely the computation in formulae (2.28) and (2.29) below, are the starting points for the characterization of intersection homology.

Let us denote by $x \in X$ a point in the stratum $S_{n-h}$ with codimension $h$ in $X$. Let us consider a neighbourhood $U$ of $x$ homeomorphic to $\mathbb{B}^{n-h} \times \mathring{c}(L_x)$. Let us denote by $\partial U \times \mathbb{B}^1$ a collar neighbourhood of the boundary of $U$ :

PROPOSITION 2.27. *Let $X$ be a locally compact stratified pseudomanifold and $\bar{p}$ any perversity. Let us consider a point $x$ in a stratum with codimension $h$ in $X$.*

*Let us denote by $U$ a neighbourhood of $x$ homeomorphic to $\mathbb{B}^{n-h} \times \mathring{c}(L_x)$, then one has:*

$$IH_i^{\bar{p}}(U) \cong IH_i^{\bar{p}}(\mathring{c}(L_x)) \cong \begin{cases} IH_i^{\bar{p}}(L_x) & i < h - p_h - 1 \\ 0 & i \geq h - p_h - 1 \end{cases} \tag{2.28}$$

$$IH_i^{\bar{p}}(U, \partial U \times \mathbb{B}^1) \cong IH_i^{\bar{p},\mathrm{cl}}(U) \cong$$

$$IH_{i-(n-h)}^{\bar{p},\mathrm{cl}}(\mathring{c}(L_x)) \cong \begin{cases} 0 & i < n - p_h \\ IH^{\bar{p}}_{i-(n-h)-1}(L_x) & i \geq n - p_h. \end{cases} \tag{2.29}$$

The link $L_x$ is compact and its homology groups, with compact and closed supports coincide.

Proof of Formulae (2.28): The Künneth formula (2.28.a) gives (with the previously defined stratification of the cone):

$$IH_i^{\bar{p}}(U) = IH_i^{\bar{p}}(\mathbb{B}^{n-h} \times \mathring{c}(L_x)) = IH_i^{\bar{p}}(\mathring{c}(L_x)).$$

Let us compute $IH_i^{\bar{p}}(\mathring{c}(L_x))$.

1) Let us assume $i < h - p_h - 1$. A $(\bar{p}, i)$-cycle $\xi$ in $\mathring{c}(L_x)$ does not contain the vertex of the cone. Indeed, for the 0-dimensional stratum $\{x\}$ of $\mathring{c}(L_x)$, one must have (formula (2.10) with $n = k = h$):

$$\dim(|\xi| \cap \{x\}) \leq i - h + p_h < -1.$$

One can deform (push) $\xi$ along the generatrices of the cone, as a $(\bar{p}, i)$-cycle in $L_x \times \{t\}$ for a fixed $t \neq 0$ in $\mathbb{B}^1$ (Figure 6 Left).

One obtains a cycle $\xi_L$ in $IC_i^{\bar{p}}(L_x)$. The cycles $\xi_L$ and $\xi$ are homologous $IC_i^{\bar{p}}(\mathring{c}(L_x))$ by a $\bar{p}$-allowable homology: $\xi - \xi_L$ is boundary of the $\bar{p}$-allowable chain $c(\xi_L) \setminus \mathring{c}(\xi)$.

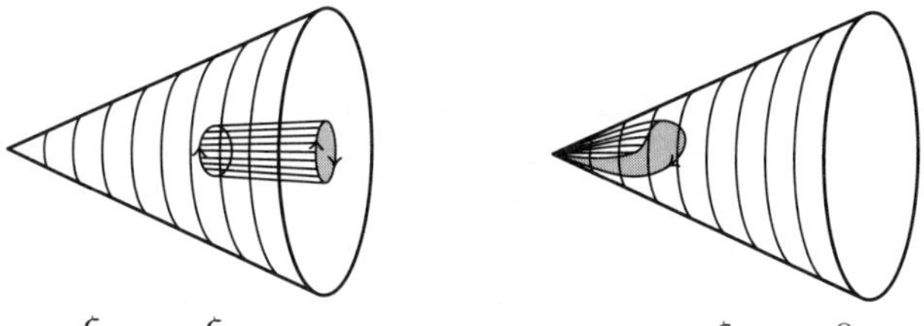

$\xi \sim \xi_L$ $\qquad\qquad$ $\xi \sim 0$

Figure 6. Cycles in the cone

2) Let us now assume $i \geq h - p_h - 1$ and let $\xi$ be a $(\bar{p}, i)$-cycle in $\mathring{c}(L_x)$. The cone over $\xi$ and vertex $\{x\}$, denoted by $c(\xi)$, is a $(\bar{p}, i+1)$-chain (Figure 6 Right). Indeed, one has in particular

$$0 = \dim(|c(\xi)| \cap \{x\}) \leq (i+1) - h + p_h.$$

That is $\xi = \partial(c(\xi))$ is a boundary in $IC_i^{\bar{p}}(\mathring{c}(L_x))$ and the result. $\square$

PROOF OF FORMULAE (2.29): The Künneth formula (2.28.d) gives (see Figure 7):

$$\begin{aligned} & IH_i^{\bar p}(U, \partial U \times \mathbb{B}^1) \\ & = IH_i^{\bar p}(\mathbb{B}^{n-h} \times \mathring{c}(L_x), (\partial \mathbb{B}^{n-h} \times \mathbb{B}^1) \times \mathring{c}(L_x) \cup \mathbb{B}^{n-h} \times (L_x \times \mathbb{B}^1)) \\ & = IH_{i-n+h}^{\bar p}(\mathring{c}(L_x), L_x \times \mathbb{B}^1). \end{aligned}$$

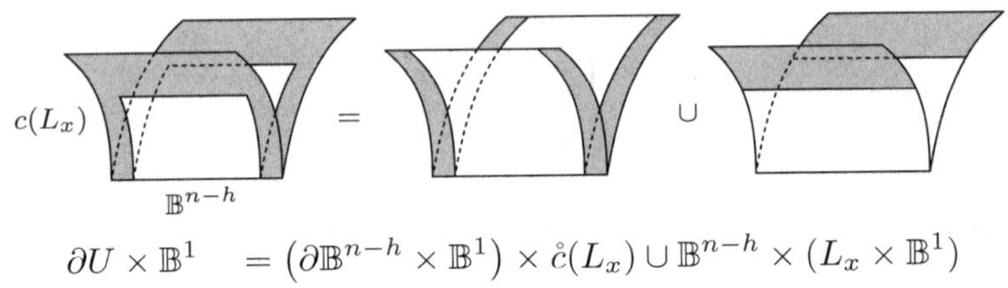


FIGURE 7. Collar neighbourhood of the boundary of $U_x$

Let us denote by $\xi$ a cycle in $IC_{i-n+h}^{\bar p}(\mathring{c}(L_x), L_x \times \mathbb{B}^1)$.

1) Let us assume $i < n - p_h$, then $|\xi|$ does not contain the vertex $\{x\}$. Indeed, one has

$$\dim(|\xi| \cap \{x\})) \leq (i - n + h) - h + p_h < 0.$$

So, one can push $\xi$ along the generatrices of the cone as a cycle lying in $L_x \times \mathbb{B}^1$ (Figure 6 Left). The class of $\xi$ in $IH_{i-n+h}^{\bar p}(\mathring{c}(L_x), L_x \times \mathbb{B}^1)$ vanishes.

Let us remark that, if $i = n - p_h - 1$, then $c(\xi)$ is a $(\bar p, i+1)$-chain with boundary $\xi$ (Figure 6 Right), that is another way to prove the result.

2) Let us assume $i \geq n - p_h$, then the long exact sequence of the pair $(\mathring{c}(L_x), L_x \times \mathbb{B}^1)$ (see Section 2.4.4) is written (with $k = i - n + h$):

$$\begin{aligned} \cdots \to IH_k^{\bar p}(\mathring{c}(L_x)) \to & IH_k^{\bar p}(\mathring{c}(L_x), L_x \times \mathbb{B}^1) \\ & \to IH_{k-1}^{\bar p}(L_x \times \mathbb{B}^1) \to IH_{k-1}^{\bar p}(\mathring{c}(L_x)) \to \cdots \end{aligned}$$

One has $k = i - n + h \geq h - p_h$, then the formula (2.28) shows that the two extreme groups vanish. One has:

$$IH_k^{\bar p}(\mathring{c}(L_x), L_x \times \mathbb{B}^1) \cong IH_{k-1}^{\bar p}(L_x \times \mathbb{B}^1) \cong IH_{k-1}^{\bar p}(L_x)$$

according to the Künneth formula (2.28.a). □

*Here is an useful and important notation for the following:*

Let us denote by $\bar p$ a perversity and $\bar q$ the complementary perversity, one recalls that $p_k + q_k = k - 2$ for all $k \geq 2$. If $j \in \mathbb{N}$, one defines

$$p^{-1}(j) = \min\{ k \mid p_k \geq j\}$$

and $p^{-1}(j) = \infty$ if $j > p_n$ (see Figure 8).

One has:

$$p_k \geq j \iff k \geq p^{-1}(j) \tag{2.29}$$

$$j \geq n - q_k \iff n - k \leq n - q^{-1}(n-j) \tag{2.29}$$

According to 2.28, one has:

$$IH_{n-j}^{\bar p}(U_x) = 0 \text{ for } n - j \geq h - p_h - 1$$

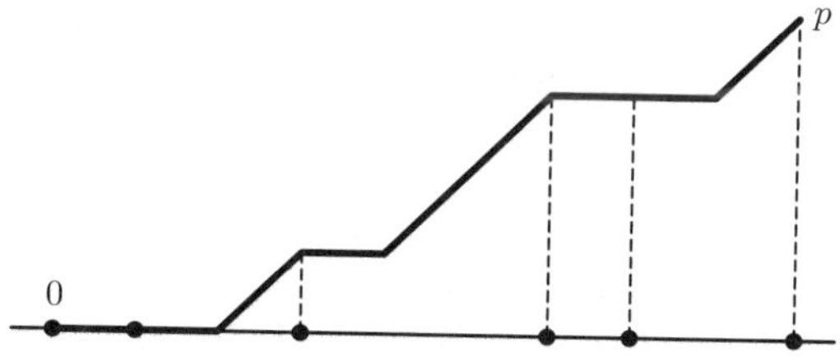


FIGURE 8. For the perversity of figure 4, one has $p^{-1}(1) = 3$, $p^{-1}(2) = 5$, $p^{-1}(3) = 6$, $p^{-1}(4) = 8$.

that is, the set of points $x \in X$ such that $IH^{\bar{p}}_{n-j}(U_x) \neq 0$ is contained in the strata $X_{n-h}$ such that $n-j < h - p_h - 1$. That can be written $n - j \leq h - p_h - 2 = q_h$ for the complementary perversity $\bar{q}$, *i.e.*, $j \geq n - q_h$ or $n - h \leq n - q^{-1}(n-j)$, according to 2.29. One has

$$\boxed{\dim\{x \in X \mid IH^{\bar{p}}_{n-j}(U_x) \neq 0\} \leq n - q^{-1}(n-j) \qquad \text{for } j \geq 1.} \tag{2.30}$$

In the same way,

$$IH^{\bar{p},\mathrm{cl}}_{n-j}(U_x) = 0 \text{ for } n - j < n - p_h$$

that is, the set of points $x \in X$ such that $IH^{\bar{p},\mathrm{cl}}_{n-j}(U_x) \neq 0$ is contained in the strata $X_{n-h}$ such that $n - j \geq n - p_h$, *i.e.*, $j \leq p_h$. According to 2.29 that is $h \geq p^{-1}(j)$, or $n - h \leq n - p^{-1}(j)$. One has:

$$\boxed{\dim\{x \in X \mid IH^{\bar{p},\mathrm{cl}}_{n-j}(U_x) \neq 0\} \leq n - p^{-1}(j) \qquad \text{for } j \leq n-1.} \tag{2.31}$$

## 3. Sheaves

The aim of the chapter is to translate the local calculus and more precisely the properties (2.30) and (2.31) into the language of sheaves and hypercohomology. That is the next step to the characterization of intersection homology. Main reference for the chapter is [**Bo**].

The reader is supposed to be familiar with the notion of sheaves (see Godement [**God**]). However, we provide two useful examples for the following. Given a topological space $X$, the set of sections of the sheaf $\mathcal{F}$ over an open subset $U$ of $X$ will be denoted by $\mathcal{F}(U)$.

EXAMPLE 3.1. [The constant sheaf] Let $X$ be a topological space and $G$ an abelian group, the *constant sheaf* $G_X = G$ is defined as follows: Given $G$ the discrete topology, one defines $G_X(\emptyset) = \{0\}$ and, if $U$ is non-empty, $G_X(U)$ is the set of all continuous maps of $U$ into $G$. The restriction maps are obvious. If $U$ is a non-empty and connected open set, every continuous map, element of $G_X(U)$ is constant, so $G_X(U) = G$. The constant sheaf we will often use are the constant sheaves $\mathbb{Q}_X$, $\mathbb{R}_X$ and $\mathbb{C}_X$.

EXAMPLE 3.2. [The orientation sheaf] Let us consider a connected $n$-dimensional manifold $X$, the *orientation sheaf* $\mathcal{O}_X$ is the sheaf associated to the presheaf (in fact, already a sheaf)

$$U \mapsto H_c^n(U; \mathbb{C}).$$

Here $\mathbb{C}$ could be any noetherian ring. If $X$ is orientable, the orientation sheaf is the constant sheaf, otherwise it is locally constant.

The orientation sheaf is an example of so-called *rank one local system.*

DEFINITION 3.3. A sheaf $\mathcal{F}$ on $X$ is locally constant, or is a *local system*, if for all $x \in X$, there is a neighborhood $U$ containing $x$ such that $\mathcal{F}|_U$ is a constant sheaf.

### 3.1. Complexes of sheaves.

DEFINITION 3.4. A *complex of sheaves* (or Differential Graded Sheaf) $\mathcal{F}^\bullet$ on a variety $X$ is a set of sheaves $\mathcal{F}^k$, $k \in \mathbb{Z}$ and sheaf maps (differential) $d^k : \mathcal{F}^k \to \mathcal{F}^{k+1}$ such that $d^{k+1} \circ d^k = 0$ for all $k$.

Let $\mathcal{F}^\bullet$ be a complex of sheaves, the shifted complex $\mathcal{F}^\bullet[n]$ is defined by $(\mathcal{F}^\bullet[n])^k = \mathcal{F}^{n+k}$ with differential $d^k_{[n]} = (-1)^n d^{n+k}$.

EXAMPLE 3.5. The complex of sheaves of chains with closed supports.

The complex of chains with closed supports on an $n$-dimensional variety is a complex of sheaves that we will denote by $\mathcal{C}^\bullet$. Firstly, for each $k$, one defines

$$\mathcal{C}^k(U) = C^{\text{cl}}_{n-k}(U).$$

The restriction maps

$$\rho_{UV} : C^{\text{cl}}_{n-k}(V) \to C^{\text{cl}}_{n-k}(U), \qquad \text{for } U \subset V$$

are the ones defined in Lemma 1.6 and the correspondence $U \mapsto C^{\text{cl}}_{n-k}(U)$ defines a presheaf which is the sheaf $\mathcal{C}^k$. Notice that the same Lemma 1.6 says that we cannot define a sheaf when using chains with compact supports.

By definition the maps $d^k : \mathcal{C}^k \to \mathcal{C}^{k+1}$ are defined by the chain maps $d_{n-k} : C^{\text{cl}}_{n-k}(U) \to C^{\text{cl}}_{n-k-1}(U)$.

The complex of sheaves of chains $\mathcal{C}^\bullet$ is sometimes denoted also by $\mathcal{C}_{n-\bullet}$ and when necessary, we will precise the space by denoting $\mathcal{C}^\bullet_X$.

EXAMPLE 3.6. The complex of sheaves of intersection chains.

In the same way the complex of sheaves of intersection chains is defined using $\bar{p}$-allowable chains with closed supports:

Let $X$ be a $n$-dimensional pseudomanifold and let $U \subset V$ be two open subsets in $X$. We have seen (2.22), that, for every perversity $\bar{p}$, one defines natural maps

$$\rho_{UV} : IC_i^{\bar{p},\text{cl}}(V) \to IC_i^{\bar{p},\text{cl}}(U).$$

The correspondence $U \mapsto IC_i^{\bar{p},\text{cl}}(U)$ defines a presheaf which is in fact a sheaf, called the sheaf of $i$-intersection chains. One denotes, with $i = n-k$,

$$\mathcal{IC}^{\bar{p},k}(U) = IC^{\bar{p},\text{cl}}_{n-k}(U).$$

The complex of sheaves $\mathcal{IC}^{\bar{p},\bullet}$ will be also called intersection complex and also denoted by $\mathcal{IC}^{\bar{p}}_{n-\bullet}$. When necessary, we will precise the space by denoting $\mathcal{IC}^{\bar{p},\bullet}_X$.

We will sometimes omit the perversity when using the middle perversity $\bar{m}$ and we will denote $\mathcal{IC}^\bullet = \mathcal{IC}^{\bar{m},\bullet}$.

DEFINITION 3.7. The $i$-th *cohomology sheaf* of a complex of sheaves $\mathcal{F}^\bullet$ will be denoted by $\mathcal{H}^i(\mathcal{F}^\bullet)$.

A map of complexes

$$\begin{array}{cccccccccccc} \mathcal{F}^\bullet : & \cdots \longrightarrow & \mathcal{F}^{i-1} & \longrightarrow & \mathcal{F}^i & \longrightarrow & \mathcal{F}^{i+1} & \longrightarrow & \mathcal{F}^{i+2} & \longrightarrow \cdots \\ f\downarrow & & \downarrow & & \downarrow & & \downarrow & & \downarrow & \\ \mathcal{G}^\bullet : & \cdots \longrightarrow & \mathcal{G}^{i-1} & \longrightarrow & \mathcal{G}^i & \longrightarrow & \mathcal{G}^{i+1} & \longrightarrow & \mathcal{G}^{i+2} & \longrightarrow \cdots \end{array}$$

is called a *quasi-isomorphism* if it induces isomorphisms $\mathcal{H}^i(f) : \mathcal{H}^i(\mathcal{F}^\bullet) \to \mathcal{H}^i(\mathcal{G}^\bullet)$ of the cohomology sheaves of the complexes.

EXAMPLE 3.8. One has a quasi-isomorphism of the "constant sheaf" complex $\mathbb{R}_X$ in the complex $\Omega^\bullet_X$ of differential forms on the differentiable manifold $X$:

$$\begin{array}{ccccccccccc} \mathbb{R}_X : & \cdots \longrightarrow & 0 & \longrightarrow & \mathbb{R}_X & \longrightarrow & 0 & \longrightarrow \cdots \longrightarrow & 0 & \longrightarrow \cdots \\ f\downarrow & & \downarrow & & \downarrow & & \downarrow & & \downarrow & \\ \Omega^\bullet_X : & \cdots \longrightarrow & 0 & \longrightarrow & \Omega^0_X & \longrightarrow & \Omega^1_X & \longrightarrow \cdots \longrightarrow & \Omega^i_X & \longrightarrow \cdots \end{array}$$

DEFINITION 3.9. [**Bo**, V] A *(right) resolution* of a sheaf $\mathcal{F}$ is the data of a complex of sheaves $\mathcal{R}^\bullet$ and a morphism $\mathcal{F} \to \mathcal{R}^\bullet$ of complexes of sheaves which induces an isomorphism $H^p(\mathcal{F}(U)) \to H^p(\mathcal{R}^\bullet(U))$ for all $p$ and all $U$.

Let us recall that one says that $I$ is an *injective* object in an abelian category if the functor Hom( , $I$) is an exact functor. By a result of N. Spaltenstein, a sheaf $\mathcal{F}$ is injective if for every open subsets $U \subset V$, the restriction map $\rho_{UV} : \mathcal{F}(V) \to \mathcal{F}(U)$ is a surjection.

THEOREM 3.10 (Godement [**God**]). *Every sheaf $\mathcal{F}$ admits an injective resolution, that is a quasi-isomorphism of complexes:*

$$\begin{array}{cccccccccccc} \mathcal{F} : & \cdots \longrightarrow & 0 & \longrightarrow & \mathcal{F} & \longrightarrow & 0 & \longrightarrow & 0 & \longrightarrow \cdots \\ f\downarrow & & \downarrow & & \downarrow & & \downarrow & & \downarrow & \\ \mathcal{I}^\bullet : & \cdots \longrightarrow & 0 & \longrightarrow & \mathcal{I}^0 & \longrightarrow & \mathcal{I}^1 & \longrightarrow & \mathcal{I}^2 & \longrightarrow \cdots \end{array}$$

*where the sheaves $\mathcal{I}^k$ are injective.*

**3.2. Čech Cohomology of Sheaves.** General references for this section are [**Bo**, II.5; V,1.4] and [**God**, II.5.2-3].

Let us consider a sheaf $\mathcal{F}$ on a topological space $X$. One denotes by $\mathcal{U} = \{U_i : i \in I\}$ a (locally finite) open covering of $X$. For every ordered subset $J_p = i_0, \dots, i_p$ of $p+1$ elements in $I$, one denotes

$$U_{J_p} = U_{i_0} \cap U_{i_1} \cap \cdots \cap U_{i_p}.$$

An element $\alpha \in C^p(\mathcal{U}, \mathcal{F})$ is a family

$$\alpha = (\alpha_{i_0,\dots,i_k}) \text{ with } \alpha_{i_0,\dots,i_p} \in \mathcal{F}(U_{J_p})$$

indexed by the subsets $J_p$ of $I$ such that $U_{J_p} \neq \emptyset$. One defines a coboundary map

$$\delta^p : C^p(\mathcal{U},\mathcal{F}) \to C^{p+1}(\mathcal{U},\mathcal{F}) \tag{3.10}$$

$$(\delta^p\alpha)_{i_0,\dots,i_{p+1}} = \sum_{k=0}^{p+1}(-1)^k\alpha_{i_0,\dots,\widehat{i_k},\dots,\alpha_{p+1}}$$

in the subsets where all elements take sense. It is an exercise to check that $\delta^{p+1}\circ\delta^p = 0$.

The corresponding cohomology groups are called *Čech cohomology groups* of the sheaf $\mathcal{F}$ and denoted by $H^p(\mathcal{U},\mathcal{F})$.

One denotes by $C^p(X,\mathcal{F})$ the direct limit of the $C^p(\mathcal{U},\mathcal{F})$ with respect to refinements of the covering $\mathcal{U}$ and by $H^p(X,\mathcal{F})$ the corresponding cohomology groups.

PROPOSITION 3.11. *Let us consider the constant sheaf* $\mathbb{R}_X$ *on the triangulable topological space* $X$, *then*

$$H^p(X) \cong H^p(X,\mathbb{R}_X).$$

**3.3. Hypercohomology.**

3.3.1. *Definition of hypercohomology.* Let us consider a sheaf complex $\mathcal{A}^\bullet$ on $X$. One denotes by $\mathcal{U}$ a locally finite open covering of $X$ and by $C^p(\mathcal{U};\mathcal{A}^q)$ the set of Čech $p$-cochains on $\mathcal{U}$ with coefficients in $\mathcal{A}^q$. One has two differentials:

$$\delta_1 : C^p(\mathcal{U};\mathcal{A}^q) \to C^{p+1}(\mathcal{U};\mathcal{A}^q)$$

induced by the one of the Čech complex (3.10) and

$$\delta_2 : C^p(\mathcal{U};\mathcal{A}^q) \to C^p(\mathcal{U};\mathcal{A}^{q+1})$$

induced by the differential $d^q : \mathcal{A}^q \to \mathcal{A}^{q+1}$ of the sheaf complex. Then

$$\delta_2^2 = \delta_1^2 = 0 \quad \text{and} \quad \delta_1\delta_2 = \delta_2\delta_1.$$

Let us denote by

$$C^{p,q} = \varinjlim_{\mathcal{U}} C^p(\mathcal{U};\mathcal{A}^q)$$

the double complex and by $\delta_1 : C^{p,q} \to C^{p+1,q}$ and $\delta_2 : C^{p,q} \to C^{p,q+1}$ the induced differentials. The total complex is denoted

$$K^i = \oplus_{p+q=i}C^{p,q} \quad \text{with differential} \quad d = \delta_1 + (-1)^p\delta_2.$$

DEFINITION 3.12. The *hypercohomology* groups of the sheaf complex $\mathcal{A}^\bullet$ are the cohomology groups of the total complex $(K^\bullet, d)$, denoted by

$$\mathbb{H}^i(X;\mathcal{A}^\bullet) = H^i(K^\bullet).$$

EXAMPLE 3.13. Let $\mathcal{A}^\bullet$ be a complex which is not zero only in degree $k$, one says *concentrated in degree* $k$, then one has:

$$\mathbb{H}^i(X;\mathcal{A}^\bullet) = H^{i-k}(X;\mathcal{A}^k). \tag{3.13}$$

The classical theory of spectral sequences (see [**God**] or any "classical book" on the subject) shows:

PROPOSITION 3.14. *If the sheaf complex $\mathcal{A}^q$ is acyclic, i.e., if for every $q$ one has*

$$H^p(X;\mathcal{A}^q)=0 \qquad \forall p>0,$$

(see [**God**, Théorème 5.2.3]) *then the hypercohomology group $\mathbb{H}^i(X;\mathcal{A}^\bullet)$ is the $i$-th cohomology group of the complex $\Gamma(X;\mathcal{A}^\bullet)$.*

Examples of acyclic complex sheaves are given by injective or "flabby" sheaves and, in the case $X$ being paracompact, by "fine" or "soft" sheaves (for the definitions of flabby and fine, see [**God**, II sections 7,1; 3.1; 3.7; 3.4] and [**God**, II, Théorème 5.2.3]). The examples we will use are soft sheaves:

DEFINITION 3.15. A sheaf $\mathcal{F}$ is called *soft* if the restriction maps

$$\Gamma(U,\mathcal{F})\to\Gamma(A,\mathcal{F})$$

are surjective for all open $U\subset X$ and closed subset $A\subset U$.

LEMMA 3.16. *The complexes $\mathcal{C}^\bullet$ and $\mathcal{IC}^\bullet$ are examples of complexes of soft sheaves.*

An alternative definition of hypercohomology groups is the following:

$$\mathbb{H}^i(X;\mathcal{A}^\bullet)=H^i(\Gamma(X;\mathcal{I}^\bullet)) \tag{3.16}$$

where $\mathcal{I}^\bullet$ is an injective resolution of the complex $\mathcal{A}^\bullet$ (cf. the isomorphism $j''^*$ in [**God**, II.5.3]).

PROPOSITION 3.17 ([**Bo**, II, §5], [**God**]). *Let $X$ be a paracompact topological space, if the sheaf complex $\mathcal{F}$ consists of injective, flabby, fine or soft sheaves, then*

$$\mathbb{H}^i(X;\mathcal{F}^\bullet)=H^i(\Gamma(X;\mathcal{F}^\bullet)). \tag{3.18}$$

DEFINITION 3.19. Considering cochains with supports in a family of supports $\Phi$, one defines *hypercohomology $\mathbb{H}^i_\Phi(X;\mathcal{A}^\bullet)$ with support in the family $\Phi$* as $\mathbb{H}^i_\Phi(X;\mathcal{A}^\bullet)=H^i(\Gamma_\Phi(X;\mathcal{I}^\bullet))$.

We will use mainly the family of compact supports and the family of closed supports, as we did previously.

3.3.2. *Examples.* Let $X$ be a $n$-dimensional algebraic complex variety, the following examples provide particular cases of hypercohomology groups:

a) Let $\mathcal{C}^\bullet$ be the *sheaf complex of singular chains with closed supports* on $X$. Elements of $\mathcal{C}^\bullet$ are soft sheaves (Lemma 3.16), hence the complex $\mathcal{C}^\bullet$ satisfies Proposition 3.17. One has:

$$\mathbb{H}^i(X;\mathcal{C}^\bullet)\cong H^i(\Gamma(X;\mathcal{C}^\bullet))=H^i(C^{\mathrm{cl}}_{n-\bullet}(X))=H^{\mathrm{cl}}_{n-i}(X)$$

where the first isomorphism is (3.18). For every family of supports $\Phi$ one has:

$$\mathbb{H}^i_\Phi(X;\mathcal{C}^\bullet)=H^\Phi_{n-i}(X). \tag{3.20}$$

In particular, one has

$$\mathbb{H}^i_c(X;\mathcal{C}^\bullet)=H_{n-i}(X).$$

b) Let us consider the *constant sheaf* $\mathbb{Q}_X$ on $X$, considered as a complex concentrated in degree 0, then, by (3.13):

$$\mathbb{H}^i(X;\mathbb{Q}_X)=H^i(X)$$

is cohomology of $X$ with closed supports. For every family of supports $\Phi$ one has:

$$\mathbb{H}^i_\Phi(X;\mathbb{Q}_X) = H^i_\Phi(X). \tag{3.21}$$

c) Let us consider $\mathcal{IC}^{\bar{p},\bullet} = \mathcal{IC}^{\bar{p},\bullet}_X$ the *sheaf complex of intersection chains* with rational coefficients, for the perversity $\bar{p}$. Elements of $\mathcal{IC}^{\bar{p},\bullet}$ are soft sheaves (Lemma 3.16), so that the complex satisfies Proposition 3.17. One has, by (3.18):

$$\mathbb{H}^i(X;\mathcal{IC}^{\bar{p},\bullet}) = IH^{\bar{p},cl}_{n-i}(X) \tag{3.22}$$

and

$$\mathbb{H}^i_c(X;\mathcal{IC}^{\bar{p},\bullet}) = IH^{\bar{p}}_{n-i}(X), \tag{3.23}$$

intersection homology with compact supports, and more generally

$$\mathbb{H}^i_\Phi(X;\mathcal{IC}^{\bar{p},\bullet}) = IH^{\bar{p},\Phi}_{n-i}(X)$$

for any family of supports $\Phi$ ([**Bo**, II,5]).

d) The dualizing complex.

The *dualizing sheaf complex* is a generalization of the orientation sheaf. On a $n$-dimensional manifold, the dualizing sheaf complex $\mathcal{D}^\bullet$ coincides, up to quasi-isomorphism, with the shifted orientation sheaf $\mathcal{O}_X[n]$, *i.e.*, with the complex whose unique element is $\mathcal{O}_X$ in degree $n$. An intrinsic definition of $\mathcal{D}^\bullet$ is given, for instance in [**Bo**, V,7.1].

For our purposes, it is enough to know that the (shifted) dualizing complex is quasi-isomorphic to $\mathcal{C}^\bullet_X$ (see [**Bo**, V,7.2]), *i.e.*,

$$\widetilde{\mathcal{D}}^\bullet_X \cong \mathcal{C}^\bullet_X. \tag{3.24}$$

Here the shifted complex $\widetilde{\mathcal{D}}^\bullet_X$ is $\widetilde{\mathcal{D}}^\bullet_X = \mathcal{D}^\bullet_X[-n] = \mathcal{D}^{\bullet-n}_X$. That means $\widetilde{\mathcal{D}}^i_X = \mathcal{D}^i_X[-n] = \mathcal{D}^{i-n}_X$ for all $i$.

We obtain, for the dualizing complex, similar results to Example (a) above. In particular, one has:

$$\mathbb{H}^i(X;\widetilde{\mathcal{D}}^\bullet_X) = H^{cl}_{n-i}(X) \qquad \text{and} \qquad \mathbb{H}^i_c(X;\widetilde{\mathcal{D}}^\bullet_X) = H_{n-i}(X).$$

More generally, one has

$$\mathbb{H}^i_\Phi(X;\widetilde{\mathcal{D}}^\bullet_X) = H^\Phi_{n-i}(X)$$

for any family of supports $\Phi$ ([**Bo**, V,7.1-2-3]).

### 3.4. Poincaré Duality "à la sauce Faisceaux".

3.4.1. *Poincaré Duality on a manifold.* The Poincaré duality can be proved by using sheaf complexes:

Let us assume that $X$ is a $n$-dimensional manifold, then one has an isomorphism of complexes of sheaves (see [**Bo**]) $\mathbb{Q}_X \cong \mathcal{C}^\bullet_X$ that induces isomorphisms of hypercohomology groups:

$$\mathbb{H}^i_\Phi(X;\mathbb{Q}_X) \cong \mathbb{H}^i_\Phi(X;\mathcal{C}^\bullet_X)$$

*i.e.*, (see 3.20, 3.21)

$$H^i_\Phi(X) \cong H^\Phi_{n-i}(X).$$

In particular, for the family of closed supports one has

$$H^i(X) \cong H^{\mathrm{cl}}_{n-i}(X)$$

and for the one of compact supports

$$H_c^i(X) \cong H_{n-i}(X).$$

3.4.2. *Poincaré Duality – Singular case.* Let us assume that $X$ is a complex algebraic variety with (real) pure dimension $n$. On the regular stratum of $X$, that is a smooth dense open subset $W \subset X$, one has quasi-isomorphisms $\mathbb{Q}_W \xrightarrow{\cong} \mathcal{C}_W^\bullet$ and $\mathcal{C}_W^\bullet \xrightarrow{\cong} \widetilde{\mathcal{D}}_W^\bullet$. They provide morphisms at the level of sheaf complexes, for any perversity $\bar{p}$:

$$\alpha_X : \mathbb{Q}_X \to \mathcal{IC}_X^{\bar{p};\bullet} \qquad \text{and} \qquad \omega_X : \mathcal{IC}_X^{\bar{p};\bullet} \to \widetilde{\mathcal{D}}_X^\bullet.$$

By taking hypercohomology, one obtain the classical comparison morphisms

$$H_c^\bullet(X) \xrightarrow{\alpha_X} IH_{n-\bullet}^{\bar{p}}(X) \quad \text{and} \quad IH_{n-\bullet}^{\bar{p}}(X) \xrightarrow{\omega_X} H_{n-\bullet}(X).$$

The composition $\omega_X \circ \alpha_X : \mathbb{Q}_X \to \widetilde{\mathcal{D}}_X^\bullet$

$$\begin{array}{ccc} \mathbb{Q}_X & \longrightarrow & \widetilde{\mathcal{D}}_X^\bullet \\ & {}_{\alpha_X}\searrow \quad \nearrow_{\omega_X} & \\ & \mathcal{IC}_X^{\bar{p};\bullet} & \end{array}$$

induces at the global level, *i.e.*, taking hypercohomology, the "classical" Poincaré duality homomorphism

$$H_c^\bullet(X) \to H_{n-\bullet}(X)$$

that is already known to be factorized by intersection homology (§4.2.1):

$$\begin{array}{ccc} H_c^\bullet(X) & \longrightarrow & H_{n-\bullet}(X) \\ & {}_{\alpha_X}\searrow \quad \nearrow_{\omega_X} & \\ & IH_{n-\bullet}^{\bar{p}}(X). & \end{array}$$

**3.5. Rewriting the properties of the intersection complex.** Let $U_x$ be a distinguished neighbourhood of a point $x$ in $X$ and let us consider the sheaf complex $\mathcal{IC}_X^{\bar{p};\bullet}$ of intersection chains with rational (or real or complex) coefficients, for the perversity $\bar{p}$. According to Examples 3.3.2 (c), one has:

$$\mathbb{H}^j(U_x; \mathcal{IC}_X^{\bar{p};\bullet}) = IH_{n-j}^{\bar{p},\mathrm{cl}}(U_x)$$
$$\mathbb{H}_c^j(U_x; \mathcal{IC}_X^{\bar{p};\bullet}) = IH_{n-j}^{\bar{p}}(U_x).$$

Properties (2.28) and (2.29) that we re-wrote (2.30) and (2.31) can be now written in terms of hypercohomology:

(3.25) $\qquad \dim\{x \in X \mid \mathbb{H}^j(U_x; \mathcal{IC}_X^{\bar{p};\bullet}) \neq 0\} \leq n - p^{-1}(j) \qquad$ for $j \geq 1$.

(3.26) $\qquad \dim\{x \in X \mid \mathbb{H}_c^j(U_x; \mathcal{IC}_X^{\bar{p};\bullet}) \neq 0\} \leq n - q^{-1}(n-j). \qquad$ for $j \leq n-1$.

## 4. Characterizations of the intersection complex

In this chapter, one translates the previous definition in terms of appropriate functors to obtain a more intrinsic notion.

The local calculus we performed for intersection homology (Proposition 2.27) is the key point for a characterization of intersection homology. In particular, we proved that if $x$ is a point in a $h$-codimensional stratum and $U_x$ a distinguished neighbourhood of $x$, then

$$IH_i^{\bar{p}}(U_x) = 0 \quad \text{for } i \geq h - p_h - 1$$
$$IH_i^{\bar{p},\mathrm{cl}}(U_x) = 0 \quad \text{for } i < n - p_h.$$

We already translated the local calculus into the language of sheaves and hypercohomology. Indeed, we already know that classical homology and cohomology theories as well as intersection homology, can be defined as hypercohomology of suitable complexes of sheaves: for instance intersection homology is hypercohomology of the complex of sheaves $\mathcal{IC}^\bullet$. Given inclusion of a point $f_x : \{x\} \hookrightarrow X$, one will define functors $f_x^*$ (section 4.2.2) and $f_x^!$ (section 4.2.5) such that (Theorem 4.14)

$$IH_i^{\bar{p}}(U_x) \cong H^{n-i}(f_x^! \mathcal{IC}_X^{\bar{p},\bullet})$$
$$IH_i^{\bar{p},\mathrm{cl}}(U_x) \cong H^{n-i}(f_x^* \mathcal{IC}_X^{\bar{p},\bullet}).$$

Hence, together Proposition 2.27 and Theorem 4.14 provide properties on cohomology of the fibers $f_x^! \mathcal{IC}_X^{\bar{p},\bullet}$ and $f_x^* \mathcal{IC}_X^{\bar{p},\bullet}$. In that way, the local calculus can be translated at the sheaf level.

The main result of this section is that hypercohomology of a complex of sheaves satisfying suitable conditions, is nothing else than intersection homology. That provides a characterisation of intersection homology.

References for this chapter are [**Bo**, **GM2**, **McP2**, **Rie**].

### 4.1. First characterization of the intersection complex.

The notion of constructible sheaf is one of the natural tools for the characterisation of intersection complex of sheaves.

Let us denote by $\mathcal{X}$ a filtration of $X$ as in (2.3).

DEFINITION 4.1. One says that the sheaf complex $\mathcal{S}^\bullet$ is *cohomologicaly locally constant* for the filtration $\mathcal{X}$ of $X$ (to shorten: $\mathcal{S}^\bullet$ is $\mathcal{X}$-clc), if $\mathcal{H}^\bullet \mathcal{S}^\bullet$ is locally constant on each stratum $S_i = X_i - X_{i-1}$ $(0 \leq i \leq n)$.

DEFINITION 4.2. One says that the sheaf complex $\mathcal{S}^\bullet$ is *constructible* relatively to the filtration (stratification) $\mathcal{X}$ of $X$ if $\mathcal{S}^\bullet$ is $\mathcal{X}$-clc and the fibers $\mathcal{H}^\bullet \mathcal{S}_x^\bullet$ are finite-dimensional.

Let us remark that Borel calls cohomologically constructible such a complex and denotes it as $\mathcal{X}$-cc.

DEFINITION 4.3. One says that the sheaf complex $\mathcal{S}^\bullet$ is *topologically constructible* if it is bounded and constructible relatively to a stratification of $X$.

One has the first characterization of the intersection complex (as a system of axioms called $[\mathrm{AX}_1]$ in [**GM2**],§3.3 and [**Bo**],§2):

THEOREM 4.4. *Let $\bar{p}$ be a perversity. The intersection complex $\mathcal{IC}_X^{\bar{p},\bullet}$ is a complex of sheaves $\mathcal{S}^\bullet$ satisfying the following properties:*

(1a) *$\mathcal{S}^\bullet$ is topologically constructible relatively to any Whitney stratification of $X$ and $\mathcal{S}^\bullet_{U_2}$ is quasi-isomorphic to the orientation sheaf $\mathcal{O}_{U_2}$,*
(1b) $\dim\{x \in X \mid \mathbb{H}^j(U_x;\mathcal{S}^\bullet) \neq 0\} \leq n - p^{-1}(j)$ *for every $j \geq 1$,*
(1c) $\dim\{x \in X \mid \mathbb{H}^j_c(U_x;\mathcal{S}^\bullet) \neq 0\} \leq n - q^{-1}(n-j)$ *for every $j \leq n-1$.*

PROOF. For $x \in U_2$, all chains are allowable, for any perversity, then one has $\mathcal{IC}^{\bar{p},\bullet}|_{U_2} = C^\bullet|_{U_2}$. Then, one has $\mathcal{H}^i(\mathcal{IC}^{\bar{p},\bullet})_x = 0$ for $i > 0$ and $\mathcal{H}^0(\mathcal{IC}^{\bar{p},\bullet})|_{U_2} = \mathcal{O}_{U_2}$ orientation sheaf on $U_2$. One concludes that $\mathcal{IC}^{\bar{p},\bullet}|_{U_2}$ is quasi-isomorphic to $\mathcal{O}_{U_2}$. Assertions (1b) and (1c) are already proved. □

THEOREM 4.5 ([**GM2, Bo**]). *Let $\bar{p}$ be a perversity and et $\mathcal{S}^\bullet$ be a fine (or soft) complex of sheaves on $X$ satisfying properties (1a), (1b) and (1c), then the cohomology groups of the complex*

$$\cdots \to \Gamma(X;\mathcal{S}^{j-1}) \to \Gamma(X;\mathcal{S}^{j}) \to \Gamma(X;\mathcal{S}^{j+1}) \to \cdots$$

*i.e., the hypercohomology groups $\mathbb{H}^j_c(X;\mathcal{S}^\bullet)$, are isomorphic to the intersection homology groups $IH^{\bar{p}}_{n-j}(X)$.*

Note that property (1c) is given in [**GM2, Bo**] in terms of attaching maps, that motivates the definition of the so-called *Deligne sheaf*, denoted by $\mathcal{P}^\bullet$ (see [**Bo**] §2.2 and [**GM2**], §3.1). The Deligne sheaf satisfies axioms $[\mathrm{AX}_1]$ and is quasi-isomorphic to the intersection complex $\mathcal{IC}_X^{\bar{p},\bullet}$. Another example is provided by the complex of $\mathcal{L}^2$-differential forms on the regular part of a Riemannian pseudomanifold with conical singularities. Cheeger [**Ch**] shown that this complex satisfies axioms $[\mathrm{AX}_1]$ for the middle perversity (see also [**CGM**] and further developments in the historical and useful paper by S. Kleiman [**Kle**]).

**4.2. Functors.** The general reference for this section is [**Bo**, VI].

We denote by $Sh(X)$ the abelian category of sheaves of $\mathbb{Q}$-vector spaces on $X$.

Let us consider a morphism $f : X \to Y$, one defines functors in the following way:

4.2.1. *Direct image $f_*$.* Let us denote, for every open subset $V$ in $Y$ and for every sheaf $\mathcal{A}$ on $X$,

$$f_*\mathcal{A}(V) = \Gamma(f^{-1}(V);\mathcal{A}).$$

The correspondence $\mathcal{A} \mapsto f_*\mathcal{A}$ defines a functor "direct image"

$$f_* : Sh(X) \to Sh(Y).$$

It is left exact and transforms injective sheaves into injective ones. If $j : X \hookrightarrow Y$ is a closed immersion and $\mathcal{A}$ a sheaf on $X$, then $j_*\mathcal{A} = \mathcal{A}^Y$ is the extension of the sheaf $\mathcal{A}$ by zero.

4.2.2. *Pull-back $f^*$.* The "pull-back" functor

$$f^* : Sh(Y) \to Sh(X)$$

is defined in the following way: Let $\pi : \mathrm{E}(\mathcal{B}) = \coprod \mathcal{B}_x \to Y$ be the "espace étalé" associated to the sheaf $\mathcal{B}$ on $Y$. By definition, $f^*(\mathcal{B})$ is the sheaf associated to the "espace étalé" $X \times_Y \mathrm{E}(\mathcal{B}) \to X$ obtained as the fibered product (Cartesian square) $f$ by $\pi$.

$$\begin{array}{ccc} X \times_Y \mathrm{E}(\mathcal{B}) & \xrightarrow{pr_2} & \mathrm{E}(\mathcal{B}) \\ \big\downarrow{\scriptstyle pr_1} & & \big\downarrow{\scriptstyle \pi} \\ X & \xrightarrow{f} & Y. \end{array}$$

For every point $x$ in $X$ one has an isomorphism at the level of stalks

$$(f^*\mathcal{B})_x = \mathcal{B}_{f(x)}.$$

The functor $f^*$ is exact, it is adjoint of the functor $f_*$. One has

$$Hom(f^*\mathcal{B}, \mathcal{A}) \cong Hom(\mathcal{B}, f_*\mathcal{A}).$$

EXAMPLE 4.6. Let $j : X \hookrightarrow Y$ be an inclusion, then $j^*\mathcal{B} = \mathcal{B}|_X$ is the restriction of the sheaf $\mathcal{B}$ to $X$.

EXAMPLE 4.7. Let us denote by $p : X \to \{\mathrm{pt}\}$ the constant map. The constant sheaf $\mathbb{Q}_X$ is equal to $\mathbb{Q}_X = p^*\mathbb{Q}_{\mathrm{pt}}$. Hence, one has a canonical isomorphism $\mathbb{Q}_X \cong f^*\mathbb{Q}_Y$ using the commutative diagram:

$$\begin{array}{ccc} X & \xrightarrow{f} & Y \\ & {\scriptstyle p}\searrow \quad \swarrow & \\ & \mathrm{pt}. & \end{array}$$

Compare with next example 4.10.

4.2.3. *The functor $f_!$.* Let $f : X \to Y$ be a continuous map between locally compact spaces, one defines the functor

$$f_! : Sh(X) \to Sh(Y)$$

direct image with proper supports, in the following way:

Let $V$ be an open subset in $Y$, one denotes $\Phi_V$ the family of closed subsets $E$ in $f^{-1}(V)$ such that the map $f|_E : E \to V$ is proper.

For a sheaf $\mathcal{A}$ on $X$, one defines the sheaf $f_!\mathcal{A}$ on $Y$ by

$$f_!\mathcal{A}(V) = \Gamma_{\Phi_V}(f^{-1}(V); \mathcal{A}) \ .$$

The functor $f_!$ is left-exact and one has a canonical morphism of functors

$$0 \longrightarrow f_! \longrightarrow f_*$$

which, if $f$ is proper, is an isomorphism.

If $Y$ is a point, then $f_!\mathcal{A} = \Gamma_c(X; \mathcal{A})$, where $c$ denotes the family of compact subsets in $X$. If $j : X \hookrightarrow Y$ is an open (or closed) immersion, then $j_!\mathcal{A} = \mathcal{A}^Y$ and the functor $j_!$ is exact. Finally, the functor $f_!$ is exact in the subcategory of injective sheaves on $X$.

4.2.4. *Derived categories.* The functor $f^!$ (next section) is not defined at the level of sheaves, but at the level of so-called *derived category* that we introduce now.

In a first step, one defines the category $K(X)$ whose objects are complexes of sheaves of $\mathbb{Q}$-vector spaces on $X$ and whose maps are homotopy classes of morphisms of complexes. $K^+(X)$ denotes the subcategory of complexes which are bounded below.

The derived category $D(X)$ [**Bo**, V;5.12] has the same objects than $K(X)$ but a morphism from $\mathcal{A}^\bullet$ to $\mathcal{B}^\bullet$ is an equivalence class of diagrams of morphisms in $K(X)$ :

$$\mathcal{A}^\bullet \xleftarrow{\cong} \mathcal{C}^\bullet \longrightarrow \mathcal{B}^\bullet$$

where $\cong$ means a quasi-isomorphism. This diagram is equivalent to

$$\mathcal{A}^\bullet \xleftarrow{\cong} \mathcal{C}'^\bullet \longrightarrow \mathcal{B}^\bullet$$

if there is $\mathcal{A}^\bullet \overset{\cong}{\leftarrow} \mathcal{D}^\bullet \to \mathcal{B}^\bullet$ and morphisms $\mathcal{C}^\bullet \leftarrow \mathcal{D}^\bullet \to \mathcal{C}'^\bullet$ such that one has a commutative diagram in $K(X)$ (*i.e.*, commutatif up to homotopy):

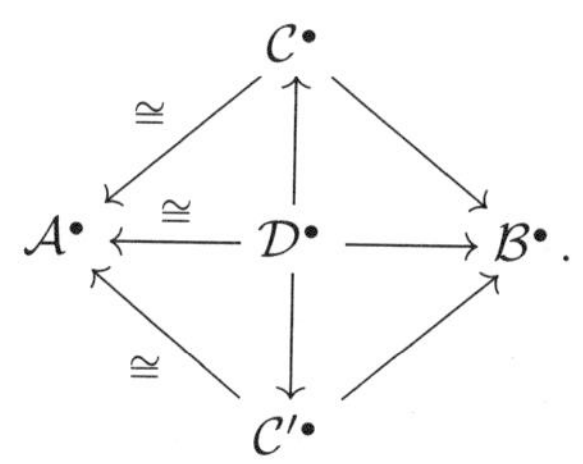

Two morphisms in $D(X)$ can be composed in the following way: a diagram

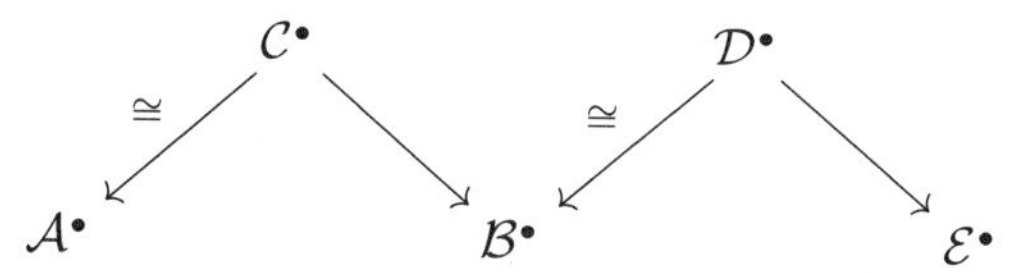

can be completed in such a way that one has a commutative diagram up-to homotpy:

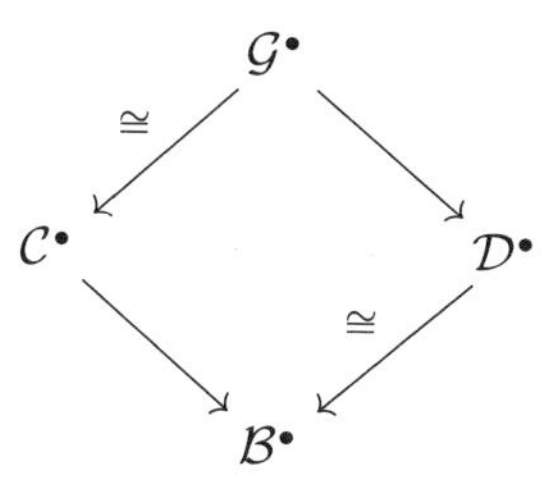

then $\mathcal{A}^\bullet \xleftarrow{\cong} \mathcal{G}^\bullet \longrightarrow \mathcal{E}^\bullet$ is the composition of the given morphisms.

One defines in the same way the derived category $D^+(X)$ of complexes bounded below.

4.2.5. *The functor $f^!$.* Let us consider a continuous map $f : X \to Y$ of locally compact spaces. Define the sheaf $\mathbb{C}_{U,X}$ by

$$\mathbb{C}_{U,X}(W) = \{s \in \mathbb{C}_X(U \cap W) : \mathrm{Supp}(s) \text{ is contained in } W\},$$

*i.e.*, $\mathbb{C}_{U,X}(W)$ is $\mathbb{C}^\ell$ where $\ell$ is the number of connected components of $W$ which are contained in $U$.

For $\mathcal{G}$ sheaf on $Y$, the presheaf defined by

$$U \mapsto \mathrm{Hom}_{Sh(Y)}(f_!\mathbb{C}_{U,X}, \mathcal{G})$$

is not necessarily a sheaf. It may have global sections while all of its stalks are zero (see [**Rie**]). That is the reason for using derived categories.

Let us consider a soft resolution of the constant sheaf $\mathbb{C}_X$ on $X$, for example the complex of chains $\mathcal{C}^\bullet = \mathcal{C}^\bullet_X$:

$$0 \to \mathbb{C}_X \to \mathcal{C}^\bullet,$$

one defines a functor

$$f^!_{\mathcal{C}^\bullet} : K^+(Y) \longrightarrow K^+(X)$$

by

$$f^!_{\mathcal{C}^\bullet}(\mathcal{G}^\bullet)(U) = Hom^\bullet(f_!(\mathcal{C}^\bullet_U); \mathcal{G}^\bullet_U)$$

The correspondence $\mathcal{G}^\bullet \mapsto f^!_{\mathcal{C}^\bullet}(\mathcal{G}^\bullet)$ defines a functor $f^!_{\mathcal{C}^\bullet} : K^+(Y) \to K^+(X)$. If $\mathcal{S}^\bullet$ is another soft resolutions of the constant sheaf $\mathbb{C}_X$, then the sheaf complexes $f^!_{\mathcal{C}^\bullet}(\mathcal{G}^\bullet)$ and $f^!_{\mathcal{S}^\bullet}(\mathcal{G}^\bullet)$ are quasi-isomorphic. One can deduce that the functor

$$f^! : D^+(Y) \longrightarrow D^+(X)$$

induced on the derived categories by the functor $f^!_{\mathcal{C}^\bullet}$ is independent, up to isomorphism, on the choice of the soft resolution $\mathcal{C}^\bullet$.

EXAMPLE 4.8. In the case of an open immersion $j : X \hookrightarrow Y$, one has $j^! = j^*$.

EXAMPLE 4.9. In the case of a closed immersion $j : X \hookrightarrow Y$, one has

$$j^!(\mathcal{G}^\bullet)(U) = \Gamma_{(X)}(V; \mathcal{G}^\bullet)$$

where $V$ is an open subset in $Y$ such that $U = V \cap X$ (see also 4.10).

Here $(X)$ denotes the family supports whose elements are closed subsets contained in $X$.

EXAMPLE 4.10. Let us consider the projection $p : X \to \{\mathrm{pt}\}$ and the sheaf $\mathbb{Q}_{\mathrm{pt}}$. The dualizing sheaf (see 3.24) satisfies [**Bo**, V,7.18]:

$$\mathcal{D}^\bullet_X \cong p^!\mathbb{Q}_{\mathrm{pt}}.$$

Hence for every map $f : X \to Y$ one has a canonical isomorphism $\mathcal{D}^\bullet_X \cong f^!\mathcal{D}^\bullet_Y$. Compare with Example 4.7.

**4.3. Local calculus "à la sauce Faisceaux".** Let us consider the inclusion of a point $x$ in $X$, $f_x \colon \{x\} \to X$, and consider the functors $f^*_x$ and $f^!_x$ (Sections 4.2.2 and 4.2.5).

PROPOSITION 4.11 ([**GM2**, P.91]). *Let $f_x \colon \{x\} \to X$ be the inclusion of a point $x$ and let $U_x$ be a distinguished neighbourhood of $x$ (definition 2.6) . One denotes by $\mathcal{S}^\bullet$ a topologically constructible sheaf complex on $X$ (definition 4.3), then one has:*

$$H^j(f^*_x\mathcal{S}^\bullet) \cong I\!H^j(U_x; \mathcal{S}^\bullet) = \mathcal{H}^j(\mathcal{S}^\bullet)_x \tag{4.12}$$

$$H^j(f^!_x\mathcal{S}^\bullet) \cong I\!H^j_c(U_x; \mathcal{S}^\bullet). \tag{4.13}$$

*These groups are respectively called the stalk homology and the costalk homology of $\mathcal{S}^\bullet$ at $x$.*

The following interpretations are taken from [**GM2**, §4]:

If a class $\xi \in \mathbb{H}^j(X;\mathcal{S}^\bullet)$ does not vanish under the homomorphism

$$\mathbb{H}^j(X;\mathcal{S}^\bullet) \to H^j(f_x^*\mathcal{S}^\bullet)$$

then any cycle representative of $\xi$ must contain the point $x$. Thus, $H^j(f_x^*\mathcal{S}^\bullet)$ represents local classes which "cannot be pulled away from the point $x$". We call

$$\{x \in X | H^j(f_x^*\mathcal{S}^\bullet) \neq 0\} \quad \text{the } \textit{local } j\textit{-support of the complex } \mathcal{S}^\bullet.$$

Similarly, a class $\eta \in \mathbb{H}^j(X;\mathcal{S}^\bullet)$ is in the image of the homomorphism

$$H^j(f_x^!\mathcal{S}^\bullet) \to \mathbb{H}^j(X;\mathcal{S}^\bullet).$$

if some cycle representative of $\eta$ is completely contained in a neighborhood of $x$. Thus $H^i(f_x^!\mathcal{S}^\bullet)$ represents local classes which are "supported near $x$". We call

$$\{x \in X | H^j(f_x^!\mathcal{S}^\bullet) \neq 0\} \quad \text{the } \textit{local } j\textit{-cosupport of the complex } \mathcal{S}^\bullet.$$

Combining formulae (3.22) and (3.23) with (4.12) and (4.13), one obtains the following theorem:

THEOREM 4.14. *Let $f_x\colon \{x\} \to X$ be the inclusion of a point $x$ in $X$, $U_x$ a distinguished neighbourhood of $x$ and let us consider the sheaf complex $\mathcal{IC}_X^{\bar{p};\bullet}$ of intersection chains with rational coefficients, for the middle perversity, one has:*

$$H^j(f_x^*\mathcal{IC}_X^{\bar{p},\bullet}) \cong \mathbb{H}^j(U_x;\mathcal{IC}_X^{\bar{p};\bullet}) = IH_{n-j}^{\bar{p},cl}(U_x) \tag{4.15}$$

$$H^j(f_x^!\mathcal{IC}_X^{\bar{p},\bullet}) \cong \mathbb{H}_c^j(U_x;\mathcal{IC}_X^{\bar{p};\bullet}) = IH_{n-j}^{\bar{p}}(U_x). \tag{4.16}$$

**4.4. Second characterization of the intersection complex.** One can now write the second characterization of the intersection complex of sheaves:

DEFINITION 4.17. [**GM2**, §4.1], [**McP**, §9] and [**Bo**, §V.4.13]. One says that the sheaf complex $\mathcal{S}^\bullet$ satisfies the axioms $[AX_2]_{\bar{p}}$ for the perversity $\bar{p}$ if one has:
(2a) $\mathcal{S}^\bullet$ is a topologically constructible complex on $X$ (definition 4.3), $\mathcal{H}^j(\mathcal{S}^\bullet) = 0$ if $j < 0$ and there is an open dense submanifold $U$ of codimension at least 2 in $X$ such that $\mathcal{S}^\bullet|_U \cong \mathbb{Q}_U$.
(2b) $\dim\{x \in X | H^j(f_x^*\mathcal{S}^\bullet) \neq 0\} \leq n - p^{-1}(j)$ for every $j > 0$.
(2c) $\dim\{x \in X | H^j(f_x^!\mathcal{S}^\bullet) \neq 0\} \leq n - q^{-1}(n-j)$ for every $j < n$.

In the definition, the constant sheaf $\mathbb{Q}_U$ can be replaced by $\mathbb{C}_U$ or, even, any (rank one) local system $\mathcal{E}$ on $U$ as in 3.3 (see [**GM2**] of [**Bo**]).

Uniqueness theorem, proved in [**GM2**, 4.1], states that up to canonical isomorphism, there exists an unique complex in $D^b(X)$ which satisfies axioms $[AX_2]_{\bar{p}}$. It is given by the sheaf $\mathcal{IC}_X^{\bar{p};\bullet}$, constructed as before with any stratification of $X$. As a corollary, the intersection homology groups $IH_*^{\bar{p}}(X)$ are topological invariant and exist independently of the choice of the stratification of $X$. One has:

THEOREM 4.18 ([**GM2**, **Bo**]). *Let $\mathcal{S}^\bullet$ be a fine (or soft) sheaf complex on $X$ satisfying Axioms $[AX_2]_{\bar{p}}$ for a perversity $\bar{p}$, then the cohomology groups of the complex*

$$\dots \to \Gamma(X;\mathcal{S}^{j-1}) \to \Gamma(X;\mathcal{S}^j) \to \Gamma(X;\mathcal{S}^{j+1}) \to \dots$$

*i.e., the hypercohomology groups $\mathbb{H}_c^j(X;\mathcal{S}^\bullet)$, are isomorphic to the intersection homology groups $IH_{n-j}^{\bar{p}}(X)$.*

## 5. Going to Perverse sheaves

Les faisceaux pervers n'étant ni des faisceaux,
ni pervers, la terminologie requiert une explication.
Beilinson, Bernstein, Deligne

In this section, $X$ will be a complex algebraic (or analytic) variety and we restrict ourselves to the middle perversity. One has $\bar{m}(2j) = \bar{m}(2j+1) = j-1$ and axioms (2b) and (2c) are written in the following way:

$$\dim\{x \in X | H^j(f_x^* \mathcal{S}^\bullet) \neq 0\} \leq n - j - 1 \qquad \text{if } j > 0 \tag{5.1}$$

$$\dim\{x \in X | H^j(f_x^! \mathcal{S}^\bullet) \neq 0\} \leq j - 1 \qquad \text{if } j < n. \tag{5.2}$$

Due to Poincaré-Verdier duality properties (see [**GM2**, **Bo**, **Ma**]) it is more convenient to consider the intersection complex of sheaves shifted by the complex dimension of $X$. In the following we denote this shifted sheaf complex by

$$\mathbf{IC}_X^\bullet = \mathcal{IC}_X^{\bar{m},\bullet}[n]$$

where $n = \dim_{\mathbb{C}} X$.

With these conventions, 5.1 and 5.2 are now written

$$\dim\{x \in X | H^{-i}(f_x^* \mathbf{IC}_X^\bullet) \neq 0\} < i \qquad \text{if } i < n \tag{5.3}$$

$$\dim\{x \in X | H^{i}(f_x^! \mathbf{IC}^\bullet)_X \neq 0\} < i \qquad \text{if } i < n. \tag{5.4}$$

Let $\mathbf{P}^\bullet$ be a complex of sheaves, we denote:

$$\begin{aligned} \operatorname{supp}^i \mathbf{P}^\bullet &= \text{ closure of } \{x \in X | H^i(f_x^* \mathbf{P}^\bullet) \neq 0\} \\ \operatorname{cosupp}^i \mathbf{P}^\bullet &= \text{ closure of } \{x \in X | H^i(f_x^! \mathbf{P}^\bullet) \neq 0\}. \end{aligned}$$

Let $j : X_{\text{sing}} \hookrightarrow X$ be the (closed) inclusion of the singular part of $X$ in $X$, the conditions (5.3) and (5.4) are equivalent to (see [**Ma**, §2]):

$$\dim \operatorname{supp}^{-i}(j^*[-1]\mathbf{IC}_X^\bullet) = \dim \overline{\{x \in X | H^{-(i+1)}(f_x^* \mathbf{IC}_X^\bullet) \neq 0\})} \leq i \quad \text{if } i < n$$

$$\dim \operatorname{cosupp}^{i}(j^![1]\mathbf{IC}_X^\bullet) = \dim \overline{\{x \in X | H^{i+1}(f_x^! \mathbf{IC}_X^\bullet) \neq 0\})} \leq i \quad \text{if } i < n.$$

The dimension is 0 if the considered set is empty set.

We now reach the final step of our promenade to the country of perverse sheaves. In fact that is the starting point of many papers on the subject. Some of them provide history and motivation for the notion of perverse sheaves. MacPherson's conference [**McP**] and Kleiman's paper [**Kle**] provide main results and historical development of the theory. Among papers on the subject, giving examples, results and applications, let us quote those of De Cataldo and Migliorini [**DM**], Klinger [**Kli**], Massey [**Ma**], Rietsch [**Rie**], Williamson [**Wi**], the list being far from complete. Many of them being available on the web, the link is provided.

The definition of perverse sheaves involves a "relaxed" version of the previous axioms:

DEFINITION 5.5. Let $X$ be a complex algebraic variety. A *perverse sheaf* for the perversity $\bar{p}$ is a complex of sheaves $\mathbf{P}^\bullet$ in the category of bounded complexes $D^b(X)$ with cohomology constructible relatively to a Whitney stratification of $X$ and satisfying

$$\dim\{x \in X | H^{-i}(f_x^* \mathbf{P}^\bullet) \neq 0\} \leq i \qquad \forall i \in \mathbb{Z}$$
$$\dim\{x \in X | H^{i}(f_x^! \mathbf{P}^\bullet) \neq 0\} \leq i \qquad \forall i \in \mathbb{Z}.$$

where $f_x$ is defined as before (4.3).

Equivalently, the conditions can be written (sse [**Ma**])

$$\dim(\mathrm{supp}^{-i}\mathbf{P}^\bullet)) \leq i$$
$$\dim(\mathrm{cosupp}^{i}\mathbf{P}^\bullet)) \leq i.$$

The category $P(X)$ of perverse sheaves on $X$ is a full subcategory of the constructible derived category $D_c(X)$. Recall that simple objects in a category are objects which do not have nontrivial sub-object or quotient. The simple objects in $P(X)$ are the intersection complexes of sheaves $\mathbf{IC}^\bullet_Y(\mathcal{L})$ associated to irreducible closed subvarieties $Y \subset X$ and irreductible local systems $\mathcal{L}$ defined on a Zariski dense open nonsingular subvariety of $Y$. The category $P(X)$ is Abelian, Noetherian and Artinian. That implies that every perverse sheaf is a finite iterated extension of simple perverse sheaves.

The main result concerning perverse sheaves is the famous *Decomposition Theorem* [**BBD**] that is described in many papers on the subject. Many operations work better in the category of perverse sheaves than in the category of sheaves, *e.g.*, perverse sheaves are preserved by duality, they are preserved by vanishing cycles and nearby cycles functors (see the warning in [**Ma**]) as well as specialization over a curve.

The definition can also be interpreted in terms of so-called $t$-structures and perverse sheaves are sometimes defined in this context. For the dictionary between the different definitions, see for instance [**Ma, Kli, Wi**].

## References


[BBD] A. Beilinson, J. Bernstein, and P. Deligne. *Faisceaux pervers.* Astérisque, 100, 1982.

[Bo] A.Borel and al. *Intersection Cohomology,* Progress in Mathematics Vol 50, Birkhäuser, 1984.

[BG] J.P.Brasselet et G.Gonzalez-Springberg *Espaces de Thom et contre-exemples de J.L. Verdier et M.Goresky,* Bol. Soc. Bras. Mat. vol.17 n°2 (1986).

[Ch] J. Cheeger *On the Hodge theory of Riemannian pseudomanifolds.* In: Geometry of the Laplace operator. Proceedings of Symposia in pure Mathematics 36, 91-146 (1980). Amer. Math. Soc. Providence, RI.

[CGM] J. Cheeger, M. Goresky and R.MacPherson *$\mathcal{L}^2$-cohomology and intersection cohomology of singular algebraic varieties,* in "Seminar on differential geometry" ed. by S. T. Yau, Princeton University Press, 1982, p. 303–340

[CGJ] D.C.Cohen, M.Goresky and L.Ji *On the Künneth Formula for Intersection Cohomology,* Trans. AMS 333 (1992),63-69.

[DM] M.A. De Cataldo and L. Migliorini *The decomposition theorem, perverse sheaves and the topology of algebraic maps.* Bull. AMS Vol 46, No 4, 2009, 535-633.

[FK] K.H.Fieseler and L. Kaup *Singular Poincaré duality and intersection homology,* Proceedings of the 1984 Vancouver Conference in Algebraic Geometry. AMS, Providence, (1986), 113-161.

[God] R. Godement *Topologie algébrique et théorie des faisceaux.* Actualites Scientifiques et Industrielles. 1252. Publications de l'Institut de Mathématique de l'Université de Strasbourg. XIII. Paris: Hermann VIII, 283 p. (1958).

[GM1] M.Goresky and R.MacPherson *Intersection homology theory,* Topology 19 (1980), 135-162.

[GM2] M.Goresky and R.MacPherson *Intersection homology theory II,* Inv. Math. 71 (1983), 77-129.

[Gu] S. Guillermou *Introduction aux faisceaux pervers.* Notes d'exposés. https://www-fourier.ujf-grenoble.fr/ guillerm/fp.ps

[Kle] S. Kleiman *The development of intersection homology theory.* A century of mathematics in America, Part II, Hist. Math. 2, Amer. Math. Soc., 1989, pp. 543-585.

[Kli] B. Klinger *Intersection homology and perverse sheaves.* http://webusers.imj-prg.fr/ bruno.klingler/cours/courspervers.pdf

[Lef] S. Lefschetz, *Transformations of Manifolds with a Boundary,* Proceedings of the National Academy of Sciences 12 (12), 737–739, 1926.

[Ma] D. Massey *Notes on perverse sheaves and vanishing cycles* http://arxiv.org/abs/math/9908107.

[McP] R.MacPherson *Global questions in the topology of singular spaces,* Proceedings of the International Congress of Mathematicians, Vol.I, pp 213-235, Warszawa, 1983.

[McP2] R.MacPherson *Intersection Homology and Perverse Sheaves,* Report December 15, 1990. Unpublished Notes: http://faculty.tcu.edu/gfriedman/notes/ih.pdf

[McC] C. Mc Crory *Stratified general position.* Algebraic and Geometric Topology Lecture Notes in Mathematics Volume 664, 1978, pp 142-146.

[Po1] H. Poincaré *Analysis Situs,* Journal de l'Ecole Polytechnique ser 2, 1 (1895) pp 1-123.

[Po2] H. Poincaré *Complément à l'Analysis Situs,* Rendiconti del Circolo matematico di Palermo, 13 (1899) pp 285-343.

[Po3] H. Poincaré *Second complément à l'Analysis Situs,* Proceedings of the London Mathematical Society, 32 (1900), pp 277-308.

[Rie] K. Rietsch *An introduction to perverse sheaves* arXiv:math/0307349v1. 26 Jul. 2003.

[Wi] G. Williamson *An illustrated guide to perverse sheaves* http://people.mpim-bonn.mpg.de/geordie/perverse_course/lectures.pdf



AIX-MARSEILLE UNIVERSITÉ, CNRS, 12M, MARSEILLE, FRANCE

*E-mail address*: jean-paul.brasselet@univ-amu.fr

Contemporary Mathematics
Volume **675**, 2016
http://dx.doi.org/10.1090/conm/675/13584

# Gauss maps and duality of sphere bundles

D. Dreibelbis

ABSTRACT. Given a sphere bundle, we define a Gauss map of the bundle, and from this we define an orthogonal sphere bundle. We define the dual of a Gauss map, and show that it is equal to either the orthogonal bundle or the singularity set of the orthogonal bundle's Gauss map. This is a direct generalization of the case where the bundles are the unit tangent and unit normal bundles of an immersed manifold. The theory is supported by several examples. We end with looking at the duals of certain submanifolds of the sphere bundle.

## 1. Introduction

Given an immersed hypersurface $M^n$ in $\mathbb{R}P^{n+1}$, it is possible to define the *dual hypersurface* $M^*$ by taking the projective dual of the tangent plane $T_pM$ for each point $p$ on $M$. This is an old concept, with classic formulas for curves in $\mathbb{C}P^2$ (such as Plücker's formulas [**9**]). The dual hypersurface generically will have singularities, and these singularities will occur at the same points where the Gauss map has singularities, which are points where the tangent plane has non-Morse contact with the hypersurface. Bruce looked at the singularities of the dual of hypersurfaces in [**1**]. Later, Bruce and Romero-Fuster looked at duality between the singularity sets of height functions and projections functions for curves and surfaces in $\mathbb{R}^3$ ([**3**]), and then Bruce and Nogueira did the same for curves and surfaces in $\mathbb{R}^4$ ([**2**]). Mochida, Romero-Fuster, and Ruas then looked at the duality of asymptotic and binormal vectors of immersed manifolds ([**11**]). In [**13**], Porteous defined a similar type of dual for surfaces in $S^3$. This is essentially the same as the projective dual, only we need to worry about orientation. If we allow ourselves to work with manifolds that have an antipodal property, then we can continue Porteous's idea without worrying about orientation. Finally Bruce and Zakalyukin showed why these duals take standard bifurcation structures in [**4**].

In this paper, we offer a definition of duals that applies to any map from a sphere bundle to a sphere $S^n$. Thc concept is a direct generalization of unit tangent bundles and unit normal bundles, which in fact is the main case we care about. We show that the dual of a sphere bundle is equal to either its orthogonal

2010 *Mathematics Subject Classification.* Primary 53A35, 57R45, 58K40.
*Key words and phrases.* Gauss map, duality, singularities, conjugate vectors, contour generators.

bundle or the singularity set of the Gauss map of the orthogonal bundle (depending on dimensions). While put in a more general setting, this is a direct generalization of the work of Bruce, Romero-Fuster, Ruas, and Mochida.

We follow our theorem with examples from low dimensions and/or codimension. We show how our theorem applies to previously known cases, and also how it can be applied to a more general situation.

Finally, we focus on submanifolds of a sphere bundle, and describe when we can assume that the dual of the submanifold is contained in the orthogonal bundle.

## 2. Second-Order Flat Geometry of Immersions

This section contains the geometry needed for immersed manifolds. In particular, we need second-order geometry that is preserved by affine transformations of the ambient space (since the topology of dualfors is preserved by such transformations). The information we need is encapsulated in the second fundamental form.

Given an immersion $\mathbf{s} : M^m \to \mathbb{R}^{n+1}$ into Euclidean space, the manifold inherits the metric from the ambient space, and from this we can define the unit tangent and unit normal bundles $UTM$ and $UNM$. Given a normal vector $\mathbf{n}$ at a point $p \in M$, we define the *second fundamental form* $\mathrm{II}_{\mathbf{n}} : T_pM \times T_pM \to \mathbb{R}$ by the map $\mathrm{II}_{\mathbf{n}}(\mathbf{v}, \mathbf{w}) = \mathbf{n} \cdot d^2\mathbf{s}(\mathbf{v}, \mathbf{w}) = (-d\mathbf{n} \cdot d\mathbf{s})(\mathbf{v}, \mathbf{w})$. If we wish, we can consider the form as the tensor $\mathrm{II} : N_pM \times T_pM \times T_pM \to \mathbb{R}$, where flat geometry is anything preserved by the natural $GL(n - m + 1) \times GL(m) \times GL(m)$ action.

For each unit normal vector $\mathbf{n}$ at $p \in M$, we can look at the height function $f_{\mathbf{n}} : M \to \mathbb{R}$ defined by $f_{\mathbf{n}}(q) = \mathbf{n} \cdot \mathbf{s}(q)$. This map necessarily has a singularity at $p$, and for most vectors it will be a Morse singularity. If it has a non-Morse singularity, then we call $\mathbf{n}$ a *binormal vector*. If we have a binormal vector, then necessarily the second fundamental form $\mathrm{II}_{\mathbf{n}}$ must be degenerate, and thus there must exist at least one tangent vector $\mathbf{v}$ such that $\mathrm{II}_{\mathbf{n}}(\mathbf{v}, \cdot) \equiv 0$. Any vector $\mathbf{v}$ satisfying this definition is called an *asymptotic vector*. For a generic immersion of $M$, the set of all triples $(p, \mathbf{n}, \mathbf{v})$ in the bundle $(UN \times UT)M$ (i.e, the fiber bundle over $M$ where the fiber at $p$ is $UN_pM \times UT_pM$), where $\mathbf{v}$ is an asymptotic vector associated with the binormal vector $\mathbf{n}$ at $p$, is an embedded manifold of dimension $n - 1$. Projecting this manifold into $UNM$, we get the set of binormal vectors as a Whitney stratified set with top dimension $n - 1$. The dimension of the set of asymptotic vectors could be less than $n - 1$, depending on the dimension of $M$.

Note that the definition of asymptotic vectors for surfaces in $\mathbb{R}^3$ is generally different: for a surface in $\mathbb{R}^3$, we say a vector $\mathbf{v} \in UT_pM$ is asymptotic if $\mathbf{v}$ is a root of the second fundamental form $\mathrm{II}_{\mathbf{n}}$. The more general definition above would only include the asymptotic vectors along the parabolic curve. Since many of our examples deal with surfaces in $\mathbb{R}^3$, we need to make the distinction.

Next, we say that two tangent vectors (or tangent directions) $\mathbf{v}, \mathbf{w} \in UT_pM$ are *conjugate* if $\mathrm{II}_{\mathbf{n}}(\mathbf{v}, \mathbf{w}) = 0$ for all normal vectors $\mathbf{n} \in UN_pM$. A few facts about conjugate vectors:

- For surfaces in $\mathbb{R}^3$, each tangent direction has a unique conjugate direction, except for asymptotic directions at a parabolic point (which are conjugate to all other tangent directions).
- Also for surfaces in $\mathbb{R}^3$, asymptotic vectors are by definition self-conjugate.

- For a generic immersion $M^m$ in $\mathbb{R}^{n+1}$, let $(UT \times UT)M$ be the fiber bundle over $M$ where the fiber at $p$ is $UT_pM \times UT_pM$. Then the set of triples $(p, \mathbf{v}, \mathbf{w}) \in (UT \times UT)M$ where $\mathbf{v}$ and $\mathbf{w}$ are conjugate at $p$ form an $(4m - n - 3)$-dimensional embedded manifold in $(UT \times UT)M$ ([**6**]).
- For an immersed manifold $M$ where the dimension equals the codimension, a vector is asymptotic if and only if it is part of a conjugate pair ([**6**]).

Many geometric concepts involve conjugate vectors, especially for surfaces in $\mathbb{R}^3$. A partial list is given in Corollary 6.2.

## 3. Duals of Gauss Maps

Let $M^{n-1}$ be an immersed hypersurface in $S^n$, which we equip with the standard Euclidean structure. We say that $M$ is *antipodal* if for all $p \in M$, the point $-p$ is also in $M$. In this case, $M$ could easily be thought of as a manifold in $\mathbb{R}P^n$ as well. We define the *dual* of $M$ as the hypersurface $M^*$, where $q \in M^*$ if and only if $q$ (thought as a vector in $\mathbb{R}^{n+1}$) is orthogonal to a point $p \in M$ as well as $T_pM$. If we consider $M$ and $M^*$ as hypersurfaces in $\mathbb{R}P^n$, then this definition is equivalent to the standard one.

If $M$ is not immersed, then we usually define $M^*$ as the closure of the dual of the immersed portion of $M$. We could also define $M^*$ by the same definition as above (which will be the result below), but then $M^*$ may not be a manifold at points dual to singular points of $M$. For example, the dual of a cusp in $S^2$ would be the union of an inflection point and its tangent great circle, as opposed to just the inflection point.

For this paper, we want to extend the notion of dual to a larger class of objects. In particular, we want to talk about duals of Gauss maps of sphere bundles.

DEFINITION 3.1. Let $M^m$ be a smooth manifold, and let $\pi : B^k \to M$ be a $k$-sphere bundle. A *Gauss map* of $B$ is a map $\Gamma : B \to S^n$, $k < n$, such that for each $p \in M$, the restriction $\Gamma|_{\pi^{-1}(p)}$ is an embedding to a great $k$-sphere in $S^n$.

If we have two sphere bundles $B_1^k$ and $B_2^{n-k-1}$ over the same manifold $M$, both equipped with Gauss maps into $S^n$, then we say they are *orthogonal* if $\Gamma_1(p, \mathbf{v})$ is orthogonal to $\Gamma_2(p, \mathbf{w})$ for all $p \in M$, $\mathbf{v} \in B_1|_p$, $\mathbf{w} \in B_2|_p$. Given a sphere bundle $B_1^k$ and a Gauss map $\Gamma_1$ into $S^n$, there is a canonical orthogonal sphere bundle:

$$B_2^{n-k-1} = \{(p, \mathbf{w}) \in M \times S^n : \mathbf{w} \text{ is orthogonal to all } \mathbf{v} \in \Gamma_1(B_1|_p)\}$$

with a canonical Gauss map $\Gamma_2$ which is simply projection into the second component. Any other orthogonal bundle of dimension $n - k - 1$ is diffeomorphic to the canonical bundle. It is clear that given a manifold $M$ immersed in Euclidean space, the unit tangent bundle $UTM$ and the unit normal bundle $UNM$, equipped with the obvious Gauss maps, are orthogonal bundles.

Now, given a sphere bundle $B$ with Gauss map $\Gamma : B \to S^n$, we can define its dual as

$$\begin{aligned}\Gamma^* \quad = \quad & \{(p, \mathbf{w}) \in M \times S^n : \mathbf{w} \text{ is orthogonal to} \\ & \Gamma(p, \mathbf{v}) \text{ and } d\Gamma|_{(p,\mathbf{v})} \text{ for some } (p, \mathbf{v}) \in B\}\end{aligned}$$

We will sometimes write $\Gamma(B)^*$ if it is necessary to emphasize the bundle as well as the Gauss map. Note that if $(p, \mathbf{w})$ is dual to a point $(p, \mathbf{v})$, then $\mathbf{w}$ must be orthogonal to the entire $k$-sphere $\Gamma(B_p)$. Thus we can easily think of $\Gamma^*$ as a subset of the canonical orthogonal sphere bundle, or we can consider it as a subset of any $(n-k-1)$-sphere bundle with a Gauss map orthogonal to $\Gamma$.

The set of points of $\Gamma^*$ associated to a point $(p, \mathbf{v}) \in B$ will always be a (possibly vacuous) great sphere, whose dimension is determined by the rank of $d\Gamma|_{(p,\mathbf{v})}$. In particular, if the rank of $d\Gamma$ changes as we travel along $B$, then we can expect that $\Gamma^*$ will not be a manifold (though it will be a Whitney stratified set).

Associated to the dual set, we define the preimage of the dual as

$$(\Gamma^*)^{-1} = \{(p, \mathbf{v}) \in B : d\Gamma \text{ has rank less than } n \text{ at } (p, \mathbf{v})\}$$

If the dimension of $B$ (as a manifold) is less than $n$, then the preimage is equal to $B$ itself. If the dimension is greater than or equal to $n$, then the preimage is the set of points where $\Gamma$ is not a submersion, and hence the preimage is the singularity set $\Sigma\Gamma$.

Finally, if $A$ is a subset of the bundle $B$ with the property that $\Gamma(A)$ satisfies the antipodal property, then we can define the dual of $A$ in the same way as we define the dual of $B$. In this case, we cannot assume that $\Gamma(A)^*$ is in the orthogonal bundle of $B$.

## 4. Singularities of a Generalized Gauss Map

Given a sphere bundle equipped with a Gauss map, we have defined its dual, which is a subset of the orthogonal bundle. Our main theorem shows that the dual of a Gauss map will be equal to the preimage of the orthogonal Gauss map.

THEOREM 4.1. *Let $M^m$ be a manifold, and let $B_1^k$ and $B_2^{n-k-1}$ be two sphere bundles on $M$ equipped with Gauss maps $\Gamma_1 : B_1 \to S^n$ and $\Gamma_2 : B_2 \to S^n$ respectively. Further, assume that the two Gauss maps are orthogonal. Then:*

$$\Gamma_1^* = (\Gamma_2^*)^{-1}$$

*and*

$$\Gamma_2^* = (\Gamma_1^*)^{-1}$$

**Proof** This is a local theorem, so we may look at a local parameterization of $\Gamma_1$ and $\Gamma_2$. In particular, we can describe $M$ in local coordinates $x_\alpha$, $\alpha = 1 \dots m$, and can find a local orthonormal frame $\mathbf{e}_1, \dots \mathbf{e}_{n+1}$ of $\mathbb{R}^{n+1}$ with

$$\Gamma_1(x_\alpha, y_\beta) = \sum_{i=1}^{k+1} a_i(x_\alpha, y_\beta)\mathbf{e}_i(x_\alpha)$$

where $\sum a_i^2 = 1$, each $a_i$ is a function of the $x_\alpha$'s as well as variables $y_\beta$, $\beta = 1 \dots k$, and finally $\Gamma_1$ is an immersion for fixed $x_\alpha$'s. Similarly, we have

$$\Gamma_2(x_\alpha, z_\gamma) = \sum_{i=k+2}^{n+1} b_i(x_\alpha, z_\gamma)\mathbf{e}_i(x_\alpha)$$

where $\sum b_i^2 = 1$, each $b_i$ is a function of the $x_\alpha$'s as well as variables $z_\beta$, $\beta = 1 \dots (n-k-1)$, and $\Gamma_2$ is an immersion for fixed $x_\alpha$'s.

First we determine the condition for a vector $\mathbf{v}$ to be in $\Gamma_1^*$. For this to happen, we need $\mathbf{v}$ to be orthogonal to $\Gamma_1(x_\alpha, y_\beta)$ and all of its derivatives, which is equivalent to saying that $\mathbf{v}$ is in the nullspace of the $(m+k+1) \times (n+1)$ matrix derived by appending $\Gamma_1(x_\alpha, y_\beta)$ as a row to $d\Gamma_1$. We can decompose the matrix as:

$$\left[\begin{array}{cc} A_{m,k+1} & B_{m,\ n-k} \\ C_{k+1,k+1} & 0_{k+1,\ n-k} \end{array}\right] \tag{4.1}$$

where $C$ is nonsingular (consisting of $\Gamma_1(x_\alpha, y_\beta)$ and its derivatives with respect to the $y_i$'s), and the first $m$ rows are the derivatives with respect to the $x_\alpha$'s. This tells us that $\mathbf{v}$ is in the dual if and only if $\mathbf{v} = \sum_{j=k+2}^{n+1} c_j \mathbf{e}_j$, where $[c_{k+2}, \ldots c_{n+1}]^T$ is in the nullspace of $B$. The entries of $B$ are given by

$$\begin{aligned} B_{\alpha j} &= (\Gamma_1)_{x_\alpha} \cdot \mathbf{e}_j \\ &= \left(\sum_{i=1}^{k+1} a_i \mathbf{e}_i\right)_\alpha \cdot \mathbf{e}_j \\ &= \left(\sum_{i=1}^{k+1} a_i \sum_{l=1}^{n+1} w_{il\alpha} \mathbf{e}_l\right) \cdot \mathbf{e}_j \\ &= \sum_{i=1}^{k+1} a_i w_{ij\alpha} \end{aligned}$$

where $w_{ij\alpha} = (\mathbf{e}_i)_{x_\alpha} \cdot \mathbf{e}_j$. So our final condition for a vector in the dual: we need the $c_j$'s to satisfy:

$$\sum_{j=k+2}^{n+1} \sum_{i=1}^{k+1} c_j a_i w_{ij\alpha} = 0 \tag{4.2}$$

for all $\alpha = 1 \ldots m$.

For the preimage of $\Gamma_2$, we look at the $(m+n-k) \times (n+1)$ matrix $d\Gamma_2$ with $\Gamma_2(x_\alpha, z_\gamma)$ appended to it as a row (this is allowed, since all derivatives of $\Gamma_2$ are orthogonal to $\Gamma_2(x_\alpha, z_\gamma)$). This matrix has the form

$$\left[\begin{array}{cc} A_{m,k+1} & B_{m,n-k} \\ 0_{n-k,k+1} & C_{n-k,n-k} \end{array}\right] \tag{4.3}$$

The matrix $C$ is nonsingular (consisting of $\Gamma_2(x_\alpha, z_\gamma)$ and its derivatives with respect to the $z_i$'s. We need the rank of this matrix to be less than $n+1$, which means that the rank of $A$ must be less than or equal to $k$. A general entry of $A$ is

$$\begin{aligned} A_{\alpha,j} &= (\Gamma_2)_{x_\alpha} \cdot \mathbf{e}_j \\ &= \left(\sum_{i=k+2}^{n+1} b_i \mathbf{e}_i\right)_\alpha \cdot \mathbf{e}_j \\ &= \left(\sum_{i=k+2}^{n+1} b_i \sum_{l=1}^{n} w_{il\alpha} \mathbf{e}_l\right) \cdot \mathbf{e}_j \\ &= \sum_{i=k+2}^{n+1} b_i w_{ij\alpha} \end{aligned}$$

The matrix $A$ has rank less than or equal to $k$ if and only if there exists a nonzero vector $\mathbf{w} = \sum_{j=1}^{k+1} d_j \mathbf{e}_j$ with

$$\sum_{j=1}^{k+1} \sum_{i=k+2}^{n+1} d_j b_i w_{ij\alpha} = 0 \tag{4.4}$$

for all $\alpha$. Since $w_{ij\alpha} = -w_{ji\alpha}$ for all indices, Equations 4.2 and 4.4 are equal. This shows that $\Gamma_1^* = (\Gamma_2^*)^{-1}$. The proof of the second equation is identical. □

The preimages are either the entire sphere bundle or its Gauss map's singularity set, depending on whether the dimension of the sphere bundles are less than $n$ or greater than or equal to $n$. We can break down the conditions explicitly:

COROLLARY 4.2. *Let $M^m$ be a manifold, and let $B_1^k$ and $B_2^{n-k-1}$ be two sphere bundles on $M$ equipped with Gauss maps $\Gamma_1 : B_1 \to S^n$ and $\Gamma_2 : B_2 \to S^n$ respectively. Further, assume that the two Gauss maps are orthogonal. Then:*

- *If $k < n - m$ and $k > m - 1$, then $\Gamma_1^* = B_2$ and $\Gamma_2^* = B_1$.*
- *If $k \geq n - m$ and $k > m - 1$, then $\Gamma_1^* = \Sigma\Gamma_2$ and $\Gamma_2^* = B_1$.*
- *If $k < n - m$ and $k \leq m - 1$, then $\Gamma_1^* = B_2$ and $\Gamma_2^* = \Sigma\Gamma_1$.*
- *If $k \geq n - m$ and $k \leq m - 1$, then $\Gamma_1^* = \Sigma\Gamma_2$ and $\Gamma_2^* = \Sigma\Gamma_1$.*

As stated previously, the motivating example for this idea comes from the tangent and normal bundles of an immersed manifold. We can restrict our theorem to this special case:

COROLLARY 4.3. *Let $M^m$ be an immersed manifold in $\mathbb{R}^{n+1}$. Then $\Gamma(UTM)^* = \Sigma(\Gamma(UNM))$. If $n \leq 2m - 1$, then we also have $\Gamma(UNM)^* = \Sigma(\Gamma(UTM))$. If $n > 2m - 1$, then $\Gamma(UNM)^* = UTM$.*

In [**11**], the authors have similar results connecting asymptotic vectors and binormal vectors. We now establish that this is the almost the same as Corollary 4.3.

THEOREM 4.4. *Let $\mathbf{s} : M^m \to \mathbb{R}^{n+1}$ be an immersion. Then $(p, \mathbf{v})$ is a singularity of the normal Gauss map $\Gamma : UNM \to S^n$ if and only if $\mathbf{v}$ is a binormal vector at $p$. Similarly, $(p, \mathbf{w})$ is a singularity of the tangential Gauss map $\Gamma : UTM \to S^n$ if and only if $\mathbf{w}$ is an asymptotic vector at $p$.*

**Proof** The first statement is a known result (see, for instance, [**5**]), so we will focus on the proof of the second statement. We can use the notation from the proof of Theorem 4.1, with $B_2 = UTM$ and $k = m - 1$. In this case, the components of the $m \times m$ submatrix $A$ are $d^2\mathbf{s}(\mathbf{v}, \cdot)$, and this matrix is singular if and only if $\mathbf{v}$ is an asymptotic vector. □

## 5. Examples

EXAMPLE 5.1. **Zero-bundles on curves.** Given an immersed curve $\mathbf{s}(t)$ in $\mathbb{R}^3$, we have the tangent bundle $\mathbf{T}(t)$, which is a 0-sphere bundle (more precisely, the 0-sphere bundle is $\mathbf{T}$ along with $-\mathbf{T}$). A simple calculation shows that the dual of $\mathbf{T}(t)$ is the binormal curve $\mathbf{B}(t)$. Another straightforward computation shows that the singularity set of the Gauss map of the normal bundle is also $\mathbf{B}(t)$. Further, the dual of $\mathbf{B}(t)$ is $\mathbf{T}(t)$, and the preimage of the dual map is also $\mathbf{T}(t)$. We note that this works even if the curve is nongeneric: if $\mathbf{s}(t)$ has a flat point at $t_0$ (a point where the curvature is zero), then $\mathbf{T}'(t_0) = \mathbf{0}$ and the entire unit normal circle is included in the dual at $t_0$. Computing the singularity set of the Gauss map of the normal bundle, we see that we also get the entire unit normal circle at $t_0$.

Because of the dimensions, the dual of the normal bundle will be the entire tangent bundle. If we work out the dual of $UNM$ explicitly, then we find that binormal vectors are the only places where the vector and its derivatives do not span all of $\mathbb{R}^3$.

The theorem does not just work for the tangent bundle, but it applies for any vector field along the curve. For instance, if we look at the dual of the binormal curve, we get the tangent curve along with the oscillating circles at torsion free points, and this is the singularity set of the oscillating plane bundle (spanned by the tangent and normal vector at each point).

We can perform the same calculations for curves in higher dimensions: given a curve in $\mathbb{R}^{n+1}$, $n > 2$, the dual of $UTM$ will be an $(n-2)$-sphere bundle consisting of all normal vectors orthogonal to the oscillating plane. This set is again the singularity set of the normal bundle, and its dual will be all of $UTM$.

EXAMPLE 5.2. **Hypersurfaces of $S^n$.** This case was studied in ([**12**]), where the Gauss map was interpreted as a wavefront set of a Legendrian immersion. An antipodal immersion $f : M^{n-1} \to S^n$ can be thought of as a Gauss map of a 0-sphere bundle $B^0 \to M$, and hence Theorem 4.1 applies. The dual in this case is equal to the usual dual, and the theorem says that we can get the same set as the singularity set of its perpendicular bundle. In particular, since the perpendicular bundle is the unit tangent bundle of $UTS^n$ restricted to $B^0$, the theorem says that the dual is the envelope of this set of tangent planes.

If $f$ is not an immersion, then the dual at a singularity will be a sphere whose dimension is dependent on the rank of the singularity. This sphere will be tangent to the rest of the dual.

EXAMPLE 5.3. **Great circle bundles in $S^3$** The idea of great circle bundles over a curve $C$ in $S^3$ (considered as a generalization of ruled surfaces) was studied in detail in [**14**]. As far as our theorem is concerned, if $B_1$ is such a sphere bundles with Gauss map $\Gamma_2$ and $B_2$ is an orthogonal bundle with Gauss map $\Gamma_2$, then our theorem states that $\Gamma_1^* = B_2$ and $\Gamma_2^* = B_1$, and we have duality directly without needing to consider a singularity set.

EXAMPLE 5.4. **Contour Generators.** Let $\mathbf{s} : M^n \to \mathbb{R}^{n+1}$ be an immersed hypersurface, and let $P \in \mathbb{R}^{n+1}$ be a point that is not on $M$. We define a 0-sphere bundle $B_1^0$ on $M$ by assigning to each point $p \in M$ the direction given by $\mathbf{s}(p) - P$. In essence, the bundle gives the direction needed to look at the point $\mathbf{s}(p)$ when viewing from the point $P$. The orthogonal bundle will be the $(n-1)$-sphere bundle $B_2$, where the sphere at point $p$ is the sphere sitting in the hyperplane orthogonal to $\mathbf{s}(p) - P$.

We say a point $p \in M$ is on the *contour generator* of $M$ with respect to $P$ if the vector $\mathbf{s}(p) - P$ is in $T_pM$. This is the set of edge points if $M$ is observed through a central projection from $P$ (see [**8**] for a nice survey on visual geometry). It is no surprise that the singularity set of the Gauss map of $B_1$ is given by the contour generator.

PROPOSITION 5.5. *Given the set up above, let $\Gamma_1 : B_1 \to S^n$ and $\Gamma_2 : B_2 \to S^n$ be the two Gauss maps. Then $\Sigma(\Gamma_1)$ is the set of all points $(p, \mathbf{v})$, where $p$ is on the contour generator and $\mathbf{v}$ is parallel to $\mathbf{s}(p) - P$. The dual $\Gamma_1^*$ is the set of normal*

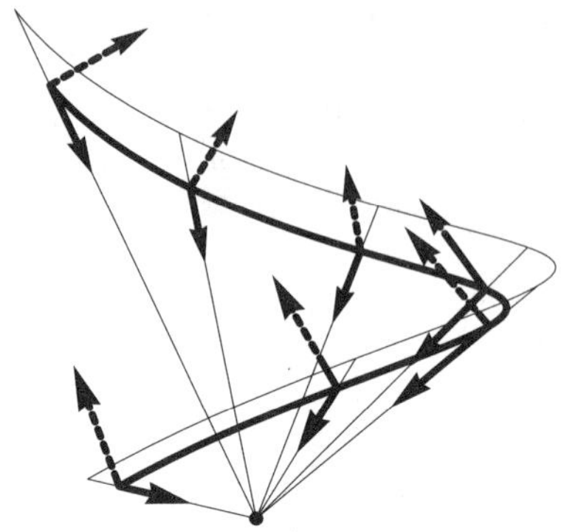

FIGURE 1. Perspective vectors are dual to the vectors normal to the cone generated by the manifold and viewing point.

*vectors of $M$ at points on the contour generator. As with Theorem 4.1, we have $\Sigma(\Gamma_2) = \Gamma_1^*$ and $\Gamma_2^* = \Sigma(\Gamma_1)$.*

**Proof** Since $\mathbf{s}$ is an immersion, a point $(p, \mathbf{v})$ is in $\Sigma(\Gamma_1)$ if and only if $\mathbf{v}$ is spanned by $d\mathbf{v}$. Since $\mathbf{v} = (\mathbf{s} - P)/||\mathbf{s} - P||$, this will occur if and only if $\mathbf{v}$ is spanned by $d(\mathbf{s} - P) = d\mathbf{s}$. So we have a singularity if and only if $p$ is on the contour generator.

For the dual, a vector $\mathbf{w}$ is in the dual at point $p$ if and only if $\mathbf{w}$ is orthogonal to $\mathbf{s}(p) - P$ and $d(\mathbf{s} - P) = d\mathbf{s}$. Again, since $\mathbf{s}$ is an immersion, this can only happen if $\mathbf{s}(p) - P$ is spanned by $d\mathbf{s}$, and the orthogonal vector much be a normal vector.

The singularity set and dual of $B_2$ could be worked out directly as well, but Theorem 4.1 already tells us that we will get the normals and tangents to the contour generator, respectively. □

The proposition gives us a nice result: given a perspective projection of a hypersurface $M$, the viewing directions and the normals along the contour generator are dual.

If the codimension of $M$ is greater than 1, then we can still define $B_1^0$ and $B_2^{n-1}$, but now dimensions dictate that $\Gamma_1^* = \Sigma(\Gamma_2)$ and $\Gamma_2^* = B_1$. In this case, the dual of $B_1$ at $P$ is the set of all normal vectors orthogonal to $\mathbf{s}(p) - P$, which could also be thought of as the normal space to the cone generated by $M$ and $P$ (See Figure 1).

EXAMPLE 5.6. **Contour Generators for Orthogonal Projections.** Given an direction $\mathbf{v}$ we could view our hypersurface from $\mathbf{v}$. This is known as an orthogonal projection, and we can still talk about the contour generator (now the set of points where $\mathbf{v}$ is tangent to $M$). Again, we can make our bundles $B_1^0$ and $B_2^{n-1}$, but our results will not be as satisfying. In this case, since $\Gamma_1$ is constant, its singularity set consists of all points, while its dual consists of all of $B_2$.

## 6. Conjugate Vectors

Since the second fundamental form involves a dot product between tangent vectors and the derivative of normal vectors, it is no surprise that conjugate vectors (where this dot product is zero) come in to play when talking about duals. In particular, we can make a strong connection between conjugate vectors and duals of curves in a hypersurface.

THEOREM 6.1. *Let* $\mathbf{s} : M^n \to \mathbb{R}^{n+1}$ *be an immersed hypersurface, and let* $C$ *be an immersed curve on* $M$ *parametrized by* $\alpha : I \to M$. *Define* $B_1^0$ *as the 0-sphere bundle created by restricting* $UNM$ *to* $C$, $B_2^{n-1}$ *as the* $(n-1)$*-sphere bundle created by restricting* $UTM$ *to* $C$, *and* $D \subset UTM$ *as the set of all pairs* $(p, \mathbf{v})$, *where* $p \in C$ *and* $\mathbf{v}$ *is conjugate to* $(\mathbf{s} \circ \alpha)'$ *at* $p$. *Then:*

(i) *The dual of* $B_1$ *is* $D$.
(ii) *The singularity set of* $\Gamma_2|_{B_2}$ *is* $D$.
(iii) *The dual of* $B_2$ *is* $B_1$.

**Proof** The Gauss map $\Gamma_1$ can be parametrized as $\mathbf{n}(\alpha(t))$, and its derivative is $d\mathbf{n}\alpha'$. The dual is then the set of tangent vectors $\mathbf{v}$ satisfying $d\mathbf{n}\alpha' \cdot \mathbf{v} = 0$, which is precisely the set of vectors conjugate to $(\mathbf{s} \cdot \alpha)'$. The proves part (i), and parts (ii) and (iii) follow from Theorem 4.1. □

This theorem works particularly well for surfaces in $\mathbb{R}^3$, where there is a unique conjugate direction for each vector, and where many geometric concepts can be connected to conjugate vectors. In particular, the theorem says that if you have a curve $C$ on a surface $M$, the dual of the normal vector field will give the vectors conjugate to the direction of $C$, and the singularity set of the tangent planes restricted to $C$ will also give the vectors conjugate to the directions of $C$. Note also in this case, the conjugate vector field is also known as the spherical osculating Darboux field $\overline{D_0}$ in the Darboux frame field ([**10**]).

COROLLARY 6.2. *Let* $M^2$ *be an immersed surface in* $\mathbb{R}^3$. *The theorem applies to the following curves and vector fields.*

- *The curve* $C$ *is the parabolic curve,* $D$ *is the asymptotic vector field along* $C$.
- *The curve* $C$ *is a contour generator,* $D$ *is the viewing direction field along* $C$.
- *The curve* $C$ *is one-half of a pair of bitangencies (see* [**11**]*),* $D$ *is the common tangent direction field along* $C$ *(Figure 2).*
- *The curve* $C$ *is an asymptotic curve,* $D$ *is the tangent vector field to the curve.*
- *the curve* $C$ *is a characteristic curve,* $D$ *is the other characteristic direction along* $C$ *(see* [**7**]*).*
- *The curve* $C$ *is an isophotic curve (a curve whose tangent vector field makes a constant angle with a fixed vector* $\mathbf{d}$*),* $D$ *is the other tangent direction making the same angle with the vector* $\mathbf{d}$ *(see* [**10**]*).*

**Proof** In all of the listed cases, the vector field is conjugate to the tangent vector field of the curve, and so Theorem 6.1 applies. □

From this corollary, we can note that asymptotic curves are the only curves on a surface where the curve's oscillating planes are equal to the surface's tangent planes.

## 7. Duals of Sub-bundles

An original motivating problem for this research was the following: given a tangent vector field on a surface in $\mathbb{R}^4$, what is its dual vector field (or given a normal vector field, what is its dual vector field)? For instance, if the tangent vector field is an asymptotic vector field, then its dual is the corresponding binormal

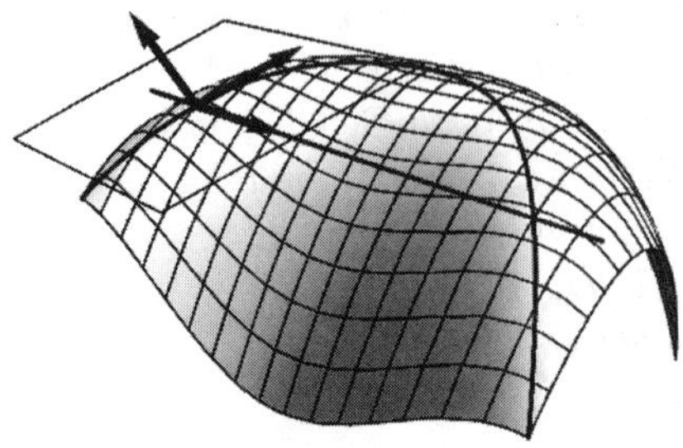

FIGURE 2. Bitangent directions are conjugate to the bitangent curve. They are dual to the normal direction and they can be thought of as the envelope of the bitangent planes.

vector field and vice versa. Given a tangent or normal vector field, we would really only be interested when its dual is a normal or tangent field. Unfortunately, the asymptotic/binormal vector fields are the only examples where the duals land in the appropriate spaces. This leads to the more general question: given an appropriate submanifold $A$ of a sphere bundle $B_1$, when is its dual in the orthogonal bundle? We now aim to answer this question for certain cases.

First, we look at the case when $A$ is a sphere bundle over $M$.

THEOREM 7.1. *Let $M^m$ be a manifold, and let $B_1^k$ and $B_2^{n-k-1}$ be two sphere bundles on $M$ equipped with Gauss maps $\Gamma_1$ and $\Gamma_2$ respectively. Assume that the two Gauss maps are orthogonal. Let $A^l$ be an $l$-sphere subbundle of $B_1$. Then $(p, \mathbf{w}) \in \Gamma_1(A)^* \cap B_2$ if and only if $(p, \mathbf{w}) \in \Gamma_1(B_1)^*$ and an associated preimage point $(p, \mathbf{v})$ is in $A$.*

**Proof** To look at $\Gamma_1(A)^*$, we proceed as we did in the proof of Theorem 4.1, only now assume that the first $l+1$ vector fields $\mathbf{e}_i$ span the subbundle $A$. With this assumption, the matrix in Equation 4.4 is simply truncated by $(k-l)$ rows in the $C$ and 0 matrices. Hence any vector from $\Gamma_2(B_2)$ in the nullspace of this matrix will necessarily be in the nullspace of Equation 4.4, and so $\Gamma_1(A)^* \cap B_2 \subseteq \Gamma_1(B_1)^*$. In addition, the existence of such a nullvector implies that there is a vector in $A$ that satisfies Equation 4.4, which means that it is a vector in the preimage of $\Gamma_1$. $\square$

EXAMPLE 7.2. If we choose a tangent vector field $A$ on an immersed surface $\mathbf{s} : M \to \mathbb{R}^4$, then we can consider $A$ as $(p, d\mathbf{s}\mathbf{v})$ for some vector field $\mathbf{v} : M \to \mathbb{R}^2$. If a vector is dual to $d\mathbf{s}\mathbf{v}$, then it must be orthogonal to $d(d\mathbf{s}\mathbf{v}) = d^2\mathbf{s}\mathbf{v} + d\mathbf{s}d\mathbf{v}$. If such a vector is a normal vector $\mathbf{n}$, then we must satisfy $\mathbf{n} \cdot d^2\mathbf{s}\mathbf{v} = 0$, which means $\mathbf{n}$ is a binormal vector and $d\mathbf{s}\mathbf{v}$ is an asymptotic vector.

Next, we consider the dual of the preimages. Ideally, we would like to say that the preimages are dual to each other. Clearly this is true if the preimage is equal to the entire sphere bundle, but if the preimage is the singularity set of $B_1$, then we can only say that the dual contains the preimage of $\Gamma_2$, and that the dual is exactly the preimage of $\Gamma_2$ when the singularity of $\Gamma_1$ is as simple as possible, i.e., a $\Sigma^{1,0}$ point. At more degenerate singularities, it is likely that we will include more vectors in the dual. We noted this problem when in the beginning, when we first defined the dual.

THEOREM 7.3. *Let $M^m$ be a manifold, and let $B_1^k$ and $B_2^{n-k-1}$ be two sphere bundles on $M$ equipped with Gauss maps $\Gamma_1$ and $\Gamma_2$ respectively. Further, assume that $m + k \geq n$. Let $P_1$ be the subset of $\Sigma\Gamma_1$ for which the singularity is of type $\Sigma^{1,0}$. Then the dual $\Gamma_1^*(P_1) \subseteq (\Gamma_2^*)^{-1}$.*

**Proof** The Gauss map $\Gamma_1$ sends a space of dimension greater than or equal to $n$ into a space of dimension $n$, and so the preimage of $\Gamma_1$ is $\Sigma\Gamma_1$. Furthermore, The set of $\Sigma^{1,0}$ points form an $n-1$ dimensional submanifold of $\Sigma\Gamma_1$, and $\Gamma_1$ is a diffeomorphism of $P_1$ into $\Gamma_1(P_1)$. Clearly at a point $(p, \mathbf{v}) \in P_1$, the span of $d\Gamma_1|_{P_1}$ is a subset of the span of $d\Gamma_1|_{B_1}$. By assumption, both spans have rank $n-1$, and so they must be the same. Hence a vector is orthogonal to $d\Gamma_1|_{P_1}$ if and only if it is orthogonal to $d\Gamma_1|_{B_1}$, and so the set of vectors dual to $B_1$ and $P_1$ are the same at $(p, \mathbf{v})$. □

EXAMPLE 7.4. For our curve in $\mathbb{R}^3$, the dual of $\mathbf{T}(t)$ was $\mathbf{B}(t)$. The theorem states that the dual of $\mathbf{B}(t)$ will be all of $\mathbf{T}(t)$, and as long as $\mathbf{B}'(t) \neq \mathbf{0}$, the dual will only include $\mathbf{T}(t)$. Unfortunately, it is possible for $\mathbf{B}(t)$ to have cusps (at torsion-free points). At these cusps, our definition of duality will include the entire circle $\cos\theta\mathbf{T} + \sin\theta\mathbf{N}$.

Theorem 7.3 is a direct generalization of the main theorems in [**11**], where they showed that the regular portions of asymptotic vectors and binormal vectors are dual to each other. In [**4**], Bruce and Zakalyukin showed that if we use the closure definition of duals, then we can make the more aesthetic claim that the set of all asymptotic vectors are dual to the set of all binormal vectors. Their approach focused on the duals as the bifurcation sets of the height and projection maps. We do not have generalizations of height and projection functions for a random sphere bundle, so it is not clear how to generalize their results. Still, we conjecture that their results can be generalized to any sphere bundles: using the closure definition of duals, the preimages of two orthogonal sphere bundles are dual.

Finally, as a comment, we could also look at a sphere bundle above an arbitrary submanifold of $M$. Indeed, this is what we did with the conjugate vectors, with good results. However, for a general theory, there needs to be some interaction between the derivatives of the manifold and the sphere bundle. We have this when the sphere bundle is the tangent bundle (thanks to the second fundamental form), but for a general theory, the results do not seem to say much.

## References


[1] J. W. Bruce. The duals of generic hypersurfaces. *Math. Scand.*, 49:36–60, 1981.

[2] J. W. Bruce and A. C. Nogueira. Surfaces in $\mathbb{R}^4$ and duality. *Quart. J. Math.*, 2(49):433–443, 1998.

[3] J. W. Bruce and M.C. Romero Fuster. Duality and orthogonal projections of curves and surfaces in euclidean 3-space. *Quarterly J. Math.*, 42(2):433–441, 1991.

[4] J. W. Bruce and V. M. Zakalyukin. On the geometry of caustics. *Amer. Math. Soc. Transl. Ser.*, 80(2):1–11, 1997.

[5] D. Dreibelbis. Singularities of the Gauss map and the binormal surface. *Adv. Geom.*, 3:453–468, 2003.

[6] D. Dreibelbis. Conjugate vectors on immersed manifolds. *Contemporary Mathematics*, 459:1–12, 2008.

[7] R. Garcia and J. Sotomayor. Harmonic mean curvature lines on surfaces immersed in $\mathbb{R}^3$. *Bull. Braz. Math. Soc. (N.S.)*, 34(2):303–331, 2003.

[8] P. Giblin. Apparent contours: an outline. In *New geometric techniques in computer vision*, number 1740, pages 1087–1102, 1998.

[9] P. Griffiths and J. Harris. *Principles of Algebraic Geometry*. Pure and Applied Mathematics. Wiley, 1978.

[10] S. Hananoi, N. Ito, and S. Izumiya. Spherical Darboux images of curves and surfaces. *Beitr. Algebra Geom.*, 56:575–585, 2015.

[11] D. K. H. Mochida, M. C. Romero-Fuster, and M. A. S. Ruas. Singularities and duality in the flat geometry of submanifolds of euclidean spaces. *Contributions to Algebra and Geometry*, 42(1):137–148, 2001.

[12] T. Nagai. The Gauss map of a hypersurface in Euclidean sphere and the spherical Legendrian duality. *Topology Appl.*, 159:545–554, 2012.

[13] I. R. Porteous. Some remarks on duality in $S^3$. In *Geometry and topology of caustics - CAUSTICS '98 (Warsaw)*, number 50, pages 217–226. Banach Center Publ., 1999.

[14] K Saji S. Izumiya, T Nagai. Great circular surfaces in the three-sphere. *Differential Geometry and its Applications*, (29):409–425, 2011.

Department of Mathematics and Statistics, University of North Florida, 1 UNF Drive, Jacksonville, Florida 32224

*E-mail address*: ddreibel@unf.edu

Contemporary Mathematics
Volume **675**, 2016
http://dx.doi.org/10.1090/conm/675/13585

# Topological formulas for closed semi-algebraic sets by Euler integration

N. Dutertre and J. A. Moya-Pérez

Abstract. This paper is divided in two parts. In the first one, we consider a polynomial function $f : \mathbb{R}^n \to \mathbb{R}$ with a finite number of critical points and we give a formula that relates its index at infinity to Euler characteristics of links at infinity, using Euler integration theory. In the second one, we generalize this result by considering a closed semi-algebraic set $X$ and a $\mathcal{C}^2$ semi-algebraic function $f : \mathbb{R}^n \to \mathbb{R}$ such that $f_{|X}$ has a finite number of critical points.

## 1. Introduction

Let $f : (\mathbb{R}^n, 0) \to (\mathbb{R}, 0)$ be an analytic function germ with an isolated critical point at 0. The Khimshiasvili formula (see [**Kh**]) states that

$$\chi(f^{-1}(\delta) \cap B^n_\epsilon) = 1 - \operatorname{sign}(-\delta)^n \deg_0 \nabla f,$$

where $0 < |\delta| \ll \epsilon \ll 1$, $B^n_\epsilon$ is a closed ball of radius $\epsilon$ centered at 0, $\nabla f$ is the gradient of $f$ and $\deg_0 \nabla f$ is the topological degree of the mapping $\frac{\nabla f}{|\nabla f|} : S^{n-1}_\epsilon \to S^{n-1}$.

As a corollary of this formula, one gets

$$\chi(\{f \le 0\} \cap S^{n-1}_\epsilon) = 1 - \deg_0 \nabla f,$$

$$\chi(\{f \ge 0\} \cap S^{n-1}_\epsilon) = 1 + (-1)^{n-1} \deg_0 \nabla f,$$

and

$$\chi(\{f = 0\} \cap S^{n-1}_\epsilon) = 2 - 2 \deg_0 \nabla f,$$

if $n$ is even.

Sekalski in [**Se**] gives a global counterpart of Khimshiasvili's formula for a polynomial function $f : \mathbb{R}^2 \to \mathbb{R}$ with a finite number of critical points. He considers the set $\Lambda_f = \{\lambda_1, \ldots, \lambda_k\}$ of critical values of $f$ at infinity, where $\lambda_1 < \lambda_2 < \cdots < \lambda_k$, and its complement $\mathbb{R} \setminus \Lambda_f = \cup_{i=0}^k ]\lambda_i, \lambda_{i+1}[$ where $\lambda_0 = -\infty$ and $\lambda_{k+1} = +\infty$.

---

2010 *Mathematics Subject Classification.* Primary 14P10, 14P25.
*Key words and phrases.* Semi-algebraic sets, Euler integration.
This work has been carried out in the framework of the Labex Archimède (ANR-11-LABEX-0033) and of the A*MIDEX project (ANR-11-IDEX-0001-02), funded by the "Investissements d'Avenir" French Government program managed by the French National Research Agency (ANR).

Denoting by $r_\infty(g)$ the number of real branches at infinity of a curve $\{g=0\}$ in $\mathbb{R}^2$, he proves that

$$\deg_\infty \nabla f = 1 + \sum_{i=1}^{k} r_\infty(f-\lambda_i) - \sum_{i=0}^{k} r_\infty(f-\lambda_i^+),$$

where for $i=0,\ldots,k$, $\lambda_i^+$ is an element of $]\lambda_i,\lambda_{i+1}[$ and $\deg_\infty \nabla f$ is the topological degree of the mapping $\frac{\nabla f}{\|\nabla f\|} : S_R^{n-1} \to S^{n-1}$, $R \gg 1$.

Gwoździewicz in [**Gw**] gives a topological proof of Sekalski's result using Euler integration. He proves that

$$\deg_\infty \nabla f = 1 + \int_{\mathbb{R}} r_\infty(f-t)d\chi_c(t),$$

where $\chi_c$ denotes the Euler characteristic with compact support that we will define later.

The first author gives global polynomial versions in $n$ variables of Khimshiasvili's result in [**Du1**], proving that if $n$ is even, then

$$\chi(f^{-1}(0)) = \deg_\infty H + \deg_\infty \nabla f - \deg_\infty L_2,$$

$$\chi(\{f \geq 0\}) - \chi(\{f \leq 0\}) = 1 - \deg_\infty K - \deg_\infty L_1,$$

and if $n$ is odd, then

$$\chi(f^{-1}(0)) = \deg_\infty K - \deg_\infty L_1,$$

$$\chi(\{f \geq 0\}) - \chi(\{f \leq 0\}) = 1 - \deg_\infty H - \deg_\infty \nabla f + \deg_\infty L_2,$$

where $H, K, L_1$ and $L_2$ are polynomial mappings defined in terms of $f$ and its partial derivatives. He generalizes this result in [**Du2**] by considering a closed semi-algebraic set $X \subset \mathbb{R}^n$ and a $\mathcal{C}^2$ semi-algebraic function $f : \mathbb{R}^n \to \mathbb{R}$ such that $f_{|X}$ has a finite number of critical points. As a particular case, he obtains again a generalization of Sekalski and Gwoździewicz's results, considering a polynomial function $f : \mathbb{R}^n \to \mathbb{R}$.

Our aim is to recover the first author's results using Euler integration, which clearly simplifies the proofs. After stating some necessary lemmas, in Section 4 we consider a polynomial function $f : \mathbb{R}^n \to \mathbb{R}$ with a finite number of critical points and we prove the known formula that relates the index at infinity to Euler characteristics of links at infinity. Using the same techniques of proof, in Section 5 we extend these results to the case of closed semi-algebraic sets.

## 2. Euler integration and the index of a critical point

Let $X \subset \mathbb{R}^n$ be a semi-algebraic set. We can write it in the following way:

$$X = \sqcup_{j=1}^{l} C_j,$$

where $C_j$ is semi-algebraically homeomorphic to $]-1,1[^{d_j}$ ($C_j$ is called a cell of dimension $d_j$). We set

$$\chi_c(X) = \sum_{j=1}^{l} (-1)^{d_j},$$

and we call it the Euler characteristic with compact support of $X$. Let us remark that if $X$ is compact, then $\chi_c(X) = \chi(X)$.

A constructible function $\varphi : X \to \mathbb{Z}$ is a $\mathbb{Z}$-valued function that can be written as a finite sum

$$\varphi = \sum_{i \in I} m_i 1_{X_i},$$

where $X_i$ is a semi-algebraic subset of $X$.

If $\varphi$ is a constructible function, the Euler integral of $\varphi$ is defined as

$$\int_X \varphi d\chi_c = \sum_{i \in I} m_i \chi_c(X_i).$$

DEFINITION 2.1. Let $f : X \to Y$ be a continuous semi-algebraic map and let $\varphi : X \to \mathbb{Z}$ be a constructible function. The push forward $f_*\varphi$ of $\varphi$ along $f$ is the function $f_*\varphi : Y \to \mathbb{Z}$ defined by

$$f_*\varphi(y) = \int_{f^{-1}(y)} \varphi d\chi_c.$$

THEOREM 2.2. *(Fubini's theorem) Let $f : X \to Y$ be a continuous semi-algebraic map and let $\varphi$ be a constructible function on $X$. Then, we have*

$$\int_Y f_*\varphi d\chi_c = \int_X \varphi d\chi_c.$$

COROLLARY 2.3. *Let $X$, $Y$ be semi-algebraic sets and let $f : X \to Y$ be a continuous semi-algebraic map. Then*

$$\chi_c(X) = \int_Y \chi_c(f^{-1}(y)) d\chi_c(y).$$

Let us consider from now on a closed semi-algebraic set $X \subset \mathbb{R}^n$. It is equipped with a finite semi-algebraic Whitney stratification $X = \sqcup_{a \in A} S_a$. Let $f : \mathbb{R}^n \to \mathbb{R}$ be a $\mathcal{C}^2$-semi-algebraic function.

DEFINITION 2.4.

(1) A point $p \in X$ is a critical point of $f_{|X}$ if it is a critical point of $f_{|S(p)}$, where $S(p)$ is the stratum that contains $p$.
(2) If $p$ is an isolated critical point of $f_{|X}$, we define the index of $f$ at $p$ by

$$\mathrm{ind}(f, X, p) = 1 - \chi(X \cap \{f = f(p) - \delta\} \cap B_\epsilon^n(p)),$$

where $0 < \delta \ll \epsilon \ll 1$.

Let us notice that if $X = \mathbb{R}^n$, by [**Kh**], $\mathrm{ind}(f, X, p) = \deg_p \nabla f$.

We suppose that $f_{|X}$ has a finite number of critical points $p_1, \ldots, p_l$. A proof of the next theorem (probably not the first one) can be found in [**Du2**], Theorem 3.1.

THEOREM 2.5. *If $X$ is compact, then*

$$\chi(X) = \sum_{i=1}^{l} \mathrm{ind}(f, X, p_i).$$

DEFINITION 2.6. Let $* \in \{\leq, =, \geq\}$.

(1) We define $\Lambda_f^*$ by

$$\Lambda_f^* = \{\alpha \in \mathbb{R} \mid \beta \mapsto \chi(\mathrm{Lk}^\infty(X \cap \{f * \beta\})) \text{ is not constant in a neighborhood of } \alpha\}.$$

Here $\mathrm{Lk}^\infty(Y) = Y \cap S_R^{n-1}$, $R \gg 1$, for a semi-algebraic set $Y$ of $\mathbb{R}^n$.

(2) We define $\widetilde{B}(f)$ by

$$\widetilde{B}(f) = f(\{p_1, \ldots, p_l\}) \cup \Lambda_f^{\leq} \cup \Lambda_f^{\geq}.$$

PROPOSITION 2.7.

(1) *The sets $\Lambda_f^*$ and $\widetilde{B}(f)$ are finite. Moreover $\Lambda_f^= \subset \Lambda_f^{\leq} \cup \Lambda_f^{\geq}$.*

(2) *If $\alpha \notin \widetilde{B}(f)$, the functions*

$$\beta \mapsto \chi(X \cap \{f * \beta\}), * \in \{\leq, =, \geq\},$$

*are constant in a neighborhood of $\alpha$.*

PROOF. See [**Du2**]. □

REMARK 2.8. Taking into account Proposition 2.7 and basic properties of $\chi_c$, if we have inclusions

$$\Lambda_f^* \subset \{c_1, c_2, \ldots, c_s\},$$

with $c_1 < c_2 < \cdots < c_s$ and

$$\widetilde{B}(f) \subset \{d_1, d_2, \ldots, d_k\},$$

with $d_1 < d_2 < \cdots < d_k$, we can express the Euler integral $\int_{\mathbb{R}} \chi(\mathrm{Lk}^\infty(X \cap \{f * t\}))dt$ as

$$\int_{\mathbb{R}} \chi(\mathrm{Lk}^\infty(X \cap \{f * t\}))dt = \sum_{i=1}^{s} \chi(\mathrm{Lk}^\infty(X \cap \{f * c_i\})) - \sum_{i=0}^{s} \chi(\mathrm{Lk}^\infty(X \cap \{f * c_i^+\})),$$

the Euler integral $\int_{\mathbb{R}} \chi(X \cap \{f * t\})dt$ as

$$\int_{\mathbb{R}} \chi(X \cap \{f * t\})dt = \sum_{j=1}^{k} \chi(X \cap \{f * d_j\}) - \sum_{j=0}^{k} \chi(X \cap \{f * d_j^+\}),$$

and the Euler integral $\int_{\mathbb{R}} \chi_c(X \cap \{f * t\})dt$ as

$$\int_{\mathbb{R}} \chi_c(X \cap \{f * t\})dt = \sum_{j=1}^{k} \chi_c(X \cap \{f * d_j\}) - \sum_{j=0}^{k} \chi_c(X \cap \{f * d_j^+\}),$$

where $c_0, d_0 = -\infty$, $c_{s+1} = +\infty$, $d_{k+1} = +\infty$, $c_i^+ \in ]c_i, c_{i+1}[$ and $d_j^+ \in ]d_j, d_{j+1}[$.

## 3. Some preliminary lemmas

In this section we state some technical lemmas that we will need in the proof of the main results. For more information, see [**Du1**] and [**Du2**].

LEMMA 3.1. *Let $f : \mathbb{R}^n \to \mathbb{R}$ a polynomial function with a finite number of critical points $p_1, p_2, \ldots, p_l$. If $\delta$ is a small regular value of $f$ and $R \gg 1$ is such that $f^{-1}(0) \cap B_R$ is a retract by deformation of $f^{-1}(0)$, then*

$$\chi(f^{-1}(\delta) \cap B_R) = \chi(f^{-1}(0)) - \mathrm{sign}(-\delta)^n \sum_{p_i \in f^{-1}(0)} \deg_{p_i} \nabla f.$$

LEMMA 3.2. *(Semi-algebraic case) Let $X$ be a closed semi-algebraic set and $f : \mathbb{R}^n \to \mathbb{R}$ a $\mathcal{C}^2$ semi-algebraic function such that $f_{|X} : X \to \mathbb{R}$ has a finite number of critical points $p_1, p_2, \ldots, p_l$. If $\delta < 0$ is a small regular value of $f$ and $R \gg 1$ is such that $X \cap f^{-1}(0) \cap B_R$ is a retract by deformation of $X \cap f^{-1}(0)$, then*

$$\chi(X \cap f^{-1}(\delta) \cap B_R) = \chi(X \cap f^{-1}(0)) - \sum_{p_i \in f^{-1}(0)} \operatorname{ind}(f, X, p_i).$$

LEMMA 3.3. *If $f_{|X}$ is proper then for any $\alpha \in \mathbb{R}$, we have*

$$\chi(X \cap \{f \geq \alpha\}) - \chi(X \cap \{f = \alpha\}) = \sum_{i: f(p_i) > \alpha} \operatorname{ind}(f, X, p_i).$$

Let $g : \mathbb{R}^n \to \mathbb{R}$ be a $\mathcal{C}^2$-semi-algebraic function such that $g^{-1}(0)$ intersects $X$ transversally. Let us suppose that $f_{|X \cap \{g \leq 0\}}$ admits an isolated critical point $p$ in $X \cap \{g = 0\}$ which is not a critical point of $f_{|X}$. If $S$ denotes the stratum of $X$ that contains $p$, this implies that

$$\nabla(f_{|S})(p) = \lambda(p) \nabla(g_{|S})(p),$$

with $\lambda(p) \neq 0$.

LEMMA 3.4. *For $0 < \delta \ll \epsilon \ll 1$, we have*

$$\chi(f^{-1}(-\delta) \cap B_\epsilon^n(p) \cap X \cap \{g \leq 0\}) = 1,$$

*if $\lambda(p) > 0$ and*

$$\chi(f^{-1}(-\delta) \cap B_\epsilon^n(p) \cap X \cap \{g \leq 0\}) = \chi(f^{-1}(-\delta) \cap B_\epsilon^n(p) \cap X \cap \{g = 0\}),$$

*if $\lambda(p) < 0$.*

REMARK 3.5. As a consequence of the last lemma and the definition of the index of a critical point $p$, we get that

$$\operatorname{ind}(f, X \cap \{g \leq 0\}, p) = 0,$$

if $\lambda(p) > 0$, and

$$\operatorname{ind}(f, X \cap \{g \leq 0\}, p) = \operatorname{ind}(f, X \cap \{g = 0\}, p),$$

if $\lambda(p) < 0$.

## 4. The index at infinity of a polynomial of $n$ variables

In this section, we give new proofs of the generalization of Sekalski and Gwoździewicz's formula for polynomial functions on $\mathbb{R}^n$.

THEOREM 4.1. *Let $f : \mathbb{R}^n \to \mathbb{R}$ a polynomial function with a finite number of critical points $p_1, p_2, \ldots, p_l$. Then,*

$$1 = \deg_\infty \nabla f - \int_{\mathbb{R}} \chi(\mathrm{Lk}^\infty(\{f \leq t\}))dt.$$

*In particular, if $n$ is even,*

$$1 = \deg_\infty \nabla f - \frac{1}{2} \int_{\mathbb{R}} \chi(\mathrm{Lk}^\infty(f^{-1}(t)))dt.$$

PROOF. Let us suppose first that $n$ is even.

By applying Euler integration, we know that $\chi_c(\mathbb{R}^n) = \int_{\mathbb{R}} \chi_c(f^{-1}(t))dt$. Moreover, we have

$$\begin{aligned}\chi_c(f^{-1}(c)) &= \chi_c(f^{-1}(c) \cap B_{R_c}) + \chi_c(f^{-1}(c) \cap \{\omega^2 > R_c^2\}) \\ &= \chi(f^{-1}(c) \cap B_{R_c}) - \chi(\mathrm{Lk}^\infty(f^{-1}(c))),\end{aligned}$$

where $\omega(x) = x_1^2 + x_2^2 + \cdots + x_n^2$ and $R_c \gg 1$ is such that $f^{-1}(c) \cap B_{R_c}$ is a deformation retract of $f^{-1}(c)$.

On the other hand, by Lemma 3.1 and taking into account that we can choose $0 < \epsilon \ll \frac{1}{R_c}$ and $\widetilde{c} = c - \epsilon$ such that $f^{-1}(\widetilde{c}) \cap B_R$ is an odd dimensional manifold with boundary, we have

$$\begin{aligned}\chi(f^{-1}(c)) &= \chi(f^{-1}(\widetilde{c}) \cap B_R) + \sum_{p \in f^{-1}(c)} \deg_p \nabla f \\ &= \frac{1}{2}\chi(f^{-1}(\widetilde{c}) \cap S_R) + \sum_{p \in f^{-1}(c)} \deg_p \nabla f \\ &= \frac{1}{2}\chi(f^{-1}(c) \cap S_R) + \sum_{p \in f^{-1}(c)} \deg_p \nabla f \\ &= \frac{1}{2}\chi(\mathrm{Lk}^\infty(f^{-1}(c))) + \sum_{p \in f^{-1}(c)} \deg_p \nabla f.\end{aligned}$$

Finally, we get

$$\begin{aligned}1 = \chi_c(\mathbb{R}^n) &= \int_{\mathbb{R}} \chi_c(f^{-1}(t))dt = \int_{\mathbb{R}} \chi(f^{-1}(t))dt - \int_{\mathbb{R}} \chi(\mathrm{Lk}^\infty(f^{-1}(t)))dt \\ &= \frac{1}{2}\int_{\mathbb{R}} \chi(\mathrm{Lk}^\infty(f^{-1}(t)))dt + \sum_p \deg_p \nabla f - \int_{\mathbb{R}} \chi(\mathrm{Lk}^\infty(f^{-1}(t)))dt \\ &= \deg_\infty \nabla f - \frac{1}{2}\int_{\mathbb{R}} \chi(\mathrm{Lk}^\infty f^{-1}(t)))dt.\end{aligned}$$

Since $n$ is even, $\chi(\mathrm{Lk}^\infty(f^{-1}(t))) = 2\chi(\mathrm{Lk}^\infty(\{f \leq t\}))$, getting the other equality.

Let us treat now the case $n$ odd. Let $\Gamma_f$ be the following polar set:

$$\Gamma_f = \{x \in \mathbb{R}^n \mid \mathrm{rank}[\nabla f(x), \nabla \omega(x)] < 2\},$$

and let $\Lambda_f$ be defined by

$$\Lambda_f = \{\alpha \in \mathbb{R} \mid \exists (x_n)_{n \in \mathbb{N}} \in \Gamma_f \text{ such that } \|x_n\| \to +\infty \text{ and } f(x_n) \to \alpha\}.$$

By Lemma 2.2 in **[Du2]**, we can assume that $\Lambda_f$ is a finite set. Furthermore, by Lemma 3.12 in **[Du2]**, $\Lambda_f^{\leq} \subset \Lambda_f$. Let us write

$$\Lambda_f \cup \tilde{B}_f = \{\alpha_1, \ldots, \alpha_r\},$$

where $\alpha_1 < \cdots < \alpha_r$.

Since $n$ is odd, $\chi(\mathrm{Lk}^\infty(f^{-1}(\alpha))) = 0$. Then, $\chi_c(f^{-1}(\alpha)) = \chi(f^{-1}(\alpha))$ and

$$-1 = \chi_c(\mathbb{R}^n) = \int_{\mathbb{R}} \chi_c(f^{-1}(t))dt = \int_{\mathbb{R}} \chi(f^{-1}(t))dt.$$

By Remark 2.8, we can write

$$-1 = \sum_{i=1}^{r} \chi(f^{-1}(\alpha_i)) - \sum_{i=0}^{r} \chi(f^{-1}(\alpha_i^+)),$$

and then

$$-1 = \deg_\infty \nabla f + \sum_{i=1}^{r} \chi(f^{-1}(\alpha_i - \delta) \cap B_{R_{\alpha_i}}) - \sum_{i=0}^{r} \chi(f^{-1}(\alpha_i^+) \cap B_{R_{\alpha_i^+}}),$$

where $\delta > 0$ is sufficiently small.

To compute the right-hand side of the above equality, we work with each difference

$$\chi(f^{-1}(\alpha_{i-1}^+) \cap B_{R_{\alpha_{i-1}^+}}) - \chi(f^{-1}(\alpha_i - \delta) \cap B_{R_{\alpha_i}}),$$

for $i = 1, \dots, r$.

Let us set $\alpha_{i-1}^+ = \alpha^-$ and $\alpha_i = \alpha$ with $\alpha^- = \alpha - \delta$, $0 < \delta \ll \frac{1}{R_{\alpha^-}}$ and $R_{\alpha^-} > R_\alpha \gg 1$. We have

$$\chi(f^{-1}(\alpha^-) \cap B_{R_{\alpha^-}}) - \chi(f^{-1}(\alpha - \delta) \cap B_{R_\alpha}) = \chi(f^{-1}(\alpha^-) \cap B_{R_{\alpha^-}}) - \chi(f^{-1}(\alpha^-) \cap B_{R_\alpha}),$$

(see Figure 1).

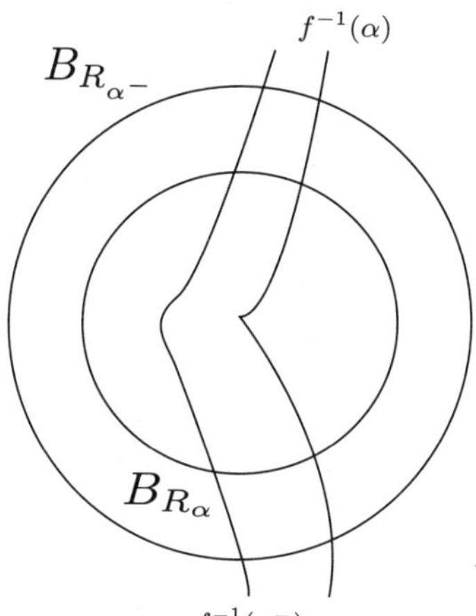


FIGURE 1

Since $n$ is odd, taking into account that the manifolds with corners $\{f \leq \alpha^-\} \cap B_{R_\alpha}$ and $\{f \leq \alpha^-\} \cap B_{R_{\alpha^-}}$ are homeomorphic after a suitable deformation to manifolds with boundary, we get that

$$\chi(\{f \leq \alpha^-\} \cap B_{R_\alpha}) = \frac{1}{2}\chi(\{f \leq \alpha^-\} \cap S_{R_\alpha}) + \frac{1}{2}\chi(\{f = \alpha^-\} \cap B_{R_\alpha}),$$

and

$$\chi(\{f \leq \alpha^-\} \cap B_{R_{\alpha^-}}) = \frac{1}{2}\chi(\{f \leq \alpha^-\} \cap S_{R_{\alpha^-}}) + \frac{1}{2}\chi(\{f = \alpha^-\} \cap B_{R_{\alpha^-}}).$$

We can choose $R_\alpha$ sufficiently big such that the critical points of $\omega_{|\{f<\alpha^-\}}$ lie in $\{\omega < R_\alpha^2\}$. This implies that the critical points of $\omega_{|\{f\leq\alpha^-\}}$ lying in $f^{-1}(\alpha^-)\cap \{\omega \geq R_\alpha^2\}$ are correct. Furthermore, by a straightforward application of the Curve Selection Lemma at infinity, these correct critical points point outwards (see Figure 2). Therefore, by Morse theory,

$$\chi(\{f\leq \alpha^-\}\cap B_{R_{\alpha^-}}) - \chi(\{f\leq\alpha^-\}\cap B_{R_\alpha}) = 0.$$

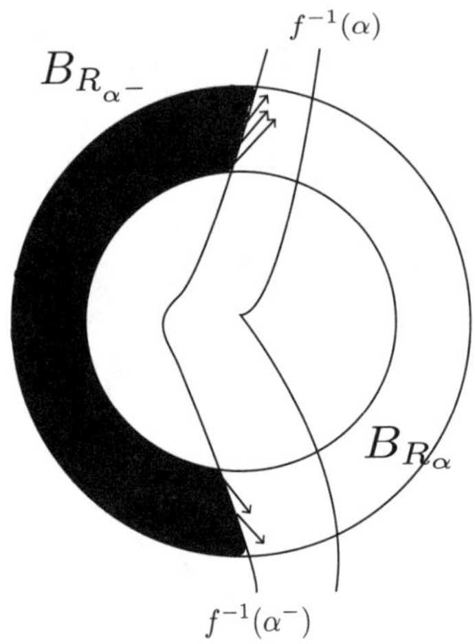


FIGURE 2

Thus,

$$\begin{aligned} 0 &= 2\Big(\chi(\{f\leq \alpha^-\}\cap B_{R_{\alpha^-}}) - \chi(\{f\leq\alpha^-\}\cap B_{R_\alpha})\Big) \\ &= \chi(\{f\leq \alpha^-\}\cap S_{R_{\alpha^-}}) - \chi(\{f\leq\alpha^-\}\cap S_{R_\alpha}) \\ &\quad +\chi(\{f= \alpha^-\}\cap B_{R_{\alpha^-}}) - \chi(\{f=\alpha^-\}\cap B_{R_\alpha}), \end{aligned}$$

and

$$\begin{aligned} &\chi(\{f= \alpha^-\}\cap B_{R_{\alpha^-}}) - \chi(\{f=\alpha^-\}\cap B_{R_\alpha}) \\ &= \chi(\mathrm{Lk}^\infty(f\leq\alpha)) - \chi(\mathrm{Lk}^\infty(f\leq\alpha^-)). \end{aligned}$$

Finally,

$$-1 = \deg_\infty \nabla f - \sum_{i=0}^{r}\chi(f^{-1}(\alpha_i^+)\cap B_{R_{\alpha_i^+}}) + \sum_{i=1}^{r}\chi(f^{-1}(\alpha_i)\cap B_{R_{\alpha_i}})$$

$$= \deg_\infty \nabla f + \sum_{i=0}^{r-1}\chi(\mathrm{Lk}^\infty(f\leq\alpha_i^+)) - \sum_{i=1}^{r}\chi(\mathrm{Lk}^\infty(f\leq\alpha_i)) - \chi(f^{-1}(\alpha_r^+)\cap B_{R_{\alpha_r^+}}),$$

with $\alpha_r^+ \in ]\alpha_r, +\infty[$.

Taking into account that $f$ does not contain any critical point in $]\alpha_r^+,+\infty[$ and that the critical points of $f_{|\{f\geq\alpha_{r^+}\}\cap S_{R_{\alpha_r^+}}}$ point outwards, by Morse theory we have $\chi(\{f\geq\alpha_r^+) = \chi(\{f=\alpha_r^+)$, and applying the Mayer-Vietoris sequence, we get that $\chi(f\leq \alpha_r^+) = 1$.

On the other hand, since $n$ is odd,

$$\chi(f^{-1}(\alpha_r^+) \cap B_{R_{\alpha+}}) + \chi(\mathrm{Lk}^\infty(f \leq \alpha_r^+)) = 2\chi(\{f \leq \alpha_r^+)\} \cap B_{R_{\alpha+}}) = 2.$$

Finally,

$$-1 = \deg_\infty \nabla f + \sum_{i=0}^{r} \chi(\mathrm{Lk}^\infty(f \leq \alpha_i^+)) - \sum_{i=1}^{r} \chi(\mathrm{Lk}^\infty(f \leq \alpha_i)) - 2,$$

obtaining that

$$1 = \deg_\infty \nabla f - \int_{\mathbb{R}} \chi(\mathrm{Lk}^\infty(f \leq t))dt.$$

□

## 5. Closed semi-algebraic sets and semi-algebraic functions

Here we give the generalization of Theorem 4.1 to $\mathcal{C}^2$ semi-algebraic functions defined on closed semi-algebraic sets.

THEOREM 5.1. ([**Du2**], Theorem 3.16) *Let $X$ be a closed semi-algebraic set and $f : \mathbb{R}^n \to \mathbb{R}$ a $\mathcal{C}^2$ semi-algebraic function such that $f_{|X} : X \to \mathbb{R}$ has a finite number of critical points $p_1, p_2, \ldots, p_l$. Then,*

$$\chi(X) = \sum_{i=1}^{l} \mathrm{ind}(f, X, p_i) - \int_{\mathbb{R}} \chi(\mathrm{Lk}^\infty(X \cap \{f \leq t\}))dt.$$

PROOF. We can equip $X$ with a finite semi-algebraic Whitney stratification, i.e. $X = \sqcup_{a\in A} S_a$. Let $\Gamma_f$ be the following polar set:

$$\Gamma_f = \{x \in \mathbb{R}^n \mid \mathrm{rank}[\nabla(f_{|S})(x), \nabla(\omega_{|S})(x)] < 2 \quad \text{where } S \text{ is the stratum that contains } x\},$$

and let $\Lambda_f$ be defined by

$$\Lambda_f = \{\alpha \in \mathbb{R} \mid \exists (x_n)_{n\in\mathbb{N}} \in \Gamma_f \text{ such that } \|x_n\| \to +\infty \text{ and } f(x_n) \to \alpha\}.$$

By Lemma 2.2 in [**Du2**], we can assume that $\Lambda_f$ is a finite set. Furthermore, by Lemma 3.12 in [**Du2**], $\Lambda_f^{\leq} \subset \Lambda_f$. By Hardt's theorem [**H**], there exists a finite set $\Delta_f \subset \mathbb{R}$ such that over each connected component of $\mathbb{R} \setminus \Delta_f$, $f$ is a semi-algebraic trivial fibration. Let us write

$$\Lambda_f \cup \tilde{B}_f \cup \Delta_f = \{b_1, \ldots, b_r\},$$

where $b_1 < \cdots < b_r$.

Let $b \in \mathbb{R}$ and let us take $R_b \gg 1$ and $0 < \delta \ll \frac{1}{R_b}$. By Lemma 3.2 we have

$$\chi(X \cap f^{-1}(b-\delta) \cap B_{R_b}) = \chi(X \cap f^{-1}(b)) - \sum_{f(p_i)=b} \mathrm{ind}(f, X, p_i).$$

We have

$$\chi(X) = \chi_c(X) + \chi(\mathrm{Lk}^\infty(X)) = \int_{\mathbb{R}} \chi_c(X \cap f^{-1}(t))dt + \chi(\mathrm{Lk}^\infty(X)).$$

By Remark 2.8, we get

$$\chi(X) = \sum_{i=1}^{r} \chi_c(X \cap f^{-1}(b_i)) - \sum_{i=0}^{r} \chi_c(X \cap f^{-1}(b_i^+)) + \chi(\mathrm{Lk}^\infty(X))$$

$$= \sum_{i=1}^{r} \chi_c(X \cap f^{-1}(b_i) \cap B_{R_{b_i}}) - \sum_{i=1}^{r} \chi_c(\mathrm{Lk}^\infty(X \cap f^{-1}(b_i)))$$

$$- \sum_{i=0}^{r} \chi_c(X \cap f^{-1}(b_i^+) \cap B_{R_{b_i^+}}) + \sum_{i=0}^{r} \chi_c(\mathrm{Lk}^\infty(X \cap f^{-1}(b_i^+))) + \chi(\mathrm{Lk}^\infty(X))$$

$$= \sum_{i=1}^{l} \mathrm{ind}(f, X, p_i) + \sum_{i=1}^{r} \chi_c(X \cap f^{-1}(b_i - \delta) \cap B_{R_{b_i}}) - \sum_{i=1}^{r} \chi_c(\mathrm{Lk}^\infty(X \cap f^{-1}(b_i)))$$

$$- \sum_{i=0}^{r} \chi_c(X \cap f^{-1}(b_i^+) \cap B_{R_{b_i^+}}) + \sum_{i=0}^{r} \chi_c(\mathrm{Lk}^\infty(X \cap f^{-1}(b_i^+))) + \chi(\mathrm{Lk}^\infty(X)),$$

where $\delta > 0$ is sufficiently small.

To compute the right-hand side of the above equality, we work with each difference

$$\chi(X \cap f^{-1}(b_{i-1}^+) \cap B_{R_{b_{i-1}^+}}) - \chi(X \cap f^{-1}(b_i - \delta) \cap B_{R_{b_i}}),$$

for $i = 1, \ldots, r$. Let us set $b_{i-1}^+ = b^-$ and $b_i = b$ with $b^- = b - \delta$, $0 < \delta \ll \frac{1}{R_{b^-}}$ and $R_{b^-} > R_b \gg 1$. We have

$$\chi(X \cap f^{-1}(b^-) \cap B_{R_{b^-}}) - \chi(X \cap f^{-1}(b - \delta) \cap B_{R_b})$$

$$= \chi(X \cap f^{-1}(b^-) \cap B_{R_{b^-}}) - \chi(X \cap f^{-1}(b^-) \cap B_{R_b}).$$

If $R_b \gg 1$ then the critical points of $-\omega_{|\{b^- \le f \le b\}}$ are contained in the interior of $B_{R_b}$ and the correct critical points of $-\omega_{|\{b^- < f < b\}}$ lying in $\{R_b^2 \le \omega \le R_{b^-}^2\}$ point outwards (see Figure 3). Then, by applying Lemmas 3.4 and 3.3, we have that

$$\chi_c(X \cap \{b^- \le f \le b\} \cap \{R^2 \le \omega \le R'^2\}) = \chi_c(X \cap \{b^- \le f \le b\} \cap \{\omega = R'^2\})$$

$$= \chi_c(X \cap \{f \le b\} \cap \{\omega = R'^2\}) - \chi_c(X \cap \{f \le b^-\} \cap \{\omega = R'^2\})$$

$$+ \chi_c(X \cap \{f = b^-\} \cap \{\omega = R'^2\})$$

$$= \chi_c(\mathrm{Lk}^\infty(X \cap \{f \le b\})) - \chi_c(\mathrm{Lk}^\infty(X \cap \{f \le b^-\})) + \chi_c(\mathrm{Lk}^\infty(X \cap \{f = b^-\})),$$

with $R = R_b$ and $R' = R_{b^-}$.

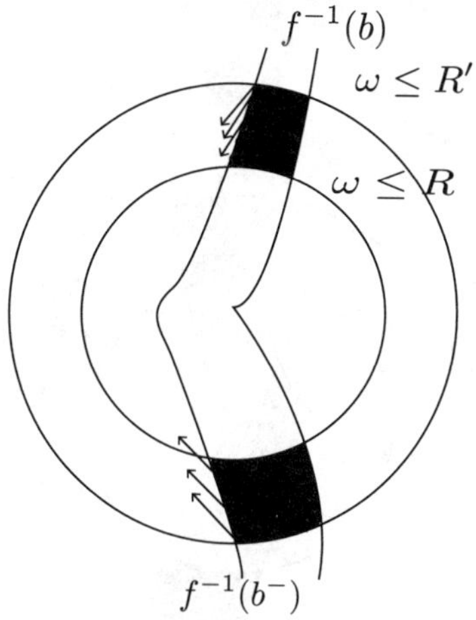


FIGURE 3

By an analogous argument, considering now the function $f_{|\{R\leq\omega\leq R'\}}$ (see Figure 4) and applying again Lemmas 3.4 and 3.3,

$$\begin{aligned}\chi_c(X\cap\{b^-\leq f\leq b\}\cap\{R^2\leq\omega\leq R'^2\}) &= \chi_c(X\cap\{f=b^-\}\cap\{R^2\leq\omega\leq R'^2\})\\ &= \chi_c(X\cap\{f=b^-\}\cap S_R)+\chi_c(X\cap\{f=b^-\}\cap\{R^2<\omega\leq R'^2\})\\ &= \chi_c(\mathrm{Lk}^\infty(X\cap\{f=b\}))+\chi_c(X\cap\{f=b^-\}\cap\{R^2<\omega\leq R'^2\}).\end{aligned}$$

Finally,

$$\begin{aligned}&\chi_c(X\cap\{f=b^-\}\cap B_{R_{b^-}})-\chi_c(X\cap\{f=b^-\}\cap B_{R_b})\\ &=\chi_c(X\cap\{f=b^-\}\cap\{R^2<\omega\leq R'^2\})=\chi_c(X\cap\{f=b^-\}\cap\{R^2\leq\omega\leq R'^2\})\\ &\qquad -\chi_c(X\cap\{f=b^-\}\cap S_R)\\ &=\chi_c(\mathrm{Lk}^\infty(X\cap\{f\leq b\}))-\chi_c(\mathrm{Lk}^\infty(X\cap\{f\leq b^-\}))\\ &\quad+\chi_c(\mathrm{Lk}^\infty(X\cap\{f=b^-\}))-\chi_c(\mathrm{Lk}^\infty(X\cap\{f=b\})).\end{aligned}$$

Returning to our initial equation and substituting, we get

$$\begin{aligned}\chi(X) &= \sum_{i=1}^{l}\mathrm{ind}(f,X,p_i)+\sum_{i=0}^{r-1}\chi_c(\mathrm{Lk}^\infty(X\cap\{f\leq b_i^+\}))\\ &-\sum_{i=0}^{r-1}\chi_c(\mathrm{Lk}^\infty(X\cap\{f\leq b_i\}))+\sum_{i=0}^{r-1}\chi_c(\mathrm{Lk}^\infty(X\cap\{f=b_i\}))-\sum_{i=0}^{r-1}\chi_c(\mathrm{Lk}^\infty(X\cap\{f=b_i^+\}))\\ &-\chi_c(X\cap\{f=b_r^+\}\cap B_{R_{r^+}})-\int_{\mathbb{R}}\chi_c(\mathrm{Lk}^\infty(X\cap f^{-1}(t))dt+\chi_c(\mathrm{Lk}^\infty(X))\\ &=\sum_{i=1}^{l}\mathrm{ind}(f,X,p_i)-\int_{\mathbb{R}}\chi_c(\mathrm{Lk}^\infty(X\cap\{f\leq t\}))dt-\chi_c(\mathrm{Lk}^\infty(X\cap\{f\leq b_r^+\}))\\ &\quad+\chi_c(\mathrm{Lk}^\infty(X\cap\{f=b_r^+\}))-\chi_c(X\cap\{f=b_r^+\}\cap B_{R_{r^+}})+\chi_c(\mathrm{Lk}^\infty(X))\end{aligned}$$

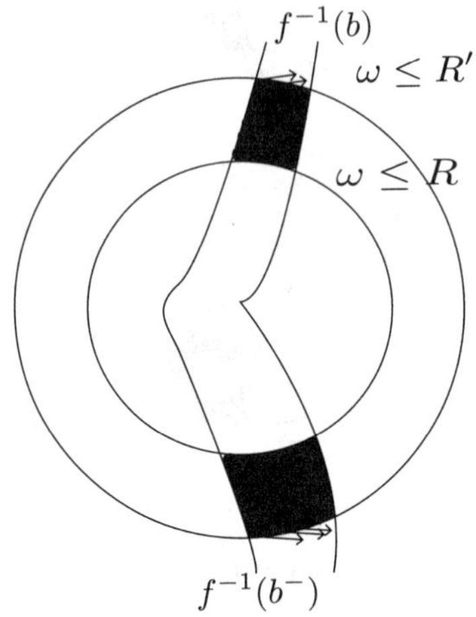


FIGURE 4

$$
\begin{aligned}
&= \sum_{i=1}^{l} \mathrm{ind}(f, X, p_i) - \int_{\mathbb{R}} \chi_c(\mathrm{Lk}^\infty(X \cap \{f \leq t\}))dt + \chi_c(\mathrm{Lk}^\infty(X \cap \{f \geq b_r^+\})) \\
&\qquad - \chi_c(X \cap \{f = b_r^+\} \cap B_{R_{r+}}) \\
&= \sum_{i=1}^{l} \mathrm{ind}(f, X, p_i) - \int_{\mathbb{R}} \chi_c(\mathrm{Lk}^\infty(X \cap \{f \leq t\}))dt + \chi_c(\mathrm{Lk}^\infty(X \cap \{f \geq b_r^+\})) \\
&\qquad - \chi_c(X \cap \{f \geq b_r^+\} \cap B_{R_{r+}}) \\
&= \sum_{i=1}^{l} \mathrm{ind}(f, X, p_i) - \int_{\mathbb{R}} \chi_c(\mathrm{Lk}^\infty(X \cap \{f \leq t\}))dt + \chi_c(\mathrm{Lk}^\infty(X \cap \{f \geq b_r^+\})) \\
&\qquad - \chi_c(\mathrm{Lk}^\infty(X \cap \{f \geq b_r^+\})) - \chi_c(X \cap \{f \geq b_r^+\}) \\
&= \sum_{i=1}^{l} \mathrm{ind}(f, X, p_i) - \int_{\mathbb{R}} \chi_c(\mathrm{Lk}^\infty(X \cap \{f \leq t\}))dt,
\end{aligned}
$$

where $\chi_c(X \cap \{f \geq b_r^+\} \cap B_{R_{r+}}) = \chi_c(X \cap \{f = b_r^+\} \cap B_{R_{r+}})$ by Lemma 3.3 and $\chi_c(X \cap \{f \geq b_r^+\}) = 0$ taking into account that $\chi_c([b_r^+, +\infty[) = 0$ and $f_{|X \cap [b_r^+, +\infty[}$ is a trivial fibration. □

COROLLARY 5.2. ([**Du2**], Theorem 3.17)
$\chi(X) = \sum_{i=1}^{l} \mathrm{ind}(-f, X, p_i) - \int_{\mathbb{R}} \chi(\mathrm{Lk}^\infty(X \cap \{f \geq t\}))dt.$

PROOF. By replacing $f$ by $-f$ and applying an analogous procedure as in the last theorem, we arrive to the desired result. □

COROLLARY 5.3. ([**Du2**], Corollary 3.18)

$$2\chi(X) - \chi(\mathrm{Lk}^\infty(X)) = \sum_{i=1}^{l} \mathrm{ind}(f, X, p_i) + \sum_{i=1}^{l} \mathrm{ind}(-f, X, p_i) - \int_{\mathbb{R}} \chi(\mathrm{Lk}^\infty(X \cap \{f = t\}))dt.$$

PROOF. It follows from Theorem 5.1 and Corollary 5.2 by applying the Mayer-Vietoris sequence. □

LEMMA 5.4. $\int_{\mathbb{R}} \chi_c(X \cap \{f \leq t\})dt = 0.$

PROOF. Let us take $b$ in $\Lambda_f \cup \tilde{B}_f \cup \Delta_f$ and $b^+ = b + \delta$, with $\delta > 0$ small enough, a regular value. Since $f_{|X \cap ]b,b^+]}$ is trivial and $\chi_c(]b, b^+]) = 0$, we conclude that

$$\chi_c(X \cap \{f \leq b^+\}) - \chi_c(X \cap \{f \leq b\}) = \chi_c(X \cap \{\alpha < f \leq b^+\}) = 0.$$

Therefore,

$$\int_{\mathbb{R}} \chi_c(X \cap \{f \leq t\})dt = \sum_{i=1}^{r} \chi_c(X \cap \{f \leq b_i\}) - \sum_{i=0}^{r} \chi_c(X \cap \{f \leq b_i^+\})$$
$$= -\chi_c(X \cap \{f \leq b_0^+\}) = 0.$$

□

COROLLARY 5.5. ([**Du2**], Theorem 3.21)
$\chi(X) = \sum_{i=1}^{l} \mathrm{ind}(f, X, p_i) - \int_{\mathbb{R}} \chi(X \cap \{f \leq t\})dt.$

PROOF. We have

$$\chi(X) = \sum_{i=1}^{l} \mathrm{ind}(f, X, p_i) - \int_{\mathbb{R}} \chi(\mathrm{Lk}^\infty(X \cap \{f \leq t\}))dt,$$

and

$$\int_{\mathbb{R}} \chi_c(X \cap \{f \leq t\})dt = \int_{\mathbb{R}} \chi_c(X \cap \{f \leq t\} \cap B_{R_t})dt - \int_{\mathbb{R}} \chi_c(\mathrm{Lk}^\infty(X \cap \{f \leq t\}))dt = 0.$$

Then,

$$\int_{\mathbb{R}} \chi(X \cap \{f \leq t\})dt = \int_{\mathbb{R}} \chi(X \cap \{f \leq t\} \cap B_{R_t})dt = \int_{\mathbb{R}} \chi_c(X \cap \{f \leq t\} \cap B_{R_t})dt$$
$$= \int_{\mathbb{R}} \chi_c(\mathrm{Lk}^\infty(X \cap \{f \leq t\}))dt = \int_{\mathbb{R}} \chi(\mathrm{Lk}^\infty(X \cap \{f \leq t\}))dt,$$

arriving to the desired result. □

COROLLARY 5.6. ([**Du2**], Theorem 3.21)
$\chi(X) = \sum_{i=1}^{l} \mathrm{ind}(-f, X, p_i) - \int_{\mathbb{R}} \chi(X \cap \{f \geq t\})dt.$

PROOF. By replacing $f$ by $-f$ and applying an analogous procedure as in the last corollary, we arrive to the desired result. □

COROLLARY 5.7. ([**Du2**], Theorem 3.20)
$\chi(X) = \sum_{i=1}^{l} \mathrm{ind}(f, X, p_i) + \sum_{i=1}^{l} \mathrm{ind}(-f, X, p_i) - \int_{\mathbb{R}} \chi(X \cap \{f = t\})dt.$

PROOF. It follows from the last two corollaries by applying the Mayer-Vietoris sequence. □

## References


[Du1] N.Dutertre, *On topological invariants associated with a polynomial with isolated critical points*, Glasgow Math J., **46** (2004), 323–334.

[Du2] N.Dutertre, *On the topology of semi-algebraic functions on closed semi-algebraic sets*, Manuscripta Math., **139** (2012), 415–441.

[Gw] J.Gwoździewicz, *The index at infinity of a polynomial in two variables*, Univ. Iagel. Acta Math, **47** (2009), 243–247.

[H] R.M. Hardt, *Topological properties of subanalytic sets*, Trans. Amer. Math. Soc., **211** (1975), 57-70.

[Se] M.Sekalski, *The degree at infinity of the gradient of a polynomial in two real variables*, Ann. Polon. Math, **87** (2005), 229–235.

[Kh] G.M.Khimshiasvili, *On the local degree of a smooth map*, Soobshch. Aka. Sauk Gruz. SSR, **85** (1977), 309–311.



Aix-Marseille Université, CNRS, Centrale Marseille, I2M, UMR 7373, 13453 Marseille, France

*E-mail address*: nicolas.dutertre@univ-amu.fr

Aix-Marseille Université, CNRS, Centrale Marseille, I2M, UMR 7373, 13453 Marseille, France

*E-mail address*: juan-antonio.moya-perez@univ-amu.fr

Contemporary Mathematics
Volume **675**, 2016
http://dx.doi.org/10.1090/conm/675/13586

# On associate families of spacelike Delaunay surfaces

Atsufumi Honda

*Dedicated to Professor María del Carmen Romero Fuster on the occasion of her 60th birthday.*

ABSTRACT. In a paper of do Carmo and Dajczer, associate families of Delaunay surfaces in the Euclidean 3-space are constructed explicitly and classified completely. Following the method of do Carmo-Dajczer, Sasahara gave a construction of some associate families of spacelike Delaunay surfaces in the Lorentz-Minkowski 3-space. In this paper, we continue the construction and give a complete classification of associate families of spacelike Delaunay surfaces. We also determine their singularity types.

## 1. Introduction

*Maximal surfaces* are spacelike surfaces of zero-mean-curvature in the Lorentz-Minkowski 3-space $L^3$. Since any complete maximal surface is a plane [**2**], maximal surfaces have singular points in general. Maximal surfaces with admissible singularities are studied intensively in the last decade [**3**, **5**–**16**, **24**–**26**, **31**].

In particular, *fold* singular points on maximal surfaces possess remarkable properties. For example, maximal surfaces with fold singularities can be extended to type-changing zero-mean-curvature surfaces analytically [**11**, **26**] (cf. [**18**–**20**], [**27**], [**25**]). Moreover, every maximal surface with fold singular points is a conjugate maximal surface with *conelike* singular points, and vice versa [**11**, **26**]. (Such a property is called 'duality of singularities'. For other singularities, see [**16**].)

Then, it is natural to ask what occurs when we consider non-maximal spacelike surfaces of *constant mean curvature* (abbreviated to *CMC*). According to the duality of singularities for maximal surfaces, we expect that the conjugates of spacelike CMC surfaces with conelike singular points may have fold singular points. In the case of maximal surfaces, we can construct conjugates by using the Weierstrass-type representation formula [**28**]. However, in the non-maximal CMC case, there is no such a formula, and hence it is not easy to construct conjugate spacelike CMC surfaces explicitly. In the case of CMC surfaces in the Euclidean 3-space, do Carmo-Dajczer [**4**] gave a method to construct associate families of Delaunay surfaces (i.e., CMC surfaces of revolution).

2010 *Mathematics Subject Classification.* Primary 53A10; Secondary 53A35, 53C50.

*Key words and phrases.* Spacelike CMC surface, constant mean curvature, spacelike Delaunay surface, surface with singularities.

The author is partially supported by Grant-in-Aid for Challenging Exploratory Research No. 26610035 of the Japan Society for the Promotion of Science and by JSPS/CAPES under the Japan-Brazil Research Cooperative Program.

In this paper, we apply their method to construct spacelike CMC surfaces which are conjugate to spacelike Delaunay surfaces with conelike singularities. More precisely, we classify the associate families of spacelike Delaunay surfaces in $L^3$ (Theorems 3.4, 3.6 and 3.7), and their singularities (Theorem 4.5). In particular, by Theorem 4.5, we have that the conjugates of spacelike Delaunay surfaces with conelike singularities do not have fold singular points but 5/2-cuspidal edges. Since Sasahara [**30**] applied the method of do Carmo-Dajczer [**4**] to the case of $L^3$ partially, we shall deal with the remaining cases. We remark that Brander-Rossman-Schmitt [**1**] gave a description of the moduli space of spacelike Delaunay surfaces and their associate families in terms of the holomorphic potentials and frames.

This paper is organized as follows. In Section 2, we review the classification of spacelike Delaunay surfaces in $L^3$. Then, in Section 3, we shall apply the method of do Carmo-Dajczer [**4**] and classify the associate families of spacelike Delaunay surfaces (Theorems 3.4, 3.6 and 3.7). Finally in Section 4, we investigate their singularities (Theorem 4.5).

## 2. Preliminaries

**2.1. Associate family.** We denote by $L^3 = (\boldsymbol{R}^3, \langle\ ,\ \rangle)$ the Lorentz-Minkowski 3-space with the Lorentz metric $\langle (x_0, x_1, x_2), (y_0, y_1, y_2)\rangle = -x_0y_0 + x_1y_1 + x_2y_2$. Then the hyperbolic plane is given by

$$H^2 = \left\{\boldsymbol{x} = (x_0, x_1, x_2) \in L^3\,;\ \langle \boldsymbol{x}, \boldsymbol{x}\rangle = -1,\ x_0 > 0\right\}.$$

Let $M^2$ be a smooth orientable 2-manifold. An immersion $f : M^2 \to L^3$ is called *spacelike* if the first fundamental form $d\sigma^2 := f^*\langle\ ,\ \rangle$ is a Riemannian metric on $M^2$. Denote by $\nu : M^2 \to H^2$ the unit normal vector field along $f$.

Let $(U; x, y)$ be an isothermal coordinate system of $(M^2, d\sigma^2)$, that is, there exists a smooth function $\tau = \tau(x, y)$ such that $d\sigma^2 = e^{2\tau}(dx^2 + dy^2)$ holds on $U$. If we denote by $II = h_{11}\,dx^2 + 2h_{12}\,dx\,dy + h_{22}\,dy^2$ the second fundamental form of $f$, the mean curvature $H$ is given by

$$H = \frac{h_{11} + h_{22}}{2e^{2\tau}}.$$

A spacelike immersion with non-zero constant mean curvature (resp. with vanishing mean curvature) is called *spacelike CMC* (resp. *maximal*). Setting $z = x + iy$, we have

$$II = Q + \bar{Q} + H\,d\sigma^2 \qquad \left(Q := q\,dz^2, \quad q := \frac{h_{11} - h_{22} - 2ih_{12}}{4}\right).$$

We call $Q$ the *Hopf differential*. The Gauss and Codazzi equations are written as

$$4\tau_{z\bar{z}} = H^2e^{2\tau} - 4|q|^2e^{-2\tau}, \tag{2.1}$$

$$2q_{\bar{z}} = H_ze^{2\tau}, \tag{2.2}$$

respectively. By the Codazzi equation (2.2), $Q$ is holomorphic if and only if $f$ is either maximal or spacelike CMC.

Let $f : M^2 \to L^3$ be a maximal or spacelike CMC immersion of mean curvature $H$. If $M^2$ is simply connected, it is known that there exists a family of immersions $\{f_\theta : M^2 \to L^3\}_{\theta\in S^1}$ such that

$$H_\theta = H, \qquad d\sigma_\theta^2 = d\sigma^2, \qquad Q_\theta = e^{i\theta}Q$$

hold for all $\theta \in S^1 := \boldsymbol{R}/2\pi\boldsymbol{Z}$, where $H_\theta$, $d\sigma_\theta^2$ and $Q_\theta$ are the mean curvature, first fundamental form and Hopf differential of $f_\theta$, respectively. We call $\{f_\theta\}_{\theta\in S^1}$ and $\theta$ the *associate family* and the *spectral parameter* of $f$, respectively. In particular, if $\theta = \pi/2$, we call $f^{\#} := f_{\pi/2}$ the *conjugate* of $f$.

REMARK 2.1. It is known that any two isometric maximal or spacelike CMC immersions which have the same mean curvature are associated: Let $f, \tilde{f} : M^2 \to L^3$ be spacelike CMC immersions of mean curvature $H$. Assume that they share the same first fundamental form $d\sigma^2 = d\tilde{\sigma}^2 = e^{2\tau}(dx^2 + dy^2)$. Then, by the Gauss equation (2.1), we have $|q| = |\tilde{q}|$, where $Q = q\,dz^2$ (resp. $\tilde{Q} = \tilde{q}\,dz^2$) is the Hopf differential of $f$ (resp. $\tilde{f}$). Therefore, $f$ and $\tilde{f}$ are associated.

**2.2. Spacelike Delaunay surfaces.** We call a spacelike CMC surface of revolution in $L^3$ a *spacelike Delaunay surface*. A spacelike Delaunay surface is called *s-Delaunay* (resp. *t-Delaunay*, *ℓ-Delaunay*), if its axis of revolution is spacelike (resp. timelike, lightlike). Without loss of generality, we may assume that the mean curvature $H$ is a positive constant. Spacelike Delaunay surfaces are classified in [**21**], [**23**], [**30**] as follows.

FACT 2.2 (Classification of spacelike $t$-Delaunay surfaces). A spacelike $t$-Delaunay surface of mean curvature $H\,(>0)$ is congruent to (cf. Figure 1)

$$f_{\mathfrak{t}}(s,t) = (\eta_{\mathfrak{t}}(s),\ \xi_{\mathfrak{t}}(s)\cos(2Ht),\ \xi_{\mathfrak{t}}(s)\sin(2Ht)) \tag{2.3}$$

$$\left(\xi_{\mathfrak{t}}(s) = \frac{U_{\mathfrak{t}}(s)}{2H}, \qquad \eta_{\mathfrak{t}}(s) = \pm\int \frac{e^{2Hs} + ke^{-2Hs} - 2}{2U_{\mathfrak{t}}(s)}ds\right)$$

for some $k \in \boldsymbol{R}$, where we set $U_{\mathfrak{t}}(s) = \sqrt{e^{2Hs} + ke^{-2Hs} - k - 1}$. The domain of $s$ is $(\frac{1}{2H}\log k, \infty)$ if $k > 1$, and $(0, \infty)$ if $k \leq 1$.

$k = 2$ $\qquad$ $k = 0.5$ $\qquad$ $k = 0.5$ (half)

FIGURE 1. Spacelike $t$-Delaunay surfaces $f_{\mathfrak{t}}(s,t)$ (cf. Fact 2.2). If $k < 1$, $f_{\mathfrak{t}}$ has self-intersection.

FACT 2.3 (Classification of spacelike $s$-Delaunay surfaces). A spacelike $s$-Delaunay surface of mean curvature $H\,(>0)$ is congruent to one of the surfaces (cf. Figures 2 and 3)

$$f_j(s,t) = (\xi_j(s)\cosh(2Ht),\ \xi_j(s)\sinh(2Ht),\ \eta_j(s)) \quad (j = \mathfrak{sc}, \mathfrak{si}, \mathfrak{cyl}), \tag{2.4}$$

where the generating curves $(\xi_j(s), \eta_j(s))$ $(j = \mathfrak{sc}, \mathfrak{si}, \mathfrak{cyl})$ are defined as

$$(2.5)\qquad \xi_{\mathfrak{sc}}(s) = \pm\frac{U_{\mathfrak{sc}}(s)}{2H}, \qquad \eta_{\mathfrak{sc}}(s) = \int \frac{e^{2Hs} + ke^{-2Hs} + 2}{2U_{\mathfrak{sc}}(s)} ds \qquad (k \geq 0),$$

$$(2.6)\qquad \xi_{\mathfrak{si}}(s) = \pm\frac{U_{\mathfrak{si}}(s)}{2H}, \qquad \eta_{\mathfrak{si}}(s) = \int \frac{e^{2Hs} + ke^{-2Hs} - 2}{2U_{\mathfrak{si}}(s)} ds \qquad (k \in \boldsymbol{R}),$$

$$(2.7)\qquad \xi_{\mathfrak{cyl}}(s) = \pm\frac{1}{2H}, \qquad \eta_{\mathfrak{cyl}}(s) = \frac{s}{2H},$$

and $U_{\mathfrak{sc}}(s) = \sqrt{1 + k + e^{2Hs} + ke^{-2Hs}}$, $U_{\mathfrak{si}}(s) = \sqrt{1 + k - e^{2Hs} - ke^{-2Hs}}$. The domain of $s$ for (2.6) is $(0, \frac{1}{2H}\log k)$ if $k > 1$, $(\frac{1}{2H}\log k, 0)$ if $0 < k \leq 1$, and $(-\infty, 0)$ if $k \leq 0$. For (2.5) and (2.7), $s$ is defined on whole $\boldsymbol{R}$.

$f_{\mathfrak{sc}}$ $(k = 10)$ $\qquad$ $f_{\mathfrak{sc}}$ $(k = 0)$ $\qquad$ $f_{\mathfrak{cyl}}$

FIGURE 2. Complete spacelike $s$-Delaunay surfaces $f_{\mathfrak{sc}}(s,t)$ and $f_{\mathfrak{cyl}}(s,t)$ (cf. Fact 2.3).

$k = -1$ $\qquad$ $k = 0$ $\qquad$ $k = 0.5$

FIGURE 3. Incomplete spacelike $s$-Delaunay surfaces $f_{\mathfrak{si}}(s,t)$ (cf. Fact 2.3).

FACT 2.4 (Classification of spacelike $\ell$-Delaunay surfaces). A spacelike $\ell$-Delaunay surface of mean curvature $H$ $(> 0)$ is congruent to (cf. Figure 4)

$$(2.8)\qquad f_j(s,t) = \left(\eta_j(s) + \xi_j(s) + \frac{t^2}{4}\xi_j(s),\ t\xi_j(s),\ \eta_j(s) - \xi_j(s) + \frac{t^2}{4}\xi_j(s)\right)$$

$(j = \mathfrak{lt}, \mathfrak{lsc}, \mathfrak{lsi}, \mathfrak{tu})$, where the generating curves $(\xi_j(s), \eta_j(s))$ $(j = \mathfrak{lt}, \mathfrak{lsc}, \mathfrak{lsi}, \mathfrak{tu})$ are defined as

$$(2.9)\qquad \xi_{\mathfrak{lt}}(s) = \pm\sqrt{e^{2Hs} - 1}, \quad \eta_{\mathfrak{lt}}(s) = \frac{\pm 1}{8H^2}\left(\frac{\xi(s)}{e^{2Hs}} - \tan^{-1}\xi(s)\right) \quad (s > 0),$$

$$(2.10)\quad \xi_{\mathfrak{lsc}}(s) = \pm\sqrt{1+e^{2Hs}}, \quad \eta_{\mathfrak{lsc}}(s) = \frac{\pm 1}{8H^2}\left(\frac{\xi(s)}{e^{2Hs}} - \coth^{-1}\xi(s)\right) \quad (s \in \boldsymbol{R}),$$

$$(2.11)\quad \xi_{\mathfrak{lsi}}(s) = \pm\sqrt{1-e^{2Hs}}, \quad \eta_{\mathfrak{lsi}}(s) = \frac{\mp 1}{8H^2}\left(\frac{\xi(s)}{e^{2Hs}} + \tanh^{-1}\xi(s)\right) \quad (s < 0),$$

$$(2.12)\quad \xi_{\mathfrak{tu}}(s) = \pm e^{Hs}, \quad \eta_{\mathfrak{tu}}(s) = \pm\frac{e^{-Hs}}{4H^2} \quad (s \in \boldsymbol{R}).$$

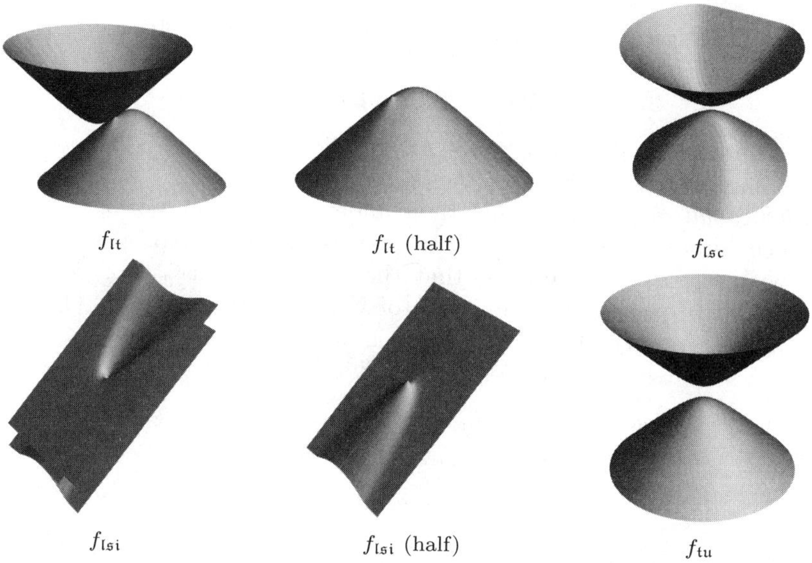


FIGURE 4. Spacelike $\ell$-Delaunay surfaces $f_{\mathfrak{lt}}(s,t)$, $f_{\mathfrak{lsc}}(s,t)$, $f_{\mathfrak{lsi}}(s,t)$ and $f_{\mathfrak{tu}}(s,t)$ (cf. Fact 2.4).

REMARK 2.5. In [**21**], $\ell$-Delaunay surfaces were classified with a parameter $a \in \boldsymbol{R}$. However, by a suitable coordinate change and an isometry of $L^3$, we have the form as in Fact 2.4.

REMARK 2.6. For each $j = \mathfrak{t}$, $\mathfrak{sc}$, $\mathfrak{si}$, the first fundamental form $d\sigma_j^2$ of $f_j(s,t)$ is given by $d\sigma_j^2 = ds^2 + U_j(s)^2 dt^2$. Similarly, we have $d\sigma_{\mathfrak{cyl}}^2 = ds^2 + dt^2$ and $d\sigma_j^2 = ds^2 + \xi_j(s)^2 dt^2$ for $j = \mathfrak{lt}$, $\mathfrak{lsc}$, $\mathfrak{lsi}$, $\mathfrak{tu}$. Moreover, if $k < 0$, a spacelike $t$-Delaunay surface $f_{\mathfrak{t}}(s,t)$ is isometric to the spacelike $s$-Delaunay surface $f_{\mathfrak{si}}(\hat{s},\hat{t})$ where $(\hat{s},\hat{t}) := (-s, -\sqrt{-k}t)$. In fact, setting $\hat{k} := 1/k$, $c := -\sqrt{-k}$,

$$\begin{aligned} d\sigma_{\mathfrak{t}}^2 &= ds^2 + U_{\mathfrak{t}}(s)^2 dt^2 \\ &= ds^2 + (e^{2Hs} - c^2 e^{-2Hs} + c^2 - 1)dt^2 \\ &= d\hat{s}^2 + \left(1 + \hat{k} - e^{2H\hat{s}} - \hat{k}e^{-2H\hat{s}}\right) d\hat{t}^2 = d\sigma_{\mathfrak{si}}^2 \end{aligned}$$

holds (cf. Theorems 3.4 and 3.7).

The isothermal coordinate system is given as follows.

LEMMA 2.7. *For a positive-valued function $U(s)$, set $d\sigma^2 := ds^2 + U(s)^2 dt^2$. Then a coordinate system $(x,y)$ satisfying $dx = ds/U(s)$, $y = t$ is isothermal.*

By a direct calculation, we have the Hopf differential (and hence the second fundamental form) of the spacelike Delaunay surfaces $f_j(s,t)$ as follows.

LEMMA 2.8. *On the spacelike Delaunay surface $f_j(s,t)$ ($j = \mathfrak{t}$, $\mathfrak{sc}$, $\mathfrak{si}$, $\mathfrak{cyl}$, $\mathfrak{lt}$, $\mathfrak{lsc}$, $\mathfrak{lsi}$, $\mathfrak{tu}$), let $(x,y)$ be the isothermal coordinate system as in Lemma 2.7. Then the Hopf differential $Q_j = q_j\,dz^2$ ($z = x+iy$) is given by*

$$q_{\mathfrak{t}} = -q_{\mathfrak{sc}} = -q_{\mathfrak{si}} = \frac{H(1-k)}{2}, \tag{2.13}$$

$$q_{\mathfrak{cyl}} = q_{\mathfrak{lt}} = -q_{\mathfrak{lsc}} = -q_{\mathfrak{lsi}} = -\frac{H}{2}, \qquad q_{\mathfrak{tu}} = 0. \tag{2.14}$$

## 3. Spacelike helicoidal CMC surfaces

In the Lorentz-Minkowski 3-space $L^3$, a helicoidal motion is defined as a non-trivial one-parameter subgroup of the isometry group of $L^3$. A surface in $L^3$ is called a *helicoidal surface* if it is invariant under a helicoidal motion.

A helicoidal motion fixes a line. If the axis is timelike (resp. spacelike), we may assume, without loss of generality, that the axis is $x_0$-axis (resp. $x_2$-axis), since every timelike (resp. spacelike) unit vector can be transformed to $(1,0,0)$ (resp. $(0,0,1)$) by an isometry of $L^3$. If the axis is lightlike, we may also assume that it is $\boldsymbol{R}\cdot(1,0,1)$. Thus, such a helicoidal surface is congruent to one of the following:

$$f(\rho,\varphi) = (h\varphi+\lambda, \rho\cos\varphi, \rho\sin\varphi), \tag{3.1}$$

$$f(\rho,\varphi) = (\rho\cosh\varphi, \rho\sinh\varphi, h\varphi+\lambda), \tag{3.2}$$

$$f(\rho,\varphi) = (\rho\sinh\varphi, \rho\cosh\varphi, h\varphi+\lambda), \tag{3.3}$$

$$f(\rho,\varphi) = (\lambda-\rho-\rho\varphi^2, -2\rho\varphi, \lambda+\rho-\rho\varphi^2) + h\left(\frac{\varphi^3}{3}+\varphi, \varphi^2, \frac{\varphi^3}{3}-\varphi\right) \tag{3.4}$$

for some $h \in \boldsymbol{R}$ and a smooth function $\lambda = \lambda(\rho)$. We call $h$ the *pitch*. Helicoidal surfaces of pitch 0 are surfaces of revolution.

Now we introduce the special coordinate system $(s,t)$ of spacelike helicoidal surfaces $f(\rho,\varphi)$. Since Sasahara [**30**] treated the cases of (3.1), (3.2) and (3.4), we consider here the case of (3.3). The first fundamental form $d\sigma^2$ of the spacelike helicoidal surface (3.3) is calculated as

$$\begin{aligned} d\sigma^2 &= (h^2-\rho^2)d\varphi^2 + 2h\frac{d\lambda}{d\rho}d\rho d\varphi + \left(1+\left(\frac{d\lambda}{d\rho}\right)^2\right)d\rho^2 \\ &= (h^2-\rho^2)\left(d\varphi + \frac{hd\rho}{h^2-\rho^2}\frac{d\lambda}{d\rho}\right)^2 + \left(1-\frac{\rho^2}{h^2-\rho^2}\left(\frac{d\lambda}{d\rho}\right)^2\right)d\rho^2. \end{aligned}$$

By the assumption that $f(\rho,\varphi)$ is spacelike, $h^2-\rho^2>0$ holds. Setting $(s,t)$ as

$$ds = \sqrt{1-\frac{\rho^2}{h^2-\rho^2}\left(\frac{d\lambda}{d\rho}\right)^2}\,d\rho, \qquad dt = d\varphi + \frac{h}{h^2-\rho^2}\frac{d\lambda}{d\rho}d\rho, \tag{3.5}$$

and $U(s) = \sqrt{h^2-\rho(s)^2}$, we have

$$d\sigma^2 = ds^2 + U(s)^2dt^2. \tag{3.6}$$

For the cases of (3.1), (3.2) and (3.4), similar arguments were applied in [**30**, Lemma 5.1]. And hence we have the following.

LEMMA 3.1. *For a spacelike helicoidal surface $f(\rho,\varphi)$ defined by either* (3.1), (3.2), (3.3) *or* (3.4), *there exists a coordinate change $(s,t) = (s(\rho),\, t(\rho,\varphi))$ and a positive-valued function $U = U(s)$ such that the first fundamental form $d\sigma^2$ is given by the form* (3.6).

We call $(s,t)$ in Lemma 3.1 the *natural coordinate system* of the spacelike helicoidal surfaces $f(\rho,\varphi)$ given by (3.1), (3.2), (3.3) and (3.4).

Now, we shall consider the converse. That is, for a given positive-valued function $U(s)$, we find a spacelike helicoidal surface whose first fundamental form $d\sigma^2$ is given by (3.6). Let us consider the case of (3.3), since the cases of (3.1), (3.2) and (3.4) are treated by Sasahara in [**30**, Lemma 5.1]. Assume there exists a local coordinate change $\alpha(s,t) = (\rho(s,t),\, \varphi(s,t))$ such that the first fundamental form $d\sigma^2$ of $\bar{f}(s,t) := f\circ\alpha(s,t)$ is given by (3.6), where $f(\rho,\varphi)$ is as in (3.3). By (3.5), it suffices to find functions $\rho(s,t), \lambda(s,t), \varphi(s,t)$ and a constant $h$ such that

$$d\rho^2 = ds^2 + \frac{\rho^2}{h^2-\rho^2}d\lambda^2, \tag{3.7}$$

$$U dt = \pm\sqrt{h^2-\rho^2}\left(d\varphi + \frac{h}{h^2-\rho^2}d\lambda\right). \tag{3.8}$$

By (3.7), we have $\rho(s,t) = \rho(s), \lambda(s,t) = \lambda(s)$. By (3.8),

$$\frac{\partial\varphi}{\partial s} = \frac{-h}{h^2-\rho(s)^2}\frac{d\lambda(s)}{ds}, \qquad \frac{\partial\varphi}{\partial t} = \pm\frac{U(s)}{\sqrt{h^2-\rho(s)^2}} \tag{3.9}$$

holds. This yields that $\dfrac{\partial^2\varphi}{\partial s\partial t} = 0$. Hence, by (3.9), there exists a constant $m$ such that

$$\frac{U(s)}{\sqrt{h^2-\rho(s)^2}} = \frac{1}{m}.$$

Therefore $\rho(s) = \sqrt{h^2-m^2U(s)^2}$ holds. By a calculation, we have $(d\rho/ds)^2 = m^4U(s)^2\dot{U}(s)^2/(h^2-m^2U(s)^2)$, where the dot means the derivative with respect to $s$. Substituting this into (3.7), we have

$$\frac{d\lambda}{ds} = mU(s)\frac{\sqrt{m^2U(s)^2(m^2\dot{U}(s)^2+1)-h^2}}{h^2-m^2U(s)^2}.$$

If we substitute this into (3.8),

$$d\varphi = \frac{dt}{m} - h\frac{\sqrt{m^2U(s)^2(m^2\dot{U}(s)^2+1)-h^2}}{mU(s)(h^2-m^2U(s)^2)}ds$$

holds. Therefore, we have the following.

PROPOSITION 3.2. For given constants $m\,(\neq 0)$, $h$ and a positive-valued function $U(s)$, let $\bar{f}(s,t)$ be the helicoidal surface defined as one of the following:

$$\bar{f}(s,t) = (\lambda + h\,\varphi,\ \rho\cos\varphi,\ \rho\sin\varphi) \tag{3.10}$$

$$\left(\begin{aligned} &\lambda(s) = \int mU(s)\Phi(s)ds, \quad \varphi(s,t) = \frac{t}{m} + h\int\frac{\Phi(s)}{mU(s)}ds, \\ &\rho(s) = \sqrt{m^2U(s)^2+h^2}, \quad \Phi(s) := \frac{\sqrt{m^2U(s)^2(m^2\dot{U}(s)^2-1)-h^2}}{m^2U(s)^2+h^2} \end{aligned}\right),$$

$$\bar{f}(s,t) = (\rho \cosh \varphi,\ \rho \sinh \varphi,\ \lambda + h\,\varphi)\,, \tag{3.11}$$

$$\left( \begin{array}{ll} \lambda(s) = \displaystyle\int mU(s)\Phi(s)ds, & \varphi(s,t) = \dfrac{t}{m} - h \displaystyle\int \dfrac{\Phi(s)}{mU(s)}ds, \\ \rho(s) = \sqrt{m^2U(s)^2 - h^2}, & \Phi(s) := \dfrac{\sqrt{m^2U(s)^2(m^2\dot{U}(s)^2+1) - h^2}}{m^2U(s)^2 - h^2} \end{array} \right),$$

$$\bar{f}(s,t) = (\rho \sinh \varphi,\ \rho \cosh \varphi,\ \lambda + h\,\varphi)\,, \tag{3.12}$$

$$\left( \begin{array}{ll} \lambda(s) = \displaystyle\int mU(s)\Phi(s)ds, & \varphi(s,t) = \dfrac{t}{m} - h \displaystyle\int \dfrac{\Phi(s)}{mU(s)}ds, \\ \rho(s) = \sqrt{h^2 - m^2U(s)^2}, & \Phi(s) := \dfrac{\sqrt{m^2U(s)^2(m^2\dot{U}(s)^2+1) - h^2}}{h^2 - m^2U(s)^2} \end{array} \right),$$

$$\bar{f}(s,t) = \left(\lambda - (\varphi^2+1)\rho,\ -2\rho\varphi,\ \lambda - (\varphi^2-1)\rho\right) + h\left(\frac{\varphi^3}{3} + \varphi, \varphi^2, \frac{\varphi^3}{3} - \varphi\right), \tag{3.13}$$

$$\left( \begin{array}{ll} \lambda(s) = \displaystyle\int mU(s)\dfrac{\Phi(s)}{4h^2}ds, & \varphi(s,t) = \dfrac{t}{m} + \displaystyle\int \dfrac{\Phi(s)}{2hmU(s)}ds, \\ \rho(s) = \dfrac{1}{2}mU(s), & \Phi(s) := m^2U(s)\dot{U}(s) + \sqrt{m^4U(s)^2\dot{U}(s)^2 - 4h^2}, \end{array} \right).$$

Then, the first fundamental form $d\sigma^2$ of $\bar{f}(s,t)$ is given by (3.6).

We denote by $[U,m,h]_T$ (resp. $[U,m,h]_{S_1}$, $[U,m,h]_{S_2}$, $[U,m,h]_L$) the spacelike helicoidal surface $\bar{f}(s,t)$ defined by (3.10) (resp. (3.11), (3.12), (3.13)).

Now, we shall find the condition under which $[U,m,h]_{S_2}$ is spacelike CMC. By (3.3), the second fundamental form $II$ of $[U,m,h]_{S_2}$ is calculated as

$$II = -\left(\rho\frac{d^2\lambda}{d\rho^2}d\rho^2 - 2h\,d\rho\,d\varphi - \rho^2\frac{d\lambda}{d\rho}d\varphi^2\right) \Big/ \sqrt{h^2 - \rho^2\left(1 + \left(\frac{d\lambda}{d\rho}\right)^2\right)}\,.$$

Substituting (3.12), we have

$$II = -\frac{m^4U(s)^3\ddot{U}(s) + h^2}{m^2U(s)^2\sqrt{\Delta}}ds^2 - \frac{2h}{m^2U(s)}ds\,dt - \frac{\sqrt{\Delta}}{m^2}dt^2, \tag{3.14}$$

and hence the mean curvature $H$ is calculated as

$$H = -\frac{m^2\left(U(s)\ddot{U}(s) + \dot{U}(s)^2\right) + 1}{2\sqrt{\Delta}}, \tag{3.15}$$

where

$$\Delta := m^2U(s)^2(m^2\dot{U}(s)^2 + 1) - h^2. \tag{3.16}$$

From now on, we assume $H$ is constant. If we put $\mu := mU(s)$, we have

$$1 + \mu\ddot{\mu} + \dot{\mu}^2 = -2H\sqrt{\mu^2(\dot{\mu}^2+1) - h^2}. \tag{3.17}$$

Setting $\beta := \sqrt{\mu^2(\dot{\mu}^2+1) - h^2}$, we have

$$2\beta\dot{\beta} = 2\mu\dot{\mu}(\dot{\mu}^2+1) + \mu^2(2\dot{\mu}\ddot{\mu}) = 2\mu\dot{\mu}(1 + \mu\ddot{\mu} + \dot{\mu}^2).$$

By (3.17), $\dot\beta = -2H\mu\dot\mu$. Therefore there exists a constant $a \in \boldsymbol{R}$ such that $\beta = -H\mu^2 + a$. Together with the definition of $\beta$, we have

$$\mu^2\dot\mu^2 = (H\mu^2 - a)^2 - \mu^2 + h^2. \tag{3.18}$$

Thus, we have the following.

LEMMA 3.3. *The mean curvature $H$ of $[U,m,h]_T$ (resp. $[U,m,h]_{S_1}$, $[U,m,h]_{S_2}$, $[U,m,h]_L$) is constant if and only if there exists a constant $a$ such that $U(s)$ satisfies*

$$m^4U(s)^2\dot U(s)^2 = H^2m^4U(s)^4 + (2aH+1)m^2U(s)^2 + a^2 + h^2, \tag{3.19}$$

$$m^4U(s)^2\dot U(s)^2 = H^2m^4U(s)^4 - (2aH+1)m^2U(s)^2 + a^2 + h^2, \tag{3.20}$$

$$m^4U(s)^2\dot U(s)^2 = H^2m^4U(s)^4 - (2aH+1)m^2U(s)^2 + a^2 + h^2, \tag{3.21}$$

$$m^4U(s)^2\dot U(s)^2 = (Hm^2U(s)^2 + a)^2 + 4h^2, \tag{3.22}$$

*respectively. Moreover, if $(x,y)$ is an isothermal coordinate system as in Lemma 2.7, the Hopf differential $Q = q\,dz^2$ $(z = x+iy)$ of $[U,m,h]_T$ (resp. $[U,m,h]_{S_1}$, $[U,m,h]_{S_2}$, $[U,m,h]_L$) is*

$$q = \frac{-a+ih}{2m^2} \qquad \left(\textit{resp.}\ \frac{a-ih}{2m^2},\quad \frac{a+ih}{2m^2},\quad \frac{-a+2ih}{2m^2}\right). \tag{3.23}$$

PROOF. Substituting $\mu = mU(s)$ into (3.18), we have (3.21). Similar arguments show (3.19), (3.20) and (3.22). Hence we shall prove the latter part. By (3.14), the second fundamental form $II$ of $[U,m,h]_{S_2}$ is represented by

$$II = -\frac{m^4U(s)^3\ddot U(s) + h^2}{m^2\sqrt{\Delta}}dx^2 - \frac{2h}{m^2}dx\,dy - \frac{\sqrt{\Delta}}{m^2}dy^2$$

in terms of the isothermal coordinate system $(x,y)$, where $\Delta$ is as in (3.16). Therefore the Hopf differential $q$ satisfies

$$4q = \frac{\Delta - h^2 - m^4U(s)^3\ddot U(s)}{m^2\sqrt{\Delta}} + \frac{2ih}{m^2}. \tag{3.24}$$

By (3.15), we have

$$-m^4U(s)^3\ddot U(s) = 2Hm^2U(s)^2\sqrt{\Delta} + m^2U(s)^2(1 + m^2\dot U(s)^2). \tag{3.25}$$

Since $\Delta$ is defined by (3.16),

$$m^2U(s)^2(1 + m^2\dot U(s)^2) = \Delta + h^2 \tag{3.26}$$

holds. Substituting (3.26) into (3.25), we have

$$-m^4U(s)^3\ddot U(s) = 2Hm^2U(s)^2\sqrt{\Delta} + \Delta + h^2. \tag{3.27}$$

On the other hand, (3.21) yields that

$$\Delta = m^4U(s)^2\dot U(s)^2 + m^2U(s)^2 - h^2 = (Hm^2U(s)^2 - a)^2. \tag{3.28}$$

By (3.24), (3.27) and (3.28), it holds that

$$\begin{aligned} 4q &= \frac{\Delta - h^2 + 2Hm^2U(s)^2\sqrt{\Delta} + \Delta + h^2}{m^2\sqrt{\Delta}} + \frac{2ih}{m^2} \\ &= \frac{2Hm^2U(s)^2 + 2\sqrt{\Delta}}{m^2} + \frac{2ih}{m^2} = 2\frac{a+ih}{m^2}. \end{aligned}$$

Similar proofs give (3.23). □

Let $f_j(s,t)$ ($j = \mathfrak{t}$, $\mathfrak{sc}$, $\mathfrak{si}$, $\mathfrak{cyl}$, $\mathfrak{lt}$, $\mathfrak{lsc}$, $\mathfrak{lsi}$, $\mathfrak{tu}$) be the spacelike Delaunay surfaces as in Facts 2.2, 2.3 and 2.4. In the rest of this section, applying Lemma 3.3, we shall find the associate family $\{f_{j,\theta}(s,t)\}_{\theta\in S^1}$ of $f_j(s,t)$ for each $j$. We remark that $f_{j,\theta}(s,t)$ is the spacelike CMC surface satisfying

$$d\sigma_{j,\theta}^2 = d\sigma_j^2, \qquad Q_{j,\theta} = e^{i\theta}Q_j$$

for each $\theta \in S^1$, where $d\sigma_{j,\theta}^2$ (resp. $d\sigma_j^2$) is the first fundamental form and $Q_{j,\theta}$ (resp. $Q_j$) is the Hopf differential of $f_{j,\theta}(s,t)$ (resp. $f_j(s,t)$). By Remark 2.1, it is enough to find spacelike CMC surfaces which are isometric to $f_j(s,t)$. Moreover, by Proposition 3.2, we can construct spacelike helicoidal surfaces whose first fundamental forms are given by (3.6).

We remark that it suffices to deal with the associate families of $f_{\mathfrak{t}}$, $f_{\mathfrak{sc}}$ and $f_{\mathfrak{si}}$ for $k \neq 1$, because of the following reasons: If $k = 1$, $f_{\mathfrak{t}}$, $f_{\mathfrak{sc}}$, $f_{\mathfrak{si}}$ and $f_{\mathfrak{tu}}$ are totally umbilic. If $k = 0$, $f_{\mathfrak{lt}}$ (resp. $f_{\mathfrak{lsc}}$, $f_{\mathfrak{lsi}}$) and $f_{\mathfrak{t}}$ (resp. $f_{\mathfrak{sc}}$, $f_{\mathfrak{si}}$) are isometric (see Remark 2.6). Moreover, each element of the associate family $\{f_{\mathfrak{cyl},\theta}\}_{\theta\in S^1}$ is congruent to $f_{\mathfrak{cyl}}(s,t)$. In fact, since $d\sigma_{\mathfrak{cyl}}^2 = dz\,d\bar{z}$ and $Q_{\mathfrak{cyl}} = -(H/2)dz^2$ by Remark 2.6 and Lemma 2.7, $f_{\mathfrak{cyl},\theta}(s,t)$ satisfies $d\sigma_{\mathfrak{cyl},\theta}^2 = dz\,d\bar{z}$ and $Q_{\mathfrak{cyl},\theta} = -(H/2)e^{i\theta}dz^2$, where $z = s+it$. Hence, setting $w := e^{i\theta/2}z$, we have $d\sigma_{\mathfrak{cyl},\theta}^2 = dw\,d\bar{w}$ and $Q_{\mathfrak{cyl},\theta} = -(H/2)dw^2$, which implies that $f_{\mathfrak{cyl},\theta}$ is congruent to $f_{\mathfrak{cyl}}$ for all $\theta$.

In the following Theorems 3.4, 3.6 and 3.7, we identify $S^1$ with the interval $(-\pi,\pi]$.

THEOREM 3.4 (Associate family of spacelike $t$-Delaunay surfaces). *Let $k$ $(\neq 1)$ be a constant and $U_{\mathfrak{t}}(s) = \sqrt{e^{2Hs}+ke^{-2Hs}-k-1}$. Then, the associate family $\{f_{\mathfrak{t},\theta}\}_{\theta\in S^1}$ of the spacelike t-Delaunay surface $f_{\mathfrak{t}}(s,t)$ is given by*

$$f_{\mathfrak{t},\theta}(s,t) = \begin{cases} \left[U_{\mathfrak{t}}(s),\ \dfrac{1}{\sqrt{2}H\sqrt{C_{k,\theta}}},\ \dfrac{(1-k)\sin\theta}{2HC_{k,\theta}}\right]_T & \left(\text{if } \begin{array}{l} k>0, \\ k=0 \text{ and } \theta\neq\pi, \\ k<0 \text{ and } |\theta|<\Theta_k \end{array}\right) \\ f_{\mathfrak{lt}}(s,t) & (\text{if } k=0 \text{ and } \theta=\pi) \\ \left[U_{\mathfrak{t}}(s),\ 1,\ \pm H\sqrt{-k}\right]_L & (\text{if } k<0 \text{ and } \theta=\pm\Theta_k) \\ \left[U_{\mathfrak{t}}(s),\ \dfrac{1}{\sqrt{2}H\sqrt{-C_{k,\theta}}},\ \dfrac{(k-1)\sin\theta}{2HC_{k,\theta}}\right]_{S_2} & (\text{if } k<0 \text{ and } |\theta|>\Theta_k), \end{cases}$$

*where we set $C_{k,\theta} = 1+k+(1-k)\cos\theta$ and $\Theta_k = 2\tan^{-1}\dfrac{1}{\sqrt{-k}}$.*

To prove Theorem 3.4, we shall prepare the following lemma.

LEMMA 3.5. *Let $C_{k,\theta}$ and $\Theta_k$ be the constants defined in Theorem 3.4. Then, $C_{k,\theta} > 0$ if and only if $k > 0$, $k = 0$ and $\theta \neq \pi$, or $k < 0$ and $|\theta| < \Theta_k$. Moreover, $C_{k,\theta} < 0$ (resp. $C_{k,\theta} = 0$) if and only if $k < 0$ and $|\theta| > \Theta_k$ (resp. $k = 0$ and $\theta = \pi$, or $k < 0$ and $\theta = \pm\Theta_k$). In particular, if $k < 0$ and $\theta = \pm\Theta_k$, then $(1-k)\sin\theta = \pm 2\sqrt{-k}$ holds.*

PROOF. By $(1+k)^2-(1-k)^2 = 4k$, $C_{k,\theta} = 1+k+(1-k)\cos\theta > 0$ for all $\theta$ if and only if $k > 0$. If $k = 0$, $C_{k,\theta} = 1+\cos\theta > 0$ if and only if $\theta \neq \pi$. If $k < 0$, set $c = \sqrt{-k}$. Then

$$\cos\theta = \frac{1-\tan^2\frac{\theta}{2}}{1+\tan^2\frac{\theta}{2}} \quad \text{and} \quad \frac{k+1}{k-1} = \frac{c^2-1}{c^2+1} = \frac{1-\frac{1}{c^2}}{1+\frac{1}{c^2}}$$

yield that $C_{k,\theta} = 0$ if and only if $\theta = \pm\Theta_k$. In this case,

$$\sin\theta = \frac{2\tan\frac{\theta}{2}}{1+\tan^2\frac{\theta}{2}} = \frac{\pm 2\sqrt{-k}}{1-k}$$

shows the last assertion. □

PROOF OF THEOREM 3.4. Since $[U_{\mathfrak{t}}(s), m, h]_\alpha$ ($\alpha = T$, $S_1$, $S_2$, $L$) is isometric to the spacelike $t$-Delaunay surface $f_{\mathfrak{t}}(s,t)$, it suffices to find the appropriate constants $m$ and $h$. By (2.13) in Lemma 2.8, the Hopf differential $Q_{\mathfrak{t}} = q_{\mathfrak{t}}\,dz^2$ of $f_{\mathfrak{t}}(s,t)$ is given by $q_{\mathfrak{t}} = H(1-k)/2$.

In the case of $C_{k,\theta} > 0$, let us consider $[U(s), m, h]_T$. By (3.23) in Lemma 3.3, the Hopf differential $Q = q\,dz^2$ of $[U(s), m, h]_T$ is given by $q = (-a+ih)/(2m^2)$. If $q = q_{\mathfrak{t}}e^{i\theta}$ holds, then we have

$$a = -m^2H(1-k)\cos\theta, \qquad h = m^2H(1-k)\sin\theta.$$

Substituting these and $U_{\mathfrak{t}}(s)$ into (3.19) in Lemma 3.3, we have $m^2 = 1/(2H^2C_{k,\theta})$.

In the case of $C_{k,\theta} < 0$, let us consider $[U(s), m, h]_{S_2}$. By (3.23) in Lemma 3.3, the Hopf differential $Q = q\,dz^2$ of $[U(s), m, h]_{S_2}$ is given by $q = (a+ih)/(2m^2)$. If $q = q_{\mathfrak{t}}e^{i\theta}$ holds, we have

$$a = m^2H(1-k)\cos\theta, \qquad h = m^2H(1-k)\sin\theta.$$

Substituting these and $U_{\mathfrak{t}}(s)$ into (3.21) in Lemma 3.3, we have $m^2 = -1/(2H^2C_{k,\theta})$.

In the case of $C_{k,\theta} = 0$ and $k = 0$, then $\theta = \pi$ and the spacelike $\ell$-Delaunay surface $f_{\mathfrak{lt}}(s,t)$ is the desired associate family of $f_{\mathfrak{t}}(s,t)$.

In the case of $C_{k,\theta} = 0$ and $k \neq 0$, we have $\theta = \pm\Theta_k$ by Lemma 3.5. Let us consider $[U(s), m, h]_L$. By (3.23) in Lemma 3.3, the Hopf differential $Q = q\,dz^2$ of $[U(s), m, h]_L$ is given by $q = (-a+2ih)/(2m^2)$. If $q = q_{\mathfrak{t}}e^{i\theta}$ holds, we have

$$a = -m^2H(1-k)\cos\theta = m^2H(1+k),$$
$$h = \frac{1}{2}m^2H(1-k)\sin\theta = \pm m^2H\sqrt{-k}$$

by Lemma 3.5. In this case, (3.22) in Lemma 3.3 holds for any $m$. Hence setting $m = 1$, we have the results. □

The proofs similar to that of Theorem 3.4 yield Theorems 3.6 and 3.7.

THEOREM 3.6 (Associate family of complete spacelike $s$-Delaunay surfaces). *Let $k$ ($\neq 1$) be a non-negative constant and $U_{\mathfrak{sc}}(s) = \sqrt{1+k+e^{2Hs}+ke^{-2Hs}}$. Then, the associate family $\{f_{\mathfrak{sc},\theta}\}_{\theta\in S^1}$ of the spacelike $s$-Delaunay surface $f_{\mathfrak{sc}}(s,t)$ is given by*

$$f_{\mathfrak{sc},\theta}(s,t) = \begin{cases} \left[U_{\mathfrak{sc}}(s), \dfrac{1}{\sqrt{2}H\sqrt{C_{k,\theta}}}, \dfrac{(1-k)\sin\theta}{2HC_{k,\theta}}\right]_{S_1} & \left(\textit{if } \begin{matrix} k>0, \\ k=0 \textit{ and } \theta\neq\pi \end{matrix}\right) \\ f_{\mathfrak{lsc}}(s,t) & (\textit{if } k=0 \textit{ and } \theta=\pi), \end{cases}$$

*where we set $C_{k,\theta} = 1+k+(1-k)\cos\theta$.*

THEOREM 3.7 (Associate family of incomplete spacelike $s$-Delaunay surfaces). *Let $k$ ($\neq 1$) be a constant and $U_{\mathfrak{si}}(s) = \sqrt{1+k-e^{2Hs}-ke^{-2Hs}}$. Then, the associate family $\{f_{\mathfrak{si},\theta}\}_{\theta\in S^1}$ of the spacelike $s$-Delaunay surface $f_{\mathfrak{si}}(s,t)$ is given by*

$$f_{\mathfrak{si},\theta}(s,t) = \begin{cases} \left[U_{\mathfrak{si}}(s),\ \dfrac{1}{\sqrt{2H\sqrt{C_{k,\theta}}}},\ \dfrac{(k-1)\sin\theta}{2HC_{k,\theta}}\right]_{S_2} & \left(\text{if } \begin{array}{l} k>0, \\ k=0 \text{ and } \theta\neq\pi, \\ k<0 \text{ and } |\theta|<\Theta_k \end{array}\right) \\ f_{\mathfrak{lsi}}(s,t) & (\text{if } k=0 \text{ and } \theta=\pi) \\ \left[U_{\mathfrak{si}}(s),\ 1,\ \mp H\sqrt{-k}\right]_L & (\text{if } k<0 \text{ and } \theta=\pm\Theta_k) \\ \left[U_{\mathfrak{si}}(s),\ \dfrac{1}{\sqrt{2H\sqrt{-C_{k,\theta}}}},\ \dfrac{(1-k)\sin\theta}{2HC_{k,\theta}}\right]_T & (\text{if } k<0 \text{ and } |\theta|>\Theta_k), \end{cases}$$

*where we set $C_{k,\theta} = 1+k+(1-k)\cos\theta$ and $\Theta_k = 2\tan^{-1}\dfrac{1}{\sqrt{-k}}$.*

## 4. Singularities of associate families

The associate families of spacelike Delaunay surfaces are given in Theorems 3.4, 3.6 and 3.7. Although each element of the associate family in Theorem 3.6 is a complete immersion, elements in the associate families in Theorems 3.4 and 3.7 have singularities. In this section, we specify the types of such singular points.

**4.1. Surfaces with singularities.** Let $f = f(u,v) : V^2 \to \boldsymbol{R}^3$ a $C^\infty$-map, where $V^2$ is a domain of $\boldsymbol{R}^2$. If $f$ is not an immersion at a point $p \in V^2$, $p$ is called *singular*. We denote by $S(f)$ the set of singular points of $f$. Then a singular point $p$ is called *cuspidal edge* (resp. *5/2-cuspidal edge*, *conelike singularity*) if there exist local diffeomorphisms $g : V^2 \to \boldsymbol{R}^2$ and $G : \boldsymbol{R}^3 \to \boldsymbol{R}^3$ such that $g(p) = (0,0)$, $G(f(p)) = (0,0,0)$ and $(G\circ f\circ g^{-1})(u,v) = (u,v^2,v^3)$ (resp. $(u,v^2,v^5)$, $(u, u\cos v, u\sin v)$).

We denote by $S^2$ the unit sphere $S^2 := \{\boldsymbol{x} \in \boldsymbol{R}^3\,;\, |\boldsymbol{x}| = 1\}$, where $|\boldsymbol{x}| = \sqrt{\boldsymbol{x}\cdot\boldsymbol{x}}$ and the dot '$\cdot$' is the standard Euclidean inner product. A smooth map $f : V^2 \to \boldsymbol{R}^3$ is called a *frontal*, if for any point $p \in V^2$, there exist a neighborhood $V_p^2$ of $p$ and a smooth map $\boldsymbol{n} : V_p^2 \to S^2$ (called a *Euclidean unit normal vector field*) such that $df_q(\boldsymbol{v})\cdot\boldsymbol{n}(q) = 0$ for each $q \in V_p^2$, $\boldsymbol{v} \in T_qV^2$. In addition, if $(f,\boldsymbol{n}) : V_p^2 \to \boldsymbol{R}^3 \times S^2$ gives an immersion, $f$ is called a *wave front* or a *front*. The function $\Lambda := \det(f_u, f_v, \boldsymbol{n})$ is called *signed area density function* with respect to the coordinate system $(u,v)$. A point $p \in V^2$ is a singular point of $f$ if and only if $\Lambda(p) = 0$. If $d\Lambda(p) \neq 0$, a singular point $p$ is called *non-degenerate*. Note that $\operatorname{rank}(df_p) = 1$ holds for a non-degenerate singular point $p$. By the implicit function theorem, there exists a regular curve $\gamma(t)$ ($|t| < \varepsilon$) on the $uv$-plane such that $\gamma(0) = p$ and the image of $\gamma$ is included in the singular point set $S(f)$, where $\varepsilon > 0$. We call $\gamma(t)$ the *singular curve* and $d\gamma/dt$ the *singular direction*. In this setting, there exists a smooth vector field $\eta(t)$ along $\gamma(t)$ such that $\eta(t)$ is a null vector (i.e., $df(\eta(t)) = \boldsymbol{0}$) for each $t$, where $\boldsymbol{0} = (0,0,0)$. Such a vector field $\eta(t)$ is called a *null vector field*. Remark that $f$ is a front at $p = \gamma(0)$ if and only if

$$(d\boldsymbol{n})_p(\eta) \neq \boldsymbol{0}. \tag{4.1}$$

Kokubu-Rossman-Saji-Umehara-Yamada [**29**] gave the following criterion.

FACT 4.1 ([**29**, Proposition 1.3]). Let $p = \gamma(0) \in V^2$ be a non-degenerate singular point of a front $f : V^2 \to \boldsymbol{R}^3$. Then, $p$ is a cuspidal edge if and only if the null vector field $\eta$ and $d\gamma/dt$ are linearly independent at $t = 0$.

**4.2. Extensions of spacelike Delaunay surfaces.** An incomplete spacelike Delaunay surface is congruent to either the $t$-Delaunay surface $f_{\mathfrak{t}}(s,t)$ defined by (2.3) with $k \neq 1$, the $s$-Delaunay surface $f_{\mathfrak{si}}(s,t)$ defined by (2.6) with $k \neq 1$, the $\ell$-Delaunay surfaces $f_{\mathfrak{lt}}(s,t)$ defined by (2.9) or $f_{\mathfrak{lsi}}(s,t)$ defined by (2.11). From now on, assume $k \neq 1$. Setting $r = U_k(s)$, these are represented as

$$(4.2)\qquad \hat{f}_{\mathfrak{t}}(r,t) = \frac{1}{2H}\left(\int_0^r \frac{\zeta^2 + k - 1}{\sqrt{(\zeta^2+k+1)^2 - 4k}}d\zeta,\ r\cos 2Ht,\ r\sin 2Ht\right),$$

$$(4.3)\qquad \hat{f}_{\mathfrak{si}}(r,t) = \frac{1}{2H}\left(r\cosh 2Ht,\ r\sinh 2Ht,\ \int_0^r \frac{\zeta^2 - k + 1}{\sqrt{(\zeta^2-k-1)^2 - 4k}}d\zeta\right),$$

$$(4.4)\qquad \hat{f}_j(r,t) = \left(\phi_j(r) + r + \frac{1}{4}t^2 r,\ t\,r,\ \phi_j(r) - r + \frac{1}{4}t^2 r\right) \quad (j = \mathfrak{lt},\ \mathfrak{lsi}),$$

respectively, where

$$\phi_{\mathfrak{lt}}(r) = \frac{1}{8H^2}\left(\frac{r}{1+r^2} - \tan^{-1} r\right), \quad \phi_{\mathfrak{lsi}}(r) = \frac{1}{8H^2}\left(\frac{r}{1-r^2} - \tanh^{-1} r\right).$$

However, as in Theorems 3.4 and 3.7, $\hat{f}_{\mathfrak{lt}}$ (resp. $\hat{f}_{\mathfrak{lsi}}$) is included in the associate family of $\hat{f}_{\mathfrak{t}}$ (resp. $\hat{f}_{\mathfrak{si}}$) with $k = 0$. Thus, it suffices to consider the associate families of $\hat{f}_{\mathfrak{t}}$ and $\hat{f}_{\mathfrak{si}}$.

**4.3. Extensions of associate families.** By Theorem 3.4, the associate family $f_{\mathfrak{t},\theta}(s,t)$ of the spacelike $t$-Delaunay surface $f_{\mathfrak{t}}(s,t)$ is either $[U_{\mathfrak{t}}, m, h]_T$, $f_{\mathfrak{lt}}$, $[U_{\mathfrak{t}}, m, h]_L$ or $[U_{\mathfrak{t}}, m, h]_{S_2}$, where $m$, $h$ are constants defined in Theorem 3.4. As in the case of spacelike $t$-Delaunay surfaces in Section 4.2, we can define the extension of $f_{\mathfrak{t},\theta}(s,t)$ by putting $r := U_{\mathfrak{t}}(s)$. Explicitly, using Theorem 3.4 and Proposition 3.2, the extension of the associate family of $\hat{f}_{\mathfrak{t}}(r,t)$ is given by

$$\hat{f}_{\mathfrak{t},\theta}(r,t) = \begin{cases} (\hat{\lambda} + h\hat{\varphi}, \hat{\rho}\cos\hat{\varphi}, \hat{\rho}\sin\hat{\varphi}) & (\text{if } C_{k,\theta} > 0) \\ \hat{f}_{\mathfrak{lt}}(r,t) & (\text{if } C_{k,\theta} = 0 \text{ and } k = 0) \\ (\hat{\lambda} - \hat{\rho} - \hat{\rho}\hat{\varphi}^2, -2\hat{\rho}\hat{\varphi}, \hat{\lambda} + \hat{\rho} - \hat{\rho}\hat{\varphi}^2) & \\ \qquad + h\left(\dfrac{\hat{\varphi}^3}{3} + \hat{\varphi}, \hat{\varphi}^2, \dfrac{\hat{\varphi}^3}{3} - \hat{\varphi}\right) & (\text{if } C_{k,\theta} = 0 \text{ and } k < 0) \\ (\hat{\rho}\sinh\hat{\varphi}, \hat{\rho}\cosh\hat{\varphi}, \hat{\lambda} + h\hat{\varphi}) & (\text{if } C_{k,\theta} < 0), \end{cases}$$

where $\hat{\rho} = \hat{\rho}(r)$, $\hat{\lambda} = \hat{\lambda}(r)$, $\hat{\varphi} = \hat{\varphi}(r,t)$ and $h$ are as follows: if $C_{k,\theta} \neq 0$,

$$\hat{\rho}(r) = \frac{\sqrt{\psi(r)}}{2H|C_{k,\theta}|}, \qquad \hat{\lambda}(r) = \int_0^r \frac{\zeta^2\Psi(\zeta)}{H}d\zeta,$$

$$\hat{\varphi}(r,t) = H\sqrt{|2C_{k,\theta}|}t + \operatorname{sgn}(C_{k,\theta})(1-k)\sin\theta \int_0^r \Psi(\zeta)d\zeta,$$

and $h = (1-k)\sin\theta/(2H|C_{k,\theta}|)$; if $C_{k,\theta} = 0$ and $k < 0$,

$$\hat{\rho}(r) = \frac{r}{2}, \qquad \hat{\lambda}(r) = -\int_0^r \frac{\zeta^2\left(\zeta^2+k+1+\sqrt{\delta_{\mathrm{t}}(\zeta)}\right)}{4kH^2\sqrt{\delta_{\mathrm{t}}(\zeta)}}d\zeta,$$

$$\hat{\varphi}(r,t) = t \pm \int_0^r \frac{\zeta^2+k+1+\sqrt{\delta_{\mathrm{t}}(\zeta)}}{2H\sqrt{-k}\sqrt{\delta_{\mathrm{t}}(\zeta)}}d\zeta \quad \text{and} \quad h = \pm H\sqrt{-k},$$

where $C_{k,\theta}$ is the constant as in Theorem 3.4 and

$$\delta_{\mathrm{t}}(r) := (r^2+k+1)^2 - 4k, \qquad \Psi(r) := \frac{\sqrt{|2C_{k,\theta}|}(r^2+(k-1)\cos\theta)}{\psi(r)\sqrt{\delta_{\mathrm{t}}(r)}},$$

$$\psi(r) := 2C_{k,\theta}r^2 + (1-k)^2\sin^2\theta.$$

LEMMA 4.2. *The extended associate family $\hat{f}_{\mathrm{t},\theta}(r,t)$ of the spacelike t-Delaunay surface $\hat{f}_{\mathrm{t}}(r,t)$ is a frontal and its singular points are non-degenerate. Moreover, these singular points are cuspidal edges if and only if $\theta \neq 0, \pm\pi/2, \pi$, where we regard $S^1$ as the interval $(-\pi,\pi]$.*

PROOF. First, we consider the case of $C_{k,\theta} > 0$. The Euclidean unit normal vector field $\boldsymbol{n}_T$ of $\hat{f}_{\mathrm{t},\theta}(r,t)$ is given by

$$\begin{aligned}\boldsymbol{n}_T(r,t) = &\frac{1}{\sqrt{2}\sqrt{\psi(r)}\sqrt{\delta_{\mathrm{t}}(r)-C_{k,\theta}r^2}}\Big(\sqrt{\delta_{\mathrm{t}}(r)}\sqrt{\psi(r)},\\ &-(k-1)\sin\theta\sqrt{\delta_{\mathrm{t}}(r)}\sin\hat{\varphi}(r,t) - r\sqrt{2C_{k,\theta}}\,p(r)\cos\hat{\varphi}(r,t),\\ &(k-1)\sin\theta\sqrt{\delta_{\mathrm{t}}(r)}\cos\hat{\varphi}(r,t) - r\sqrt{2C_{k,\theta}}\,p(r)\sin\hat{\varphi}(r,t)\Big),\end{aligned}$$

where $p(r) = r^2 + (k-1)\cos\theta$. Since the signed area density function $\Lambda(r,t)$ is

$$\Lambda(r,t) = \det\left((\hat{f}_{\mathrm{t},\theta})_r, (\hat{f}_{\mathrm{t},\theta})_t, \boldsymbol{n}_T\right) = \frac{r\sqrt{\delta_{\mathrm{t}}(r)-C_{k,\theta}r^2}}{H\sqrt{C_{k,\theta}\delta_{\mathrm{t}}(r)}},$$

the singular point set is $\{(r,t)\,;\,\Lambda(r,t) = 0\} = \{(0,t)\}$. Since the derivative $d\Lambda_{(0,t)} = dr/(H\sqrt{C_{k,\theta}})$ does not vanish, the singular points are non-degenerate. The singular curve is $\gamma(t) = (0,t)$ and the null vector field is

$$\eta = H|k-1|\sin\theta\,\partial_r + \cos\theta\,\partial_t,$$

where $\partial_r := \partial/\partial r$, $\partial_t := \partial/\partial t$. Then

$$\det\left(\frac{d}{dt}\gamma, \eta\right) = \begin{vmatrix} 0 & H|k-1|\sin\theta \\ 1 & \cos\theta \end{vmatrix} = -H|k-1|\sin\theta$$

yields that $d\gamma/dt$ and $\eta$ are linearly independent if $\theta \neq 0, \pi$. Hence, by Fact 4.1, the singular point $(0,t)$ is a cuspidal edge if and only if $\hat{f}_{\mathrm{t},\theta}$ is a front at $(0,t)$. Since

$$\left|(d\boldsymbol{n}_T)_{(0,t)}(\eta)\right|^2 = H^2C_{k,\theta}\cos^2\theta,$$

$\hat{f}_{\mathrm{t},\theta}$ is a front, that is, $(d\boldsymbol{n}_T)_{(0,t)}(\eta) \neq \boldsymbol{0}$ holds (cf. (4.1)) if and only if $\theta \neq \pm\pi/2$. A similar argument is valid for the case of $C_{k,\theta} < 0$.

In the case of $C_{k,\theta} = 0$ and $k = 0$, since $C_{0,\theta} = 1 + \cos\theta$, we have $\theta = \pi$. By calculating the unit normal, we can check that $\hat{f}_{\mathrm{t},\pi}(r,t) = \hat{f}_{\mathrm{lt}}(r,t)$ is a frontal and all the singular points are non-degenerate. Moreover, by a direct calculation, we have that the singular direction $d\gamma/dt$ and the null vector field $\eta$ are linearly dependent on the singular point set $S(\hat{f}_{\mathfrak{lt}})$, and hence Fact 4.1 yields that the singular points are not cuspidal edges.

Similarly, in the case of $C_{k,\theta} = 0$ and $k < 0$, by calculating the unit normal $\boldsymbol{n}_L$, we can check that $\hat{f}_{\mathfrak{t},\theta}(r,t)$ is a frontal and all the singular points are non-degenerate. By a direct calculation, we have that the singular curve $\gamma$ is given by $\gamma(t) = (0,t)$ and

$$\eta = -2H|k-1|\sqrt{-k}\partial_r - (k+1)\partial_t$$

gives the null vector field, which implies that $d\gamma/dt$ and $\eta$ are linearly independent. Hence, by Fact 4.1, the singular point $(0,t)$ is a cuspidal edge if and only if $\hat{f}_{\mathfrak{t},\theta}$ is a front at $(0,t)$. Since

$$\left|(d\boldsymbol{n}_L)_{(0,t)}(\eta)\right|^2 = \frac{2(k+1)^2}{(1+t^2)^2},$$

$\hat{f}_{\mathfrak{t},\theta}$ is a front, that is, $(d\boldsymbol{n}_L)_{(0,t)}(\eta) \neq \boldsymbol{0}$ holds (cf. (4.1)) if and only if $k \neq -1$. Therefore, $C_{-1,\theta} = 2\cos\theta$ implies that $\hat{f}_{\mathfrak{t},\theta}$ is a front at $(0,t)$ if and only if $\theta \neq \pm\pi/2$. We remark that, by the condition $C_{k,\theta} = 0$ and $k < 0$, we have $\theta \neq 0,\ \pi$ (cf. Lemma 3.5). Thus, we have the desired result. □

As in the case of spacelike $t$-Delaunay surfaces, by Theorem 3.7, the associate family $\{\hat{f}_{\mathfrak{si},\theta}(r,t)\}_{\theta\in S^1}$ of the extended $s$-Delaunay surface $\hat{f}_{\mathfrak{si}}(r,t)$ is given by

$$\hat{f}_{\mathfrak{si},\theta}(r,t) = \begin{cases} (\hat{\rho}\sinh\hat{\varphi}, \hat{\rho}\cosh\hat{\varphi}, \hat{\lambda} + h\hat{\varphi}) & (\text{if } C_{k,\theta} > 0) \\ \hat{f}_{\mathfrak{lsi}}(r,t) & (\text{if } C_{k,\theta} = 0 \text{ and } k = 0) \\ (\hat{\lambda} - \hat{\rho} - \hat{\rho}\hat{\varphi}^2, -2\hat{\rho}\hat{\varphi}, \hat{\lambda} + \hat{\rho} - \hat{\rho}\hat{\varphi}^2) & \\ \qquad +h\left(\dfrac{\hat{\varphi}^3}{3} + \hat{\varphi}, \hat{\varphi}^2, \dfrac{\hat{\varphi}^3}{3} - \hat{\varphi}\right) & (\text{if } C_{k,\theta} = 0 \text{ and } k < 0) \\ (\hat{\lambda} + h\hat{\varphi}, \hat{\rho}\cos\hat{\varphi}, \hat{\rho}\sin\hat{\varphi}) & (\text{if } C_{k,\theta} < 0), \end{cases}$$

where $\hat{\rho} = \hat{\rho}(r)$, $\hat{\lambda} = \hat{\lambda}(r)$, $\hat{\varphi} = \hat{\varphi}(r,t)$ and $h$ are as follows: if $C_{k,\theta} \neq 0$,

$$\hat{\rho}(r) = \frac{\sqrt{\psi(r)}}{2HC_{k,\theta}}, \qquad \hat{\lambda}(r) = -\operatorname{sgn}(C_{k,\theta})\int_0^r \frac{\zeta^2\Psi(\zeta)}{H}d\zeta,$$

$$\hat{\varphi}(r,t) = H\sqrt{|2C_{k,\theta}|}t + \operatorname{sgn}(C_{k,\theta})(1-k)\sin\theta\int_0^r \Psi(\zeta)d\zeta,$$

and $h = (k-1)\sin\theta/(2H|C_{k,\theta}|)$; if $C_{k,\theta} = 0$ and $k < 0$,

$$\hat{\rho}(r) = \frac{r}{2}, \qquad \hat{\lambda}(r) = -\int_0^r \frac{\zeta^2\left(\zeta^2 - k - 1 + \sqrt{\delta_{\mathfrak{si}}(\zeta)}\right)}{4kH^2\sqrt{\delta_{\mathfrak{si}}(\zeta)}}d\zeta,$$

$$\hat{\varphi}(r,t) = t \pm \int_0^r \frac{\zeta^2 - k - 1 + \sqrt{\delta_{\mathfrak{si}}(\zeta)}}{2H\sqrt{-k}\sqrt{\delta_{\mathfrak{si}}(\zeta)}}d\zeta \quad \text{and} \quad h = \pm H\sqrt{-k},$$

where $C_{k,\theta}$ is the constant as in Theorem 3.4 and

$$\delta_{\mathfrak{si}}(r) = (r^2 - k - 1)^2 - 4k, \qquad \Psi(r) = \frac{\sqrt{|2C_{k,\theta}|}(r^2 + (1-k)\cos\theta)}{\psi(r)\sqrt{\delta_{\mathfrak{si}}(r)}},$$

$$\psi(r) = -2C_{k,\theta}r^2 + (1-k)^2\sin^2\theta.$$

By a proof similar to that of Lemma 4.2, we have the following.

LEMMA 4.3. *The extended associate family $\hat{f}_{\mathfrak{si},\theta}(r,t)$ of the spacelike s-Delaunay surface $\hat{f}_{\mathfrak{si}}(r,t)$ is a frontal and its singular points are non-degenerate. Moreover, these singular points are cuspidal edges if and only if $\theta \neq 0, \pm\pi/2, \pi$, where we regard $S^1$ as the interval $(-\pi,\pi]$.*

By Lemmas 4.2 and 4.3, singular points of the associate family of spacelike Delaunay surfaces are cuspidal edges, except for the case of $\theta = \pm\pi/2$ and $\theta = 0, \pi$. In [**22**], a criterion for 5/2-cuspidal edges is given to prove the following.

FACT 4.4 ([**22**]). The singular point set of $\hat{f}_j^{\#}(r,t)$ consists of 5/2-cuspidal edges for each $j = \mathfrak{t}, \mathfrak{lt}, \mathfrak{si}, \mathfrak{lsi}$, where $\hat{f}_j^{\#}(r,t)$ is the conjugate of the extended spacelike Delaunay surface $\hat{f}_j(r,t)$ (i.e., $\hat{f}_j^{\#}(r,t) := \hat{f}_{j,\pi/2}(r,t)$).

Thus, we have the following.

THEOREM 4.5. *Singular points of the associate family $\hat{f}_{\mathfrak{t},\theta}(r,t)$ (resp. $\hat{f}_{\mathfrak{si},\theta}(r,t)$) of extended spacelike Delaunay surfaces $\hat{f}_{\mathfrak{t}}(r,t)$ (resp. $\hat{f}_{\mathfrak{si}}(r,t)$) are cuspidal edges if $\theta \neq 0, \pm\pi/2, \pi$, conelike singularities if $\theta = 0, \pi$, and 5/2-cuspidal edges if $\theta = \pm\pi/2$, where we regard $S^1$ as the interval $(-\pi,\pi]$.*

PROOF. By Lemmas 4.2, 4.3 and Fact 4.4, it suffices to show that $\hat{f}_{\mathfrak{t},\theta}(r,t)$ and $\hat{f}_{\mathfrak{si},\theta}(r,t)$ of $\theta = 0, \pi$, that is, the extended spacelike Delaunay surfaces $\hat{f}_{\mathfrak{t}}(r,t)$, $\hat{f}_{\mathfrak{si}}(r,t)$, $\hat{f}_{\mathfrak{lt}}(r,t)$ and $\hat{f}_{\mathfrak{lsi}}(r,t)$ defined by (4.2), (4.3) and (4.4) have conelike singularities. Since we can prove similarly, we only deal with $\hat{f}_{\mathfrak{t}}(r,t)$ here.

Set $G_1$ and $g$ as $G_1(x_0,x_1,x_2) := 2H(x_0,x_1,x_2)$ and $g(r,t) := (r, 2Ht)$. Then, we have $G_1 \circ \hat{f}_{\mathfrak{t}} \circ g^{-1}(u,v) = (\phi(u),\ u\cos v,\ u\sin v)$, where

$$\phi(u) := \int_0^u \frac{\zeta^2+k-1}{\sqrt{(\zeta^2+k+1)^2-4k}}\,d\zeta.$$

Since $\phi(u)$ is an odd function, by the Malgrange preparation theorem and Whitney's lemma (cf. [**17**]), there exists a smooth function $\tilde{\phi}$ such that $\phi(u) = u\,\tilde{\phi}(u^2)$. Then, $\phi'(0) = \operatorname{sgn}(k-1)\ (\neq 0)$ yields that the map

$$G_2(x_0,x_1,x_2) := \left(\frac{x_0}{\tilde{\phi}(x_1^2+x_2^2)},\ x_1,\ x_2\right)$$

defines a local diffeomorphism of $\boldsymbol{R}^3$ around the origin. Setting $G := G_2 \circ G_1$, we have $G \circ \hat{f}_{\mathfrak{t}} \circ g^{-1}(u,v) = (u,\ u\cos v,\ u\sin v)$, which gives the desired result. □

ACKNOWLEDGEMENTS. The author would like to express his sincere gratitude to Professor Miyuki Koiso for fruitful discussions and helpful advices. He also thanks the referee for careful reading and valuable comments and Masashi Yasumoto for pointing out typographical errors.

## References

[1] David Brander, Wayne Rossman, and Nicholas Schmitt, *Holomorphic representation of constant mean curvature surfaces in Minkowski space: consequences of non-compactness in loop group methods*, Adv. Math. **223** (2010), no. 3, 949–986, DOI 10.1016/j.aim.2009.09.006. MR2565553

[2] Eugenio Calabi, *Examples of Bernstein problems for some nonlinear equations*, Global Analysis (Proc. Sympos. Pure Math., Vol. XV, Berkeley, Calif., 1968), Amer. Math. Soc., Providence, R.I., 1970, pp. 223–230. MR0264210

[3] L. Desideri, *The Plateau problem for polygonal boundary curves in Minkowski 3-space*, preprint (arXiv:1012.3582).

[4] Manfredo P. do Carmo and Marcos Dajczer, *Helicoidal surfaces with constant mean curvature*, Tôhoku Math. J. (2) **34** (1982), no. 3, 425–435, DOI 10.2748/tmj/1178229204. MR676120

[5] Isabel Fernández, *The number of conformally equivalent maximal graphs*, J. Geom. Anal. **20** (2010), no. 4, 955–969, DOI 10.1007/s12220-010-9133-4. MR2683771

[6] Isabel Fernández and Francisco J. López, *Periodic maximal surfaces in the Lorentz-Minkowski space* $\mathbb{L}^3$, Math. Z. **256** (2007), no. 3, 573–601, DOI 10.1007/s00209-006-0087-y. MR2299572

[7] Isabel Fernandez and Francisco J. Lopez, *On the uniqueness of the helicoid and Enneper's surface in the Lorentz-Minkowski space* $\mathbb{R}^3_1$, Trans. Amer. Math. Soc. **363** (2011), no. 9, 4603–4650, DOI 10.1090/S0002-9947-2011-05133-0. MR2806686

[8] Isabel Fernández, Francisco J. López, and Rabah Souam, *The space of complete embedded maximal surfaces with isolated singularities in the 3-dimensional Lorentz-Minkowski space*, Math. Ann. **332** (2005), no. 3, 605–643, DOI 10.1007/s00208-005-0642-6. MR2181764

[9] Isabel Fernández, Francisco J. López, and Rabah Souam, *The moduli space of embedded singly periodic maximal surfaces with isolated singularities in the Lorentz-Minkowski space* $\mathbb{L}^3$, Manuscripta Math. **122** (2007), no. 4, 439–463, DOI 10.1007/s00229-007-0079-1. MR2300054

[10] S. Fujimori, Y. W. Kim, S.-E. Koh, W. Rossman, H. Shin, H. Takahashi, M. Umehara, K. Yamada, and S.-D. Yang, *Zero mean curvature surfaces in* $\mathbf{L}^3$ *containing a light-like line* (English, with English and French summaries), C. R. Math. Acad. Sci. Paris **350** (2012), no. 21-22, 975–978, DOI 10.1016/j.crma.2012.10.024. MR2996778

[11] S. Fujimori, Y. W. Kim, S.-E. Koh, W. Rossman, H. Shin, M. Umehara, K. Yamada, and S.-D. Yang, *Zero mean curvature surfaces in Lorentz-Minkowski 3-space and 2-dimensional fluid mechanics*, Math. J. Okayama Univ. **57** (2015), 173–200. MR3289302

[12] S. Fujimori, Y. W. Kim, S.-E. Koh, W. Rossman, H. Shin, M. Umehara, K. Yamada, and S.-D. Yang, *Zero mean curvature surfaces in Lorentz-Minkowski 3-space which change type across a light-like line*, Osaka J. Math. **52** (2015), no. 1, 285–297. MR3326612

[13] Shoichi Fujimori and Francisco J. López, *Nonorientable maximal surfaces in the Lorentz-Minkowski 3-space*, Tohoku Math. J. (2) **62** (2010), no. 3, 311–328, DOI 10.2748/tmj/1287148614. MR2742011

[14] Shoichi Fujimori, Wayne Rossman, Masaaki Umehara, Kotaro Yamada, and Seong-Deog Yang, *New maximal surfaces in Minkowski 3-space with arbitrary genus and their cousins in de Sitter 3-space*, Results Math. **56** (2009), no. 1-4, 41–82, DOI 10.1007/s00025-009-0443-4. MR2575851

[15] Shoichi Fujimori, Wayne Rossman, Masaaki Umehara, Kotaro Yamada, and Seong-Deog Yang, *Embedded triply periodic zero mean curvature surfaces of mixed type in Lorentz-Minkowski 3-space*, Michigan Math. J. **63** (2014), no. 1, 189–207, DOI 10.1307/mmj/1395234364. MR3189473

[16] Shoichi Fujimori, Kentaro Saji, Masaaki Umehara, and Kotaro Yamada, *Singularities of maximal surfaces*, Math. Z. **259** (2008), no. 4, 827–848, DOI 10.1007/s00209-007-0250-0. MR2403743

[17] M. Golubitsky and V. Guillemin, *Stable mappings and their singularities*, Springer-Verlag, New York-Heidelberg, 1973. Graduate Texts in Mathematics, Vol. 14. MR0341518

[18] Chao Hao Gu, *The extremal surfaces in the 3-dimensional Minkowski space*, Acta Math. Sinica (N.S.) **1** (1985), no. 2, 173–180, DOI 10.1007/BF02560031. MR858565

[19] Chao Hao Gu, *A global study of extremal surfaces in 3-dimensional Minkowski space*, Differential geometry and differential equations (Shanghai, 1985), Lecture Notes in Math., vol. 1255, Springer, Berlin, 1987, pp. 26–33, DOI 10.1007/BFb0077678. MR895395

[20] Chao Hao Gu, *Extremal surfaces of mixed type in Minkowski space* $\mathbf{R}^{n+1}$, Variational methods (Paris, 1988), Progr. Nonlinear Differential Equations Appl., vol. 4, Birkhäuser Boston, Boston, MA, 1990, pp. 283–296. MR1205160

[21] Jun-ich Hano and Katsumi Nomizu, *Surfaces of revolution with constant mean curvature in Lorentz-Minkowski space*, Tohoku Math. J. (2) **36** (1984), no. 3, 427–437, DOI 10.2748/tmj/1178228808. MR756026

[22] A. Honda, M. Koiso and K. Saji, *Fold singularities on spacelike CMC surfaces in Lorentz-Minkowski space*, preprint (arXiv:1509.03050), to appear in Hokkaido Mathematical Journal.

[23] T. Ishihara and F. Hara, *Surfaces of revolution in the Lorentzian 3-space*, J. Math. Tokushima Univ. **22** (1988), 1–13.
[24] Y. Kawakami, *On the maximal number of exceptional values of Gauss maps for various classes of surfaces*, Math. Z. **274** (2013), 1249–1260.
[25] Y. W. Kim, S.-E. Koh, H. Shin and S.-D. Yang, *Spacelike maximal surfaces, timelike minimal surfaces, and Björling representation formulae*, J. Korean Math. Soc. **48** (2011), 1083–1100.
[26] Y. W. Kim and S.-D. Yang, *Prescribing singularities of maximal surfaces via a singular Björling representation formula*, J. Geom. Phys. **57** (2007), 2167–2177.
[27] V. A. Klyachin, *Surfaces of zero mean curvature of mixed type in Minkowski space*, Izv. Ross. Akad. Nauk Ser. Mat. **67** (2003), 5–20.
[28] O. Kobayashi, *Maximal surfaces in the 3-dimensional Minkowski space* $L^3$, Tokyo J. Math. **6** (1983), 297–309.
[29] M. Kokubu, W. Rossman, K. Saji, M. Umehara and K. Yamada, *Singularities of flat fronts in hyperbolic space*, Pacific J. Math. **221** (2005), 303–351.
[30] N. Sasahara, *Spacelike helicoidal surfaces with constant mean curvature in Minkowski 3-space*, Tokyo J. Math. **23** (2000), 477–502.
[31] M. Umehara and K. Yamada, *Maximal surfaces with singularities in Minkowski space*, Hokkaido Math. J. **35** (2006), 13–40.

National Institute of Technology, Miyakonojo College, Yoshio, Miyakonojo 885-8567, Japan
*E-mail address*: atsufumi@cc.miyakonojo-nct.ac.jp

Contemporary Mathematics
Volume **675**, 2016
http://dx.doi.org/10.1090/conm/675/13587

# Generalized distance-squared mappings of $\mathbb{R}^{n+1}$ into $\mathbb{R}^{2n+1}$

S. Ichiki and T. Nishimura

ABSTRACT. We classify generalized distance-squared mappings of $\mathbb{R}^{n+1}$ into $\mathbb{R}^{2n+1}$ ($n \geq 1$) having generic central points (Theorem 1). Theorem 1 provides a complete classification of mappings of a restrictive class with respect to $\mathcal{A}$-equivalence. Generic points in Theorem 1 depend on matrices $A$ constructed by coefficients of generalized distance-squared mappings, which seems to be not the best formulated. We show also that generic points must depend on matrices $A$, namely, there does not exist a universal bad set $\Sigma \subset (\mathbb{R}^{n+1})^{2n+1}$ in the case of this dimension-pair (Theorem 2).

## 1. Introduction

For any positive integers $k, n$, let $p_i = (p_{i0}, p_{i1}, \ldots, p_{in})$ $(0 \leq i \leq k)$ (resp., $A = (a_{ij})_{0\leq i\leq k, 0\leq j\leq n}$) be a point of $\mathbb{R}^{n+1}$ (resp., a $(k+1) \times (n+1)$ matrix with non-zero entries). Let $G_{(p_0,p_1,\ldots,p_k,A)} : \mathbb{R}^{n+1} \to \mathbb{R}^{k+1}$ be the mapping defined by

$$G_{(p_0,p_1,\ldots,p_k,A)}(x) = \left( \sum_{j=0}^{n} a_{0j}(x_j - p_{0j})^2, \sum_{j=0}^{n} a_{1j}(x_j - p_{1j})^2, \ldots, \sum_{j=0}^{n} a_{kj}(x_j - p_{kj})^2 \right),$$

where $x = (x_0, x_1, \ldots, x_n) \in \mathbb{R}^{n+1}$. The mapping $G_{(p_0,p_1,\ldots,p_k,A)}$ is called a *generalized distance-squared mapping*, and the $(k+1)$-tuple of points $(p_0, \ldots, p_k) \in (\mathbb{R}^{n+1})^{k+1}$ is called the *central point* of the generalized distance-squared mapping $G_{(p_0,p_1,\ldots,p_k,A)}$. Each component of a generalized distance-squared mapping is merely a quadratic function, and thus its behavior is very simple. A generalized distance-squared mapping is merely a finite tuple of quadratic functions, and thus it should be a simple object. Nevertheless, the behavior of generalized distance-squared mappings seems to be wrapped in mystery. Classifying generalized distance-squared mappings with respect to $\mathcal{A}$-equivalence can be an important clue to solve the mystery. Therefore, it is natural to classify them. In the case of $n = k = 1$, a recognizable classification is known for $G_{(p_0,p_1,\ldots,p_k,A)}$ (see Proposition 1 below). The purpose of this paper is to obtain a classification in the case of $k = 2n$.

---

2010 *Mathematics Subject Classification.* Primary 57R45, 58C25, 58K50.
*Key words and phrases.* Generalized distance-squared mapping, central point, stable mapping, $\mathcal{A}$-equivalence, normal form of Whitney umbrella, bad set, universal bad set .
The first author is a Research Fellow of the Japan Society for the Promotion of Science.
The second author was partially supported by JSPS-CAPES under the JAPAN-BRAZIL research cooperative program.

Two mappings $f_i : \mathbb{R}^{n+1} \to \mathbb{R}^{k+1}$ $(i = 1, 2)$ are said to be *$\mathcal{A}$-equivalent* if there exist $C^\infty$ diffeomorphisms $h : \mathbb{R}^{n+1} \to \mathbb{R}^{n+1}$ and $H : \mathbb{R}^{k+1} \to \mathbb{R}^{k+1}$ such that the identity $f_1 = H \circ f_2 \circ h$ holds.

PROPOSITION 1 ([**3**]). *Let $((x_0, y_0), (x_1, y_1))$ be the standard coordinates of $(\mathbb{R}^2)^2$ and let $\Sigma$ be the hypersurface in $(\mathbb{R}^2)^2$ defined by $(x_0 - x_1)(y_0 - y_1) = 0$. Let $(p_0, p_1)$ be a point in $(\mathbb{R}^2)^2 - \Sigma$ and let $A_k$ be a $2 \times 2$ matrix of rank $k$ with non-zero entries (k=1, 2). Then, the following hold:*

(1) *The mapping $G_{(p_0,p_1,A_1)}$ is proper and stable, and it is $\mathcal{A}$-equivalent to $(x, y) \mapsto (x, y^2)$.*
(2) *The mapping $G_{(p_0,p_1,A_2)}$ is proper and stable, and it is not $\mathcal{A}$-equivalent to $G_{(p_0,p_1,A_1)}$.*
(3) *Let $B_2$ be a $2 \times 2$ matrix of rank 2 with non-zero entries and let $(q_0, q_1)$ be a point in $(\mathbb{R}^2)^2 - \Sigma$. Then, $G_{(p_0,p_1,A_2)}$ is $\mathcal{A}$-equivalent to $G_{(q_0,q_1,B_2)}$.*

In the case of $k = 2n$, only a few partial classification result for $G_{(p_0,\dots,p_k,A)}$ is known. A *distance-squared mapping* $D_{(p_0,p_1,\dots,p_k)}$ (resp., *Lorentzian distance-squared mapping* $L_{(p_0,p_1,\dots,p_k)}$) is the mapping $G_{(p_0,p_1,\dots,p_k,A)}$ satisfying that each entry of $A$ is 1 (resp., $a_{i0} = -1$ and $a_{ij} = 1$ if $j \neq 0$).

PROPOSITION 2 ([**1**, **2**]). *There exists a proper algebraic set $\Sigma \subset (\mathbb{R}^{n+1})^{2n+1}$ such that for any $p = (p_0, p_1, \dots, p_{2n}) \in (\mathbb{R}^{n+1})^{2n+1} - \Sigma$, both the distance-squared mapping $D_{(p_0,p_1,\dots,p_{2n})}$ and the Lorentzian distance-squared mapping $L_{(p_0,p_1,\dots,p_{2n})}$ are $\mathcal{A}$-equivalent to the following inclusion:*

$$(x_0, x_1, \dots, x_n) \mapsto (x_0, x_1, \dots, x_n, 0, \dots, 0).$$

THEOREM 1. *Let $A = (a_{ij})_{0\leq i\leq 2n, 0\leq j\leq n}$ be a $(2n + 1) \times (n + 1)$ matrix with non-zero entries. Then, the following two hold:*

(1) *Suppose that the rank of $A$ is $n+1$. Then, there exists a proper algebraic set $\Sigma_A \subset (\mathbb{R}^{n+1})^{2n+1}$ such that for any $p = (p_0, p_1, \dots, p_{2n}) \in (\mathbb{R}^{n+1})^{2n+1} - \Sigma_A$, $G_{(p,A)}$ is $\mathcal{A}$-equivalent to the following mapping:*

$$(x_0, x_1, \dots, x_n) \mapsto (x_0^2, x_0 x_1, \dots, x_0 x_n, x_1, \dots, x_n).$$

(2) *Suppose that the rank of $A$ is less than $n + 1$. Then, there exists a proper algebraic set $\Sigma_A \subset (\mathbb{R}^{n+1})^{2n+1}$ such that for any $p = (p_0, p_1, \dots, p_{2n}) \in (\mathbb{R}^{n+1})^{2n+1} - \Sigma_A$, $G_{(p,A)}$ is $\mathcal{A}$-equivalent to the following inclusion:*

$$(x_0, x_1, \dots, x_n) \mapsto (x_0, x_1, \dots, x_n, 0, \dots, 0).$$

The proper algebraic set $\Sigma_A$ given in Theorem 1 is called a *bad set*. The mapping given in the assertion (1) of Theorem 1 was firstly given in [**5**] and is called *the normal form of Whitney umbrella*. It is easily seen that the normal form of Whitney umbrella is not $\mathcal{A}$-equivalent to the inclusion $(x_0, x_1, \dots, x_n) \mapsto (x_0, x_1, \dots, x_n, 0, \dots, 0)$. Moreover, since these two mappings are proper, it is easily shown that they are stable by Mather's characterization theorem of stable proper mappings given in [**4**]. Thus, Theorem 1 may be regarded as a result of Proposition 1 type. On the other hand, it is desirable to improve Theorem 1 so that the bad set $\Sigma_A$ given in Theorem 1 does not depend on the given matrix $A$. However, contrary to the case of $n = k = 1$, in this case it is impossible to expect the existence of such a universal bad set $\Sigma$ which is independent of matrices $A$.

THEOREM 2. *There does not exist a subset* $\Sigma \subset (\mathbb{R}^{n+1})^{2n+1}$ *with Lebesgue measure zero such that for any point* $p = (p_0, p_1, \ldots, p_{2n}) \in (\mathbb{R}^{n+1})^{2n+1} - \Sigma$ *and any* $(2n+1) \times (n+1)$ *matrix* $A$ *with non-zero entries such that the rank of* $A$ *is* $n+1$, $G_{(p,A)}$ *is* $\mathcal{A}$*-equivalent to the following mapping:*

$$(x_0, x_1, \ldots, x_n) \mapsto (x_0^2, x_0x_1, \ldots, x_0x_n, x_1, \ldots, x_n).$$

Notice that the Lebesgue measure zero subset $\Sigma$ given in Theorem 2 does not depend on a matrix $A$ although the bad set $\Sigma_A$ in Theorem 1 depends on $A$. In this sense, $\Sigma$ is called a *universal bad set.*

The assertion (1) of Theorem 1, the assertion (2) of Theorem 1 and Theorem 2 are proved in Sections 2, 3 and 4 respectively.

## 2. Proof of the assertion (1) of Theorem 1

Set $A_1 = (a_{ij})_{0 \le i,j \le n}$ and $A_2 = (a_{ij})_{n+1 \le i \le 2n, 0 \le j \le n}$. Taking permutations of coordinates of the target space if necessary, without loss of generality, from the first we may assume that $\mathrm{rank}(A_1) = n+1$.

**2.1. STEP 1.** The purpose of this step is to delete quadratic terms as many as possible by a linear transformation $H_1 : \mathbb{R}^{2n+1} \to \mathbb{R}^{2n+1}$ of the following type.

$$H_1(X_0, \ldots, X_{2n}) = (X_0, \ldots, X_{2n}) \begin{pmatrix} \lambda_{0,0} & \cdots & \lambda_{0,n} & \lambda_{0,n+1} & \cdots & \lambda_{0,2n} \\ \vdots & \ddots & \vdots & \ddots & \vdots & \vdots \\ \lambda_{n,0} & \cdots & \lambda_{n,n} & \lambda_{n,n+1} & \cdots & \lambda_{n,2n} \\ 0 & \cdots & 0 & 1 & 0 & 0 \\ \vdots & \ddots & \vdots & 0 & \ddots & 0 \\ 0 & \cdots & 0 & 0 & 0 & 1 \end{pmatrix}.$$

Set $\Lambda_1 = (\lambda_{i,j})_{0 \le i,j \le n}$ and $\Lambda_2 = (\lambda_{i,j})_{0 \le i \le n, n+1 \le j \le 2n}$. Two matrices $\Lambda_1$ and $\Lambda_2$ are obtained as the solutions of the following linear equations, where $E_{n+1}$ is the $(n+1) \times (n+1)$ unit matrix and $M^T$ stands for the transposed matrix of a matrix $M$.

$$A_1^T \Lambda_1 = E_{n+1}, \quad A_1^T \Lambda_2 = -A_2^T.$$

Set $H_1 \circ G_{(p,A)} = (\varphi_{1,0}, \varphi_{1,1}, \ldots, \varphi_{1,2n})$. Then, $\varphi_{1,i}$ $(0 \le i \le 2n)$ may be expressed as follows.

$$\varphi_{1,i}(x_0, x_1, \ldots, x_n) = \begin{cases} x_i^2 + \sum_{j=0}^n b_{ij} x_j + c_i & (0 \le i \le n) \\ \sum_{j=0}^n b_{ij} x_j + c_i & (n+1 \le i \le 2n), \end{cases}$$

where $c_i$ stands for the constant term and $b_{ij}$ is as follows.

$$b_{ij} = \begin{cases} -2 \sum_{k=0}^n \lambda_{k,i} a_{kj} p_{kj} & (0 \le i \le n) \\ -2 \left(\sum_{k=0}^n \lambda_{k,i} a_{kj} p_{kj} + a_{ij} p_{ij}\right) & (n+1 \le i \le 2n). \end{cases}$$

**2.2. STEP 2.** The purpose of this step is to delete constant terms by the parallel transformation $H_2 : \mathbb{R}^{2n+1} \to \mathbb{R}^{2n+1}$ defined by

$$H_2(X_0, X_1, \ldots, X_{2n}) = (X_0 - c_0, X_1 - c_1, \ldots, X_{2n} - c_{2n}).$$

Set $H_2 \circ H_1 \circ G_{(p,A)} = (\varphi_{2,0}, \varphi_{2,1}, \ldots, \varphi_{2,2n})$. Then, $\varphi_{2,i}$ $(0 \leq i \leq 2n)$ may be expressed as follows.

$$\varphi_{2,i}(x_0, x_1, \ldots, x_n) = \begin{cases} x_i^2 + \sum_{j=0}^n b_{ij}x_j & (0 \leq i \leq n) \\ \sum_{j=0}^n b_{ij}x_j & (n+1 \leq i \leq 2n). \end{cases}$$

**2.3. STEP 3.** The purpose of this step is to construct the bad set $\Sigma_A$. The desirable bad set $\Sigma_A$ has the property that for any $p \in (\mathbb{R}^{n+1})^{2n+1} - \Sigma_A$, $H_2 \circ H_1 \circ G_{(p,A)}$ is transformed to the following type form:

$$\left(x_0^2 + d_{0,0}x_0, \cdots, x_n^2 + d_{n,n}x_n, d_{n+1,0}x_0 + d_{n+1,1}x_1, \cdots, d_{2n,0}x_0 + d_{2n,n}x_n\right),$$

by composing a linear transformation $H_3 : \mathbb{R}^{2n+1} \to \mathbb{R}^{2n+1}$ of the following type.

$$\begin{aligned} &H_3(X_0, X_1, \ldots, X_{2n}) \\ &= (X_0, X_1, \ldots, X_{2n}) \begin{pmatrix} 1 & 0 & 0 & 0 & \cdots & 0 \\ 0 & \ddots & 0 & \vdots & \vdots & \vdots \\ 0 & 0 & 1 & 0 & \cdots & 0 \\ \gamma_{n+1,0} & \cdots & \gamma_{n+1,n} & \gamma_{n+1,n+1} & \cdots & \gamma_{n+1,2n} \\ \vdots & \vdots & \vdots & \vdots & \vdots & \vdots \\ \gamma_{2n,0} & \cdots & \gamma_{2n,n} & \gamma_{2n,n+1} & \cdots & \gamma_{2n,2n} \end{pmatrix}, \end{aligned}$$

where $d_{ii}$ $(0 \leq i \leq n)$ are constants and $d_{ij}$ $(n+1 \leq i \leq 2n)$ are non-zero constants.

For any $j$ $(0 \leq j \leq n)$, set $\mathbf{b}_j = (b_{n+1,j}, \ldots, b_{2n,j})^T$. Moreover, for any $j$ $(0 \leq j \leq n)$, set

$$B_j = (\mathbf{b}_0, \ldots, \mathbf{b}_{j-1}, \hat{\mathbf{b}_j}, \mathbf{b}_{j+1}, \ldots, \mathbf{b}_n),$$

where $\hat{\mathbf{b}_j}$ stands for deleting $\mathbf{b}_j$. Then, $B_j$ is an $n \times n$ matrix for any $j$ $(0 \leq j \leq n)$.

DEFINITION 1. (1) For any $j$ $(0 \leq j \leq n)$, $\Sigma_{B_j}$ is the set consisting of points $p \in (\mathbb{R}^{n+1})^{2n+1}$ such that $\det B_j = 0$.

(2)

$$\Sigma_A = \bigcup_{j=0}^n \Sigma_{B_j}.$$

LEMMA 2.1. *The set $\Sigma_A$ is a proper algebraic set.*

PROOF. By the construction, it is clear that $\Sigma_A$ is an algebraic set. In order to show the properness, it is sufficient to find a point $p = (p_0, p_1, \ldots, p_{2n}) \in (\mathbb{R}^{n+1})^{2n+1} - \Sigma_A$.

For any $i$ $(0 \le i \le n)$, set $p_i = (0, 0, \ldots, 0) \in \mathbb{R}^{n+1}$. Moreover, set as follows.

$$\begin{array}{rcl} p_{n+1} & = & \left(\dfrac{1}{a_{n+1,0}}, \dfrac{1}{a_{n+1,1}}, 0, 0, \ldots, 0, 0\right) \\ p_{n+2} & = & \left(\dfrac{1}{a_{n+2,0}}, 0, \dfrac{1}{a_{n+2,2}}, 0, \ldots, 0, 0\right) \\ \vdots & \vdots & \qquad\qquad \vdots \\ p_{2n} & = & \left(\dfrac{1}{a_{2n,0}}, 0, 0, 0, \ldots, 0, \dfrac{1}{a_{2n,n}}\right). \end{array}$$

Then, we have the following:

$$\begin{array}{rcl} \mathbf{b}_0 & = & (1, 1, 1, \ldots, 1, 1) \\ \mathbf{b}_1 & = & (1, 0, 0, \ldots, 0, 0) \\ \mathbf{b}_2 & = & (0, 1, 0, \ldots, 0, 0) \\ \vdots & \vdots & \qquad \vdots \\ \mathbf{b}_n & = & (0, 0, 0, \ldots, 0, 1). \end{array}$$

Thus, we have that $\det B_j \ne 0$ for any $j$ $(0 \le j \le n)$. It follows that $p = (p_0, \ldots, p_{2n}) \notin \Sigma_A$. □

Set $H_3 \circ H_2 \circ H_1 \circ G_{(p,A)} = (\varphi_{3,0}, \varphi_{3,1}, \ldots, \varphi_{3,2n})$.

LEMMA 2.2. *Let $p$ be a point of $(\mathbb{R}^{n+1})^{2n+1} - \Sigma_A$. Then, the following hold:*

(1) *For any $j$ $(0 \le j \le n)$, $\varphi_{3,j}$ may be expressed as follows where $d_{j,j}$ is a constant.*

$$\varphi_{3,j}(x_0, x_1, \ldots, x_n) = x_j^2 + d_{j,j}x_j.$$

(2) *For any $j$ $(n+1 \le j \le 2n)$, $\varphi_{3,j}$ may be expressed as follows where $d_{j,0}, d_{j,j-n}$ are non-zero constants.*

$$\varphi_{3,j}(x_0, x_1, \ldots, x_n) = d_{j,0}x_0 + d_{j,j-n}x_{j-n}.$$

PROOF. Since the given point $p$ is outside the bad set $\Sigma_A$, it follows that $\det B_j \ne 0$ for any $j$ $(0 \le j \le n)$. Thus, for any $j$ $(0 \le j \le n)$, there exists the unique solution $(\gamma_{n+1,j}, \ldots, \gamma_{2n,j})^T$ for the following linear equations.

$$(-b_{j,0}, \ldots, -b_{j,j-1}, -\hat{b}_{j,j}, -b_{j,j+1}, \ldots, -b_{j,n})^T = B_j^T(\gamma_{n+1,j}, \ldots, \gamma_{2n,j})^T,$$

where $-\hat{b}_{j,j}$ stands for deleting $-b_{j,j}$. Therefore, the assertion (1) follows.

Next, we show the assertion (2). For any $j$ $(1 \le j \le n)$, set

$$\widetilde{B}_j = (\mathbf{b}_1 \ldots, \hat{\mathbf{b}}_j, \ldots, \mathbf{b}_n)$$

where $\hat{\mathbf{b}}_j$ stands for deleting $\mathbf{b}_j$. Then, $\widetilde{B}_j$ is an $n \times (n-1)$ matrix for any $j$ $(1 \le j \le n)$. Since $\det B_j \ne 0$ for any $j$ $(0 \le j \le n)$, it follows that $\operatorname{rank}(\widetilde{B}_j) = n-1$ for any $j$ $(1 \le j \le n)$. Thus, for any $j$ $(1 \le j \le n)$, the solution space for the linear equation

$$(0, \ldots, 0)^T = \widetilde{B}_j^T(\gamma_{n+1,n+j}, \ldots, \gamma_{2n,n+j})^T$$

is one-dimensional. Moreover, since $\det B_j \ne 0$ for any $j$ $(0 \le j \le n)$, both $\mathbf{b}_0$ and $\mathbf{b}_j$ are not perpendicular to non-zero solution vector $(\gamma_{n+1,n+j}, \ldots, \gamma_{2n,n+j})$ for the above linear equations for any $j$ $(1 \le j \le n)$. Thus, the assertion (2) follows. □

**2.4. STEP 4.** The purpose of this step is to complete the square for the first $(n+1)$ components of $H_3 \circ H_2 \circ H_1 \circ G_{(p,A)}$. Set

$$h_1(x_0, x_1, \ldots, x_n) = (x_0 - \frac{1}{2}d_{0,0}, x_1 - \frac{1}{2}d_{1,1}, \ldots, x_n - \frac{1}{2}d_{n,n})$$

and $H_3 \circ H_2 \circ H_1 \circ G_{(p,A)} \circ h_1 = (\varphi_{4,0}, \varphi_{4,1}, \ldots, \varphi_{4,2n})$. Then, $\varphi_{4,i}$ $(0 \le i \le 2n)$ may be expressed as follows, where $d_{i,0}, d_{i,i-n}$ are non-zero real numbers obtained in STEP 3 and $\widetilde{d}_i$ are some constants.

$$\varphi_{4,i}(x_0, x_1, \ldots, x_n) = \begin{cases} x_i^2 + \widetilde{d}_i & (0 \le i \le n) \\ d_{i,0}x_0 + d_{i,i-n}x_{i-n} + \widetilde{d}_i & (n+1 \le i \le 2n). \end{cases}$$

**2.5. STEP 5.** The purpose of this step is to delete constant terms by the parallel transformation $H_4 : \mathbb{R}^{2n+1} \to \mathbb{R}^{2n+1}$ defined by

$$H_4(X_0, X_1, \ldots, X_{2n}) = (X_0 - \widetilde{d}_0, X_1 - \widetilde{d}_1, \ldots, X_{2n} - \widetilde{d}_{2n}).$$

Set $H_4 \circ H_3 \circ H_2 \circ H_1 \circ G_{(p,A)} \circ h_1 = (\varphi_{5,0}, \varphi_{5,1}, \ldots, \varphi_{5,2n})$. Then, $\varphi_{5,i}$ $(0 \le i \le 2n)$ may be expressed as follows.

$$\varphi_{5,i}(x_0, x_1, \ldots, x_n) = \begin{cases} x_i^2 & (0 \le i \le n) \\ d_{i,0}x_0 + d_{i,i-n}x_{i-n} & (n+1 \le i \le 2n). \end{cases}$$

**2.6. STEP 6.** The purpose of this step is to simplify the last $n$ components of $H_4 \circ H_3 \circ H_2 \circ H_1 \circ G_{(p,A)} \circ h_1$. Set

$$\widetilde{x}_0 = x_0, \widetilde{x}_1 = d_{n+1,0}x_0 + d_{n+1,1}x_1, \ldots, \widetilde{x}_n = d_{i,0}x_0 + d_{2n,n}x_n.$$

Then, since $d_{i,i-n} \neq 0$ for any $i$ $(n+1 \le i \le 2n)$, the mapping $(x_0, x_1, \ldots, x_n) \mapsto (\widetilde{x}_0, \widetilde{x}_1, \ldots, \widetilde{x}_n)$ is a linear transformation of $\mathbb{R}^{n+1}$. Thus, setting

$$h_2(x_0, x_1, \ldots, x_n) = \left( x_0, \frac{1}{d_{n+1,1}}(x_1 - d_{n+1,0}x_0), \ldots, \frac{1}{d_{2n,n}}(x_n - d_{2n,0}x_0) \right)$$

and composing $h_2$ to $H_4 \circ H_3 \circ H_2 \circ H_1 \circ G_{(p,A)} \circ h_1$, the desired form may be obtained as follows:

$$\begin{aligned} & H_4 \circ H_3 \circ H_2 \circ H_1 \circ G_{(p,A)} \circ h_1 \circ h_2(x_0, x_1, \ldots, x_n) \\ = & \left( x_0^2, \frac{1}{d_{n+1,1}^2}(x_1 - d_{n+1,0}x_0)^2, \ldots, \frac{1}{d_{2n,n}^2}(x_n - d_{2n,0}x_0)^2, x_1, \ldots, x_n \right). \end{aligned}$$

**2.7. STEP 7.** This is the final step. Notice that $d_{i,0} \neq 0$ for any $i$ $(n+1 \le i \le 2n)$. Thus, the following coordinate transformation $H_5$ is well-defined.

$$\begin{aligned} & H_5(X_0, X_1, \ldots, X_{2n}) \\ = & \left( X_0, -\frac{d_{n+1,1}^2}{2d_{n+1,0}} \left( X_1 - \frac{d_{n+1,0}^2}{d_{n+1,1}^2}X_0 - \frac{1}{d_{n+1,1}^2}X_{n+1}^2 \right), \ldots, \right. \\ & \left. -\frac{d_{2n,n}^2}{2d_{2n,0}} \left( X_n - \frac{d_{2n,0}^2}{d_{2n,n}^2}X_0 - \frac{1}{d_{2n,n}^2}X_{2n}^2 \right), X_{n+1}, \ldots, X_{2n} \right). \end{aligned}$$

Then, the desired normal form of Whitney umbrella may be obtained as follows:

$$\begin{aligned} & H_5 \circ H_4 \circ H_3 \circ H_2 \circ H_1 \circ G_{(p,A)} \circ h_1 \circ h_2(x_0, x_1, \ldots, x_n) \\ = & (x_0^2, x_0x_1, \ldots, x_0x_n, x_1, \ldots, x_n). \end{aligned}$$

□

## 3. Proof of the assertion (2) of Theorem 1

Set $A_3 = (a_{ij})_{0\le i,j\le n-1}$. Taking permutations of coordinates of the target space if necessary, without loss of generality, from the first we may assume that $\text{rank}(A_3) = \text{rank}(A) \le n$.

**3.1. STEP 1.** The purpose of this step is to delete quadratic terms as many as possible by a linear transformation $H_1 : \mathbb{R}^{2n+1} \to \mathbb{R}^{2n+1}$ of the following type.

$$H_1(X_0, X_1, \ldots, X_{2n})$$
$$= (X_0, X_1, \ldots, X_{2n}) \begin{pmatrix} 1 & 0 & \cdots & 0 & 0 & \lambda_{0,n} & \cdots & \lambda_{0,2n} \\ 0 & 1 & \ddots & 0 & 0 & \lambda_{1,n} & \cdots & \lambda_{1,2n} \\ \vdots & \ddots & \ddots & \vdots & \vdots & \vdots & \vdots & \vdots \\ 0 & 0 & \cdots & 1 & 0 & \lambda_{n-2,n} & \cdots & \lambda_{n-2,2n} \\ 0 & 0 & \cdots & 0 & 1 & \lambda_{n-1,n} & \cdots & \lambda_{n-1,2n} \\ 0 & 0 & \cdots & 0 & 0 & 1 & 0 & 0 \\ 0 & 0 & \cdots & 0 & 0 & 0 & \ddots & 0 \\ 0 & 0 & \cdots & 0 & 0 & 0 & 0 & 1 \end{pmatrix}.$$

For any $i$ $(1 \le i \le 2n+1)$, let $\mathbf{a}_{i-1}$ be the $i$-th row vector of $A$. Then, by the assumption of $\text{rank}(A_3) = \text{rank}(A) \le n$, the following claim holds:

CLAIM 3.1. *For any $i$ $(n \le i \le 2n+1)$, there exist $\alpha_{0i}, \ldots, \alpha_{(n-1)i} \in \mathbb{R}$ such that the following equality holds, where $\mathbf{0}$ is the $(n+1)$ dimensional zero row vector:*

$$\sum_{j=0}^{n-1} \alpha_{ji}\mathbf{a}_j + \mathbf{a}_i = \mathbf{0}.$$

Set $H_1 \circ G_{(p,A)} = (\varphi_{1,0}, \varphi_{1,1}, \ldots, \varphi_{1,2n})$. Then, $\varphi_{1,i}$ $(0 \le i \le 2n)$ may be expressed as follows:

$$\varphi_{1,i}(x_0, x_1, \ldots, x_n) = \begin{cases} \sum_{j=0}^{n} a_{ij}(x_j - p_{ij})^2 & (0 \le i \le n-1) \\ \sum_{j=0}^{n} b_{ij}x_j + c_i & (n \le i \le 2n), \end{cases}$$

where $b_{ij}, c_i$ stands for some constants.

**3.2. STEP 2.** The purpose of this step is to delete constant terms $c_i$ by the parallel transformation $H_2 : \mathbb{R}^{2n+1} \to \mathbb{R}^{2n+1}$ defined by

$$H_2(X_0, X_1, \ldots, X_{2n}) = (X_0, X_1, \ldots, X_{n-1}, X_n - c_n, \ldots, X_{2n} - c_{2n}).$$

Set $H_2 \circ H_1 \circ G_{(p,A)} = (\varphi_{2,0}, \varphi_{2,1}, \ldots, \varphi_{2,2n})$. Then, $\varphi_{2,i}$ $(0 \le i \le 2n)$ may be expressed as follows.

$$\varphi_{2,i}(x_0, x_1, \ldots, x_n) = \begin{cases} \sum_{j=0}^{n} a_{ij}(x_j - p_{ij})^2 & (0 \le i \le n-1) \\ \sum_{j=0}^{n} b_{ij}x_j & (n \le i \le 2n). \end{cases}$$

**3.3. STEP 3.** The purpose of this step is to construct the bad set $\Sigma_A$. Set $B = (b_{ij})_{n\le i\le 2n, 0\le j\le n}$. Then, $B$ is an $(n+1)\times(n+1)$ matrix.

DEFINITION 2. $\Sigma_A$ is the set consisting of points $p \in (\mathbb{R}^{n+1})^{2n+1}$ such that $\det B = 0$.

LEMMA 3.1. *The set $\Sigma_A$ is a proper algebraic set.*

PROOF. Since it is clear that $\Sigma_A$ is an algebraic set. In order to show properness, it is sufficient to find a point $p = (p_0, p_1, \ldots, p_{2n}) \in (\mathbb{R}^{n+1})^{2n+1} - \Sigma_A$.

For any $i$ $(0 \le i \le n-1)$, set $p_i = (0, 0, \ldots, 0) \in \mathbb{R}^{n+1}$. Moreover, set as follows.

$$\begin{aligned} p_n &= \left(\frac{1}{a_{n,0}}, 0, 0, \ldots, 0, 0\right) \\ p_{n+1} &= \left(0, \frac{1}{a_{n+1,1}}, 0, \ldots, 0, 0\right) \\ \vdots \;\; & \;\;\vdots \qquad\qquad \vdots \\ p_{2n} &= \left(0, 0, 0, \ldots, 0, \frac{1}{a_{2n,n}}\right). \end{aligned}$$

Then, it is not difficult to see that $\det B = \det E_{n+1} = 1$. It follows that $p = (p_0, \ldots, p_{2n}) \notin \Sigma_A$. $\square$

Take a point $p \in \left(\mathbb{R}^{n+1}\right)^{2n+1} - \Sigma_A$. Then, by the construction of $\Sigma_A$, the matrix $B$ has its inverse matrix $B^{-1}$. Let $H_3 : \mathbb{R}^{2n+1} \to \mathbb{R}^{2n+1}$ be the linear transformation defined as follows:

$$\begin{aligned} & H_3(X_0, X_1, \ldots, X_{2n}) \\ = \; & (X_0, X_1, \ldots, X_{2n}) \left(\begin{array}{c|c} E_n & 0 \\ \hline 0 & \left(B^T\right)^{-1} \end{array}\right), \end{aligned}$$

where $E_n$ is the $n \times n$ unit matrix.

Set $H_3 \circ H_2 \circ H_1 \circ G_{(p,A)} = (\varphi_{3,0}, \varphi_{3,1}, \ldots, \varphi_{3,2n})$. Then, $\varphi_{3,i}$ $(0 \le i \le 2n)$ may be expressed as follows.

$$\varphi_{3,i}(x_0, x_1, \ldots, x_n) = \begin{cases} \sum_{j=0}^{n} a_{ij}(x_j - p_{ij})^2 & (0 \le i \le n-1) \\ x_{i-n} & (n \le i \le 2n). \end{cases}$$

**3.4. STEP 4.** The purpose of this step is to delete remaining constant terms by the parallel transformation $H_4 : \mathbb{R}^{2n+1} \to \mathbb{R}^{2n+1}$ defined by

$$\begin{aligned} & H_4(X_0, X_1, \ldots, X_{2n}) \\ = \; & (X_0 - \sum_{j=0}^{n} a_{0j}p_{0j}^2, X_1 - \sum_{j=0}^{n} a_{1j}p_{1j}^2, \ldots, X_{n-1} - \sum_{j=0}^{n} a_{(n-1)j}p_{(n-1)j}^2, \\ & \quad X_n, X_{n+1} \ldots, X_{2n}). \end{aligned}$$

Set $H_4 \circ H_3 \circ H_2 \circ H_1 \circ G_{(p,A)} = (\varphi_{4,0}, \varphi_{4.1}, \ldots, \varphi_{4,2n})$. Then, $\varphi_{4,i}$ $(0 \le i \le 2n)$ may be expressed as follows.

$$\varphi_{4,i}(x_0, x_1, \ldots, x_n) = \begin{cases} \sum_{j=0}^{n} a_{ij}(x_j^2 - 2p_{ij}x_j) & (0 \le i \le n-1) \\ x_{i-n} & (n \le i \le 2n). \end{cases}$$

**3.5. STEP 5.** This is the final step. Let $H_5 : \mathbb{R}^{2n+1} \to \mathbb{R}^{2n+1}$ be the coordinate transformation defined by

$$\begin{aligned} & H_5(X_0, X_1, \ldots, X_{2n}) \\ = & \left( X_n, X_{n+1}, \ldots, X_{2n}, X_0 - \sum_{j=0}^{n} a_{0j} \left(X_{n+j}^2 - 2p_{0j}X_{n+j}\right), \right. \\ & X_1 - \sum_{j=0}^{n} a_{1j} \left(X_{n+j}^2 - 2p_{1j}X_{n+j}\right), \ldots, \\ & \left. X_{n-1} - \sum_{j=0}^{n} a_{(n-1)j} \left(X_{n+j}^2 - 2p_{(n-1)j}X_{n+j}\right) \right). \end{aligned}$$

Then, we have the following:

$$H_5 \circ H_4 \circ H_3 \circ H_2 \circ H_1 \circ G_{(p,A)}(x_0, x_1, \ldots, x_n) = (x_0, \ldots, x_n, 0, \ldots, 0).$$

□

**3.6. REMARK.** As a by-product of the proof of the assertion (2) of Theorem 1 given in this section, we have the following:

THEOREM 3. *Let $n$ be a positive integer and $k$ be an integer such that $k > 2n$. Let $A = (a_{ij})_{0\le i\le k, 0\le j\le n}$ be a $(k+1) \times (n+1)$ matrix with non-zero entries. Then, there exists a proper algebraic set $\Sigma_A \subset (\mathbb{R}^{n+1})^{k+1}$ such that for any $p = (p_0, \ldots, p_k) \in (\mathbb{R}^{n+1})^{k+1} - \Sigma_A$, $G_{(p,A)}$ is $\mathcal{A}$-equivalent to the inclusion $(x_0, \ldots, x_n) \mapsto (x_0, \ldots, x_n, 0, \ldots, 0)$.*

## 4. Proof of Theorem 2

It is sufficient to show the following proposition.

PROPOSITION 3. *There exists an open dense subset $X$ of $(\mathbb{R}^{n+1})^{2n+1}$ such that for any point $p \in X$, there exists a $(2n+1) \times (n+1)$ full-rank matrix $A$ with non-zero entries satisfying the property that $G_{(p,A)}$ is $\mathcal{A}$-equivalent to the inclusion.*

Proof of Proposition 3. Take an arbitrary point $p \in (\mathbb{R}^{n+1})^{2n+1}$ and an arbitrary small positive number $\varepsilon$. The point $p$ and the positive real number $\varepsilon$ are fixed in the following arguments. It will be shown that there exist a sufficiently small positive number $\delta$ and a point $q \in (\mathbb{R}^{n+1})^{2n+1}$ such that $D_\delta(q) \subset D_\varepsilon(p)$ and $D_\delta(q) \subset X$; where for a positive number $a$ and $x_0 \in (\mathbb{R}^{n+1})^{2n+1}$, $D_a(x_0)$ stands for the open disc centered at $x_0$ with radius $a$, namely, $D_a(x_0) = \{x \in (\mathbb{R}^{n+1})^{2n+1} \mid ||x - x_0|| < a\}$.

The same terminologies as in Section 2 are used in the following arguments. It is clear that there exists an orthogonal $(n+1)\times(n+1)$ matrix with non-zero entries. Let $A_1 = (a_{ij})_{0\le i,,j\le n}$ be an orthogonal $(n+1) \times (n+1)$ matrix with non-zero entries and fix it. Then, it is clear that there exists a point $\widetilde{p} \in (\mathbb{R}^{n+1})^{2n+1}$ such that the following three are satisfied.

(a) $||p - \widetilde{p}|| < \varepsilon/2$.
(b) $-\sum_{k=0}^{n} a_{k\ell}a_{kj}\widetilde{p}_{kj} \neq 0$ for any $j, \ell$ $(0 \le j, \ell \le n, j \neq \ell)$.
(c) $-\sum_{k=0}^{n} a_{kj}^2\widetilde{p}_{kj} + \widetilde{p}_{ij} \neq 0$ for any $i, j$ $(n+1 \le i \le 2n, 0 \le j \le n)$.

Here, $\widetilde{p} = (\widetilde{p}_0, \widetilde{p}_1, \ldots, \widetilde{p}_n)$ and $\widetilde{p}_i = (\widetilde{p}_{i0}, \widetilde{p}_{i1}, \ldots, \widetilde{p}_{in})$ for any $i$ $(0 \leq i \leq n)$. Take one point $\widetilde{p}$ satisfying the above (a), (b) and (c).

Next, we modify the point $\widetilde{p}$ to obtain the new point $q$ such that for any point of a sufficiently small open neighborhood of the new point $q$, an $n \times (n+1)$ matrix $A_2 = (a_{ij})_{n+1 \leq i \leq 2n, 0 \leq j \leq n}$ with non-zero entries can be constructed. Since $A_1$ is an orthogonal matrix, we have that $(A_1^T)^{-1} = A_1$. Thus, for the matrix $\Lambda_2 = (\lambda_{i,j})_{0 \leq i \leq n, n+1 \leq j \leq 2n}$ defined in Section 2, the following equality holds:

$$\Lambda_2 = -A_1 A_2^T.$$

Therefore, for any $i, k$ $(n+1 \leq i \leq 2n, 0 \leq k \leq n)$, we have the following:

$$\lambda_{k,i} = -\sum_{\ell=0}^{n} a_{k\ell} a_{i\ell}.$$

By substituting the above equalities, we have the following:

$$\begin{aligned} b_{ij} &= -2\left(\sum_{k=0}^{n} \lambda_{k,i} a_{kj} \widetilde{p}_{kj} + a_{ij}\widetilde{p}_{ij}\right) \\ &= -2\left(\sum_{k=0}^{n}\left(-\sum_{\ell=0}^{n} a_{k\ell}a_{i\ell}\right) a_{kj}\widetilde{p}_{kj} + a_{ij}\widetilde{p}_{ij}\right) \\ &= -2\left(\sum_{\ell=0,\ell\neq j}^{n}\left(-\sum_{k=0}^{n} a_{k\ell}a_{kj}\widetilde{p}_{kj}\right) a_{i\ell} + \left(-\sum_{k=0}^{n} a_{kj}^2\widetilde{p}_{kj} + \widetilde{p}_{ij}\right) a_{ij}\right). \end{aligned}$$

For any $i, j, \ell$ $(n+1 \leq i \leq 2n, 0 \leq j, \ell \leq n)$, set

$$\begin{aligned} c_{i\ell} &= -\sum_{k=0}^{n} a_{k\ell}a_{kj}\widetilde{p}_{kj}, \\ d_{ij} &= -\sum_{k=0}^{n} a_{kj}^2\widetilde{p}_{kj} + \widetilde{p}_{ij}. \end{aligned}$$

Then, for any $i, j$ $(n+1 \leq i \leq 2n, 0 \leq j \leq n)$, $b_{ij}$ can be simply expressed as follows:

$$b_{ij} = -2\left(\sum_{\ell=0,\ell\neq j}^{n} c_{i\ell}a_{i\ell} + d_{ij}a_{ij}\right).$$

Notice that by the properties (b) and (c) above, it follows that $c_{i\ell}, d_{ij}$ are non-zero constants for any $i, j, \ell$ $(n+1 \leq i \leq 2n, 0 \leq j, \ell \leq n)$. Thus, for any $i, j$ $(n+1 \leq i \leq 2n, 0 \leq j \leq n)$, the following linear equation with respect to variables $a_{i0}, a_{i1}, \ldots, a_{in}$ has a solution $(a_{i0}, a_{i1}, \ldots, a_{in})$ satisfying $a_{i0}a_{i1}\cdots a_{in} \neq 0$.

$$b_{ij} = 0.$$

Next, for any $i$ $(n+1 \leq i \leq 2n)$, we investigate the following simultaneous equations:

$$(*) \qquad b_{i0} = b_{i1} = \cdots = \widehat{b_{i(i-n)}} = \cdots = b_{in} = 0,$$

where $\widehat{b_{i(i-n)}}$ stands for deleting $b_{i(i-n)}$. For any $i, j$ $(n+1 \leq i \leq 2n, 0 \leq j \leq n)$, let $\mathrm{grad}(b_{ij})$ be the gradient vector of the linear function $b_{ij}$. For any $i$, set $B(i)$ be the $(n+1) \times (n+1)$ matrix whose $j$-th column vector is $\mathrm{grad}(b_{ij})$:

$$B(i) = (\mathrm{grad}(b_{i0}), \mathrm{grad}(b_{i1}), \ldots, \mathrm{grad}(b_{in})).$$

Let $B(i,\widetilde{x})$ be the $(n+1)\times(n+1)$ matrix obtained by replacing the constants $\widetilde{p}_{kj},\widetilde{p}_{ij}$ of $B(i)$ with real variables $\widetilde{x}_{kj},\widetilde{x}_{ij}$ $(n+1\le i\le 2n, 0\le j,k\le n)$. Moreover, for any $j,\ell$ $(1\le j,\ell\le n+1)$, let $B(i)_{j,\ell}$ (resp., $B(i,\widetilde{x})_{j,\ell}$) be the $(j,\ell)$-cofactor of $B(i)$ (resp., $B(i,\widetilde{x})$). Then, it is clear that for any $i$ $(n+1\le i\le 2n)$, the following algebraic equation defines a proper algebraic set in the space $(\mathbb{R}^{n+1})^{2n+1}$.

$$\det B(i,\widetilde{x})\cdot B(i,\widetilde{x})_{1,i-n+1}\cdots B(i,\widetilde{x})_{n+1,i-n+1}=0.$$

Therefore, there exists a point $q\in(\mathbb{R}^{n+1})^{2n+1}$ satisfying the following four properties:

(a') $||\widetilde{p}-q||<\varepsilon/2$.
(b') $-\sum_{k=0}^{n} a_{k\ell}a_{kj}q_{kj}\neq 0$ for any $0\le j,\ell\le n$.
(c') $-\sum_{k=0}^{n} a_{kj}^2 q_{kj}+q_{ij}\neq 0$ for any $i,j$ $(n+1\le i\le 2n, 0\le j\le n)$.
(d) $\det B(i,q)\cdot B(i,q)_{1,i-n+1}\cdots B(i,q)_{n+1,i-n+1}\neq 0$ for any $i$ $(n+1\le i\le 2n)$.

By the properties (a) and (a'), it follows that $||p-q||<\varepsilon$. Moreover, all functions appeared so far are continuous. Hence, it follows that there exists a sufficiently small positive number $\delta$ such that the following four properties hold:

(a") $D_\delta(q)\subset D_\varepsilon(p)$.
(b") $-\sum_{k=0}^{n} a_{k\ell}a_{kj}\widetilde{q}_{kj}\neq 0$ for any $0\le j,\ell\le n$ and any $\widetilde{q}\in D_\delta(q)$.
(c") $-\sum_{k=0}^{n} a_{kj}^2\widetilde{q}_{kj}+\widetilde{q}_{ij}\neq 0$ for any $i,j$ $(n+1\le i\le 2n, 0\le j\le n)$ and any $\widetilde{q}\in D_\delta(q)$.
(d') $\det B(i,\widetilde{q})\cdot B(i,\widetilde{q})_{1,i-n+1}\cdots B(i,\widetilde{q})_{n+1,i-n+1}\neq 0$ for any $i$ $(n+1\le i\le 2n)$ and any $\widetilde{q}\in D_\delta(q)$.

Take one point $\widetilde{q}\in D_\delta(q)$ and fix it. By the properties (b"), (c"), (d'), the simultaneous equations (*) has a solution $(a_{i0},a_{i1},\ldots,a_{in})$ satisfying $a_{i0}a_{i1}\cdots a_{in}\neq 0$. Take a solution $(a_{i0},a_{i1},\ldots,a_{in})$ satisfying $a_{i0}a_{i1}\cdots a_{in}\neq 0$, in fact it is sufficient to take

$$(a_{i0},a_{i1},\ldots,a_{in})=\left(B(i,\widetilde{q})_{1,i-n+1},B(i,\widetilde{q})_{2,i-n+1},\ldots,B(i,\widetilde{q})_{n+1,i-n+1}\right).$$

Then, for any $i$ $(n+1\le i\le 2n)$, $\varphi_{2,i}$ has the following form, where $\varphi_{2,i}$ is the $i$-th component function of $H_2\circ H_1\circ G_{(\widetilde{q},A)}$ defined in STEP 2 of Section 2.

$$\varphi_{2,i}(x_0,x_1,\ldots,x_n)=\begin{cases} x_i^2+\sum_{j=0}^{n} b_{ij}x_j & (0\le i\le n)\\ b_{i(i-n)}x_{i-n} & (n+1\le i\le 2n).\end{cases}$$

Since $\det B(i,\widetilde{q})\neq 0$ by the above property (d'), it follows that $b_{i(i-n)}\neq 0$ for any $i$ $(n+1\le i\le 2n)$. Therefore, $H_2\circ H_1\circ G_{(\widetilde{q},A)}$ is $\mathcal{A}$-equivalent to the following mapping:

$$(**)\qquad (x_0,x_1,\ldots,x_n)\mapsto(x_0^2+b_{00}x_0,b_{10}x_0,\ldots,b_{n0}x_0,x_1,\ldots,x_n).$$

If one of $b_{10},\ldots,b_{n0}$ is non-zero, then this mapping is $\mathcal{A}$-equivalent to the inclusion. By the form of $b_{ij}$ $(0\le i,j\le n)$ given in Step 2 of Section 2 and the fact that $A_1$ is an orthogonal matrix, it can be seen that all of $b_{10},\ldots,b_{n0}$ are actually non-zero. Hence, the mapping (**) is non-singular. □

**4.1. REMARK.**

(1) Contrary to our expectation, the following Theorem 4 can not be a by-product of the proof of Theorem 2.

THEOREM 4. *Let $n$ be a positive integer and $k$ be an integer such that $k > 2n$. Then, there does not exist a subset $\Sigma \subset (\mathbb{R}^{n+1})^{k+1}$ with Lebesgue measure zero such that for any point $p = (p_0, p_1, \ldots, p_k) \in (\mathbb{R}^{n+1})^{k+1} - \Sigma$ and any $(k+1) \times (n+1)$ matrix $A$ with non-zero entries, $G_{(p,A)}$ is $\mathcal{A}$-equivalent to the inclusion $(x_0, \ldots, x_n) \mapsto (x_0, \ldots, x_n, 0, \ldots, 0)$.*

In order to prove Theorem 4, it is sufficient to show the following Proposition 4 since there are no stable and singular mappings for the dimension-pair $(n+1, k+1)$.

PROPOSITION 4. *Let $\Sigma$ be the proper algebraic set defined by*

$$\Pi_{0\le i,\ell\le k, 0\le j\le n}(x_{ij} - x_{\ell j}) = 0,$$

*where $x = (x_0, \ldots x_k) \in (\mathbb{R}^{n+1})^{k+1}$, $x_i = (x_{i0}, \ldots, x_{in})$. Then, for any $p \in (\mathbb{R}^{n+1})^{k+1} - \Sigma$, there exists a $(k+1) \times (n+1)$ full-rank matrix $A$ with non-zero entries satisfying the property that $G_{(p,A)}$ is singular.*

Proof of Proposition 4. We may assume that $A_1 = (a_{ij})_{0\le i,j\le n}$ is of full rank. Set $\widetilde{p} = (p_0, \ldots p_n) \in (\mathbb{R}^{n+1})^{n+1}$. Then, the rank of $JG_{(\widetilde{p},A_1)}$ is less than $n+1$ at (for instance) $x = p_0$. Since $p \in (\mathbb{R}^{n+1})^{k+1} - \Sigma$, it follows non-zero component $a_{ij}$ $(n+1 \le i \le k, 0 \le j \le n)$ can be chosen so that the rank of $JG_{(p,A)}$ at $x = p_0$ is equal to the rank of $JG_{(\widetilde{p},A_1)}$ at $x = p_0$. □

(2) The proof of Theorem 2 has one more advantage. It makes clear the reason why we can expect the existence of a universal bad set $\Sigma$ in the case $n = k = 1$.

## Acknowledgements

The authors are grateful to the reviewers for careful reading of the first draft of their paper and giving invaluable comments. They thank also to M. A. S. Ruas and R. Oset Sinha for their kind suggestions and advices. The first author is supported by JSPS KAKENHI Grant number 16J06911.

## References

[1] Shunsuke Ichiki and Takashi Nishimura, *Distance-squared mappings*, Topology Appl. **160** (2013), no. 8, 1005–1016, DOI 10.1016/j.topol.2013.03.011. MR3043130

[2] Shunsuke Ichiki and Takashi Nishimura, *Recognizable classification of Lorentzian distance-squared mappings*, J. Geom. Phys. **81** (2014), 62–71, DOI 10.1016/j.geomphys.2014.03.005. MR3194215

[3] S. Ichiki, T. Nishimura, R. Oset Sinha, and M. A. S. Ruas, *Generalized distance-squared mappings of the plane into the plane*, Adv. Geom. **16** (2016), no. 2, 189–198, DOI 10.1515/advgeom-2015-0044. MR3489595

[4] John N. Mather, *Stability of $C^\infty$ mappings. V. Transversality*, Advances in Math. **4** (1970), 301–336 (1970). MR0275461

[5] H. Whitney, *The singularities of a smooth $n$-manifold in $(2n-1)$-space*, Ann. of Math., **45** (1944), 247–293.

GRADUATE SCHOOL OF ENVIRONMENT AND INFORMATION SCIENCES, YOKOHAMA NATIONAL UNIVERSITY, YOKOHAMA 240-8501, JAPAN
*E-mail address*: ichiki-shunsuke-jb@ynu.jp

RESEARCH INSTITUTE OF ENVIRONMENT AND INFORMATION SCIENCES, YOKOHAMA NATIONAL UNIVERSITY, YOKOHAMA 240-8501, JAPAN
*E-mail address*: nishimura-takashi-yx@ynu.jp

Contemporary Mathematics
Volume **675**, 2016
http://dx.doi.org/10.1090/conm/675/13588

# Caustics of world hyper-sheets in the Minkowski space-time

Shyuichi Izumiya

ABSTRACT. In the Minkowski space-time, a world hyper-sheet is a timelike hypersurface formed by a one-parameter family of spacelike submanifolds. Recently, Bousso and Randall introduced the notion of caustics of world hyper-sheets in order to define the notion of holographic domains in the space-time. Here, we give a mathematical framework for describing the caustics of world hyper-sheets in the Minkowski space-time. As a consequence, we investigate the singularities of the caustics of world hyper-sheets and whose geometrical meanings. Although the Minkowski space-time has zero gravity, this framework gives a simple toy model for general cases.

## 1. Introduction

In this paper we investigate geometrical properties of caustics of world hyper-sheets in the Minkowski space-time as an application of the theory of graph-like Legendrian unfoldings [**19**]. Caustics appear in several area in physics (i.e. geometrical optics [**24**], the theory of underwater acoustics [**3**] and the theory of gravitational lensings[**25**] , and so on) and mathematics (i.e. classical differential geometry [**26**] and theory of differential equations [**8**, **11**], and so on [**2**]). The notion of caustics originally belongs to geometrical optics. We can observe the caustic formed by the rays reflected at a mirror. One of the examples of caustics in the classical differential geometry is the evolute of a curve in the Euclidean plane which is given by the envelope of normal lines emanate from the curve. The ray in the Euclidean plane is considered to be a line, so that the evolute is the caustic in the sense of geometrical optics. Moreover, the singular points of the evolute correspond to the vertices of the original curve. The vertex is the point at where the curve has higher order contact with the osculating circle (i.e. the point where the curvature has an extremum). Therefore, the singularities of the evolute provide an important geometrical information of the curve. We have the notion of evolutes for general hypersurfaces in the Euclidean space similar to the plane curve case. In particular, there are detailed investigations on evolutes for surfaces in Euclidean 3-space [**13**, **26**]. The evolute of a hypersurface can be defined in the Minkowski space-time analogous to the Euclidean case. If we consider a timelike hypersurface in the Minkowski space-time, the psuedo-normal line is directed by a spacelike vector, whose speed exceeds the speed of the ray. Therefore the evolute of a timelike hypersurface is not a caustic in the sense of physics. In the Minkowski space-time, the

2010 *Mathematics Subject Classification.* Primary 58K35; Secondary 53Z05, 78A10, 78A05.
*Key words and phrases.* Wave front propagations, caustics, Minkowski space-time.

ray emanate from a codimension two spacelike submanifold is a pseudo-normal line of the submanifold whose direction vector is lightlike, so the family of rays forms a lightlike hypersurface (i.e. a *light-sheet* in physics). The set of critical values of the light sheet is called a *lightlike focal set* along the spacelike submanifold. Actually, the notion of light-sheets plays an important role in physics which provides models of several kinds of horizons in space-times [**7**].

On the other hand, a *world hyper-sheet* in the Minkowski space-time is a timelike hypersurface formed by a one-parameter family of spacelike submanifolds of codimension two in the ambient space. Each spacelike submanifold in the world hyper-sheet is called a *momentary space*. We consider the family of lightlike hypersurfaces along monetary spaces in the world hyper-sheet. In [**4, 5**] Bousso and Randall considered that the locus of the singularities (the lightlike focal sets) of lightlike hypersurfaces along momentary spaces form a caustic in the Minkowski space-time. This construction is originally from the theoretical physics (the string theory, the brane world scenario, the cosmology, and so on). We call it a *BR-caustic* of the world hyper-sheet. Moreover, we have no notion of the time constant in the relativity theory. Hence everything that is moving depends on the time. Therefore, we consider world hyper-sheets in the relativity theory. Even if we consider a fixed light source (i.e. a shining surface), it must be a world hyper-sheet in the 4-dimensional Minkowski space-time. So the caustic in the Euclidean 3-space is a slice of the BR-caustic of the world hyper-sheet with a spacelike hyperplane.

In this paper we investigate the geometrical properties of BR-caustics as an application of the theory of graph-like Legendrian unfoldings [**19**]. In §2 we give the basic notions related to the Minkowski space-time. Basic geometrical framework for world hyper-sheets is given in §3. The light-sheet along a momentary space is introduced in §4 and some calculations are given by using Lorentz distance squared functions. In §5 the calculations in §4 are interpreted from the view point of contact with lightcones. We briefly review the theory of graph-like Legendrian unfoldings in §6. The notion of unfolded lightsheets is introduced as a special case of the graph-like wave front in §7. The caustic and the Maxwell set of the graph-like wave front are naturally induced in the theory of graph-like Legendrian unfoldings. In §8 the BR-caustic and the BR-Maxwell set are defined as the caustic and the Maxwell set of the graph-like wave front with respect to the distance squared function. We give a classification of the caustics of world sheets in the 3-dimensional Minkowski space-time in §9. As a consequence, the local classification of BR-cautics in §9 is different from the local classification of the evolutes of timelike surfaces in the 3-dimensional Minkowski space-time.

## 2. The Minkowski space-time

We now introduce some basic notions on the $(n+1)$-dimensional Minkowski space-time. For basic concepts and properties, see [**23**]. Let

$$\mathbb{R}^{n+1} = \{(x_0, x_1, \dots, x_n) \mid x_i \in \mathbb{R}\ (i = 0, 1, \dots, n)\ \}$$

be an $(n+1)$-dimensional cartesian space. For any $\boldsymbol{x} = (x_0, x_1, \dots, x_n)$, $\boldsymbol{y} = (y_0, y_1, \dots, y_n) \in \mathbb{R}^{n+1}$, the *pseudo scalar product* of $\boldsymbol{x}$ and $\boldsymbol{y}$ is defined to be $\langle \boldsymbol{x}, \boldsymbol{y} \rangle = -x_0y_0 + \sum_{i=1}^n x_iy_i$. We call $(\mathbb{R}^{n+1}, \langle , \rangle)$ the $(n+1)$*-dimensional Minkowski space-time* (or briefly, the *Lorentz-Minkowski* $(n+1)$*-space*). We write $\mathbb{R}^{n+1}_1$ instead of $(\mathbb{R}^{n+1}, \langle , \rangle)$. We say that a non-zero vector $\boldsymbol{x} \in \mathbb{R}^{n+1}_1$ is *spacelike, lightlike* *or timelike* if $\langle \boldsymbol{x}, \boldsymbol{x}\rangle > 0$, $\langle \boldsymbol{x}, \boldsymbol{x}\rangle = 0$ or $\langle \boldsymbol{x}, \boldsymbol{x}\rangle < 0$, respectively. The norm of the vector $\boldsymbol{x} \in \mathbb{R}^{n+1}_1$ is defined to be $\|\boldsymbol{x}\| = \sqrt{|\langle \boldsymbol{x}, \boldsymbol{x}\rangle|}$. We have the canonical projection $\pi : \mathbb{R}^{n+1}_1 \longrightarrow \mathbb{R}^n$ defined by $\pi(x_0, x_1, \ldots, x_n) = (x_1, \ldots, x_n)$. Here we identify $\{\mathbf{0}\} \times \mathbb{R}^n$ with $\mathbb{R}^n$ and it is considered as the Euclidean $n$-space whose scalar product is induced by the pseudo scalar product $\langle , \rangle$. For a vector $\boldsymbol{v} \in \mathbb{R}^{n+1}_1$ and a real number $c$, we define a *hyperplane with pseudo normal* $\boldsymbol{v}$ by

$$HP(\boldsymbol{v}, c) = \{\boldsymbol{x} \in \mathbb{R}^{n+1}_1 \mid \langle \boldsymbol{x}, \boldsymbol{v}\rangle = c\, \}.$$

We call $HP(\boldsymbol{v}, c)$ a *spacelike hyperplane*, a *timelike hyperplane* or a *lightlike hyperplane* if $\boldsymbol{v}$ is timelike, spacelike or lightlike, respectively.

We now define the *Hyperbolic n-space* by

$$H^n_+(-1) = \{\boldsymbol{x} \in \mathbb{R}^{n+1}_1 | \langle \boldsymbol{x}, \boldsymbol{x}\rangle = -1, x_0 > 0\}$$

and the *de Sitter n-space* by

$$S^n_1 = \{\boldsymbol{x} \in \mathbb{R}^{n+1}_1 | \langle \boldsymbol{x}, \boldsymbol{x}\rangle = 1\, \}.$$

We define

$$LC(\boldsymbol{\lambda}) = \{\boldsymbol{x} = (x_0, x_1, \ldots, x_n) \in \mathbb{R}^{n+1}_1 \mid \langle \boldsymbol{x} - \boldsymbol{\lambda}, \boldsymbol{x} - \boldsymbol{\lambda}\rangle = 0\}$$

and we call it *the lightcone* with the vertex $\boldsymbol{\lambda} \in \mathbb{R}^{n+1}_1$. We write $LC^* = LC(\mathbf{0}) \setminus \{\mathbf{0}\}$, which is called an *open lightcone* at the origin.

For any $\boldsymbol{x}_1, \boldsymbol{x}_2, \ldots, \boldsymbol{x}_n \in \mathbb{R}^{n+1}_1$, we define a vector $\boldsymbol{x}_1 \wedge \boldsymbol{x}_2 \wedge \cdots \wedge \boldsymbol{x}_n$ by

$$\boldsymbol{x}_1 \wedge \boldsymbol{x}_2 \wedge \cdots \wedge \boldsymbol{x}_n = \begin{vmatrix} -\boldsymbol{e}_0 & \boldsymbol{e}_1 & \cdots & \boldsymbol{e}_n \\ x^1_0 & x^1_1 & \cdots & x^1_n \\ x^2_0 & x^2_1 & \cdots & x^2_n \\ \vdots & \vdots & \cdots & \vdots \\ x^n_0 & x^n_1 & \cdots & x^n_n \end{vmatrix},$$

where $\boldsymbol{e}_0, \boldsymbol{e}_1, \ldots, \boldsymbol{e}_n$ is the canonical basis of $\mathbb{R}^{n+1}_1$ and $\boldsymbol{x}_i = (x^i_0, x^i_1, \ldots, x^i_n)$. We can easily check that $\langle \boldsymbol{x}, \boldsymbol{x}_1 \wedge \boldsymbol{x}_2 \wedge \cdots \wedge \boldsymbol{x}_n\rangle = \det(\boldsymbol{x}, \boldsymbol{x}_1, \ldots, \boldsymbol{x}_n)$, so that $\boldsymbol{x}_1 \wedge \boldsymbol{x}_2 \wedge \cdots \wedge \boldsymbol{x}_n$ is pseudo orthogonal to any $\boldsymbol{x}_i$ $(i = 1, \ldots, n)$.

## 3. World hyper-sheets in the Minkowski space-time

We briefly review the basic geometrical framework for the study of world hyper-sheets in the $(n+1)$-dimensional Minkowski space-time in [**18**]. Let $\mathbb{R}^{n+1}_1$ be a time-oriented space (cf. [**23**]). We choose $\boldsymbol{e}_0 = (1, 0, \ldots, 0)$ as the future timelike vector field. In the theory of relativity, we do not have the notion of time constant, so that everything that is moving depends on the time. Therefore, we consider world sheets. Although we have the notion of world sheets with general codimension, we stick to the case when the codimension one, that are called world hyper-sheets in the Minkowski space-time. A world sheet is defined to be a timelike submanifold foliated by codimension one spacelike submanifolds. Here, we only investigate the local situation, so that we consider a one-parameter family of spacelike submanifolds. Let $\boldsymbol{X} : U \times I \longrightarrow \mathbb{R}^{n+1}_1$ be a timelike embedding, where $U \subset \mathbb{R}^{n-1}$ is an open subset and $I$ is an open interval. We write $W = \boldsymbol{X}(U \times I)$ and identify $W$ and $U \times I$ through the embedding $\boldsymbol{X}$. The embedding $\boldsymbol{X}$ is said to be *timelike* if the tangent space $T_pW$ of $W$ is a timelike hyperplane at any point $p \in W$. We write that $\mathcal{S}_t = \boldsymbol{X}(U \times \{t\})$ for each $t \in I$. We have a foliation of $W$ defined by $\mathcal{S} = \{\mathcal{S}_t\}_{t \in I}$.

We say that $W = \boldsymbol{X}(U \times I)$ (or, $(W, \mathcal{S})$) is a *world hyper-sheet* if $W$ is a time-orientable timelike hypersurface and each $\mathcal{S}_t$ is spacelike. Here, we say that $\mathcal{S}_t$ is *spacelike* if the tangent space $T_p\mathcal{S}_t$ consists only spacelike vectors (i.e. spacelike subspace of $\mathbb{R}^{n+1}_1$) for any point $p \in \mathcal{S}_t$. Each $\mathcal{S}_t$ is called a *momentary space* of $W$. For any $p = \boldsymbol{X}(\overline{u}, t) \in W \subset \mathbb{R}^{n+1}_1$, we have

$$T_pW = \langle \boldsymbol{X}_{u_1}(\overline{u}, t), \dots, \boldsymbol{X}_{u_{n-1}}(\overline{u}, t), \boldsymbol{X}_t(\overline{u}, t)\rangle_{\mathbb{R}},$$

where we write $(\overline{u}, t) = (u_1, \dots, u_{n-1}, t) \in U \times I$, $\boldsymbol{X}_t = \partial \boldsymbol{X}/\partial t$ and $\boldsymbol{X}_{u_j} = \partial \boldsymbol{X}/\partial u_j$. We also have

$$T_p\mathcal{S}_t = \langle \boldsymbol{X}_{u_1}(\overline{u}, t), \dots, \boldsymbol{X}_{u_{n-1}}(\overline{u}, t)\rangle_{\mathbb{R}}.$$

Since $W$ is time-orientable, there exists a timelike vector field $\boldsymbol{v}(\overline{u}, t)$ on $W$ [**23**, Lemma 32]. Moreover, we can choose that $\boldsymbol{v}$ is *future directed* which means that $\langle \boldsymbol{v}(\overline{u}, t), \boldsymbol{e}_0\rangle < 0$. Since $\operatorname{codim} W = 1$, we have $\operatorname{codim} \mathcal{S}_t = 2$. Moreover, $\mathcal{S}_t$ is spacelike, so that we can apply the method developed in [**14**]. We consider the unit normal spacelike vector of $W$ defined by

$$\boldsymbol{n}^S(\overline{u}, t) = \frac{\boldsymbol{X}_{u_1}(\overline{u}, t) \wedge \cdots \wedge \boldsymbol{X}_{u_{n-1}}(\overline{u}, t) \wedge \boldsymbol{X}_t(\overline{u}, t)}{\|\boldsymbol{X}_{u_1}(\overline{u}, t) \wedge \cdots \wedge \boldsymbol{X}_{u_{n-1}}(\overline{u}, t) \wedge \boldsymbol{X}_t(\overline{u}, t)\|}.$$

For any $t \in I$, Let $N_p(\mathcal{S}_t)$ be the pseudo-normal space of $\mathcal{S}_t$ at $p = \boldsymbol{X}(\overline{u}, t)$ in $\mathbb{R}^{n+1}_1$. Since $\mathcal{S}_t$ is a codimension one in $W$, $N_p(\mathcal{S}_t)$ is a two dimensional Lorentz space. There exists a unique timelike unit vector field $\boldsymbol{n}^T(\overline{u}, t) \in N_p(\mathcal{S}_t) \cap T_pW$ such that it is future directed (i.e. $\langle \boldsymbol{n}^T(\overline{u}, t), \boldsymbol{e}_0\rangle < 0$). We now define maps $\mathbb{LG}^{\pm}(\mathcal{S}_t) : \mathcal{S}_t \longrightarrow LC^*$ by $\mathbb{LG}^{\pm}(\mathcal{S}_t)(p) = \boldsymbol{n}^T(\overline{u}, t) \pm \boldsymbol{n}^S(\overline{u}, t)$, where $p = \boldsymbol{X}(\overline{u}, t)$. We call each one of $\mathbb{LG}^{\pm}(\mathcal{S}_t)$ a *momentary lightcone Gauss map*. These maps lead us to the notion of curvatures (cf. [**18**]). We have linear maps $d\mathbb{LG}^{\pm}(\mathcal{S}_t)_p : T_p\mathcal{S}_t \longrightarrow T_{\widetilde{p}}LC^* \subset T_{\widetilde{p}}\mathbb{R}^{n+1}_1$, where $p = \boldsymbol{X}(\overline{u}, t)$ and $\widetilde{p} = \boldsymbol{n}^T(\overline{u}, t) \pm \boldsymbol{n}^S(\overline{u}, t)$. With the identification $T_{\widetilde{p}}\mathbb{R}^{n+1}_1 \equiv \mathbb{R}^{n+1}_1 \equiv T_p\mathbb{R}^{n+1}_1$, we have the canonical decomposition $T_p\mathbb{R}^{n+1} = T_p\mathcal{S}_t \oplus N_p(\mathcal{S}_t)$. Let $\Pi^t : T_p\mathbb{R}^{n+1} = T_p\mathcal{S}_t \oplus N_p(\mathcal{S}_t) \longrightarrow T_p\mathcal{S}_t$ be the canonical projection. Then we have linear transformations

$$S^{\pm}_{\ell}(\mathcal{S}_t)_p = -\Pi^t \circ d\mathbb{LG}^{\pm}(\mathcal{S}_t)_p : T_p\mathcal{S}_t \longrightarrow T_p\mathcal{S}_t.$$

Each one of the above mappings is called a *momentary lightcone shape operator* of $\mathcal{S}_t$ at $p = \boldsymbol{X}(\overline{u}, t)$. Let $\{\kappa^{\pm}_i(\mathcal{S}_t)(p)\}_{i=1}^{n-1}$ be the set of eigenvalues of $S^{\pm}_{\ell}(\mathcal{S}_t)_p$, which are called *momentary lightcone principal curvatures* of $\mathcal{S}_t$ at $p = \boldsymbol{X}(\overline{u}, t)$. Then *momentary lightcone Gauss-Kronecker curvatures* of $\mathcal{S}_t$ at $p = \boldsymbol{X}(\overline{u}, t)$ are defined to be

$$K^{\pm}_{\ell}(\mathcal{S}_t)(p) = \det S^{\pm}_{\ell}(\mathcal{S}_t)_p.$$

We obtain now the lightcone Weingarten formulae. Since $\mathcal{S}_t$ is a spacelike submanifold, we have a Riemannian metric (the *first fundamental form* ) on $\mathcal{S}_t$ defined by $ds^2 = \sum_{i=1}^{n-1} g_{ij}du_idu_j$, where $g_{ij}(\overline{u}, t) = \langle \boldsymbol{X}_{u_i}(\overline{u}, t), \boldsymbol{X}_{u_j}(\overline{u}, t)\rangle$ for any $(\overline{u}, t) \in U \times I$. *Lightcone second fundamental invariants* are defined to be $h[\pm]_{ij}(\overline{u}, t) = \langle -(\boldsymbol{n}^T \pm \boldsymbol{n}^S)_{u_i}(\overline{u}, t), \boldsymbol{X}_{u_j}(\overline{u}, t)\rangle$ for any $(\overline{u}, t) \in U \times I$. The following lightcone Weingarten formulae are given as special cases of the formulae in [**14**]:

(a) $(\boldsymbol{n}^T \pm \boldsymbol{n}^S)_{u_i} = \langle \boldsymbol{n}^S, \boldsymbol{n}^T_{u_i}\rangle(\boldsymbol{n}^T \pm \boldsymbol{n}^S) - \sum_{j=1}^{n-1} h^j_i[\pm]\boldsymbol{X}_{u_j}$

(b) $\Pi^t \circ (\boldsymbol{n}^T + \boldsymbol{n}^S)_{u_i} = -\sum_{j=1}^{n-1} h^j_i[\pm]\boldsymbol{X}_{u_j}$.

Here $\left(h^j_i[\pm]\right) = \left(h_{ik}[\pm]\right)\left(g^{kj}\right)$ and $\left(g^{kj}\right) = \left(g_{kj}\right)^{-1}$.

It follows that the momentary lightcone principal curvatures are the eigenvalues of $\left(h_i^j[\pm]\right)$.

## 4. Light sheets along momentary spaces

We define hypersurfaces $\mathbb{LH}^{\pm}_{\mathcal{S}_{t_0}} : U \times \{t_0\} \times \mathbb{R} \longrightarrow \mathbb{R}^{n+1}_1$ by

$$\mathbb{LH}^{\pm}_{\mathcal{S}_{t_0}}(p,\mu) = \mathbb{LH}^{\pm}_{\mathcal{S}_{t_0}}(\overline{u},t_0,\mu) = \boldsymbol{X}(\overline{u},t_0) + \mu\mathbb{LG}^{\pm}(\mathcal{S}_{t_0})(\overline{u},t_0),$$

where $p = \boldsymbol{X}(\overline{u},t_0)$. We call $\mathbb{LH}^{\pm}_{\mathcal{S}_{t_0}}$ *light sheets* along $\mathcal{S}_{t_0}$. A hypersurface $H \subset \mathbb{R}^{n+1}_1$ is, generally, called a *lightlike hypersurface* if it is tangent to a lightcone at any point. The light sheet along $\mathcal{S}_{t_0}$ is a lightlike hypersurface. We also define $\mathbb{LH}^{\pm}_{W} : U \times I \times \mathbb{R} \longrightarrow \mathbb{R}^{n+1}_1 \times I$ by

$$\mathbb{LH}^{\pm}_{W}(\overline{u},t,\mu) = (\mathbb{LH}^{\pm}_{\mathcal{S}_t}(\overline{u},t,\mu),t),$$

which are called *unfolded light sheets* of $(W,\mathcal{S})$.

We introduce the notion of Lorentz distance-squared functions on a world hyper-sheet, which is useful for the study of singularities of light sheets. We define a family of functions $G : W \times \mathbb{R}^{n+1}_1 \longrightarrow \mathbb{R}$ on $W = \boldsymbol{X}(U \times I)$ by

$$G(p,\boldsymbol{\lambda}) = G(\overline{u},t,\boldsymbol{\lambda}) = \langle \boldsymbol{X}(\overline{u},t) - \boldsymbol{\lambda}, \boldsymbol{X}(\overline{u},t) - \boldsymbol{\lambda}\rangle,$$

where $p = \boldsymbol{X}(\overline{u},t)$. We call $G$ a *Lorentz distance-squared function* on the world hyper-sheet $(W,\mathcal{S})$. For any fixed $(t_0,\boldsymbol{\lambda}_0) \in I \times \mathbb{R}^{n+1}_1$, we write $g(\overline{u}) = G_{(t_0,\boldsymbol{\lambda}_0)}(\overline{u}) = G(\overline{u},t_0,\boldsymbol{\lambda}_0)$ and have the following proposition.

PROPOSITION 4.1. *Let $\mathcal{S}_{t_0}$ be a momentary space of $(W,\mathcal{S})$ and $G : W \times \mathbb{R}^{n+1}_1 \to \mathbb{R}$ the Lorentz distance-squared function on $(W,\mathcal{S})$. Suppose that $p_0 = \boldsymbol{X}(\overline{u}_0,t_0) \neq \boldsymbol{\lambda}_0$. Then we have the following:*
*(1) $g(\overline{u}_0) = \partial g/\partial u_i(\overline{u}_0) = 0$ $(i = 1,\dots,n-1)$ if and only if $p_0 - \boldsymbol{\lambda}_0 = \mu\mathbb{LG}^{\pm}(\mathcal{S}_{t_0})(p_0)$ for some $\mu \in \mathbb{R} \setminus \{0\}$.*
*(2) $g(\overline{u}_0) = \partial g/\partial u_i(\overline{u}_0) = \det \mathcal{H}(g)(\overline{u}_0) = 0$ $(i = 1,\dots,n-1)$ if and only if*

$$p_0 - \boldsymbol{\lambda}_0 = \mu\mathbb{LG}^{\pm}(\mathcal{S}_{t_0})(p_0)$$

*for $\mu \in \mathbb{R}\setminus\{0\}$ such that $-1/\mu$ is one of the non-zero momentary lightcone principal curvatures $\{\kappa_i^{\pm}(\mathcal{S}_t)(p)\}_{i=1}^{n-1}$.*
*Here, $\det \mathcal{H}(g)(\overline{u}_0)$ is the determinant of the Hessian matrix of $g$ at $\overline{u}_0$.*

*Proof.* (1) The condition $g(\overline{u}) = \langle \boldsymbol{X}(\overline{u},t_0) - \boldsymbol{\lambda}_0, \boldsymbol{X}(\overline{u},t_0) - \boldsymbol{\lambda}_0\rangle = 0$ means that $\boldsymbol{X}(\overline{u},t_0) - \boldsymbol{\lambda}_0 \in LC^*$. We can observe that $dg(\overline{u}) = \langle d\boldsymbol{X}(\overline{u},t_0), \boldsymbol{X}(\overline{u},t_0) - \boldsymbol{\lambda}_0\rangle = 0$ if and only if $\boldsymbol{X}(\overline{u},t_0) - \boldsymbol{\lambda}_0 \in N_pM$. Hence $g(\overline{u}_0) = dg(\overline{u}_0) = 0$ if and only if $p_0 - \boldsymbol{\lambda}_0 \in N_pM \cap LC^*$. This is equivalent to the condition that $p_0 - \boldsymbol{\lambda}_0 = \mu\mathbb{LG}^{\pm}(\mathcal{S}_{t_0})(p_0)$ for some $\mu \in \mathbb{R} \setminus \{0\}$.

(2) We can show that

$$\frac{\partial g}{\partial u_i} = 2\langle \boldsymbol{X}_{u_i}, \boldsymbol{X} - \boldsymbol{\lambda}_0\rangle$$

and

$$\frac{\partial^2 g}{\partial u_i \partial u_j} = 2\left\{\boldsymbol{X}_{u_iu_j}, \boldsymbol{X} - \boldsymbol{\lambda}_0\rangle + \langle \boldsymbol{X}_{u_i}, \boldsymbol{X}_{u_j}\rangle\right\}.$$

With the condition $p_0 - \boldsymbol{\lambda}_0 = \mu\mathbb{LG}^{\pm}(\mathcal{S}_{t_0})(p_0)$, we have

$$\frac{\partial^2 g}{\partial u_i \partial u_j} = 2\left\{\langle \boldsymbol{X}_{u_i u_j}, \mu\mathbb{LG}^{\pm}(\mathcal{S}_{t_0})(p_0)\rangle + g_{ij}(\overline{u}_0, t_0)\right\}.$$

Therefore, we have

$$\left(\frac{\partial^2 g}{\partial u_i \partial u_j}\right)(g^{k\ell}) = \left(2\left\{\mu h^i_j[\pm] + \delta^i_j\right\}\right).$$

It follows that $\det \mathcal{H}(g)(p_0) = 0$ if and only if $-1/\mu$ is an eigenvalue of $(h^i_j[\pm](p_0))$. □

Inspired by the above result, we define

$$\mathbb{LF}^{\pm}_{\mathcal{S}_{t_0}} = \bigcup_{i=1}^{n-1}\left\{\boldsymbol{X}(u,t_0) + \frac{1}{\kappa^{\pm}_i(\mathcal{S}_t)(p)}\mathbb{LG}^{\pm}(\mathcal{S}_{t_0})(p) \mid u \in U, p = \boldsymbol{X}(u,t_0)\right\},$$

which are called *lightlike focal sets* of $\mathcal{S}_{t_0}$. Moreover, *unfolded lightcone focal sets* of $(W, \mathcal{S})$ are defined to be

$$\mathbb{LF}^{\pm}_{(W,\mathcal{S})} = \bigcup_{t\in I} \mathbb{LF}^{\pm}_{\mathcal{S}_t} \times \{t\} \subset \mathbb{R}^{n+1}_1 \times I.$$

Each one of $\mathbb{LF}^{\pm}_{(W,\mathcal{S})}$ is the critical value set of $\mathbb{LH}^{\pm}_W$, respectively.

## 5. Contact with lightcones

In this section we interpret the results of Proposition 4.1 from the view point of the contact with lightcones.

We consider the relationship between the contact of a one parameter family of submanifolds with a submanifold and $S.P$-$\mathcal{K}$-equivalence among functions (cf. [**9**]). Let $U_i \subset \mathbb{R}^r$, $(i = 1, 2)$ be open sets and $g_i : (U_i \times I, (\overline{u}_i, t_i)) \longrightarrow (\mathbb{R}^n, \boldsymbol{y}_i)$ immersion germs. We define $\overline{g}_i : (U_i \times I, (\overline{u}_i, t_i)) \longrightarrow (\mathbb{R}^n \times I, (\boldsymbol{y}_i, t_i))$ by $\overline{g}_i(\overline{u}, t) = (g_i(\overline{u}), t)$. We write that $(\overline{Y}_i, (\boldsymbol{y}_i, t_i)) = (\overline{g}_i(U_i \times I), (\boldsymbol{y}_i, t_i))$. Let $f_i : (\mathbb{R}^n, \boldsymbol{y}_i) \longrightarrow (\mathbb{R}, 0)$ be submersion germs and write that $(V(f_i), \boldsymbol{y}_i) = (f_i^{-1}(0), \boldsymbol{y}_i)$. We say that *the contact of* $\overline{Y}_1$ *with the trivial family of* $V(f_1)$ *at* $(\boldsymbol{y}_1, t_1)$ is of the *same type in the strict sense* as *the contact of* $\overline{Y}_2$ *with the trivial family of* $V(f_2)$ *at* $(\boldsymbol{y}_2, t_2)$ if there is a diffeomorphism germ $\Phi : (\mathbb{R}^n \times I, (\boldsymbol{y}_1, t_1)) \longrightarrow (\mathbb{R}^n \times I, (\boldsymbol{y}_2, t_2))$ of the form $\Phi(\boldsymbol{y}, t) = (\phi_1(\boldsymbol{y}, t), t + (t_2 - t_1))$ such that $\Phi(\overline{Y}_1) = \overline{Y}_2$ and $\Phi(V(f_1) \times I) = V(f_2) \times I$. In this case we write $SK(\overline{Y}_1, V(f_1) \times I; (\boldsymbol{y}_1, t_1)) = SK(\overline{Y}_2, V(f_2) \times I; (\boldsymbol{y}_2, t_2))$. We can show the following proposition which is analogous to Montldi's theorem on contact between submanifolds in [**22**]:

PROPOSITION 5.1. *With the same notations as in the above paragraphs,*

$$SK(\overline{Y}_1, V(f_1) \times I; (\boldsymbol{y}_1, t_1)) = SK(\overline{Y}_2, V(f_2) \times I; (\boldsymbol{y}_2, t_2))$$

*if and only if* $f_1 \circ g_1$ *and* $f_2 \circ g_2$ *are* $S.P$-$\mathcal{K}$*-equivalent (i.e. there exists a diffeomorphism germ* $\Psi : (U_1 \times I, (\overline{u}_1, t_1)) \longrightarrow (U_2 \times I, (\overline{u}_2, t_2))$ *of the form* $\Psi(\overline{u}, t) = (\psi_1(\overline{u}, t), t - (t_2 - t_1))$ *and a function germ* $\lambda : (U_1 \times I, (\overline{u}_1, t_1)) \longrightarrow \mathbb{R}$ *with* $\lambda(\overline{u}_1, t_1) \neq 0$ *such that* $(f_2 \circ g_2) \circ \Phi(\overline{u}, t) = \lambda(\overline{u}, t) f_1 \circ g_1(\overline{u}, t)$*).*

Since the proof of Proposition 5.1 is given by the arguments just along the line of the proof of the original theorem in [**22**], we omit the proof here.

We now consider a function $\mathfrak{g}_{\boldsymbol{\lambda}} : \mathbb{R}^{n+1}_1 \longrightarrow \mathbb{R}$ defined by $\mathfrak{g}_{\boldsymbol{\lambda}}(\boldsymbol{x}) = \langle \boldsymbol{x} - \boldsymbol{\lambda}, \boldsymbol{x} - \boldsymbol{\lambda}\rangle$, where $\boldsymbol{\lambda} \in \mathbb{R}^{n+1}_1 \setminus W$. For any $\boldsymbol{\lambda}_0 \in \mathbb{R}^{n+1}_1$, we have a lightcone $\mathfrak{g}^{-1}_{\boldsymbol{\lambda}_0}(0) = LC(\boldsymbol{\lambda}_0)$. Moreover, we consider the lightlike vectors $\boldsymbol{\lambda}^{\pm}_0 = \mathbb{LH}^{\pm}_{\mathcal{S}_{t_0}}(p_0, \mu_0)$, where $p_0 = \boldsymbol{X}(\overline{u}_0, t_0)$. Then we have

$$\mathfrak{g}_{\boldsymbol{\lambda}^{\pm}_0} \circ \boldsymbol{X}(\overline{u}_0, t_0) = G((u_0, t_0), \mathbb{LH}^{\pm}_{\mathcal{S}_{t_0}}(p_0, \mu_0)) = 0.$$

By Proposition 4.1, we also have relations that

$$\frac{\partial \mathfrak{g}_{\boldsymbol{\lambda}^{\pm}_0} \circ \boldsymbol{X}}{\partial u_i}(\overline{u}_0, t_0) = \frac{\partial G}{\partial u_i}((u_0, t_0), \mathbb{LH}^{\pm}_{\mathcal{S}_{t_0}}(p_0, \mu_0)) = 0.$$

for $i = 1, \dots, n-1$. These relations mean that the lightcones $\mathfrak{g}^{-1}_{\boldsymbol{\lambda}^{\pm}_0}(0) = LC(\boldsymbol{\lambda}^{\pm}_0)$ are tangent to $\mathcal{S}_{t_0} = \boldsymbol{X}(U \times \{t_0\})$ at $p_0 = \boldsymbol{X}(\overline{u}_0, t_0)$. Each one of the lightcones $LC(\boldsymbol{\lambda}^{\pm}_0)$ is said to be a *tangent lightcone* of $\mathcal{S}_{t_0} = \boldsymbol{X}(U \times \{t_0\})$ at $p_0 = \boldsymbol{X}(\overline{u}_0, t_0)$, which we write $TLC(\mathcal{S}_{t_0}, \boldsymbol{\lambda}^{\pm}_0)$, where $\boldsymbol{\lambda}^{\pm}_0 = \mathbb{LH}^{\pm}_{\mathcal{S}_{t_0}}(p_0, \mu_0)$. Then we have the following simple lemma.

LEMMA 5.2. *Let $\boldsymbol{X} : U \times I \longrightarrow \mathbb{R}^{n+1}_1$ be a world hyper-sheet. Consider two points $p_i = \boldsymbol{X}(\overline{u}_i, t_0)$, $(i = 1, 2)$. Then*

$$\mathbb{LH}^{\pm}_{\mathcal{S}_{t_0}}(p_1, \mu_1) = \mathbb{LH}^{\pm}_{\mathcal{S}_{t_0}}(p_2, \mu_2)$$

*if and only if*

$$TLC(\mathcal{S}_{t_0}, \mathbb{LH}^{\pm}_{\mathcal{S}_{t_0}}(p_1, \mu_1)) = TLP(\mathcal{S}_{t_0}, \mathbb{LH}^{\pm}_{\mathcal{S}_{t_0}}(p_2, \mu_2)).$$

As a consequence, we have tools for the study of the contact between momentary spaces and families of lightcones. We write that $g_{\boldsymbol{\lambda}}(\overline{u}, t) = G(\overline{u}, t, \boldsymbol{\lambda})$. Then we have $g_{\boldsymbol{\lambda}}(\overline{u}, t) = \mathfrak{g}_{\boldsymbol{\lambda}} \circ \boldsymbol{X}(\overline{u}, t)$, so that we have the following proposition as a corollary of Proposition 5.1.

PROPOSITION 5.3. *Let $\boldsymbol{X}_i : (U \times I, (\overline{u}_i, t_0)) \longrightarrow (\mathbb{R}^{n+1}_1, p_i)$, $(i = 1, 2)$, be world hypersheet germs and $\boldsymbol{\lambda}^{\pm}_i = \mathbb{LH}^{\pm}_{\mathcal{S}_{t_0}}(p_i, \mu_i))$ and $W_i = \boldsymbol{X}_i(U \times I)$. Then the following conditions are equivalent:*
(1) $SK(\overline{W}_1, TLC(\mathcal{S}_{t_0}, \boldsymbol{\lambda}^{\pm}_1) \times I; (p_1, t_0)) = SK(\overline{W}_2, TLC(\mathcal{S}_{t_0}, \boldsymbol{\lambda}^{\pm}_2) \times I; (p_2, t_0))$,
(2) $g_{1,\boldsymbol{\lambda}^{\pm}_1}$ *and* $g_{2,\boldsymbol{\lambda}^{\pm}_2}$ *are S.P-$\mathcal{K}$-equivalent.*
*Here,* $g_{i,\boldsymbol{\lambda}^{\pm}_i}(\overline{u}, t) = G_i(\overline{u}, t, \boldsymbol{\lambda}^{\pm}_i) = \langle \boldsymbol{X}_i(\overline{u}, t) - \boldsymbol{\lambda}^{\pm}_i, \boldsymbol{X}_i(\overline{u}, t) - \boldsymbol{\lambda}^{\pm}_i\rangle$, $(i = 1, 2)$.

## 6. Graph-like wave fronts

In this section we briefly review the theory of graph-like Legendrian unfoldings. Graph-like Legendrian unfoldings belong to a special class of big Legendrian submanifolds (for detail, see [**10, 15–17, 30**]). Recently there appeared a survey article [**19**] on the theory of graph-like Legendrian unfoldings. Let $\mathcal{F} : (\mathbb{R}^k \times (\mathbb{R}^m \times \mathbb{R}), 0) \to (\mathbb{R}, 0)$ be a function germ. We say that $\mathcal{F}$ is a *graph-like Morse family of hypersurfaces* if $(\mathcal{F}, d_q\mathcal{F}) : (\mathbb{R}^k \times (\mathbb{R}^m \times \mathbb{R}), 0) \to (\mathbb{R} \times \mathbb{R}^k, 0)$ is a non-singular and $(\partial\mathcal{F}/\partial t)(0) \neq 0$, where

$$d_q\mathcal{F}(q, x, t) = \left(\frac{\partial \mathcal{F}}{\partial q_1}(q, x, t), \dots, \frac{\partial \mathcal{F}}{\partial q_k}(q, x, t)\right).$$

Moreover, we say that $\mathcal{F}$ is *non-degenerate* if $(\mathcal{F}, d_q\mathcal{F})|_{\mathbb{R}^k \times (\mathbb{R}^m \times \{0\})}$ is non-singular. For a graph-like Morse family of hypersurfaces $\mathcal{F}$, $\Sigma_*(\mathcal{F}) = (\mathcal{F}, d_q\mathcal{F})^{-1}(0)$ is a smooth $m$-dimensional submanifold germ of $(\mathbb{R}^k \times (\mathbb{R}^m \times \mathbb{R}), 0)$. We now consider the space of 1-jets $J^1(\mathbb{R}^m, \mathbb{R})$ with the canonical coordinates $(x_1, \ldots, x_m, t, p_1, \ldots, p_m)$ such that the canonical contact form is $\theta = dt - \sum_{i=1}^m p_i dx_i$. We define a mapping $\Pi : J^1(\mathbb{R}^m, \mathbb{R}) \longrightarrow T^*\mathbb{R}^m$ by $\Pi(x, t, p) = (x, p)$, where

$$(x, t, p) = (x_1, \ldots, x_m, t, p_1, \ldots, p_m).$$

Here, $T^*\mathbb{R}^m$ is a symplectic manifold with the canonical symplectic structure $\omega = \sum_{i=1}^m dp_i \wedge dx_i$ (cf. [**1**]). We define a mapping $\mathscr{L}_{\mathcal{F}} : (\Sigma_*(\mathcal{F}), 0) \to J^1(\mathbb{R}^m, \mathbb{R})$ by

$$\mathscr{L}_{\mathcal{F}}(q, x, t) = \left( x, t, -\frac{\dfrac{\partial \mathcal{F}}{\partial x_1}(q, x, t)}{\dfrac{\partial \mathcal{F}}{\partial t}(q, x, t)}, \ldots, -\frac{\dfrac{\partial \mathcal{F}}{\partial x_m}(q, x, t)}{\dfrac{\partial \mathcal{F}}{\partial t}(q, x, t)}, \right).$$

It is easy to show that $\mathscr{L}_{\mathcal{F}}(\Sigma_*(\mathcal{F}))$ is a Legendrian submanifold germ (cf., [**1**]), which is called a *graph-like Legendrian unfolding germ.* We call

$$\overline{\pi}|_{\mathscr{L}_{\mathcal{F}}(\Sigma_*(\mathcal{F}))} : \mathscr{L}_{\mathcal{F}}(\Sigma_*(\mathcal{F})) \longrightarrow \mathbb{R}^m \times \mathbb{R}$$

a *graph-like Legendrian map germ*, where $\overline{\pi} : J^1(\mathbb{R}^m, \mathbb{R}) \longrightarrow \mathbb{R}^m \times \mathbb{R}$ is the canonical projection. We also call $W(\mathscr{L}_{\mathcal{F}}(\Sigma_*(\mathcal{F}))) = \overline{\pi}(\mathscr{L}_{\mathcal{F}}(\Sigma_*(\mathcal{F})))$ a *graph-like wave front* of $\mathscr{L}_{\mathcal{F}}(\Sigma_*(\mathcal{F}))$. We say that $\mathcal{F}$ is a *graph-like generating family* of $\mathscr{L}_{\mathcal{F}}(\Sigma_*(\mathcal{F}))$. Moreover, we call $W_t(\mathscr{L}_{\mathcal{F}}(\Sigma_*(\mathcal{F}))) = \pi_1(\pi_2^{-1}(t) \cap W(\mathscr{L}_{\mathcal{F}}(\Sigma_*(\mathcal{F})))$ a *momentary front* for each $t \in (\mathbb{R}, 0)$, where $\pi_1 : \mathbb{R}^m \times \mathbb{R} \longrightarrow \mathbb{R}^m$ and $\pi_2 : \mathbb{R}^m \times \mathbb{R} \longrightarrow \mathbb{R}$ are the canonical projections. The *discriminant set of the family* $\{W_t(\mathscr{L}_{\mathcal{F}}(\Sigma_*(\mathcal{F})))\}_{t\in(\mathbb{R},0)}$ is defined by the union of the caustic $C_{\mathscr{L}_{\mathcal{F}}(\Sigma_*(\mathcal{F}))} = \pi_1(\Sigma(W(\mathscr{L}_{\mathcal{F}}(\Sigma_*(\mathcal{F})))))$ and the Maxwell stratified set $M_{\mathscr{L}_{\mathcal{F}}(\Sigma_*(\mathcal{F}))} = \pi_1(SI_{W(\mathscr{L}_{\mathcal{F}}(\Sigma_*(\mathcal{F})))})$, where $\Sigma(W(\mathscr{L}_{\mathcal{F}}(\Sigma_*(\mathcal{F})))$ is the critical value set of $\overline{\pi}|_{\mathscr{L}_{\mathcal{F}}(\Sigma_*(\mathcal{F}))}$ and $SI_{W(\mathscr{L}_{\mathcal{F}}(\Sigma_*(\mathcal{F})))}$ is the closure of the self intersection set of $W(\mathscr{L}_{\mathcal{F}}(\Sigma_*(\mathcal{F})))$.

We now define equivalence relations among graph-like Legendrian unfoldings. Let $\mathcal{F} : (\mathbb{R}^k \times (\mathbb{R}^m \times \mathbb{R}), 0) \to (\mathbb{R}, 0)$ and $\mathcal{G} : (\mathbb{R}^k \times (\mathbb{R}^m \times \mathbb{R}), 0) \to (\mathbb{R}, 0)$ be graph-like Morse families of hypersurfaces. We say that $\mathscr{L}_{\mathcal{F}}(\Sigma_*(\mathcal{F}))$ and $\mathscr{L}_{\mathcal{G}}(\Sigma_*(\mathcal{G}))$ are *Legendrian equivalent* if there exist a diffeomorphism germ $\Phi : (\mathbb{R}^m \times \mathbb{R}, \overline{\pi}(p)) \longrightarrow (\mathbb{R}^m \times \mathbb{R}, \overline{\pi}(p'))$ and a contact diffeomorphism germ $\widehat{\Phi} : (J^1(\mathbb{R}^m, \mathbb{R}), p) \longrightarrow (J^1(\mathbb{R}^m, \mathbb{R}), p')$ such that $\overline{\pi} \circ \widehat{\Phi} = \Phi \circ \overline{\pi}$ and $\widehat{\Phi}(\mathscr{L}_{\mathcal{F}}(\Sigma_*(\mathcal{F}))) = (\mathscr{L}_{\mathcal{G}}(\Sigma_*(\mathcal{G})))$, where $p = \mathscr{L}_{\mathcal{F}}(0)$ and $p' = \mathscr{L}_{\mathcal{G}}(0)$. We also say that $\mathscr{L}_{\mathcal{F}}(\Sigma_*(\mathcal{F}))$ and $\mathscr{L}_{\mathcal{G}}(\Sigma_*(\mathcal{G}))$ are *$S.P^+$-Legendrian equivalent* if these are Legendrian equivalent by a diffeomorphism germ $\Phi : (\mathbb{R}^m \times \mathbb{R}, \overline{\pi}(p)) \longrightarrow (\mathbb{R}^m \times \mathbb{R}, \overline{\pi}(p'))$ of the form $\Phi(x, t) = (\phi_1(x), t + \alpha(x))$ and a contact diffeomorphism germ $\widehat{\Phi} : (J^1(\mathbb{R}^m, \mathbb{R}), p) \longrightarrow (J^1(\mathbb{R}^m, \mathbb{R}), p')$ with $\overline{\pi} \circ \widehat{\Phi} = \Phi \circ \overline{\pi}$. Moreover, graph-like wave fronts $W(\mathscr{L}_{\mathcal{F}}(\Sigma_*(\mathcal{F})))$ and $W(\mathscr{L}_{\mathcal{G}}(\Sigma_*(\mathcal{G})))$ are *$S.P^+$-diffeomorphic* if there exists a diffeomorphism germ $\Phi : (\mathbb{R}^m \times \mathbb{R}, \overline{\pi}(p)) \longrightarrow (\mathbb{R}^m \times \mathbb{R}, \overline{\pi}(p'))$ of the form $\Phi(x, t) = (\phi_1(x), t + \alpha(x))$ such that $\Phi(W(\mathscr{L}_{\mathcal{F}}(\Sigma_*(\mathcal{F})))) = W(\mathscr{L}_{\mathcal{G}}(\Sigma_*(\mathcal{G})))$ as set germs. By definition, if $\mathscr{L}_{\mathcal{F}}(\Sigma_*(\mathcal{F}))$ and $\mathscr{L}_{\mathcal{G}}(\Sigma_*(\mathcal{G}))$ are $S.P^+$-Legendrian equivalent, then $W(\mathscr{L}_{\mathcal{F}}(\Sigma_*(\mathcal{F})))$ and $W(\mathscr{L}_{\mathcal{G}}(\Sigma_*(\mathcal{G})))$ are $S.P^+$-diffeomorphic. The converse assertion holds generically [**19, 20**].

PROPOSITION 6.1 ([**20**]). *With the same notations as those of the above, we suppose that the sets of critical points of* $\overline{\pi}|_{\mathscr{L}_{\mathcal{F}}(\Sigma_*(\mathcal{F}))}, \overline{\pi}|_{\mathscr{L}_{\mathcal{G}}(\Sigma_*(\mathcal{G}))}$ *are nowhere dense respectively. Then* $\mathscr{L}_{\mathcal{F}}(\Sigma_*(\mathcal{F}))$ *and* $\mathscr{L}_{\mathcal{G}}(\Sigma_*(\mathcal{G}))$ *are $S.P^+$-Legendrian equivalent if and only if* $W(\mathscr{L}_{\mathcal{F}}(\Sigma_*(\mathcal{F})))$ *and* $W(\mathscr{L}_{\mathcal{G}}(\Sigma_*(\mathcal{G})))$ *are $S.P^+$-diffeomorphic.*

We remark that if $W(\mathscr{L}_{\mathcal{F}}(\Sigma_*(\mathcal{F})))$ and $W(\mathscr{L}_{\mathcal{G}}(\Sigma_*(\mathcal{G})))$ are $S.P^+$-diffeomorphic by a diffeomorphism germ $\Phi : (\mathbb{R}^m \times \mathbb{R}, \overline{\pi}(p)) \longrightarrow (\mathbb{R}^m \times \mathbb{R}, \overline{\pi}(p'))$, then

$$\Phi(C_{\mathscr{L}_{\mathcal{F}}(\Sigma_*(\mathcal{F}))} \cup M_{\mathscr{L}_{\mathcal{F}}(\Sigma_*(\mathcal{F}))}) = C_{\mathscr{L}_{\mathcal{G}}(\Sigma_*(\mathcal{G}))} \cup M_{\mathscr{L}_{\mathcal{G}}(\Sigma_*(\mathcal{G}))}.$$

For a graph-like Morse family hypersurfaces $\mathcal{F} : (\mathbb{R}^k \times (\mathbb{R}^m \times \mathbb{R}), 0) \to (\mathbb{R}, 0)$, by the implicit function theorem, there exist function germs $F : (\mathbb{R}^k \times \mathbb{R}^m, 0) \to (\mathbb{R}, 0)$ and $\lambda : (\mathbb{R}^k \times (\mathbb{R}^m \times \mathbb{R}), 0) \longrightarrow \mathbb{R}$ with $\lambda(0) \neq 0$ such that $\mathcal{F}(q, x, t) = \lambda(q, x, t)(F(q, x) - t)$. We have shown in [**19**] that $\mathcal{F}$ is a graph-like Morse family of hypersurfaces if and only if $F$ is a Morse family of functions. Here we say that $F : (\mathbb{R}^k \times \mathbb{R}^m, 0) \longrightarrow (\mathbb{R}, 0)$ is a *Morse family of functions* if

$$dF_q = \left(\frac{\partial F}{\partial q_1}, \dots, \frac{\partial F}{\partial q_k}\right) : (\mathbb{R}^k \times \mathbb{R}^m, 0) \longrightarrow \mathbb{R}^k$$

is non-singular. We consider a graph-like Morse family of hypersurfaces $\mathcal{F}(q, x, t) = \lambda(q, x, t)(F(q, x) - t)$. In this case,

$$\Sigma_*(\mathcal{F}) = \{(q, x, F(q, x)) \in (\mathbb{R}^k \times (\mathbb{R}^m \times \mathbb{R}), 0) \mid (q, x) \in C(F)\},$$

where

$$C(F) = \left\{(q, x) \in (\mathbb{R}^k \times \mathbb{R}^m, 0) \;\middle|\; \frac{\partial F}{\partial q_1}(q, x) = \cdots = \frac{\partial F}{\partial q_k}(q, x) = 0\right\}.$$

Moreover, we define a map germ $L(F) : (C(F), 0) \longrightarrow T^*\mathbb{R}^m$ by

$$L(F)(q, x) = \left(x, \frac{\partial F}{\partial x_1}(q, x), \dots, \frac{\partial F}{\partial x_m}(q, x)\right)$$

It is known that $L(F)(C(F))$ is a Lagrangian submanifold germ (cf., [**1**]) for the canonical symplectic structure. In this case $F$ is said to be a *generating family* of the Lagrangian submanifold germ $L(F)(C(F))$. We remark that $\Pi(\mathscr{L}_{\mathcal{F}}(\Sigma_*(\mathcal{F}))) = L(F)(C(F))$ and the graph-like wave front $W(\mathscr{L}_{\mathcal{F}}(\Sigma_*(\mathcal{F})))$ is the graph of $F|C(F)$. Here we call $\pi|_{L(F)(C(F))} : L(F)(C(F)) \longrightarrow \mathbb{R}^m$ a *Lagrangian map germ*, where $\pi : T^*\mathbb{R}^m \longrightarrow \mathbb{R}^m$ is the canonical projection. Then the set of critical values of $\pi|_{L(F)(C(F))}$ is called a *caustic* of $L(F)(C(F)) = \Pi(\mathscr{L}_{\mathcal{F}}(\Sigma_*(\mathcal{F})))$ in the theory of Lagrangian singularities, which is denoted by $C_{L(F)(C(F))}$. By definition, we have $C_{L(F)(C(F))} = C_{\mathscr{L}_{\mathcal{F}}(\Sigma_*(\mathcal{F}))}$.

Let $\mathcal{F}, \mathcal{G} : (\mathbb{R}^k \times (\mathbb{R}^m \times \mathbb{R}), 0) \to (\mathbb{R}, 0)$ be graph-like Morse families of hypersurfaces. We say that $\Pi(\mathscr{L}_{\mathcal{F}}(\Sigma_*(\mathcal{F})))$ and $\Pi(\mathscr{L}_{\mathcal{G}}(\Sigma_*(\mathcal{G})))$ are *Lagrangian equivalent* if there exist a diffeomorphism germ $\Psi : (\mathbb{R}^m, \pi \circ \Pi(p)) \longrightarrow (\mathbb{R}^m, \pi \circ \Pi(p'))$ and a symplectic diffeomorphism germ $\widehat{\Psi} : (T^*\mathbb{R}^m, \Pi(p)) \longrightarrow (T^*\mathbb{R}^m, \Pi(p'))$ such that $\pi \circ \widehat{\Psi} = \Psi \circ \pi$ and $\widehat{\Psi}(\Pi(\mathscr{L}_{\mathcal{F}}(\Sigma_*(\mathcal{F})))) = \Pi(\mathscr{L}_{\mathcal{G}}(\Sigma_*(\mathcal{G})))$, where $p = \mathscr{L}_{\mathcal{F}}(0)$ and $p' = \mathscr{L}_{\mathcal{G}}(0)$. By definition, if $\Pi(\mathscr{L}_{\mathcal{F}}(\Sigma_*(\mathcal{F})))$ and $\Pi(\mathscr{L}_{\mathcal{G}}(\Sigma_*(\mathcal{G})))$ are Lagrangian equivalent, then the caustics $C_{\mathscr{L}_{\mathcal{F}}(\Sigma_*(\mathcal{F}))}$ and $C_{\mathscr{L}_{\mathcal{G}}(\Sigma_*(\mathcal{G}))}$ are diffeomorphic as set germs. The converse assertion, however, does not hold (cf. [**20**]). Recently, we have shown the following theorem (cf. [**16, 19, 20**])

THEOREM 6.2. *With the same notations as above,* $\Pi(\mathscr{L}_{\mathcal{F}}(\Sigma_*(\mathcal{F})))$ *and* $\Pi(\mathscr{L}_{\mathcal{G}}(\Sigma_*(\mathcal{G})))$ *are Lagrangian equivalent if and only if* $\mathscr{L}_{\mathcal{F}}(\Sigma_*(\mathcal{F}))$ *and* $\mathscr{L}_{\mathcal{G}}(\Sigma_*(\mathcal{G}))$ *are* $S.P^+$*-Legendrian equivalent.*

We have the following corollary of Proposition 6.1 and Theorem 6.2.

COROLLARY 6.3. *Suppose that both of the sets of critical points of* $\overline{\pi}|_{\mathscr{L}_{\mathcal{F}}(\Sigma_*(\mathcal{F}))}$ *and* $\overline{\pi}|_{\mathscr{L}_{\mathcal{G}}(\Sigma_*(\mathcal{G}))}$ *are nowhere dense. Then* $\Pi(\mathscr{L}_{\mathcal{F}}(\Sigma_*(\mathcal{F})))$ *and* $\Pi(\mathscr{L}_{\mathcal{G}}(\Sigma_*(\mathcal{G})))$ *are Lagrangian equivalent if and only if* $W(\mathscr{L}_{\mathcal{F}}(\Sigma_*(\mathcal{F})))$ *and* $W(\mathscr{L}_{\mathcal{G}}(\Sigma_*(\mathcal{G})))$ *are* $S.P^+$*-diffeomorphic.*

There are the notions of Lagrangian stability of Lagrangian submanifold germs and $S.P^+$-Legendrian stability of graph-like Legendrian unfolding germs, respectively. Here we do not use the exact definitions of those notions of stability, so that we omit to give the definitions. For detailed properties of such stabilities, see [**1**, **19**]. We have the following corollary of Theorem 6.2.

COROLLARY 6.4. *The graph-like Legendrian unfolding* $\mathscr{L}_{\mathcal{F}}(\Sigma_*(\mathcal{F}))$ *is* $S.P^+$*-Legendrian stable if and only if the Lagrangian submanifold* $\Pi(\mathscr{L}_{\mathcal{F}}(\Sigma_*(\mathcal{F})))$ *is Lagrangian stable.*

Let $\mathcal{F} : (\mathbb{R}^k \times (\mathbb{R}^m \times \mathbb{R}), 0) \to (\mathbb{R}, 0)$ be a graph-like Morse family of hypersurfaces. We define $\overline{f} : (\mathbb{R}^k \times \mathbb{R}, 0) \longrightarrow (\mathbb{R}, 0)$ by $\overline{f}(q,t) = \mathcal{F}(q,0,t)$. For graph-like Morse families of hypersurfaces $\mathcal{F} : (\mathbb{R}^k \times (\mathbb{R}^m \times \mathbb{R}), 0) \to (\mathbb{R}, 0)$ and $\mathcal{G} : (\mathbb{R}^k \times (\mathbb{R}^m \times \mathbb{R}), 0) \to (\mathbb{R}, 0)$, we say that $\overline{f}$ and $\overline{g}$ are *$S.P$-$\mathcal{K}$-equivalent* if there exist a function germ $\nu : (\mathbb{R}^k \times \mathbb{R}, 0) \longrightarrow \mathbb{R}$ with $\nu(0) \neq 0$ and a diffeomorphism germ $\phi : (\mathbb{R}^k \times \mathbb{R}, 0) \longrightarrow (\mathbb{R}^k \times \mathbb{R}, 0)$ of the form $\phi(q,t) = (\phi_1(q,t), t)$ such that $\overline{f}(q,t) = \nu(q,t)\overline{g}(\phi(q,t))$. Although we do not give the definition of $S.P^+$-Legendrian stability, we give a corresponding notion for graph-like Morse family of hypersurfaces. We say that $\mathcal{F}$ is an *infinitesimally $S.P^+$-$\mathcal{K}$-versal unfolding* of $\overline{f}$ if

$$\mathcal{E}_{k+1} = \left\langle \frac{\partial \overline{f}}{\partial q_1}, \dots, \frac{\partial \overline{f}}{\partial q_k}, \overline{f} \right\rangle_{\mathcal{E}_{k+1}} + \left\langle \frac{\partial \overline{f}}{\partial t} \right\rangle_{\mathbb{R}} + \left\langle \frac{\partial \mathcal{F}}{\partial x_1}|_{\mathbb{R}^k \times \{0\} \times \mathbb{R}}, \dots, \frac{\partial \mathcal{F}}{\partial x_m}|_{\mathbb{R}^k \times \{0\} \times \mathbb{R}} \right\rangle_{\mathbb{R}},$$

where $\mathcal{E}_{k+1}$ is the local $\mathbb{R}$-algebra of $C^\infty$-function germs $(\mathbb{R}^k \times \mathbb{R}, 0) \longrightarrow \mathbb{R}$. It is known the following theorem in [**12**, **30**].

THEOREM 6.5. *For a graph-like Morse family of hypersurfaces* $\mathcal{F} : (\mathbb{R}^k \times (\mathbb{R}^m \times \mathbb{R}), 0) \to (\mathbb{R}, 0)$*, the graph-like Legendrian unfolding* $\mathscr{L}_{\mathcal{F}}(\Sigma_*(\mathcal{F}))$ *is* $S.P^+$*-Legendrian stable if and only if* $\mathcal{F}$ *is an infinitesimally* $S.P^+$*-*$\mathcal{K}$*-versal unfolding of* $\overline{f}$.

In [**19**] we have shown the following theorem.

THEOREM 6.6. *Let* $\mathcal{F}, \mathcal{G} : (\mathbb{R}^k \times (\mathbb{R}^m \times \mathbb{R}), 0) \to (\mathbb{R}, 0)$ *be graph-like Morse families of hypersurfaces such that* $\mathscr{L}_{\mathcal{F}}(\Sigma_*(\mathcal{F})), \mathscr{L}_{\mathcal{G}}(\Sigma_*(\mathcal{G}))$ *are* $S.P^+$*-Legendrian stable. Then the following conditions are equivalent:*
(1) $\mathscr{L}_{\mathcal{F}}(\Sigma_*(\mathcal{F}))$ *and* $\mathscr{L}_{\mathcal{G}}(\Sigma_*(\mathcal{G}))$ *are* $S.P^+$*-Legendrian equivalent,*
(2) $\overline{f}$ *and* $\overline{g}$ *are* $S.P$*-*$\mathcal{K}$*-equivalent,*
(3) $\Pi(\mathscr{L}_{\mathcal{F}}(\Sigma_*(\mathcal{F})))$ *and* $\Pi(\mathscr{L}_{\mathcal{G}}(\Sigma_*(\mathcal{G})))$ *are Lagrangian equivalent,*
(4) $W(\mathscr{L}_{\mathcal{F}}(\Sigma_*(\mathcal{F})))$ *and* $W(\mathscr{L}_{\mathcal{G}}(\Sigma_*(\mathcal{G})))$ *are* $S.P^+$*-diffeomorphic.*

## 7. Unfolded light sheets of world hyper-sheets

In this section we investigate unfolded lightcone focal sets of world hyper-sheets as an application of the theory of graph-like Legendrian unfoldings. We show the following proposition.

PROPOSITION 7.1. *Let* $G : U \times I \times (\mathbb{R}^{n+1}_1 \setminus W) \to \mathbb{R}$ *be a Lorentz distance-squared function on a world hyper-sheet* $(W, \mathcal{S})$*. For any point* $(\overline{u}_0, t_0, \boldsymbol{\lambda}_0) \in \Sigma_*(G)$*,* $G$ *is a non-degenerate graph-like Morse family of hypersurfaces around* $(\overline{u}_0, t_0, \boldsymbol{\lambda}_0)$*.*

*Proof.* We write that

$$\boldsymbol{X}(\overline{u},t)=(X_0(\overline{u},t),X_1(\overline{u},t),\dots,X_n(\overline{u},t)) \text{ and } \boldsymbol{\lambda}=(\lambda_0,\lambda_1,\dots,\lambda_n).$$

By definition, we have

$$G(\overline{u},t,\boldsymbol{\lambda})=-(X_0(\overline{u},t)-\lambda_0)^2+(X_1(\overline{u},t)-\lambda_1)^2+\cdots+(X_n(\overline{u},t)-\lambda_n)^2.$$

We now prove that the mapping

$$\Delta^*G(\overline{u},t_0,\boldsymbol{\lambda})=\left(G(\overline{u},t_0,\boldsymbol{\lambda}),\frac{\partial G}{\partial u_1}(\overline{u},t_0,\boldsymbol{\lambda}),\dots,\frac{\partial G}{\partial u_{n-1}}(\overline{u},t_0,\boldsymbol{\lambda})\right)$$

is non-singular at $(\overline{u}_0,t_0,\boldsymbol{\lambda}_0)\in\Sigma_*(G)$. The Jacobian matrix of $\Delta^*G|_{U\times\{t_0\}\times\mathbb{R}^{n+1}_1}$ is given by

$$\left(\begin{array}{c|cccc} \mathsf{A} & \begin{matrix}2(X_0-\lambda_0) & -2(X_1-\lambda_1) & \cdots & -2(X_n-\lambda_n)\\ 2X_{0u_1} & -2X_{1u_1} & \cdots & -2X_{nu_1}\\ \vdots & \vdots & \vdots & \vdots\\ 2X_{0u_{n-1}} & -2X_{1u_{n-1}} & \cdots & -2X_{nu_{n-1}}\end{matrix}\end{array}\right),$$

where $\mathsf{A}$ is an $n\times(n-1)$-matrix defined by

$$\mathsf{A}=\begin{pmatrix} 2\langle \boldsymbol{X}-\boldsymbol{\lambda},\boldsymbol{X}_{u_j}\rangle\\ 2(\langle \boldsymbol{X}_{u_i},\boldsymbol{X}_{u_j}\rangle+\langle \boldsymbol{X}-\boldsymbol{\lambda},\boldsymbol{X}_{u_iu_j}\rangle)\end{pmatrix},$$

for $i=1,\dots,n-1$; $j=1,\dots,n-1$.

Since $\boldsymbol{X}$ is an immersion, the rank of the matrix

$$\begin{pmatrix} 2X_{0u_1} & -2X_{1u_1} & \cdots & -2X_{nu_1}\\ \vdots & \vdots & \ddots & \vdots\\ 2X_{0u_{n-1}} & -2X_{1u_{n-1}} & \cdots & -2X_{nu_{n-1}}\end{pmatrix}.$$

is equal to $n-1$. Since $\boldsymbol{X}-\boldsymbol{\lambda}$ is lightlike and $T_p\mathcal{S}_{t_0}$ is spacelike, then $\boldsymbol{X}-\boldsymbol{\lambda},\boldsymbol{X}_{u_1},\dots,\boldsymbol{X}_{u_{n-1}}$ are linearly independent at $(u_0,t_0,\boldsymbol{\lambda}_0)\in\Sigma_*(G)$. This means that the rank of the matrix

$$\begin{pmatrix} 2(X_0-\lambda_0) & -2(X_1-\lambda_1) & \cdots & -2(X_n-\lambda_n)\\ 2X_{0u_1} & -2X_{1u_1} & \cdots & -2X_{nu_1}\\ \vdots & \vdots & \ddots & \vdots\\ 2X_{0u_{n-1}} & -2X_{1u_{n-1}} & \cdots & -2X_{nu_{n-1}}\end{pmatrix}$$

is equal to $n$. Therefore the Jacobi matrix of $\Delta^*G|_{U\times\{t_0\}\times\mathbb{R}^{n+1}_1}$ is non-singular at $(u_0,t_0,\boldsymbol{\lambda}_0)\in\Sigma_*(G)$.

On the other hand, we have

$$\frac{\partial G}{\partial t}(\overline{u},t,\boldsymbol{\lambda})=2\langle \boldsymbol{X}_t(\overline{u},t),\boldsymbol{X}(\overline{u},t)-\boldsymbol{\lambda}\rangle.$$

For any $(\overline{u}_0,t_0,\boldsymbol{\lambda}_0)\in\Sigma_*(G)$, there exists $\mu\neq 0$ such that $\boldsymbol{\lambda}_0=\boldsymbol{X}(\overline{u}_0,t_0)+\mu\mathbb{LG}^{\pm}(\mathcal{S}_{t_0})(\overline{u}_0,t_0)$. Since $\boldsymbol{n}^S(\overline{u}_0,t_0)$ is the unit normal vector of $W$, we have

$$\langle \boldsymbol{X}_t(\overline{u}_0,t_0),\mathbb{LG}^{\pm}(\mathcal{S}_{t_0})(\overline{u}_0,t_0)\rangle=\langle \boldsymbol{X}_t(\overline{u}_0,t_0),\boldsymbol{n}^T(\overline{u}_0,t_0)\rangle.$$

Moreover,

$$\{\boldsymbol{X}_t(\overline{u}_0,t_0),\boldsymbol{X}_{u_1}(\overline{u}_0,t_0),\dots,\boldsymbol{X}_{u_{n-1}}(\overline{u}_0,t_0)\}$$

is a basis of $T_pW$ and $\boldsymbol{n}^T(\overline{u}_0, t_0) \in N_p(\mathcal{S}_{t_0}) \cap T_pW$, where $p = \boldsymbol{X}(\overline{u}_0, t_0)$. It follows that $\langle \boldsymbol{X}_t(\overline{u}_0, t_0), \boldsymbol{n}^T(\overline{u}_0, t_0)\rangle \neq 0$. Then we have

$$\begin{aligned}\frac{\partial G}{\partial t}(\overline{u}_0, t_0, \boldsymbol{\lambda}_0) &= 2\langle \boldsymbol{X}_t(\overline{u}_0, t_0), -\mu\mathbb{LG}^{\pm}(\mathcal{S}_{t_0})(\overline{u}_0, t_0)\rangle \\ &= -\mu\langle \boldsymbol{X}_t(\overline{u}_0, t_0), \boldsymbol{n}^T(\overline{u}_0, t_0)\rangle \neq 0.\end{aligned}$$

This completes the proof. □

By Proposition 4.1, we have

$$\Sigma_*(G) = \{(\overline{u}, t, \mathbb{LH}^{\pm}_{\mathcal{S}_t}(p, \mu)) \in U \times I \times \mathbb{R}^{n+1}_1 \mid p = \boldsymbol{X}(\overline{u}, t), \mu \in \mathbb{R} \setminus \{0\}\}.$$

We define a map $\mathscr{L}_G : \Sigma_*(G) \longrightarrow J^1(\mathbb{R}^{n+1}_1, I)$ by

$$\mathscr{L}_G(\overline{u}, t, \mathbb{LH}^{\pm}_{\mathcal{S}_t}(p, \mu)) = \left(\mathbb{LH}^{\pm}_{\mathcal{S}_t}(p, \mu), t, \frac{2}{\langle \boldsymbol{X}_t(\overline{u}, t), \boldsymbol{n}^T(\overline{u}, t)\rangle}\overline{\mathbb{LG}^{\pm}(\mathcal{S}_t)(\overline{u}, t)}\right),$$

where we define $\overline{\boldsymbol{x}} = (-x_0, x_1, \dots x_n)$ for $\boldsymbol{x} = (x_0, x_1, \dots, x_n) \in \mathbb{R}^{n+1}_1$. By the construction of the graph-like Legendrian unfolding from a graph-like Morse family of hypersurfaces, $\mathscr{L}_G(\Sigma_*(G))$ is a graph-like Legendrian unfolding in $J^1(\mathbb{R}^{n+1}_1, I)$. Therefore, the graph-like wave front $W(\mathscr{L}_G(\Sigma_*(G)))$ is equal to

$$\{(\mathbb{LH}^{\pm}_{\mathcal{S}_t}(p, \mu), t) \in \mathbb{R}^{n+1}_1 \times I \mid p = \boldsymbol{X}(\overline{u}, t), (\overline{u}, t) \in U \times I, \mu \in \mathbb{R} \setminus \{0\}\}.$$

This means that $W(\mathscr{L}_G(\Sigma_*(G))) = \mathbb{LH}^+_W(U \times I \times (\mathbb{R} \setminus \{0\})) \cup \mathbb{LH}^-_W(U \times I \times (\mathbb{R} \setminus \{0\}))$. By Proposition 4.1, the set of singularities of $W(\mathscr{L}_G(\Sigma_*(G)))$ is the union of the critical value sets of $\mathbb{LH}^{\pm}_W$ which is the union of unfolded lightcone focal sets $\mathbb{LF}^+_W \cup \mathbb{LF}^-_W$. Therefore, we have shown the following proposition.

PROPOSITION 7.2. *Let $(W, \mathcal{S})$ be a world hyper-sheet in $\mathbb{R}^{n+1}_1$ and $G : W \times (\mathbb{R}^{n+1}_1 \setminus W) \longrightarrow \mathbb{R}$ the Lorentz distance squared function. Then we have the graph-like legendrian unfolding $\mathscr{L}_G(\Sigma_*(G)) \subset J^1(\mathbb{R}^{n+1}_1, I)$ such that*

$$W(\mathscr{L}_G(\Sigma_*(G))) = \mathbb{LH}^+_W(U \times I \times (\mathbb{R} \setminus \{0\})) \cup \mathbb{LH}^-_W(U \times I \times (\mathbb{R} \setminus \{0\})).$$

We write

$$\mathbb{LH}^{\pm}_{(W,\mathcal{S})} = \mathbb{LH}^{\pm}_W(U \times I \times (\mathbb{R} \setminus \{0\})).$$

We also call $\mathbb{LH}^+_{(W,\mathcal{S})} \cup \mathbb{LH}^-_{(W,\mathcal{S})}$ an *unfolded light sheet* of $(W, \mathcal{S})$. On the other hand, we have the corresponding Lagrangian submanifold $\Pi(\mathscr{L}_G(\Sigma_*(G))) \subset T^*\mathbb{R}^{n+1}_1$. We now consider the natural question what are the caustic $C_{\mathscr{L}_G(\Sigma_*(G))}$ and the Maxwell set $M_{\mathscr{L}_G(\Sigma_*(G))}$? Moreover, are there any meanings of $C_{\mathscr{L}_G(\Sigma_*(G))}$ and $M_{\mathscr{L}_G(\Sigma_*(G))}$ in physics?

## 8. Caustics of world hyper-sheets

In [**4, 5**] Bousso and Randall gave an idea of caustics of world sheets in order to define the notion of holographic domains. The family of light sheets $\{\mathbb{LH}^{\pm}_{\mathcal{S}_t}(U \times \{t\}) \times \mathbb{R}\}_{t \in J}$ sweeps out a region in $\mathbb{R}^{n+1}_1$. A *caustic* of a world sheet is the union of the sets of critical values of light sheets along momentary spaces $\{\mathcal{S}_t\}_{t \in I}$. A *holographic domain* of the world sheet is the region where the light-sheets sweep out until *caustics*. So this means that the boundary of the holographic domain consists the caustic of the world sheet. The set of critical values of the light sheet of a momentary space is the lightlike focal set of the momentary space. Therefore the notion of caustics in the sense of Bousso-Randall is formulated as follows: *Caustics of a world sheet* $(W, \mathcal{S})$ are defined to be

$$C^{\pm}(W,\mathcal{S}) = \bigcup_{t\in I} \mathbb{LF}^{\pm}_{\mathcal{S}_t} = \pi_1(\mathbb{LF}^{\pm}_{(W,\mathcal{S})}),$$

where $\pi_1 : \mathbb{R}^{n+1}_1 \times I \longrightarrow \mathbb{R}^{n+1}_1$ is the canonical projection. We call $C^{\pm}(W,\mathcal{S})$ *BR-caustics* of $(W,\mathcal{S})$. We write that $C(W,\mathcal{S}) = \pi_1(\mathbb{LF}^+_W \cup \mathbb{LF}^-_W)$ and call it a *total BR-caustic* of $(W,\mathcal{S})$. By definition, we have $\Sigma(W(\mathscr{L}_G(\Sigma_*(G))) = \mathbb{LF}^+_{(W,\mathcal{S})} \cup \mathbb{LF}^-_{(W,\mathcal{S})}$, so that we have the following proposition.

PROPOSITION 8.1. *Let $(W,\mathcal{S})$ be a world hyper-sheet in $\mathbb{R}^{n+1}_1$ and $G : U \times I \times (\mathbb{R}^{n+1}_1 \setminus W) \longrightarrow \mathbb{R}$ the Lorentz distance squared function. Then we have $C(W,\mathcal{S}) = C_{\mathscr{L}_G(\Sigma_*(G))}$.*

In [**4**, **5**] the authors did not consider the Maxwell set of a world sheet. However, the notion of Maxwell sets plays an important role in the cosmology which has been called a *crease set* by Penrose (cf. [**27**, **28**]). Actually, the topological shape of the event horizon is determined by the crease set of light sheets. Here, we write $M(W,S) = M_{\mathscr{L}_G(\Sigma_*(G))}$ and call it a *BR-Maxwell set* of the world sheet $(W,\mathcal{S})$.

Let $\boldsymbol{X}_i : (U_i \times I_i, (\overline{u}_i, t_i)) \longrightarrow (\mathbb{R}^{n+1}_1, p_i)$, $(i = 1, 2)$ be germs of timelike embeddings such that $(W_i, \mathcal{S}_i)$ are world hyper-sheet germs, where $W_i = \boldsymbol{X}_i(U)$. For $\boldsymbol{\lambda}_i = \mathbb{LH}^+_{\mathcal{S}_i}(p_i, \overline{u}_i)$ or $\boldsymbol{\lambda}_i = \mathbb{LH}^-_{\mathcal{S}_i}(p_i, \overline{u}_i)$, let $G_i : (U_i \times I_i \times (\mathbb{R}^{n+1}_1 \setminus W_i), (\overline{u}_i, t_i, \boldsymbol{\lambda}_i)) \longrightarrow \mathbb{R}$ be Lorentz distance squared function germs. We also write that $g_{i,\boldsymbol{\lambda}_i}(\overline{u}, t) = G_i(\overline{u}, t, \boldsymbol{\lambda}_i)$. Since

$$W(\mathscr{L}_{G_i}(\Sigma_*(G_i))) = \mathbb{LH}^+_{(W_i,\mathcal{S}_i)} \cup \mathbb{LH}^-_{(W_i,\mathcal{S}_i)},$$

we can apply Theorem 6.2 and Corollary 6.3 to our case. Then we have the following theorem.

THEOREM 8.2. *Suppose that the set of critical points of $\overline{\pi}|_{\mathscr{L}_{G_i}(\Sigma_*(G_i))}$ is nowhere dense for each $i = 1, 2$, respectively. Then the following conditions are equivalent:*
(1) $(\mathbb{LH}^+_{(W_1,\mathcal{S}_1)} \cup \mathbb{LH}^-_{(W_1,\mathcal{S}_1)}, \boldsymbol{\lambda}_1)$, $(\mathbb{LH}^+_{(W_2,\mathcal{S}_2)} \cup \mathbb{LH}^-_{(W_2,\mathcal{S}_2)}, \boldsymbol{\lambda}_2)$ *are $S.P^+$-diffeomorphic,*
(2) $\mathscr{L}_{G_1}(\Sigma_*(G_1))$, $\mathscr{L}_{G_2}(\Sigma_*(G_2))$ *are $S.P^+$-Legendrian equivalent,*
(3) $\Pi(\mathscr{L}_{G_1}(\Sigma_*(G_1)))$, $\Pi(\mathscr{L}_{G_2}(\Sigma_*(G_2))$ *are Lagrangian equivalent.*

We remark that conditions (2) and (3) are equivalent without any assumptions (cf. Theorem 6.2). Moreover, if we assume that $\mathscr{L}_{G_i}(\Sigma_*(G_i))$ is $S.P^+$-Legendrian stable, then we can apply Proposition 5.3 and Theorem 6.6 and show the following theorem.

THEOREM 8.3. *Suppose that $\mathscr{L}_{G_i}(\Sigma_*(G_i))$ is $S.P^+$-Legendrian stable for each $i = 1, 2$, respectively. Then the following conditions are equivalent:*
(1) $(\mathbb{LH}^+_{(W_1,\mathcal{S}_1)} \cup \mathbb{LH}^-_{(W_1,\mathcal{S}_1)}, \boldsymbol{\lambda}_1)$, $(\mathbb{LH}^+_{(W_2,\mathcal{S}_2)} \cup \mathbb{LH}^-_{(W_2,\mathcal{S}_2)}, \boldsymbol{\lambda}_2)$ *are $S.P^+$-diffeomorphic,*
(2) $\mathscr{L}_{G_1}(\Sigma_*(G_1))$, $\mathscr{L}_{G_2}(\Sigma_*(G_2))$ *are $S.P^+$-Legendrian equivalent,*
(3) $\Pi(\mathscr{L}_{G_1}(\Sigma_*(G_1)))$, $\Pi(\mathscr{L}_{G_2}(\Sigma_*(G_2))$ *are Lagrangian equivalent,*
(4) $g_{1,\boldsymbol{\lambda}_1}$, $g_{2,\boldsymbol{\lambda}_2}$ *are $S.P$-$\mathcal{K}$-equivalent,*
(5) $SK(\overline{W}_1, TLC(\mathcal{S}_{t_0}, \boldsymbol{\lambda}_1) \times I; (p_1, t_0)) = SK(\overline{W}_2, TLC(\mathcal{S}_{t_0}, \boldsymbol{\lambda}_2) \times I; (p_2, t_0))$.

By definition and Proposition 8.1, we have the following proposition.

PROPOSITION 8.4. *If* $\Pi(\mathscr{L}_{G_1}(\Sigma_*(G_1)))$ *and* $\Pi(\mathscr{L}_{G_2}(\Sigma_*(G_2))$ *are Lagrangian equivalent, then total BR-caustics*

$$C(W_1, \mathcal{S}_1),\ C(W_2, \mathcal{S}_2)$$

*and BR-Maxwell sets*

$$M(W_1, \mathcal{S}_1),\ M(W_2, \mathcal{S}_2)$$

*are diffeomorphic as set germs, respectively.*

## 9. World sheets in $\mathbb{R}^3_1$

In this section we consider world hyper-sheets in the 3-dimensional Minkowski space-time as an example. Let $(W, \mathcal{S})$ be a world hyper-sheet in $\mathbb{R}^3_1$, which is parameterized by a timelike embedding $\mathbf{\Gamma} : J \times I \longrightarrow \mathbb{R}^3_1$ such that $\mathcal{S}_t = \mathbf{\Gamma}(J \times \{t\})$ for $t \in I$. In this case we call $(W, \mathcal{S})$ a *world sheet* and $\mathcal{S}_t$ a *momentary curve*. We assume that $s \in J$ is the arc-length parameter. Since $W$ is a timelike surface, we have the unit pseudo-normal vector field $\boldsymbol{n}(s,t)$ of $W$ in $\mathbb{R}^3_1$. Then $\boldsymbol{t}(s,t) = \gamma_t'(s)$ is the unit spacelike tangent vector of $\mathcal{S}_t$. We also define $\boldsymbol{b}(s,t) = \boldsymbol{n}(s,t) \wedge \boldsymbol{t}(s,t)$, which is the unit timelike normal vector of $\mathcal{S}_t$ in $TW$. We choose the orientation of $\mathcal{S}_t$ such that $\boldsymbol{b}(s,t)$ is *future directed* (i.e. $\langle \boldsymbol{e}_0, \boldsymbol{b}(s,t)\rangle < 0$). Therefore, $\{\boldsymbol{b}(s,t), \boldsymbol{n}(s,t), \boldsymbol{t}(s,t)\}$ is a pseudo-orthonormal frame along $W$. On this moving frame, we can show the following *Frenet-Serret type formulae* for $\mathcal{S}_t$:

$$\begin{cases} \dfrac{\partial \boldsymbol{b}}{\partial s}(s,t) &= \tau_g(s,t)\boldsymbol{n}(s,t) - \kappa_g(s,t)\boldsymbol{t}(s,t), \\ \dfrac{\partial \boldsymbol{n}}{\partial s}(s,t) &= \tau_g(s,t)\boldsymbol{b}(s,t) - \kappa_n(s,t)\boldsymbol{t}(s,t), \\ \dfrac{\partial \boldsymbol{t}}{\partial s}(s,t) &= -\kappa_g(s,t)\boldsymbol{b}(s,t) + \kappa_n(s,t)\boldsymbol{n}(s,t), \end{cases}$$

where

$$\kappa_g(s,t) = \langle \frac{\partial \boldsymbol{t}}{\partial s}(s,t), \boldsymbol{b}(s,t)\rangle,$$
$$\kappa_n(s,t) = \langle \frac{\partial \boldsymbol{t}}{\partial s}(s,t), \boldsymbol{n}(s,t)\rangle,$$
$$\tau_g(s,t) = \langle \frac{\partial \boldsymbol{b}}{\partial s}(s,t), \boldsymbol{n}(s,t)\rangle.$$

We call $\kappa_g(s,t)$ a *geodesic curvature*, $\kappa_n(s,t)$ a *normal curvature* and $\tau_g(s,t)$ a *geodesic torsion* of $\mathcal{S}_t$ respectively. It is known that

1) $\mathcal{S}_{t_0}$ is an asymptotic curve of $W$ if and only if $\kappa_n(s,t_0) \equiv 0$,
2) $\mathcal{S}_{t_0}$ is a geodesic of $W$ if and only if $\kappa_g(s,t_0) \equiv 0$,
3) $\mathcal{S}_{t_0}$ is a line of principal curvature of $W$ if and only if $\tau_g(s,t_0) \equiv 0$.

Then $\boldsymbol{b}(s,t_0) \pm \boldsymbol{n}(s,t_0)$ are lightlike. It follows that we have the light sheets $\mathbb{LS}^{\pm}_{\mathcal{S}_{t_0}} : J \times \{t_0\} \times \mathbb{R} \longrightarrow \mathbb{R}^3_1$ along $\mathcal{S}_{t_0}$ defined by $\mathbb{LS}^{\pm}_{\mathcal{S}_{t_0}}((s,t_0),u) = \mathbf{\Gamma}(s,t_0) + u(\boldsymbol{b}(s,t_0) \pm \boldsymbol{n}(s,t_0))$. Here, we use the notation $\mathbb{LS}^{\pm}_{\mathcal{S}_{t_0}}$ instead of $\mathbb{LH}^{\pm}_{\mathcal{S}_{t_0}}$ because the images of these mappings are lightlike surfaces. We adopt $\boldsymbol{n}^T = \boldsymbol{b}$ and $\boldsymbol{n}^S = \boldsymbol{n}$. By the Frenet-Serret type formulae, we have

$$\frac{\partial(\boldsymbol{n}^T \pm \boldsymbol{n}^S)}{\partial s}(s,t) = \frac{\partial(\boldsymbol{b} \pm \boldsymbol{n})}{\partial s}(s,t) = \tau_g(s,t))(\boldsymbol{n} \pm \boldsymbol{b})(s,t) - (\kappa_g(s,t) \pm \kappa_n(s,t))\boldsymbol{t}(s,t).$$

Therefore, we have $\kappa^{\pm}(\mathcal{S}_t)(s,t) = \kappa_g(s,t) \pm \kappa_n(s,t)$. It follows that the lightlike focal sets $\mathbb{LF}^{\pm}_{\mathcal{S}_{t_0}}$ are equal to

$$\left\{\boldsymbol{\Gamma}(s,t_0) + \frac{1}{\kappa_g(s,t_0) \pm \kappa_n(s,t_0)}(\boldsymbol{b} \pm \boldsymbol{t})(s,t_0) \mid s \in J, \kappa_g(s,t_0) \pm \kappa_n(s,t_0) \neq 0\right\}.$$

We consider the Lorentz distance squared function $G : J \times I \times \mathbb{R}^3_1 \longrightarrow \mathbb{R}$. In [**21**], by the calculation of the first and second derivative of $G$ with respect to $s$, we have introduced an invariant defined by

$$\sigma^{\pm}(s,t) = ((\kappa_n \pm \kappa_g)\tau_g \mp (\kappa_n' \pm \kappa_g'))(s,t).$$

A geometrical meaning of these invariants is given as follows [**21**].

PROPOSITION 9.1. *The following conditions are equivalent:*
(1) $\sigma^{\pm}(s,t_0) \equiv 0$,
(2) $\{\boldsymbol{\lambda}^{\pm}\} = \mathbb{LF}^{\pm}_{\mathcal{S}_{t_0}}$ *is a point,*
(3) $\mathcal{S}_{t_0} \subset LC_{\boldsymbol{\lambda}^{\pm}}$.

Moreover, as an application of the matrix criterion for $\mathcal{R}$-versality in [**6**, Section 6.10], we have shown the following proposition in [**21**].

PROPOSITION 9.2. (1) *The light sheet* $\mathbb{LS}^{\pm}_{\mathcal{S}_{t_0}}(J \times \{t_0\} \times \mathbb{R})$ *at* $\boldsymbol{\lambda}_0 \in \mathbb{LF}^{\pm}_{\mathcal{S}_{t_0}}$ *is local diffeomorphic to the cuspialedge* $\boldsymbol{CE}$ *if* $\sigma^{\pm}(s_0,t_0) \neq 0$,

(2) *The light sheet* $\mathbb{LS}^{\pm}_{\mathcal{S}_{t_0}}(J \times \{t_0\} \times \mathbb{R})$ *at* $\boldsymbol{\lambda}_0 \in \mathbb{LF}^{\pm}_{\mathcal{S}_{t_0}}$ *is local diffeomorphic to the swallowtail* $\boldsymbol{SW}$ *if* $\sigma^{\pm}(s_0,t_0) = 0$ *and* $\partial\sigma^{\pm}/\partial s(s_0,t_0) \neq 0$.
*Here,* $\boldsymbol{CE} = \{(u,v^2,v^3) \mid (u,v) \in \mathbb{R}^2\}$, $\boldsymbol{SW} = \{(3u^4 + u^2v, 4u^3 + 2uv, v) \mid (u,v) \in \mathbb{R}^2\}$.

On the other hand, we now classify $S.P^+$-Legendrian stable graph-like Legendrian unfoldings $\mathscr{L}_G(\Sigma_*(G))$ by $S.P^+$-Legendrian equivalence. By Theorems 6.5 and 6.6, it is enough to classify $\overline{f}$ by $S.P$-$\mathcal{K}$-equivalence under the condition that

$$\dim_{\mathbb{R}} \frac{\mathcal{E}_{1+1}}{\left\langle \frac{\partial \overline{f}}{\partial q}, \overline{f} \right\rangle_{\mathcal{E}_{1+1}} + \left\langle \frac{\partial \overline{f}}{\partial t} \right\rangle_{\mathbb{R}}} \leq 3.$$

In [**9, 12**] we have the following proposition.

PROPOSITION 9.3. *With the above condition,* $\overline{f} : (\mathbb{R} \times \mathbb{R}, 0) \longrightarrow (\mathbb{R}, 0)$ *with* $\partial\overline{f}/\partial t(0) \neq 0$ *is* $S.P$-$\mathcal{K}$*-equivalent to one of the following germs:*
(1) $q$,
(2) $\pm t \pm q^2$,
(3) $\pm t + q^3$,
(4) $\pm t \pm q^4$,
(5) $\pm t + q^5$.

The infinitesimally $S.P^+$-$\mathcal{K}$-versal unfolding $\mathcal{F} : (\mathbb{R} \times (\mathbb{R}^3 \times \mathbb{R}), 0) \longrightarrow (\mathbb{R}, 0)$ of each germ in the above list is given as follows (cf. [**12**, Theorem 4.2]):
(1) $q$
(2) $\pm t \pm q^2$,
(3) $\pm t + q^3 + x_0 q$,
(4) $\pm t \pm q^4 + x_0 q + x_1 q^2$,
(5) $\pm t + q^5 + x_0 q + x_1 q^2 + x_2 q^3$.
By Theorem 6.6, we have the following classification.

THEOREM 9.4. *Let $(W, \mathcal{S})$ be a world sheet in $\mathbb{R}^3_1$ parametrized by a timelike embedding $\boldsymbol{\Gamma} : J \times I \longrightarrow \mathbb{R}^3_1$ and $G : J \times I \times \mathbb{R}^3_1 \longrightarrow \mathbb{R}$ be the Lorentz distance squared function of $(W, \mathcal{S})$. Suppose that the corresponding graph-like Legendrian unfolding $\mathscr{L}_G(\Sigma_*(G)) \subset J^1(\mathbb{R}^3_1, I)$ is $S.P^+$-Legendrian stable. Then the germ of the unfolded light sheet $\mathbb{LH}^+_{(W,\mathcal{S})} \cup \mathbb{LH}^-_{(W,\mathcal{S})}$ at any point is $S.P^+$-diffeomorphic to one of the following set germs in $(\mathbb{R}^3 \times \mathbb{R}, 0)$:*

(1) $\{((u, v, w), 0) \mid (u, v, w) \in (\mathbb{R}^3, 0)\ \}$,
(2) $\{(-u^2, v, w), \pm 2u^3) \mid (u, v, w) \in (\mathbb{R}^3, 0)\ \}$,
(3) $\{(\mp 4u^3 - 2vu, v, w), 3u^3 \pm vu^2) \mid (u, v, w) \in (\mathbb{R}^3, 0)\ \}$,
(4) $\{((5u^4 + 2vu + 3wu^2, v, w), \pm(4u^4 + vu^2 + 2wu^3)) \mid (u, v, w) \in (\mathbb{R}^3, 0)\ \}$.

*Proof.* For any $(s_0, t_0, \boldsymbol{\lambda}_0) \in J \times I \times \mathbb{R}^3_1$, the germ of $\mathscr{L}_G(\Sigma_*(G)) \subset J^1(\mathbb{R}^3_1, I)$ at $z_0 = \mathscr{L}_G(s_0, t_0, \boldsymbol{\lambda}_0)$ is $S.P^+$-Legendrian stable. It follows that the germ of $g_{\boldsymbol{\lambda}_0}$ at $(s_0, t_0)$ is $S.P$-$\mathcal{K}$-equivalent to one of the germs in the list of Proposition 9.3. By Theorem 6.6, the graph-like Legendrian unfolding $\mathscr{L}_G(\Sigma_*(G))$ is $S.P^+$-Legendrian equivalent to the graph-like Legendrian unfolding $\mathscr{L}_{\mathcal{F}}(\Sigma_*(\mathcal{F}))$ where $\mathcal{F}$ is the infinitesimally $S.P$-$\mathcal{K}$-versal unfolding of one of the germs in the list of Proposition 9.3. It is also equivalent to the condition that the germ of the graph-like wave front $W(\mathscr{L}_{\mathcal{F}}(\Sigma_*(\mathcal{F})))$ is $S.P^+$-diffeomorphic to the corresponding graph-like wave front of one of the normal forms. For each normal form, we can obtain the graph-like wave front. We only show that case (5) in Proposition 9.3. In this case we consider $\mathcal{F}(q, x_0, x_1, x_2, t) = \pm t + q^5 + x_0 q + x_1 q^2 + x_2 q^3$. Then we have

$$\frac{\partial \mathcal{F}}{\partial q} = 5q^4 + x_0 + 2x_1 q + 3x_1 q^2,$$

so that the condition $\mathcal{F} = \partial \mathcal{F}/\partial q = 0$ is equivalent to the condition that

$$x_0 = -(5q^4 + x_0 + 2x_1 q + 3x_1 q^2),\ t_0 = \pm(4q^5 + x_1 q^2 + 2x_2 q^3).$$

If we put $u = q, v = x_0, w = x_1$, then $W(\mathscr{L}_{\mathcal{F}}(\Sigma_*(\mathcal{F})))$ is equal to

$$\{((-(5u^4 + 2vu + 3wu^2), v, w), \pm(4u^4 + vu^2 + 2wu^3))|(u, v, w) \in (\mathbb{R}^3, 0)\}.$$

It is $S.P^+$-diffeomorphic to the set germ of (4). We have similar calculations for other cases. We only remark here that we obtain the germ of (1) for both the germs of (1) and (2) in Proposition 9.3. Since $W(\mathscr{L}_{\mathcal{F}}(\Sigma_*(\mathcal{F}))) = \mathbb{LH}^+_{(W,\mathcal{S})} \cup \mathbb{LH}^-_{(W,\mathcal{S})}$, this completes the proof. □

As a corollary, we have a local classification of BR-caustics in this case.

COROLLARY 9.5. *With the same assumption for the world sheet $(W, \mathcal{S})$ as Theorem 9.4, the total BR-caustic $C(W, \mathcal{S})$ of $(W, \mathcal{S})$ at a singular point is locally diffeomorphic to the cuspidaledge* $\boldsymbol{CE}$ *or the swallowtail* $\boldsymbol{SW}$.

*Proof.* The total BR-caustic $C(W, \mathcal{S})$ of $(W, \mathcal{S})$ is the set of the critical values of $\pi_1 \circ \overline{\pi}|_{\mathscr{L}_G(\Sigma_*(G))}$. Therefore, it is enough to calculate the set of critical values of $\pi_1 \circ \overline{\pi}|_{\mathscr{L}_{\mathcal{F}}(\Sigma_*(\mathcal{F}))}$ for each normal form $\mathcal{F}$ in Proposition 9.3. For the germ (5) in Proposition 9.3, by the proof of Theorem 9.4 we have

$$\Sigma_*(\mathcal{F}) = \{(u, 5u^4 + 2vu + 3wu^2, v, w) \in (\mathbb{R} \times (\mathbb{R}^3 \times \mathbb{R}), 0)|(u, v, w) \in (\mathbb{R}^3, 0)\}.$$

It follows that

$$\pi_1 \circ \overline{\pi} \circ \mathscr{L}_{\mathcal{F}}(u, 5u^4 + 2vu + 3wu^2, v, w) = (5u^4 + 2vu + 3wu^2, v, w).$$

Then the Jacobi matrix of $f(u,v,w) = (5u^4 + 2vu + 3wu^2, v, w)$ is

$$J_f = \begin{pmatrix} 20u^3 + 2v + 6wu & 0 & 0 \\ 2u & 1 & 0 \\ 3u^2 & 0 & 1 \end{pmatrix},$$

so that the set of critical values of $f$ is given by

$$\{(-(15u^4 + 3wu^2), -10u^3 - 3wu, w) \in (\mathbb{R}^3, 0) \mid (u,w) \in (\mathbb{R}^2, 0)\}.$$

If we consider a linear transformation $\psi : (\mathbb{R}^3, 0) \longrightarrow \mathbb{R}^3, 0)$ defined by

$$\psi(x_0, x_1, x_2) = (-\frac{1}{5}x_0, -\frac{2}{5}x_1, \frac{3}{5}x_2),$$

then we have $\psi(-(15u^4 + 3wu^2), -10u^3 - 3uw, w) = (3u^4 + \frac{3}{5}wu^2, 4u^3 + \frac{6}{5}wu, \frac{3}{5}w)$. If we put $U = u, V = \frac{3}{5}w$, then we have $(3U^4 + VU^2, 4U^3 + 2VU, V)$, which is the parametrization of $\boldsymbol{SW}$. By the arguments similar to the above, we can show that the set of critical values of $\pi_1 \circ \overline{\pi}|_{\mathscr{L}_{\mathcal{F}}(\Sigma_*(\mathcal{F}))}$ is a regular surface for the germ (3) and is diffeomorphic to $\boldsymbol{CE}$ for the germ (4) in Proposition 9.3, respectively. This completes the proof. □

REMARK 9.6. Since a world sheet $(W, \mathcal{S})$ is a timelike surface in $\mathbb{R}^3_1$, we can define the evolute of $(W, \mathcal{S})$ by

$$Ev_{(W,\mathcal{S})} = \bigcup_{i=1}^{2} \left\{ \boldsymbol{X}(u,t) + \frac{1}{\kappa_i(u,t)} \boldsymbol{n}^S(u,t) \mid u \in U, \kappa_i(u,t) \neq 0 \right\},$$

where $\kappa_i(s,t)$ $(i = 1, 2)$ are the principal curvatures of $W$ at $p = \boldsymbol{X}(u,t)$. The evolute of a timelike surface, in general, has singularities. The singularities of the evolute of a generic timelike surface in $\mathbb{R}^3_1$ is classified into $\boldsymbol{CE}$, $\boldsymbol{SW}$, $\boldsymbol{PY}$ or $\boldsymbol{PU}$ (cf. [**29**]), where $\boldsymbol{PY} = \{(u^2 - v^2 + 2uv, -2uv + 2uw, w) | w^2 = u^2 + v^2\}$ is the *pyramid* and $\boldsymbol{PU} = \{(3u^2 + wv, 3v^2 + wu, w) | w^2 = 36uv\}$ is the *purse*. So the singularities of BR-caustics of world sheets are different from those of the evolutes of timelike surfaces. Since the singularities of BR-caustics are only corank one singularities, the pyramid and the purse never appear. Moreover, the normal line of a timelike surface is a spacelike line, so that it is not a ray in the sense of the relativity theory. Therefore, the evolute of a timelike surface in the Minkowski space-time is not a caustic in the sense of physics.

## References


[1] V. I. Arnol'd, S. M. Guseĭn-Zade, and A. N. Varchenko, *Singularities of differentiable maps. Vol. I*, Monographs in Mathematics, vol. 82, Birkhäuser Boston, Inc., Boston, MA, 1985. The classification of critical points, caustics and wave fronts; Translated from the Russian by Ian Porteous and Mark Reynolds. MR777682

[2] V. I. Arnol'd, *Singularities of caustics and wave fronts*, Mathematics and its Applications (Soviet Series), vol. 62, Kluwer Academic Publishers Group, Dordrecht, 1990. MR1151185

[3] L. M. Brekhovskikh, *Waves in layered media*, Applied Mathematics and Mechanics, vol. 16, Academic Press, Inc. [Harcourt Brace Jovanovich, Publishers], New York-London, 1980. Translated from the second Russian edition by Robert T. Beyer. MR693455

[4] Raphael Bousso, *The holographic principle*, Rev. Modern Phys. **74** (2002), no. 3, 825–874, DOI 10.1103/RevModPhys.74.825. MR1925130

[5] Raphael Bousso and Lisa Randall, *Holographic domains of anti-de Sitter space*, J. High Energy Phys. **4** (2002), No. 57, 26, DOI 10.1088/1126-6708/2002/04/057. MR1911390

[6] J. W. Bruce and P. J. Giblin, *Curves and singularities*, 2nd ed., Cambridge University Press, Cambridge, 1992. A geometrical introduction to singularity theory. MR1206472

[7] Subrahmanyan Chandrasekhar, *The mathematical theory of black holes*, International Series of Monographs on Physics, vol. 69, The Clarendon Press, Oxford University Press, New York, 1983. Oxford Science Publications. MR700826

[8] Lars Hörmander, *Fourier integral operators. I*, Acta Math. **127** (1971), no. 1-2, 79–183. MR0388463

[9] Shyūichi Izumiya, *Generic bifurcations of varieties*, Manuscripta Math. **46** (1984), no. 1-3, 137–164, DOI 10.1007/BF01185199. MR735518

[10] Shyūichi Izumiya, *Perestroikas of optical wave fronts and graphlike Legendrian unfoldings*, J. Differential Geom. **38** (1993), no. 3, 485–500. MR1243783

[11] Shyūichi Izumiya, *Geometric singularities for Hamilton-Jacobi equation*, Progress in differential geometry, Adv. Stud. Pure Math., vol. 22, Math. Soc. Japan, Tokyo, 1993, pp. 89–100. MR1274941

[12] Shyūichi Izumiya, *Completely integrable holonomic systems of first-order differential equations*, Proc. Roy. Soc. Edinburgh Sect. A **125** (1995), no. 3, 567–586, DOI 10.1017/S0308210500032686. MR1359492

[13] Shyuichi Izumiya, *Differential geometry from the view point of Lagrangian or Legendrian singularity theory*, Singularity theory, World Sci. Publ., Hackensack, NJ, 2007, pp. 241–275, DOI 10.1142/9789812707499_0008. MR2342914

[14] Shyuichi Izumiya and María del Carmen Romero Fuster, *The lightlike flat geometry on spacelike submanifolds of codimension two in Minkowski space*, Selecta Math. (N.S.) **13** (2007), no. 1, 23–55, DOI 10.1007/s00029-007-0033-9. MR2330586

[15] Shyuichi Izumiya and Masatomo Takahashi, *Spacelike parallels and evolutes in Minkowski pseudo-spheres*, J. Geom. Phys. **57** (2007), no. 8, 1569–1600, DOI 10.1016/j.geomphys.2007.01.008. MR2316715

[16] Shyuichi Izumiya and Masatomo Takahashi, *Caustics and wave front propagations: applications to differential geometry*, Geometry and topology of caustics—CAUSTICS '06, Banach Center Publ., vol. 82, Polish Acad. Sci. Inst. Math., Warsaw, 2008, pp. 125–142, DOI 10.4064/bc82-0-9. MR2463603

[17] Shyuichi Izumiya and Masatomo Takahashi, *Pedal foliations and Gauss maps of hypersurfaces in Euclidean space*, J. Singul. **6** (2012), 84–97. MR2971307

[18] S. Izumiya, *Geometry of world sheets in Lorentz-Minkowski space*, to appear in RIMS Kôkyûroku Bessatsu (2015)

[19] Shyuichi Izumiya, *The theory of graph-like Legendrian unfoldings and its applications*, J. Singul. **12** (2015), 53–79. MR3317140

[20] S. Izumiya, *Geometric interpretation of Lagrangian equivalence*, preprint (2014).

[21] S. Izumiya, *Caustics of world sheets in Lorenz-Minkowski 3-space*, to appear in the Proceedings of ICGA 2014.

[22] James A. Montaldi, *On contact between submanifolds*, Michigan Math. J. **33** (1986), no. 2, 195–199, DOI 10.1307/mmj/1029003348. MR837577

[23] Barrett O'Neill, *Semi-Riemannian geometry*, Pure and Applied Mathematics, vol. 103, Academic Press, Inc. [Harcourt Brace Jovanovich, Publishers], New York, 1983. With applications to relativity. MR719023

[24] J. F. Nye, *Natural focusing and fine structure of light*, Institute of Physics Publishing, Bristol, 1999. Caustics and wave dislocations. MR1684422

[25] Arlie O. Petters, Harold Levine, and Joachim Wambsganss, *Singularity theory and gravitational lensing*, Progress in Mathematical Physics, vol. 21, Birkhäuser Boston, Inc., Boston, MA, 2001. With a foreword by David Spergel. MR1836154

[26] I. R. Porteous, *The normal singularities of a submanifold*, J. Differential Geometry **5** (1971), 543–564. MR0292092

[27] R. Penrose, *Null hypersurface initial data for classical fields of arbitrary spin and for general relativity*, Gen. Relativity Gravitation **12** (1980), no. 3, 225–264, DOI 10.1007/BF00756234. MR574333

[28] Masaru Siino and Tatsuhiko Koike, *Topological classification of black holes: generic Maxwell set and crease set of a horizon*, Internat. J. Modern Phys. D **20** (2011), no. 6, 1095–1122, DOI 10.1142/S0218271811019220. MR2804243

[29] Farid Tari, *Caustics of surfaces in the Minkowski 3-space*, Q. J. Math. **63** (2012), no. 1, 189–209, DOI 10.1093/qmath/haq030. MR2889187

[30] V. M. Zakalyukin, *Envelopes of wave front families, and control theory* (Russian), Trudy Mat. Inst. Steklov. **209** (1995), no. Osob. Gladkikh Otobrazh. s Dop. Strukt., 133–142. MR1422221

DEPARTMENT OF MATHEMATICS, HOKKAIDO UNIVERSITY, SAPPORO 060-0810, JAPAN
*E-mail address*: izumiya@math.sci.hokudai.ac.jp

Contemporary Mathematics
Volume **675**, 2016
http://dx.doi.org/10.1090/conm/675/13589

# On genericity of a linear deformation of an isolated singularity

Masayuki Kawashima

ABSTRACT. In this paper, we study genericity and the number of cusps of linear deformations of an isolated singularity which is obtained by adding linear terms consisting of the conjugates of complex variables. To study this, we use higher differentials and mixed polynomials of join-type.

## 1. Introduction

Let $X$ and $Y$ be smooth manifolds of dimension $2n$ and 2 respectively and let $f : X \to Y$ be a smooth map with an isolated singularity. A smooth map $f$ is called *a generic map* if $f$ has only fold and cusp singularities. It is well known that the set of generic maps is a dense subset in the set of smooth maps $C^\infty(X, Y)$ with Whitney topology. We are interested in genericity and the number of cusps of a deformation $f_t$ of $f$. It is a difficult to know them even if $f$ is a polynomial map and $f_t$ is a linear deformation on $\mathbb{R}^{2n} \to \mathbb{R}^2$ where a linear deformation is a deformation given by a linear form on $\mathbb{R}^{2n}$. Polynomials which have complex and complex conjugate variables are called *mixed polynomials*. They are investigated in [**Oka08, Oka10, RSV02**] to study Milnor fibrations in the real analytic category.

DEFINITION 1.1. Let $q_1, \dots, q_n$ be positive integers such that $q_1 \geq \dots \geq q_n \geq 2$. We say that a linear deformation $f_t$ is type $(q_1, \dots, q_n)$ if $f_t$ is a deformation of the following type:

$$f_t(u_1, \dots, u_n) = u_1^{q_1} + \dots + u_n^{q_n} + a_1\bar{u}_1 + \dots + a_n\bar{u}_n$$

where $a_1, \dots, a_n$ are analytic functions with variable $t \in \mathbb{R}$ which vanish at $t = 0$. Namely $f_t$ is a mixed linear deformation of $f(u_1, \dots, u_n) = u_1^{q_1} + \dots + u_n^{q_n}$.

The linear deformation of type (2) $u^2 + a\bar{u}$ is $C^\infty$-equivalent to $u^2 + 2\bar{u}$. It is known that the map $u^2 + 2\bar{u}$ is a stable map with 3-cusps as a real analytic map from $\mathbb{R}^2$ to $\mathbb{R}^2$.

The type $(p, q)$ is studied by K. Inaba, M. Ishikawa, N. T. Thang and author in [**IIKN**]. For this type, the deformation $f_t$ is a generic map for generic choice of coefficients and the number of cusps is not constant and it is dependent on coefficients of complex conjugates. They showed that the minimum and maximum number of cusps is $(p + 1)(q - 1)$ and $(p - 1)(q + 1)$ respectively.

---

2010 *Mathematics Subject Classification.* Primary 57R45, 58C27, 14B05.
*Key words and phrases.* Stable map, higher differential, mixed polynomial.

To study linear deformations, we focus on their forms. Namely each variable is separated.

DEFINITION 1.2. Let $f(\boldsymbol{u}, \bar{\boldsymbol{u}})$ be a mixed polynomial of $n$-variables. We say that $f$ is a *join-type* if there exist mixed polynomials in one variable $f_1(u_1, \bar{u}_1), \dots, f_n(u_n, \bar{u}_n)$ such that

$$f(\boldsymbol{u}, \bar{\boldsymbol{u}}) = f_1(u_1, \bar{u}_1) + \cdots + f_n(u_n, \bar{u}_n).$$

Let $p = (p_1, \dots, p_n)$ be a singular point of join-type $f = f_1 + \cdots + f_n$. We say that $p$ is an *inner cusp* if each $p_j$ is a cusp singularity of $f_j$ for all $j = 1, \dots, n$. It is difficult to know whether $p$ is a cusp singularity of $f$ if $p$ is an inner cusp.

In this paper, we study genericity and the number of cusps which come from inner cusps of a linear deformation of type $(q, \dots, q)$.

THEOREM 1.3. *Let $f_t : \mathbb{C}^n \to \mathbb{C}$ be a linear deformation of type $(q, \dots, q)$. Suppose that coefficients of linear terms are generic. Then we have the following:*

(1) *The map $f_t$ is a generic map.*
(2) *The number of inner cusps of $f_t$ equals with $(q+1)(q-1)^{n-1}$.*

To prove Theorem 1.3, we use the theory of H. Levine about higher differentials [**Lev65**] for mixed functions of join-type.

In Section 2 and 3, we recall briefly the definitions and its properties of mixed functions, higher differentials and its examples.

In Section 4 and 5, we prove main Theorem (1) and (2) respectively using higher differentials.

## 2. Mixed polynomials

Let $f(\boldsymbol{u}, \bar{\boldsymbol{u}})$ be an $n$-variables mixed polynomial where $\boldsymbol{u} = (u_1, \dots, u_n)$ and $\bar{\boldsymbol{u}} = (\bar{u}_1, \dots, \bar{u}_n)$. Putting $u_i := x_i + \sqrt{-1}\, y_i$, we can consider that $f$ is a $2n$-variables polynomial. Thus $f$ is regarded as a real analytic function $(Q, R) : \mathbb{R}^{2n} \to \mathbb{R}^2$ where $Q$ and $R$ are real and imaginary parts of $f$ such that

$$f(\boldsymbol{u}, \bar{\boldsymbol{u}}) = Q\left(\frac{\boldsymbol{u} + \bar{\boldsymbol{u}}}{2}, \frac{\boldsymbol{u} - \bar{\boldsymbol{u}}}{2\sqrt{-1}}\right) + \sqrt{-1} R\left(\frac{\boldsymbol{u} + \bar{\boldsymbol{u}}}{2}, \frac{\boldsymbol{u} - \bar{\boldsymbol{u}}}{2\sqrt{-1}}\right).$$

A point $p_0 \in \mathbb{R}^{2n}$ is called *a singular point of $f$* if the gradient vectors of $Q$ and $R$ are linearly dependent over $\mathbb{R}$ at $p_0$. It is not easy to check dependency of these two gradient vectors. In [**Oka08**], M. Oka proved the following useful criterion.

LEMMA 2.1 ([**Oka08**]). *The following conditions are equivalent:*

(1) *A point $p_0 \in \mathbb{C}^n$ is a singular point of $f$.*
(2) *There exists $\alpha \in S^1$ such that*

$$\overline{\frac{\partial f}{\partial u_i}}(p_0) = \alpha \frac{\partial f}{\partial \bar{u}_i}(p_0), \quad i = 1, \dots, n.$$

We denote the set of singular points and the set of singular values of $f$ by $S(f)$ and by $C(f)$ respectively.

Let $f$ be a mixed function of join-type. We consider the set of singular points of $f$. As $f$ is a join-type, we have equalities $\frac{\partial f}{\partial u_j} = \frac{\partial f_j}{\partial u_j}$ and $\frac{\partial f}{\partial \bar{u}_j} = \frac{\partial f_j}{\partial \bar{u}_j}$ for all $j$. Using Lemma 2.1, $S(f)$ is written as

$$S(f) = \left\{ (p_1, \dots, p_n) \in \mathbb{C}^n \mid \exists e^{i\theta} \in S^1 \left[ \overline{\frac{\partial f_j}{\partial u_j}}(p_j) = \frac{\partial f_j}{\partial \bar{u}_j}(p_j) e^{i\theta}, j = 1, \dots, n \right] \right\}.$$

Note that even though each $p_j$ is a singular point of $f_j$ for $j = 1, \dots, n$, the $n$-ple $(p_1, \dots, p_n)$ may not be a singular point of $f$. Thus $S(f)$ is a proper subset of the product $S(f_1) \times \dots \times S(f_n)$.

## 3. H. Levine's Theory

In this section, we recall H. Levine's Theory with respect to the higher differentials and genericity of smooth maps [**Lev65**]. Let $X$ and $Y$ be smooth manifolds of dimension $n$ and 2 respectively.

### 3.1. Generic maps.

First of all, we recall the definition of generic maps.

DEFINITION 3.1. Let $f : X \to Y$ be a smooth map. We call $f$ *generic map* if for any $p \in X$, there exist local coordinates $(x_1, \dots, x_n)$ and $(y_1, y_2)$ centered at $p$ and $f(p)$ respectively such that $f$ is locally described in one of the following forms:

(1) $(x_1, \dots, x_n) \mapsto (x_1, x_2)$,

(2) $(x_1, \dots, x_n) \mapsto (x_1, \sum_{i=2}^n \pm x_i^2)$,

(3) $(x_1, \dots, x_n) \mapsto (x_1, x_2^3 + x_1 x_2 + \sum_{i=3}^n \pm x_i^2)$.

The points in cases (2) and (3) are called fold and cusp respectively.

We define a subset $S_i(f)$ of $X$ as $S_i(f) := \{p \in X \mid \operatorname{rank} df_p = 2 - i\}$ for $i = 0, 1, 2$. Note that $S_0(f)$ is the set of regular points of $f$ and the union $S_1(f) \cup S_2(f)$ is the set of singular points $S(f)$.

### 3.2. Higher Differential.

To show genericity of smooth maps, we use *higher differentials* which are defined by H. Levine in [**Lev65**]. In our situations, we can define *the second differential* $d^2 f$ and *the third differential* $d^3 f$ on a neighborhood $U$ at a singular point $p \in S(f)$. Refer their precise definitions and properties in [**Lev65, IIKN**].

LEMMA 3.2 (H. Levine). *Let $f : X \to Y$ be a smooth map. Then $f$ is a generic map if and only if the following conditions hold:*

(a) *The second differential $d^2 f_p : T_p X \to L_p^* \otimes G_p$ is surjective for all $p \in X$ where $L_p := \ker df_p$ and $G_p := \operatorname{coker} df_p$.*

(b) *The third differential $d^3 f_p : T_p \to R_p^* \otimes K_p$ is surjective for all $p \in X$ where $T_p := \ker d^2 f_p$, $R_p := \ker d^2 f_p|_{L_p}$ and $K_p := \operatorname{coker} d^2 f_p|_{L_p}$.*

REMARK 3.3. The condition $(a)$ implies $S_2(f) = \emptyset$. Let $p \in X$ be a regular point. Then we have three equalities: $\dim G_p = 0$, $T_p = T_p X$ and $\dim K_p = 0$. Hence the condition $(b)$ is trivially satisfied. If $p \in S_1(f) \setminus S_{1,1}(f)$, then the dimension of $R_p$ is 0 and hence $(b)$ is also trivially satisfied where $S_{1,1}(f) := \{p \in X \mid \operatorname{rank} d^2 f_p|_{L_p} = n - 2\}$. Thus the non-trivial part of the condition $(b)$ is the case $p \in S_{1,1}(f)$.

Using some local coordinates at a singular point, we can rewrite H. Levine's Theorem:

(a)' The represented matrix $M_p$ of $d^2 f_p$ has rank $n-1$.

(b)' Let $H_p$ and $S_p$ be the represented matrix of $d^2 f_p|_{L_p}$ and $d^3 f_p$ respectively. Then $p \in S_{1,1}(f)$ if and only if rank $H_p = n-2$. Under the condition $(a)'$, the dimension of $T_p$ is 1 and $S_p$ is a constant. Therefore $d^3 f_p$ is surjective if and only if $S_p$ is not 0.

The following lemma gives a criterion to determine if a singular point is a fold, a cusp, or none of them.

LEMMA 3.4. *Suppose that $p \in S_1(f)$ and $d^2 f_p : T_pX \to L_p^* \otimes G_p$ is surjective. Then we have the following.*

(1) *$p$ is a fold singularity of $f$ if and only if $p \notin S_{1,1}(f)$.*

(2) *$p$ is a cusp singularity of $f$ if and only if $p \in S_{1,1}(f)$ and $d^3 f_p : T_p \to R_p^* \otimes K_p$ is surjective. In other words, $p$ is a cusp singularity if and only if* $\operatorname{rank} H_p = n-2$ *and* $S_p \neq 0$.

3.2.1. *Example 1. The case* $\dim X = \dim Y = 2$. Let $f = (Q, R) : X \to Y$ be a smooth map and $p \in S(f)$. Suppose that there exist local coordinates $(x_1, x_2)$ and $(y_1, y_2)$ at $p$ and $f(p)$ respectively such that $\operatorname{grad} Q(p) = (1, 0)$ and $\operatorname{grad} R(p) = (0, 0)$. Using these coordinates, we have

$$df_p = dx_1 \otimes \frac{\partial}{\partial y_1}, \quad L_p = \left\langle \frac{\partial}{\partial x_2} \right\rangle, \quad G_p = \left\langle \frac{\partial}{\partial y_2} \right\rangle.$$

Then the the second differential $d^2 f_p : T_pX \to L_p^* \otimes G_p$ is defined as

$$d^2 f_p\left(\frac{\partial}{\partial x_1}\right) = R_{1,2}\, dx_2 \otimes \frac{\partial}{\partial y_2}, \quad d^2 f_p\left(\frac{\partial}{\partial x_2}\right) = R_{2,2}\, dx_2 \otimes \frac{\partial}{\partial y_2}$$

where $R_{i,j} = \frac{\partial^2 R}{\partial x_i \partial x_j}(p)$. As the represented matrix $H_p$ of $d^2 f_p|_{L_p}$ is $R_{2,2}$, we have the condition $p \in S_{1,1}(f)$ if and only if $R_{2,2} = 0$.

Next we consider the third differential $d^3 f_p : T_p \to R_p^* \otimes K_p$ under the conditions $p \in S_{1,1}(f)$ and $d^2 f_p$ is surjective. Note that the equalities hold: $T_p = R_p = \langle \frac{\partial}{\partial x_2} \rangle$ and $K_p = L_p^* \otimes G_p = \langle dx_2 \otimes \frac{\partial}{\partial y_2} \rangle$.

Finally, the third differential $d^3 f_p$ is written as

$$d^3 f_p\left(\frac{\partial}{\partial x_2}\right) = (R_{2,2,2} - 3Q_{2,2}R_{1,2}) dx_2 \otimes dx_2 \otimes \frac{\partial}{\partial y_2}.$$

Hence the represented matrix $S_p$ of $d^3 f_p$ is $R_{2,2,2} - 3Q_{2,2}R_{1,2}$. Thus we have the following criterion for the case $\dim X = \dim Y = 2$.

LEMMA 3.5. *Suppose that $p$ is a singular point with $d^2 f_p$ is surjective. If $R_{2,2} \neq 0$, then $p$ is a fold singularity. If $R_{2,2} = 0$ and $R_{2,2,2} - 3Q_{2,2}R_{1,2} \neq 0$, then $p$ is a cusp singularity.*

3.2.2. *Example 2. The case join-type.* Let $f : \mathbb{R}^{2n} \to \mathbb{R}^2$ be a join-type and let $Q$ and $R$ be the real and imaginary parts of $f$ respectively.

Suppose that there exist local coordinates $(x_1, x_2, \ldots, x_{2n})$ and $(y_1, y_2)$ at each singular point $p \in S(f)$ and the image $f(p)$ such that

$$\operatorname{grad} Q(p) = (1, 0, \ldots, 1, 0), \quad \operatorname{grad} R(p) = (0, 0, \ldots, 0, 0).$$

Under these coordinates, we have

$$df_p = \sum_{k=1}^{n} dx_{2k-1} \otimes \frac{\partial}{\partial y_1}, \quad L_p = \left\langle \frac{\hat{\partial}}{\partial x_2}, \frac{\hat{\partial}}{\partial x_3} \cdots, \frac{\hat{\partial}}{\partial x_{2n}} \right\rangle, \quad G_p = \left\langle \frac{\partial}{\partial y_2} \right\rangle$$

where $\frac{\hat{\partial}}{\partial x_{2k}} := \frac{\partial}{\partial x_{2k}}$ and $\frac{\hat{\partial}}{\partial x_{2k+1}} := \frac{\partial}{\partial x_{2k+1}} - \frac{\partial}{\partial x_{2k-1}}$ for $k = 1, \ldots, n-1$.

Now we consider surjectivity of $d^2 f_p : T_p X \to L_p^* \otimes G_p$. Let $H(R,k)$ be the Hesse matrix of $R$ with respect to variables $x_{2k-1}$ and $x_{2k}$ at $p$:

$$H(R,k) := \begin{pmatrix} R_{2k-1,2k-1} & R_{2k-1,2k} \\ R_{2k-1,2k} & R_{2k,2k} \end{pmatrix}.$$

We say that $R$ is *non-degenerate with respect to* $x_{2k-1}$ *and* $x_{2k}$ if the determinant of $H(R,k)$ is not zero.

LEMMA 3.6. *Suppose that $R$ is non-degenerate with respect to $x_{2k-1}$ and $x_{2k}$ for all $k$. Then $d^2 f_p$ is surjective.*

PROOF. Our proof is a simple calculation under the assumption non-degeneracy of $R$. □

Next we consider the determinant of the represented matrix $H_p = (h_{ij})$ of $d^2 f_p|_{L_p}$. It is obtained by the determinant of $H(R,k)$.

LEMMA 3.7. *Suppose that $R$ is non-degenerate with respect to $x_{2k-1}$ and $x_{2k}$ for all $k$. Then the determinant of $H_p$ is*

$$\det H_p = T \sum_{k=1}^{n} \frac{R_{2k,2k}}{T_{2k-1}}$$

*where* $T_{2k-1} := \det H(R,k)$ *and* $T = \prod_{k=1}^{n} T_{2k-1}$.

Therefore $p$ is a fold singularity if and only if $R$ is non-degenerate with respect to $x_{2k-1}$ and $x_{2k}$ for all $k$ and $\det H_p \neq 0$.

Let $p = (p_1, \ldots, p_n) \in S_1(f)$ be an inner cusp singularity. By Lemma 3.5, the conditions

$$R_{2k,2k} = 0, \quad S_{p_k} := R_{2k,2k,2k} - 3Q_{2k,2k} R_{2k-1,2k} \neq 0$$

hold for our local coordinates $(x_{2k}, x_{2k+1})$ at $p_k$ and $(y_1, y_2)$ at $f(p)$. Note that if $p$ is an inner cusp singularity, then $\operatorname{rank} H_p = n-2$ and $p \in S_{1,1}(f)$. By simple calculations, we have the following:

LEMMA 3.8. *Under our assumptions, the represented matrix $S_p$ of $d^3 f_p$ is given by*

$$S_p = R_{2n-1,2n} R^2 \sum_{k=1}^{n} \frac{S_{p_k}}{R^3_{2k-1,2k}} \quad \textit{where} \quad R = \prod_{s=1}^{n} R_{2s-1,s}.$$

Thus if $p$ is an inner cusp, then the represented matrix $S_p$ of the third differential of join-type is written by using the represented matrix $S_{p_k}$ of the third differential of each variable.

## 4. Genericity linear deformations of type $(q, \ldots, q)$

In this section, we will show that the first part of Theorem 1.3. Consider a linear deformation of type $(q, \ldots, q)$:

$$f_t(u_1, \ldots, u_n) = u_1^q + \cdots + u_n^q + a_1\bar{u}_1 + \cdots + a_n\bar{u}_n, \quad a_1, \ldots, a_n \in \mathbb{C}.$$

Hereafter we omit the variable $t$ and we write $f_t$ by $g$.

**4.1. The one variable case.** First we consider the linear deformation of the one variable case:

$$g(u, \bar{u}) = u^q + a\bar{u}, \quad a := r_a e^{i\theta_a} \in \mathbb{C}.$$

Now we use Lemma 2.1 to determine singularities of $g$. Then $u \in \mathbb{C}$ is a singular point of $g$ if and only if $u$ satisfies the equation $u^{q-1} = \bar{a}e^{i\theta}$ for some $e^{i\theta} \in S^1$. Thus the set of singular points $S(g)$ of $g$ is written as

$$S(g) = \left\{u \in \mathbb{C} \mid \exists e^{i\theta} \in S^1 \left[qu^{q-1} = \bar{a}e^{i\theta}\right]\right\}.$$

Using the polar coordinate $(r, \theta)$ of $\mathbb{C}$, $g$ is written as $g(r, \theta) = r^q e^{iq\theta} + r r_a e^{-i(\theta-\theta_a)}$. To use Levine's theory, we consider the gradient vectors of the real part $Q$ and the imaginary part $R$ of $g$ at a singular point $p_0 = (c_a, \theta_0)$ where $c_a := \sqrt[q-1]{\frac{r_a}{q}}$.

Lemma 4.1. *The gradient vectors of $Q$ and $R$ at $p_0$ are the following:*

$$\operatorname{grad} Q(p_0) = (\kappa_1, \kappa_2) = (2r_a \cos\Theta_1 \cos\Theta_2, -2r_a c_a \sin\Theta_1 \cos\Theta_2),$$
$$\operatorname{grad} R(p_0) = (\kappa_3, \kappa_4) = (2r_a \cos\Theta_1 \sin\Theta_2, -2r_a c_a \sin\Theta_1 \sin\Theta_2)$$

*where* $\Theta_1 := \dfrac{q+1}{2}\theta_0 - \dfrac{\theta_a}{2}$ *and* $\Theta_2 := \dfrac{q-1}{2}\theta_0 + \dfrac{\theta_a}{2}$.

Proof. Calculating their partial derivatives and using the formula $e^{iA} + e^{-iB} = 2\cos\left(\frac{A+B}{2}\right)e^{\frac{A-B}{2}i}$, we have

$$g_r(p_0) = 2a\cos\left(\frac{q+1}{2}\theta_0 - \frac{\theta_a}{2}\right)e^{\left(\frac{q-1}{2}\theta_0 + \frac{\theta_a}{2}\right)i} = 2a\cos\Theta_1 e^{\Theta_2 i},$$

$$g_\theta(p_0) = 2ac_a ii\sin\left(\frac{q+1}{2}\theta_0 - \frac{\theta_a}{2}\right)e^{\left(\frac{q-1}{2}\theta_0 + \frac{\theta_a}{2}\right)i} = -2ac_a\sin\Theta_1 e^{\Theta_2 i}.$$

Using these two equalities, we have the assertion. □

**The case $\kappa_1 \neq 0$.** Now we consider the case $\kappa_1 \neq 0$. We take new coordinates $(r_1, \theta_1)$ and $(\hat{y}_1, \hat{y}_2)$ such that

$$(r, \theta) = \left(\frac{1}{\kappa_1}r_1 - \frac{\kappa_2}{\kappa_1}\theta_1, \theta_1\right), \quad (\hat{y}_1, \hat{y}_2) = (y_1, y_2 - sy_1)$$

where $s := R_{r_1}(p_0) = \tan\Theta_2$. Using these coordinates, we have $\operatorname{grad} Q(p_0) = (1, 0)$ and $\operatorname{grad}\hat{R}(p_0) = (0, 0)$ where $\hat{R} := R - sQ$.

Lemma 4.2. *Under our situations, the second derivative $\hat{R}_{\theta_1\theta_1}$ is written as*

$$\hat{R}_{\theta_1\theta_1} = c_a r_a (q-1)\frac{\sin\Theta_1}{\cos^2\Theta_1\cos\Theta_2}.$$

*Moreover there are no common zero points of the two equations $\hat{R}_{\theta_1\theta_1} = 0$ and $S_{p_0} = 0$.*

PROOF. The equality of $\hat{R}_{\theta_1\theta_1}$ is obtained by a simple calculation. Now we show the second assertion. If $\hat{R}_{\theta_1\theta_1} = 0$, then $\sin\Theta_1 = 0$ and $\cos\Theta_1 = \pm 1$. Then we have $S_{p_0} = \pm\dfrac{c_a r_a(q^2-1)}{2\cos\Theta_2} \neq 0$ where the sign of $S_{p_0}$ is the sign of $\cos\Theta_1$. □

Under the assumption $\kappa_1 \neq 0$, Lemma 3.5 and 4.2 show that the linear deformation $g$ is generic and the cusp points corresponding to $\Theta_1 = n\pi$ where $n \in \mathbb{Z}$. As $\Theta_1 = \dfrac{q+1}{2}\theta_0 - \dfrac{\theta_a}{2}$, the singular point $p_0$ is cusp if and only if

$$\theta_0 = \frac{2n\pi}{q+1} + \frac{\theta_a}{q+1}, \qquad n = 0, \dots, q.$$

Therefore if $\kappa_1 \neq 0$, then $g$ has $(q+1)$-cusp singularities.

**The case $\kappa_1 = 0$.** Now we consider the case $\kappa_1 = 2r_a\cos\Theta_1\cos\Theta_2 = 0$. There are two cases: $\cos\Theta_1 = 0$ or $\cos\Theta_2 = 0$. Note that the case $\cos\Theta_1 = \cos\Theta_2 = 0$ does not occur.

If $\cos\Theta_1 = 0$, then $\sin\Theta_1 = \pm 1$ and $\kappa_2 = \mp 2ac_a\cos\Theta_2 \neq 0$. Then the point $p_0$ is a fold point. In fact, we can take new coordinates $(\theta, r) = (\frac{\theta_2}{\kappa_2}, r_2)$. Then the gradient vectors are $\operatorname{grad} Q(p_0) = (1,0)$ and $\operatorname{grad}\hat{R}(p_0) = (0,0)$ where $\hat{R} = R - \tan\Theta_2 Q$. Then we have

$$\hat{R}_{r_2,r_2} = \frac{ac_a(q-1)}{\cos\Theta_2}\sin\left(\frac{2n+1}{2}\pi\right) \neq 0, \qquad n \in \mathbb{Z}.$$

This shows that $p_0$ is a fold point.

If $\cos\Theta_2 = 0$, then $\kappa_1 = \kappa_2 = 0$ and $\kappa_3 \neq 0$. By the same argument as above, we can show that the point $p_0$ is a fold point.

**4.2. The join-type case.** Let $g(\boldsymbol{u}, \bar{\boldsymbol{u}}) = g_1(u_1, \bar{u}_1) + \cdots + g_n(u_n, \bar{u}_n)$ be a mixed polynomial of join-type and each component is defined by

$$g_k(u_k, \bar{u}_k) = u_k^q + a_k\bar{u}_k, \quad a_k := r_{a_k}e^{i\theta_{a_k}} \in \mathbb{C}, \quad k = 1, \dots, n.$$

Using Lemma 2.1 and the argument of the one variable case, the set of singular points $S(g)$ of $g$ is written as

$$S(g) = \left\{(u_1, \dots, u_n) \in \mathbb{C}^n \mid \forall k,\ qu_k^{q-1} = r_{a_k}e^{i(\theta-\theta_{a_k})} \text{ for some } e^{i\theta} \in S^1\right\}.$$

We use the polar coordinates $(r_1, \theta_1, \dots, r_n, \theta_n)$ in $\mathbb{C}^n$ where $u_k = (r_k, \theta_k)$. Take a singular point $p_0 := (c_1, \theta_1, \dots, c_n, \theta_n) \in S(g)$ where $c_k := \sqrt[q-1]{\frac{r_{a_k}}{q}}$. Each argument $\theta_k$ satisfies the relation

$$(\star)_{k,m} \qquad \theta_k = \theta_1 + \frac{\theta_{a_1} - \theta_{a_k} + 2m\pi}{q-1}$$

where $m \in J := \{0, \dots, q-2\}$. Hence $S(g)$ has $(q-1)^{n-1}$-components. In fact, we take $\boldsymbol{m} := (m_2, \dots, m_n) \in J^{n-1}$ and put

$$S(g, \boldsymbol{m}) = \{(c_1, \theta_1, \dots, c_n, \theta_n) \in S(g) \mid \theta_k \text{ satisfies } (\star)_{k,m_k} \text{ for } k = 2, \dots, n\}.$$

Thus we have $S(g) = \bigcup_{\boldsymbol{m}\in J^{n-1}} S(g, \boldsymbol{m})$.

Let $Q := Q^{(1)} + \cdots + Q^{(n)}$ and $R := R^{(1)} + \cdots + R^{(n)}$ be the decomposition into each variable. The gradient vector of $Q$ is given by

$$\operatorname{grad} Q = \left(\operatorname{grad} Q^{(1)}, \ldots, \operatorname{grad} Q^{(n)}\right) \quad \text{where}$$
$$\operatorname{grad} Q^{(k)} = \left(\kappa_1^{(k)}, \kappa_2^{(k)}\right) = \left(2r_{a_k} \cos\Theta_1^{(k)} \cos\Theta_2^{(k)}, -2r_{a_k} c_k \sin\Theta_1^{(k)} \cos\Theta_2^{(k)}\right),$$

where $\Theta_1^{(k)} := \dfrac{q+1}{2}\theta_k - \dfrac{\theta_{a_k}}{2}$ and $\Theta_2^{(k)} := \dfrac{q-1}{2}\theta_k + \dfrac{\theta_{a_k}}{2}$ for $k = 1, \ldots, n$.

**The case $\kappa_1^{(k)} \neq 0$ for all $k = 1, \ldots, n$.** Suppose that $\kappa_1^{(k)} \neq 0$ for all $k = 1, \ldots, n$. In this case, we can take new coordinates $(r'_k, \theta'_k)$ and $(\hat{y}_1, \hat{y}_2)$ such that

$$(r_k, \theta_k) = \left(\frac{1}{\kappa_1^{(k)}} r'_k - \frac{\kappa_2^{(k)}}{\kappa_1^{(k)}} \theta'_k, \theta'_k\right), \qquad (\hat{y}_1, \hat{y}_2) = (y_1, y_2 - s y_1)$$

where $s = \tan\Theta_2$. Using these coordinates, the gradient vectors are

$$\operatorname{grad} Q^{(k)}(p_0) = (1, 0), \quad \operatorname{grad} \hat{R}^{(k)}(p_0) = (0, 0)$$

where $\hat{R}^{(k)} := R^{(k)} - sQ^{(k)}$ for all $k$.

LEMMA 4.3. *The function $\hat{R}$ is a non-degenerate function with respect to $(r'_k, \theta'_k)$. Therefore the second derivative $d^2 g_p$ is surjective.*

PROOF. We denote the Hesse matrix of $\hat{R}$ with respect to $(r'_k, \theta'_k)$ by $H(\hat{R}, k)$:

$$H(\hat{R}, k) = \begin{pmatrix} \dfrac{q(q-1)c_k^{q-2} \sin\Theta_1^{(k)}}{4r_{a_k}^2 \cos^2\Theta_1^{(k)} \cos^3\Theta_2^{(k)}}, & \dfrac{q-1}{2\cos^2\Theta_1^{(k)} \cos^2\Theta_2^{(k)}} \\[2ex] \dfrac{q-1}{2\cos^2\Theta_1^{(k)} \cos^2\Theta_2^{(k)}}, & \dfrac{r_{a_k} c_k (q-1) \sin\Theta_1^{(k)}}{\cos^2\Theta_1^{(k)} \cos\Theta_2^{(k)}} \end{pmatrix}.$$

Hence the determinant of $H(\hat{R}, k)$ is

$$\det H(\hat{R}, k) = \frac{(q-1)^2}{4\cos^2\Theta_1^{(k)} \cos^4\Theta_2^{(k)}} \neq 0.$$

□

Next we consider the determinant of the represented matrix $H_p$ of $d^2 f|_{L_p}$. By Lemma 3.7, it is given as

$$\begin{aligned} \det H_p &= T \sum_{k=1}^{n} \frac{\hat{R}_{\theta'_k, \theta'_k}}{\det H(\hat{R}, k)} \quad \text{where} \quad T = \prod_{k=1}^{n} \det H(\hat{R}, k) \\ &= T \sum_{k=1}^{n} \frac{c_k r_{a_k} (q-1) \sin\Theta_1^{(k)}}{\cos^2\Theta_1^{(k)} \cos\Theta_2^{(k)}} \cdot \frac{4\cos^2\Theta_1^{(k)} \cos^4\Theta_2^{(k)}}{(q-1)^2} \\ &= \frac{4T}{q-1} \sum_{k=1}^{n} c_k r_{a_k} \sin\Theta_1^{(k)} \cos^3\Theta_2^{(k)}. \end{aligned}$$

PROPOSITION 4.4. *Suppose that $a_1, \dots, a_n$ are generic and $p$ is an inner fold singularity. Then $p$ is a fold singularity of $g$.*

PROOF. If $p = (p_1, \dots, p_n)$ is an inner fold, then $\hat{R}_{\theta'_k,\theta'_k} \neq 0$ for all $k$. That is $\sin \Theta_1^{(k)} \neq 0$. Hence $\det H_p \neq 0$ under the assumption $a_1, \dots, a_n$ are generic. □

Next we consider the case $p$ is an inner cusp. If $p$ is an inner cusp, by the previous arguments, then we have $\hat{R}_{r'_k,\theta'_k} \neq 0$, $\hat{R}_{\theta'_k,\theta'_k} = 0$ and $S_{p_k} = \pm \dfrac{c_k r_{a_k}(q^2-1)}{2\cos\Theta_2^{(k)}}$ where the sign of $S_p$ is the sign of $\cos\Theta_1^{(k)}$. Now we consider the third derivative. By Lemma 3.8, we have

$$\begin{aligned} d^3 g_p(t) &= \hat{R}_{r'_n,\theta'_n} \hat{R}^2 \sum_{k=1}^{n} \frac{S_{p_k}}{\hat{R}^3_{r'_k,\theta'_k}} \\ &= \pm \frac{4(q+1)}{(q-1)^2} \hat{R}_{r'_n,\theta'_n} \hat{R}^2 \sum_{k=1}^{n} c_k r_k \cos^5 \Theta_2^{(k)}. \end{aligned}$$

LEMMA 4.5. *Suppose that $a_1, \dots, a_n$ are generic and $p$ is an inner cusp singularity. then $p$ is a cusp singularity of $g$.*

PROOF. As $p$ is an inner cusp, then $\hat{R}_{\theta'_k,\theta'_k} = 0$ and hence $\operatorname{rank} H_p = n-2$ and $p \in S_{1,1}(f)$. By the above arguments, $S_p$ is not zero as $a_1, \dots, a_n$ are generic. □

**The case $\kappa_1^{(k)} = 0$ for some $k \in \{1, \dots, n\}$.** Now we consider the case $\kappa_1^{(k)} \neq 0$ for some $k \in \{1, \dots, n\}$. For general $a_1, \dots, a_n$, by similar arguments as the one variable case, we can show that the corresponding points are fold points.

Therefore, by Lemma 4.4 and 4.5, if $a_1, \dots, a_n$ are generic, then our deformation $g$ is generic.

## 5. The number of cusp points

In this section, we will prove the second part of Theorem 1.3. Namely we consider the number of cusp points of our linear deformation $g$ under the assumption that $g$ is a generic map. It is sufficient to count the number of cusps on the set of singular values. We recall a smooth curve $h : I \to \mathbb{C}$ has $(2,3)$-cusp at $t_0 \in I$ if $\gamma$ is $C^\infty$-equivalent to the curve $c(t) := t^2 + \sqrt{-1}\, t^3$. For a $(2,3)$-cusps, there is the following useful criterion in [**Ish93**].

LEMMA 5.1 ([**Ish93**]). *Let $\gamma : I \to \mathbb{C}$ be a smooth curve. Then $\gamma$ has a $(2,3)$-cusp at $t_0$ if and only if $\gamma'(t_0) = 0$, $\gamma''(t_0) \neq 0$ and $\Im\left(\overline{\gamma''} \cdot \gamma'''\right)(t_0) \neq 0$ where $\Im(z)$ is the imaginary part of $z$.*

Hereafter we call a $(2,3)$-cusp by a cusp simply. There is a typical example of a $(q+1)$-cuspidal curve. We consider the curve $h_q : S^1 \to \mathbb{C}$ which is defined by $h_q(\theta) = e^{qi\theta} + qe^{-i\theta}$. We can check easily that $h$ has $(q+1)$-cusps using Lemma 5.1.

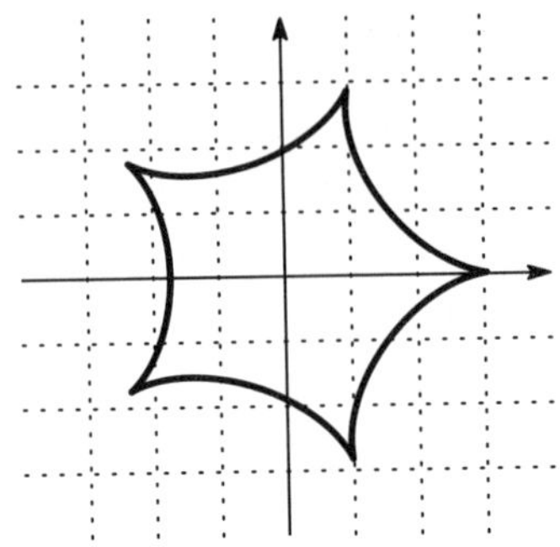

FIGURE 1. The graph of $h_4(t)$

**5.1. The set of singular values.** Now we consider the set of singular values of our linear deformation $g$ on $S(g)$.

PROPOSITION 5.2. *Let $g$ be a linear deformation of type $(q, \dots, q)$ into generic maps. Then the set of singular values $C(g)$ of $g$ in $\mathbb{C}$ is a scaling and rotation of the curve given by the map $h_q : S^1 \to \mathbb{C}$.*

PROOF. Take a singular point $p_0 = (c_1, \theta_1, \dots, c_n, \theta_n) \in S(g, \boldsymbol{m})$. As $\theta_k$ satisfies $\theta_k = \theta_1 + \frac{\theta_{a_1} - \theta_{a_k} + 2m_k\pi}{q-1}$, we have

$$\begin{aligned} g(p_0) &= \sum_{k=1}^{n} \left( B_k e^{i(q\theta_1 + \theta_{a_1})} + \bar{B}_k q e^{-i\theta_1} \right), \quad B_k = \frac{c_k r_{a_k}}{q} e^{\frac{\theta_{a_1} - q\theta_{a_k} + 2m_k\pi}{q-1}} \\ &= B e^{i(q\theta_1 + \theta_{a_1})} + \bar{B} q e^{-i\theta_1}, \quad B = \sum_{k=1}^{n} B_k \\ &= |B| \left( e^{i(q\theta_1 + \theta_{a_1} + \theta_B)} + q e^{-i(\theta_1 + \theta_B)} \right) \\ &= |B| e^{i\beta} h_q(\theta_1 + \alpha) \end{aligned}$$

where $\alpha = \frac{2}{q+1}\theta_B + \frac{1}{q+1}\theta_{a_1}$ and $\beta = -\frac{q-1}{q+1}\theta_B + \frac{1}{q+1}\theta_{a_1}$. Hence $g(S(g, \boldsymbol{m}))$ coincide with the image $h_q(S^1)$ which is the $(q+1)$-cuspidal simple closed curve. □

## Acknowledgment

I would like to express my deepest gratitude to seminar members Inaba Kazumasa, Masaharu Ishikawa and Nguyen Tat Thang. Moreover I am supported by JSPS-CAPES under the JAPAN-BRAZIL research cooperative program.

## References

[IIKN] K. Inaba, M. Ishikawa, M. Kawashima, and T. T. Nguyen, *On linear deformations of brieskorn singularities of two variables into generic maps*, Tohoku Mathematical Journal, to appear.

[Ish93] Goo Ishikawa, *Determinacy of the envelope of the osculating hyperplanes to a curve*, Bull. London Math. Soc. **25** (1993), no. 6, 603–610, DOI 10.1112/blms/25.6.603. MR1245089

[Lev65] Harold I. Levine, *Elimination of cusps*, Topology **3** (1965), no. suppl. 2, 263–296. MR0176484

[Oka08] Mutsuo Oka, *Topology of polar weighted homogeneous hypersurfaces*, Kodai Math. J. **31** (2008), no. 2, 163–182, DOI 10.2996/kmj/1214442793. MR2435890

[Oka10] Mutsuo Oka, *Non-degenerate mixed functions*, Kodai Math. J. **33** (2010), no. 1, 1–62, DOI 10.2996/kmj/1270559157. MR2732230

[RSV02] Maria Aparecida Soares Ruas, José Seade, and Alberto Verjovsky, *On real singularities with a Milnor fibration*, Trends in singularities, Trends Math., Birkhäuser, Basel, 2002, pp. 191–213. MR1900787

Department of Mathematics, Tokyo University of Science 1-3 Kagurazaka, Shinjuku, Tokyo 162-8601 Japan

*E-mail address*: `kawashima@ma.kagu.tus.ac.jp`

Contemporary Mathematics
Volume **675**, 2016
http://dx.doi.org/10.1090/conm/675/13590

# Topological classification of simple Morse Bott functions on surfaces

J. Martínez-Alfaro, I. S. Meza-Sarmiento, and R. Oliveira

ABSTRACT. We present a global topological classification of Morse Bott functions on orientable closed surfaces. The invariant is based on the Reeb graph of the function and the topological type of the singular level sets. Connection with other known invariants it is shown. We also prove a realization Theorem of the Morse Bott functions.

## 1. Introduction

The characterization of a set of functions that satisfies a determined property involves two major questions. To get an invariant for particular functions of the set and how is the set of all the invariants admissible for these functions.

With respect to the first question, the classification problem of singular points of smooth map germs is an important problem in Singularity Theory, nevertheless non-local results are not abundant. The classification of functions on surfaces up different types of equivalence can be found, for instance, in [**3**, **24**] considering smooth equivalence and in [**15**, **27**, **28**] considering topological equivalence.

In particular, Arnold [**2**], Kulinich [**15**] and Sharko [**28**] classified Morse functions on surfaces using Reeb graphs with some additional information and Prishyak [**22**] classified smooth functions with isolated critical points on closed surfaces.

In this paper we present a topological classification of Morse Bott functions on orientable closed surfaces. In order to obtain this classification we construct an invariant that is based on the Reeb graph of the function and on the topological type of the singular level sets. This is a kind of additional information that, as far as we know, it was not used until now. In Section 3 of the paper we will show how the topological type of the singular level sets of a function is related to the order of the vertex of the Reeb graph associated to $f$ and induced by the values of $f$.

Given a Morse Bott function $f$ and denoting by $\mathcal{R}^\eta(f)$ the ordered Reeb graph associated to $f$ the first main result of this paper is:

THEOREM 1.1. *Two simple Morse Bott functions* $f$, $g : \Sigma \to \mathbb{R}$ *are conjugate if and only if* $\mathcal{R}^\eta(f)$ *and* $\mathcal{R}^\eta(g)$ *are isomorphic.*

The definition of conjugated functions is given in Section 2.

2010 *Mathematics Subject Classification.* Primary 37J35, 37D15, 34C25.
*Key words and phrases.* Topological invariant, Morse Bott functions, Reeb graph.

With respect to the second problem stated at the beginning of the introduction, the natural approach will be to give necessary and sufficient conditions for a connected oriented and finite graph $G$ to be associated to Morse Bott functions on orientable closed surfaces. Following this ideas the second main result of this paper is a realization theorem for Morse Bott functions.

THEOREM 1.2. *Let $\Sigma$ be an orientable closed surface. A connected directed, finite graph $G$ can be the Reeb graph associated to a Morse Bott function $f : \Sigma \to \mathbb{R}$ on $\Sigma$ if and only if the following conditions are satisfied:*

(a) **local conditions**: *the sink and source vertices of $G$ have degree 1 or 2; the interior vertices have degree 3.*

(b) ***global conditions***:

  (b.1) *the cycle rank of $G$ coincides with the genus of the surface $\Sigma$,*

  (b.2) *the direction of the graph defines a linear ordering of the vertices.*

(c) ***Conditions for orientability***: *$G$ possesses no loops nor oriented cycles*

A final remark is that the classification of functions is related with some topological questions such as homology theory and, at the same time, with problems pertaining to areas outside topology since Morse Bott functions or Morse functions are associated with other concepts such as polynomials, flows and computer graphic recognition. See for instance [**2, 17, 21**].

## 2. Basic concepts

Denote by $\mathbb{M}^n$ a smooth manifold of dimension $n$. Let $f$ be a twice continuously differentiable function from $\mathbb{M}^n$ to $M^m$. A point $p \in \mathbb{M}^n$ is called a *singular point* if $rank(\mathrm{df}(p))$ is not maximum. Otherwise it is a *regular point.* A point $b \in \mathbb{R}$ is called a *singular value* of $f$ if $f^{-1}(b)$ contains a singular point of $f$. The *singular set* of $f$, denoted by $Sing(f)(\mathbb{M}^n)$, is the set of all singular points of $f$ on $\mathbb{M}^n$. In this paper we assume that $\mathbb{M}^m$ is $\mathbb{R}$ unless otherwise indicated.

For each $a \in \mathbb{R}$ consider the level set $I_a(f) = f^{-1}(a)$. $I_a(f)$ is a union of connected components, $i_a^k(f)$, $k = 1, \dots, m(a)$, called *fibers.* A *singular fiber* is a connected component of a level set $I_a(f)$ which contains a singular point of $f$ and it is denoted by $s_a(f)$.

If all nearby fibers around a singular fiber are homeomorphic to it then this fiber is reducible. See [**1**] for details.

From now on, we assume that $f$ is a *simple* or *non resonant* function, which means that there is a unique connected component of singular points in the singular level. It is contained in a singular fiber $s_a(f) \subset I_a(f)$ for each $a \in \mathbb{R}$.

If $f : \mathbb{M}^n \to \mathbb{R}$ has isolated singular values then the *Reeb graph* of $f$, denoted by $\mathcal{R}_f$, is the graph obtained by contracting each fiber $I_a(f)$ to a point, the vertices correspond to the singular fibers $s_a(f)$ of $f$ (see [**23**]). The Reeb graph is also known as the Kronrod-Reeb graph [**28**]. See Figure 1 for an example where $f$ is the height function on the 2-torus.

As the target of $f$ is $\mathbb{R}$, the order in $\mathbb{R}$ defines a unique direction on the Reeb graph. This direction, in turn, induces an order in the set of singular fibers.

Associated with the quotient map $\pi : \mathbb{M}^n \to \mathcal{R}_f$ and the inclusion map $\iota : \mathcal{R}_f \to \mathbb{M}^n$ there are two known functions

(i) Stein factor of $f$ ($\overline{f}$), such that $f = \overline{f} \circ \pi$ (see [**24**] for instance).

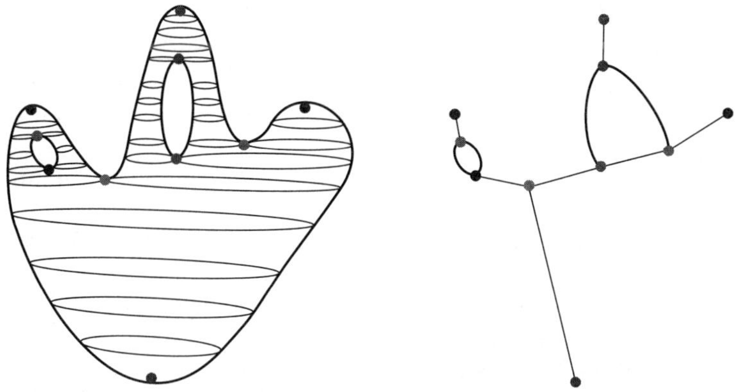

FIGURE 1. Left: the 2-torus; Right: the Reeb graph of the height function.

(ii) The restriction of $f$ to $\mathcal{R}_f$ ($f_{\mathcal{R}}$), given by $f_{\mathcal{R}} = f \circ \iota$.

Both functions are closely related since the Reeb graph can be considered as a sub-complex of the surface, (see [**13**]).

DEFINITION 2.1 ([**7**]). Let $f : \mathbb{M}^n \to \mathbb{R}$ be a smooth function on a $n$-dimensional manifold. A smooth submanifold $S \subset Sing(f)(\mathbb{M}^n)$ is a nondegenerate singular submanifold of $f$ if

(i) $\partial S = \emptyset$
(ii) $S$ is compact and connected
(iii) $\forall s \in S$, we have $T_s S = ker\ (Hess_s f)$.

The function $f$ is a Morse Bott function ($\mathcal{MB}$ function from now on) if the set $Sing(f)(\mathbb{M}^n)$ consists of isolated points and nondegenerate singular submanifolds.

Let $p \in Sing(f)(\mathbb{M}^n)$. By the Morse Bott Lemma([**4**]) there exists a local chart of $\mathbb{M}^n$ around $p$ and a local splitting of the normal bundle of $S$, $N_p(S) = N_p^+(S) \oplus N_p^-(S)$ so that if $p = (s, x, y)$, $s \in S$, $x \in N_p^+(S)$, $y \in N_p^-(S)$ then

$$T_p(\mathbb{M}^n) = T_p(S) \oplus N_p^+(S) \oplus N_p^-(S) \quad \text{and}$$
$$f(p) = f(S) + |x|^2 - |y|^2.$$

The dimension of $N_p^-(S)$ is the index of $S$ and if $p$ is not an isolated singularity of $f$ then $f$ is locally a Morse function on the image of $N_p(S)$ under the exponential map.

It also follows from the Morse Bott Lemma that Morse functions are $\mathcal{MB}$ functions with isolated singular points. Moreover, if $\mathbb{M}^n$ is compact then the function has a finite number of isolated singular points.

Another particular case of $\mathcal{MB}$ functions consist of Round Bott functions ([**14**]) where all singular submanifolds are circles.

Denote by $\Sigma(g, m)$ a connected orientable surface $\Sigma$ of genus $g \geq 0$ with $m$ punctures. Considering the dimension of the singular submanifolds and its index, the singular set of a $\mathcal{MB}$ function function $f$, denoted by $Sing(f)(\Sigma(g,0))$, can be subdivided in three subsets

(i) $Cir(f)(\Sigma(g,0))$: points in singular submanifolds that are homeomorphic to $S^1$. On these circles the function assumes extremal values. We call such singular submanifolds singular circles.
(ii) $Cen(f)(\Sigma(g,0))$: isolated singular points which are extremum points of $f$ called center points.
(iii) $Sad(f)(\Sigma(g,0))$: isolated singularities of index 1 of $f$ called saddle points.

**2.1. Morse Bott foliations and its singularities.** Let $\mathcal{F}(\Sigma, f)$ be the foliation on $\Sigma(g,0)$ defined by the level sets of $f$ and let $Sing(\mathcal{F}(\Sigma, f))$ be the set of singularities of $\mathcal{F}(\Sigma, f)$. We shall omit some arguments in the previous notations if they are irrelevant or can be determined from context. A Morse Bott foliation ($\mathcal{MB}$ foliation from now on) is a foliation defined by the level sets of a $\mathcal{MB}$ function.

As $f$ is simple then $\mathcal{F}(\Sigma, f)$ is also simple.

Two $\mathcal{MB}$ foliations on $\Sigma$ are *topologically equivalent* if there exists a homeomorphism on $\Sigma$ that sends the leaves of one foliation to the leaves of the other.

As $\Sigma$ is an orientable surface, the singular set of a $\mathcal{MB}$ foliation $Sing(\mathcal{F}(\Sigma))$ consist of (see [**25**], [**26**] for more details): $Cen(\mathcal{F}(\Sigma))$ center points and $Sad(\mathcal{F}(\Sigma))$ saddle points. The singular circles of $f$ are not singularities of the foliation.

**2.2. Equivalence of Morse Bott functions.** Let $\Sigma(g,0)$ be a compact connected orientable surface of genus $g$ and $f_1$, $f_2$: $\Sigma(g,0) \to \mathbb{R}$, smooth functions. Functions $f_1$, $f_2$ are *topologically equivalent* if there are homeomorphisms $k : \Sigma(g,0) \to \Sigma(g,0)$, $l : \mathbb{R} \to \mathbb{R}$ such that $f_2 \circ k = l \circ f_1$. The choice of $k$ and $l$ is not unique and $k$ sends level sets of $f_1$ to level sets of $f_2$ since they are related by the following equality $I_{l(a)}(f_2) = k(I_a(f_1))$.

We say that $f_1$ and $f_2$ are topologically conjugated if they are topologically equivalent and $l$ preserves the orientation. See for instance [**22, 27–29**]. In this case if $f_1$ and $f_2$ are topologically conjugated then $f_1$ and $-f_2$ are not necessarily topologically conjugated.

In the case of a $\mathcal{MB}$ function the homeomorphism $k$ could send a singular circle and its neighborhood to a regular cylinder, i.e. it does not preserve the singular circle, so the definition of topologically equivalent must be adapted here for the case of $\mathcal{MB}$ functions.

DEFINITION 2.2. Two $\mathcal{MB}$ functions $f_1$ and $f_2$ from $\Sigma$ to $\mathbb{R}$ are conjugated if there exist homeomorphisms $k : \Sigma(g,0) \to \Sigma(g,0)$, $l : \mathbb{R} \to \mathbb{R}$ such that $l$ preserves orientation, $f_2 = l \circ f_1 \circ k^{-1}$ and $k$ sends singular fibers of $f_1$ to singular fibers of $f_2$.

As $f$ is simple we get the following result.

PROPOSITION 2.3. *If $f_1$ and $f_2$ are simple conjugated $\mathcal{MB}$ functions such that $f_2 = l \circ f_1 \circ k^{-1}$, then the sets $s_a(f_1)$ and $s_{l(a)}(f_2)$ are homeomorphic.*

An *essential singular value* of a $\mathcal{MB}$ function is a value $a \in \mathbb{R}$ such that $I_a(f)$ is topologically distinct from $I_b(f)$ for any $b$ in a neighborhood of $a$. In the case of Morse functions all singular values are essential.

We will use the notion of *slicing of a Morse function* defined in [**20**, **21**] and adapted here for the case of $\mathcal{MB}$ functions. This definition will be useful in this paper. A *slicing* of a $\mathcal{MB}$ function $f$ with $n$ isolated singular fibers is an increasing sequence of real numbers

$$-\infty = a_0 < a_1 < \cdots < a_n < \infty$$

such that for every $i = 1, \ldots, n$ the intervals $(a_{i-1}, a_i)$ contain precisely one singular value of $f$.

The notion of equivalent Morse functions on $\Sigma(g, 0)$ given in [**21**] uses the idea of the slicing as follows. Two Morse functions $f_1$ and $f_2$ on $\Sigma(g, 0)$ are topologically equivalent if there exist a slicing of $f_1$, $a_0 < a_1 < \cdots < a_n$, a slicing of $f_2$, $b_0 < b_1 < \cdots < b_n$ and homeomorphisms $\Phi_i$, $i = 1, .., n$ preserving orientation

$$\Phi_i : M_{a_i}(f_1) \to M_{b_i}(f_2),\ i = 1, \ldots, n,$$

where $M_c(f_j) = \{p \in \Sigma(g) : f_j(p) \leq c\}$, $j = 1, 2$ (see [**19**]). Both notions of equivalence for Morse function (the one introduced in [**21**] and the previous one) are distinct, see for more details Theorems 2.5 and 2.6 in [**6**].

Then, if $f_1$ and $f_2$ are conjugated $\mathcal{MB}$ functions on $\Sigma(g, 0)$, the existence of the homeomorphisms $k$ from $\Sigma(g, 0)$ to $\Sigma(g, 0)$ and $l$ from $\mathbb{R}$ to $\mathbb{R}$ such that $k$ sends singular fibers of $f_1$ to singular fibers of $f_2$ gives us the existence of homeomorphisms $\Phi_i$, $i = 1, .., n$. In fact, each set $M_c(f_j)$ is the union of level sets of $f_1$ and $f_2$ respectively, and $k$ generates a bijection between these sets.

## 3. Construction of the invariant

The main objective of this section is to define a topological invariant to classify Morse Bott functions on compact, connected and orientable surfaces. To do this we will use the classification of the singular level sets of Morse Bott functions that are closed curves and eights. This classification is done in the paper [**17**].

### 3.1. Classification of circles and separatrix eights.

DEFINITION 3.1. An embedded circle on $\Sigma(g, 0)$ will be the image of an embedding $\phi : S^1 \to \Sigma(g, 0)$.

DEFINITION 3.2. A separatrix eight $\mathfrak{B}$, or in short an eight, is the image of an immersion of $S^1$ into $\Sigma$, $\psi : S^1 \to \Sigma(g, 0)$, homeomorphic to two circles glued by a point $p$. A component $s_i$, will be any of the two circles.

Denote by $\mathfrak{N}\mathfrak{B}$ a closed regular neighborhood of $\mathfrak{B} = s_1 \underset{p}{\cup} s_2$ (for details about regular neighborhoods see [**10**, **12**]). Then:

LEMMA 3.3 ([**17**]). *A closed regular neighborhood of* $\mathfrak{B}$ *is homeomorphic to* $\Sigma(0, 3)$ *or* $\Sigma(1, 1)$. *The* $\mathfrak{B}$ *whose regular neighborhood is* $\Sigma(1, 1)$ *is not an admissible singular fiber of a* $\mathcal{MB}$ *function.*

DEFINITION 3.4 ([**17**]). We will say that $\mathfrak{B}$ is a toroidal eight, if $\mathfrak{N}\mathfrak{B}$ is homeomorphic to $\Sigma(1, 1)$ and a planar eight, if $\mathfrak{N}\mathfrak{B}$ is homeomorphic to $\Sigma(0, 3)$.

DEFINITION 3.5. We will say that two circles (or two eights) are topologically equivalent if there is a homeomorphism on $\Sigma(g, 0)$ that sends one of them in the other.

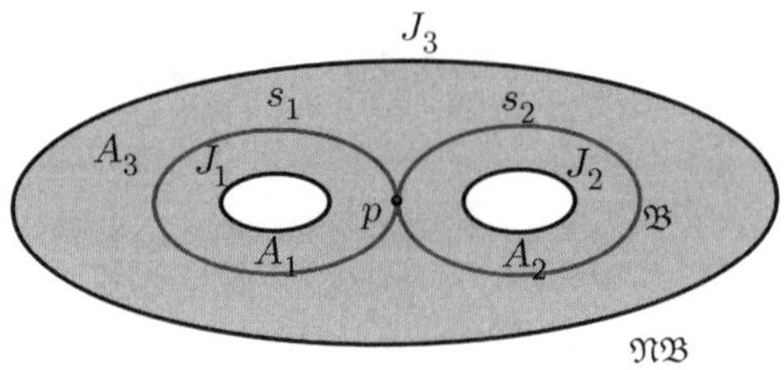


FIGURE 2. The closed regular neighborhood of a planar eight $\mathfrak{B}$.

Let $E(a)$ be the largest integer not greater than $a$ and $C(a)$ is the smallest integer not less than $a$.

THEOREM 3.6 ([**17**]). *Let $\Sigma(g,0)$ be an orientable closed surface. The number of non-equivalent embedding of $S^1$ on $\Sigma$ is*

(i) *1 if $g=0$,*
(ii) *$E\left(\frac{g}{2}\right)+2$ with representant $l_0, l_1, \dots, l_{E(\frac{g}{2})}, l_K$ if $g>0$.*

THEOREM 3.7 ([**17**]). *Let $\Sigma(g,0)$ be an orientable, closed surface with $g \geq 0$. Then, the number of topological types of eights on $\Sigma(g,0)$ is*

(1) *$3g+1$, if $g=0,1$,*
(2) *$E\left(\frac{g}{2}\right)C\left(\frac{g}{2}\right)+E\left(\frac{g}{2}\right)+2g+3$, if $g \geq 2$.*

If $f$ is a $\mathcal{MB}$ function then $\mathfrak{B}$ is a planar eight so $\mathfrak{N}\mathfrak{B}$, the closed regular neighborhood of $\mathfrak{B}$, has three boundary curves (see Figure 2). Two boundary curves $J_1$ and $J_2$ of $\mathfrak{N}\mathfrak{B}$ are contractible to $s_1$ and $s_2$ respectively and we will note these type of curves by $J_s$. The third boundary curve, $J_3$, is contractible to $\mathfrak{B}$ and will be noted by $J_{\mathfrak{B}}$. Considering the Reeb graph of $f$, in a saddle singularity the edge containing $J_{\mathfrak{B}}$ curves bifurcates in two edges of $J_s$ circles.

Foliations defined by Morse functions may differ from foliations defined by a $\mathcal{MB}$ functions as shown in the next result.

PROPOSITION 3.8. *Let $f$ be a Morse function on $\Sigma$ and $\mathcal{F}(\Sigma,f)$ the foliation induced by the level sets of $f$. Then two components of an eight $\mathfrak{B}$ of $\mathcal{F}(\Sigma,f)$ cannot be connected by a family of closed curves. Moreover, two regular cylinders connecting two eights only contain circles of the type $J_s$.*

PROOF. Assume that two components of an eight $\mathfrak{B}$ are connected by a family of closed curves. This family can be parameterized by an open interval $]a,b[$ with $a$ and $b$ assigned to the components of the eight. As $f(a)=f(b)$, $f$ must have a least a singular value on $]a,b[$. The singular level set will be a circle but this is not an admissible singularity in Morse foliations.

Given an eight $\mathfrak{B}$, in $\mathfrak{N}\mathfrak{B}$ the signs of $f(\mathfrak{B})-f(J_s)$ and $f(\mathfrak{B})-f(J_{\mathfrak{B}})$ are always opposite. This sign induces an order in the level sets of $\mathfrak{N}\mathfrak{B}$. Suppose that there are two regular cylinders connecting $\mathfrak{B}_1$ and $\mathfrak{B}_2$ and one of the cylinders are bounded by all $\mathfrak{B}_2$; therefore it contains $J_{\mathfrak{B}}$ circles. Consider a parametrization on the second connecting cylinder as in the last case and $f(\mathfrak{B}_1)=f(a)<f(\mathfrak{B}_2)=f(b)$. Let $c_2$ be a $J_s$ circle near $\mathfrak{B}_2$; assume that it corresponds to the parameter $b+\epsilon$.

Assume that by the ordering in $\mathfrak{N}\mathfrak{B}_2$, $f(c_2)>f(b)$. Then $f$ decreases from $f(b+\epsilon)$ to $f(b)$, but $f$ must also be an increasing function in part of the cylinder since $f(a) < f(b)$. Therefore $f$ must have at least one singularity along this second cylinder. Other assumptions on the ordering on $\mathfrak{NB}_1$ and $\mathfrak{NB}_2$ conclude similarly. □

**3.2. Invariant for topologically equivalent $\mathcal{MB}$ functions.** First of all we recall some basic terminology from topological graph theory (more details in [**11**]). A directed graph $G$ (or *digraph*) consists of a finite nonempty set $V$ of points together with a prescribed collection $E$ of ordered pairs of distinct points. The elements of $V$ are called *vertices* and the elements of $E$ are *directed edges or arcs.* By definition, a directed graph is simple if it has no loops or multiple edges. Denote by $e = (u, v)$ an edge of $G$. Then, the edge goes from $u$ to $v$ and it is *incident* with $u$ and $v$. We also say that $e$ is adjacent to $v$ and $v$ is adjacent from $u$. The *outdegree* of a vertex $v$, denoted by $e^-$, is the number of vertices adjacent from it, and the *indegree*, denoted by $e^+$, is the number of vertices adjacent to it. The sum of the indegree and the outdegree of $v$ is called the *degree* of $v$. A *source vertex* is a vertex with indegree 0, a *sink vertex* has outdegree 0 and an *interior vertex* has nonzero indegree and nonzero outdegree.

Let $\nu > 0$ be the number of critical values of a simple $\mathcal{MB}$ function $f$ and $\mathcal{R}_f$ the Reeb graph of $f$. Let $\eta$ be the function that associates to each vertex a natural number between 1 and $\nu$ following the order induced by the slicing of $f$. From now on, $\mathcal{R}^\eta(f)$ denotes the pair $(\mathcal{R}_f, \eta)$, i.e. the ordered Reeb graph of $f$. A vertex of $\mathcal{R}_f$ is a *saddle vertex* if it is a vertex associated to a saddle point of $f$.

DEFINITION 3.9. Let $\xi$ be the function that associates to each saddle vertex of $\mathcal{R}_f$ the edge that contains the $J_\mathcal{B}$ circle. See Figure 3.

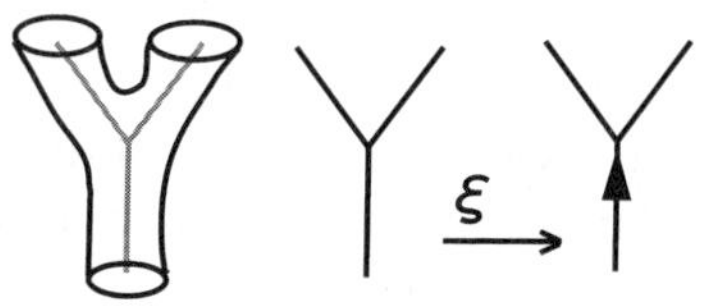


FIGURE 3. The function $\xi$.

In [**17**] the authors proved:

PROPOSITION 3.10. *The Reeb graph of a $\mathcal{MB}$ function and $\xi$ determines the topological type of the eights.*

PROPOSITION 3.11. *Given a $\mathcal{MB}$ function $f$, $\eta$ determines the function $\xi$ on the Reeb graph $\mathcal{R}_f$.*

PROOF. Given a saddle point $s$ there exists an interval $(a_i, a_{i+1})$ in the slicing of $f$ such that $s$ corresponds to a real value in the interval $(a_i, a_{i+1})$ and this real number splits the interval $(a_i, a_{i+1})$ in two components. Each point of an edge bounded by $s$ and near $s$ is associated with a real value in the interval $(a_i, a_{i+1})$. Therefore given $p_1, p_2$ and $p_3$ points on different edges, two of them have associated real numbers in the same component of the interval $(a_i, a_{i+1})$. The other component is associated to the edge bounded by $s$ which contains the curves of type $J_\mathcal{B}$. □

The values of $f$ on the vertices make $\mathcal{R}_f$ a directed graph. Conversely, given a directed graph, we can not compare final vertices as the values of a function are not specified.

PROPOSITION 3.12. *Let $\Gamma$ be a connected subgraph of the Reeb graph associated to a $\mathcal{MB}$ function $f$ having only saddle vertices. The function $\xi$ and the direction of one edge of $\Gamma$ determines the direction in all $\Gamma$. In other words, the function $\xi$ determines a linear ordering of the vertices of $\Gamma$ that coincides with the order defined by $\eta$ or with the inverse order, depending on the choice of the direction of one edge of $\Gamma$.*

PROOF. Given a saddle vertex $s$, the function $\xi$ distinguishes what is the $s$-adjacent edge that contains the curve $J_{\mathcal{B}}$. So the function $\xi$ and the direction of an edge adjacent to a saddle vertex determines the direction on the other two adjacent edges. If $\Gamma$ is a subgraph having only saddle vertices, choosing one of the two possible directions to one edge of $\Gamma$ and fixing the function $\xi$ we obtain a directed subgraph which has the same *direction* (order) of $\eta$ or the inverse order, depending on the chosen direction. This concludes the proof of this proposition. □

Two graphs are *isomorphic* if there exists a one-to-one correspondence between their vertices and edges which preserves adjacency.

The graph of a $\mathcal{MB}$ foliation on orientable closed surface $\Sigma$ is defined in [**17**] as follows.

DEFINITION 3.13. Let $\mathcal{F}$ be a $\mathcal{MB}$ foliation and $f$ such that $\mathcal{F} = \mathcal{F}(f)$. Then the Graph $\Theta(\mathcal{F})$ of the $\mathcal{MB}$ foliation is:

a)A circle, in the case of a regular foliation by circles on the torus.

b)The graph obtained from the Reeb graph of $f$ transforming the union of each vertex $v$ associated to singular circles of $f$ and the two incident edges in a new edge.

$\Theta(\mathcal{F})$ does not depend on the particular function $f$ such that $\mathcal{F} = \mathcal{F}(f)$. This construction is related to the construction described in [**8**], section 1.3 and page 13. This graph $\Theta(\mathcal{F})$ carries the information about the surface $\Sigma$ since the number of independent cycles in $\Theta(\mathcal{F})$ is the genus of $\Sigma$.

PROPOSITION 3.14 ([**17**]). *Let $\xi$ be a function on $\Theta(\mathcal{F})$ that associates to each saddle vertex the edge that contains the $J_{\mathfrak{B}}$ circles. The graph and $\xi$ determine the topological type of the eights.*

By $\Theta_\xi(\mathcal{F})$ we denoted the pair formed by $\Theta(\mathcal{F})$ and the function $\xi$ introduced in Proposition 3.14.

We assume here that $\Theta_\xi(\mathcal{F}_1)$ and $\Theta_\xi(\mathcal{F}_2)$ are isomorphic if there exists an isomorphism from $\Theta(\mathcal{F}_1)$ onto $\Theta(\mathcal{F}_2)$ that preserves the assignments of the functions $\xi$.

THEOREM 3.15 ([**17**]). *$\Theta_\xi(\mathcal{F})$ is a complete topological invariant for $\mathcal{MB}$ foliations on orientable closed surfaces.*

DEFINITION 3.16. We say that $\mathcal{R}^\eta(f)$ and $\mathcal{R}^\eta(g)$ are isomorphic if $\mathcal{R}_f$ and $\mathcal{R}_g$ are isomorphic and the assignments of the function $\eta$ are preserved.

THEOREM 3.17 (Completeness Theorem). *Two simple $\mathcal{MB}$ functions $f : \Sigma \to \mathbb{R}$ and $g : \Sigma \to \mathbb{R}$ are conjugate if and only if $\mathcal{R}^\eta(f)$ and $\mathcal{R}^\eta(g)$ are isomorphic.*

PROOF. (Necessity). Let $f$ and $g$ be simple conjugate $\mathcal{MB}$ functions and $a_0 < a_1 < \cdots < a_n$ a slicing of $f$ and $b_0 < b_1 < \cdots < b_n$ a slicing of $g$. Then there exist a homeomorphism preserving the orientation $\Phi_i : M_{a_i}(f) \to M_{b_i}(g)$, $i = 1, \ldots, n$, and $\mathcal{R}^\eta(f)$ and $\mathcal{R}^\eta(g)$ are isomorphic.

(Sufficiency). The functions $f$ and $g$ induce Morse Bott foliations $\mathcal{F}(f)$ and $\mathcal{F}(g)$ respectively on $\Sigma$. The isomorphism from $\mathcal{R}^\eta(f)$ onto $\mathcal{R}^\eta(g)$ defines a isomorphism between $\Theta_\xi(\mathcal{F}(f))$ and $\Theta_\xi(\mathcal{F}(g))$. By Theorem 3.15, we conclude that there is a homeomorphism $k$ from $\Sigma$ into $\Sigma$ that conjugates $\mathcal{F}(f)$ and $\mathcal{F}(g)$. Since the singular circles are reducible singularities we can choose $k$ in such a way that it sends singular circles of one foliation to singular circles of the other foliation.

The function $f \circ k^{-1}$ is conjugate to $f$ and defines on the surface the same foliation as $g$. Therefore $f \circ k^{-1}$ and $g$ are conjugate and by transitivity, $g$ and $f$ also. □

## 4. Examples

In this section we present several example of Morse Bott functions.

EXAMPLE 4.1. Let $\Sigma = S^2$ and embed $S^2$ in $\mathbb{R}^3$ as the unit sphere. Consider $f : S^2 \to \mathbb{R}$, $f(x, y, z) = z$ and $g : S^2 \to \mathbb{R}$, $g(x, y, z) = -z^2$.

The functions $f$ and $g$ are Morse Bott functions. The nondegenerate critical submanifolds of $f$ are two critical points the north pole and the south pole, and the nondegenerate critical submanifolds of $g$ are the north pole, the south pole and the equator points $z = 0$. As they have non isomorphic Reeb graphs, from Theorem 3.17, $f$ and $g$ are not conjugated.

The Reeb graph is not enough to characterize $\mathcal{MB}$ functions on orientable connected and compact surfaces.

EXAMPLE 4.2. Arnold in [**2**] present two examples of Morse functions with the same ordered Reeb graph which are non conjugated. For both examples the function considered is the height function $f$. In the first example Arnold studied the twin peak mountain Elbrouz and in the second one, the volcanic Vesuvius mountain. See the Figures 4, 5 and with its respective Reeb graphs. The graphs in the previous examples are isomorphic but the functions are not equivalent because the ordering induced by the singular values of the height function $f$ are distinct. For the first graph we have

$$f(D) = 1, f(C) = 2, f(B) = 3, f(A) = 4$$

and for the second one

$$f(D) = 1, f(B) = 2, f(C) = 3, f(A) = 4.$$

See for more details [**2**].

There are other special examples of Morse Bott functions, for instance, Wigner's functions. These functions are quasi-probability distribution functions in phase-space. See [**16, 30**]. According to [**30**] Wigner's functions have been useful in describing transport in quantum optics; nuclear physics; and quantum computing, decoherence and chaos. We present here some particular cases.

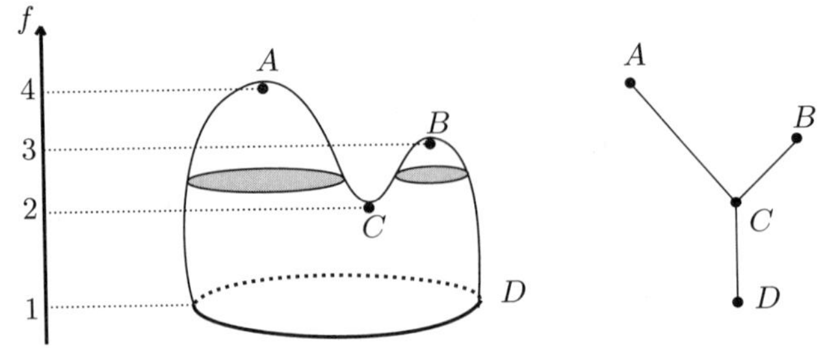


FIGURE 4. The twin peak mountain Elbrouz.

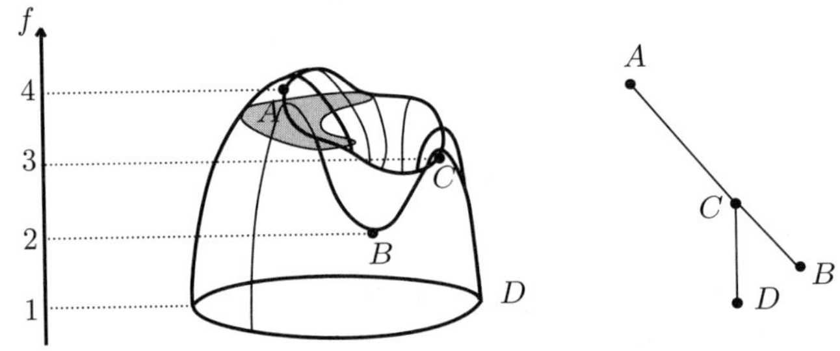


FIGURE 5. The volcanic Vesuvius mountain.

EXAMPLE 4.3. Let us consider

$$f_n(x,p) = \frac{(-1)^n}{\pi \cdot h} e^{-2H/h} \cdot L_n(4H/h),$$

where $L_n(z) = \dfrac{1}{n!} e^z \partial_z^n (e^{-z} z^n)$ are Laguerre polynomials, for $n = 0, 1, 2$, so that for $z = 4H/h$, $L_0 = 1$, $L_1 = 1 - \dfrac{4H}{h}$, $L_2 = \dfrac{8H^2}{h^2} - \dfrac{8H}{h} + 1$. When we suppose $H = \dfrac{p^2 + x^2}{2}$ and $h = 1$ the Wigner's functions are:

(i) $f_0(x,p) = e^{(-p^2-x^2)}(2p^2 + 2x^2)/\pi$;
(ii) $f_1(x,p) = e^{(-p^2-x^2)}(-2p^2 - 2x^2 + 4p^4 + 8x^2p^2 + 4x^4)/\pi$;
(iii) $f_2(x,p) = e^{(-p^2-x^2)}(4p^6 + 12x^2p^4 + 12x^4p^2 - 8p^4 - 16x^2p^2 + 2p^2 + 4x^6 - 8x^4 + 2x^2)/\pi$.

These function are Morse Bott functions. All these functions have the origin as nondegenerate critical point. Also, they have one, two and three nondegenerate critical submanifolds homeomorphic to circles, respectively (see Figures 6, 7 and 8), so it follows from the Theorem 3.17 that these functions are not conjugated.

## 5. Realization theorem

In this section we describe sufficient and necessary conditions for a connected oriented and finite graph $G$ to be associated with a Morse Bott function defined on an orientable closed surface $\Sigma$.

A *walk* in a graph is an alternating sequence of vertices and edges, $v_0$, $e_1$, $v_1$, $e_2$, $\ldots x_n$, $v_n$ in which each edge is either $e_i = (v_{i-1}, v_i)$ or $e_i = (v_i, v_{i-1})$. A walk

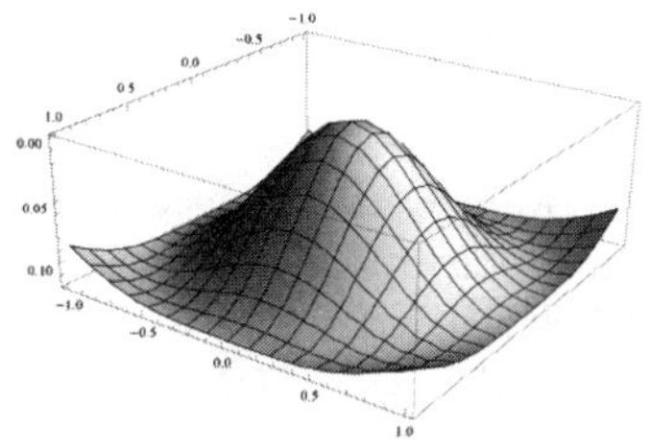

FIGURE 6. The function $f_0(x, p)$.

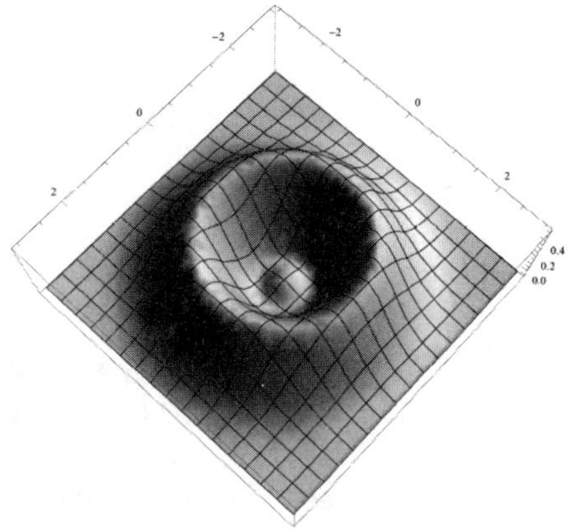


FIGURE 7. The function $f_1(x, p)$.

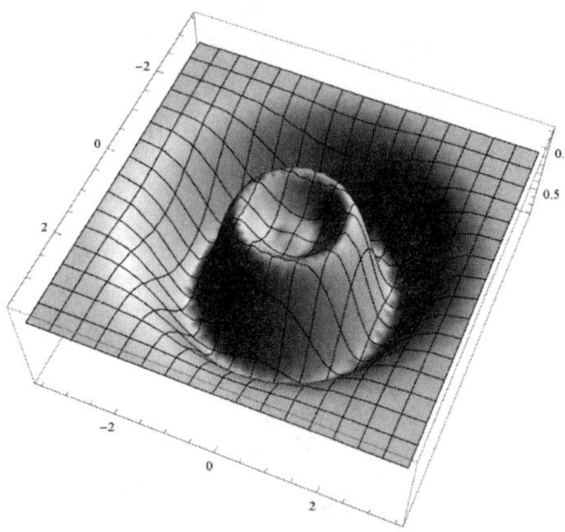


FIGURE 8. The function $f_2(x, p)$.

is a *path* if its vertices (and thus necessarily all the edges) are distinct. A walk is a *cycle* if $v_0 = v_n$ and its edges are distinct. The first vertex of the first edge of a path is the *origin* and the second vertex of the last edge is the *final vertex*. Both the origin and final vertex are called *endpoints* of the path. A graph is *connected* if every pair of vertices are joined by a walk. An oriented cycle is a cycle with all the edges being oriented in the same direction. If $G$ is a connected graph, then the *cycle rank* is $m(G) = \#E - \#V + 1$.

Let be a $f$ Morse function, we will say that $\mathcal{R}_f$ is canonical (see [**9**]) if:

- it has exactly a maximal and a minimal vertex.
- the cycles, if any, have length two.

A *path graph* or *linear graph* is a connected simple graph that contains only vertices of degree 2 and 1.

Let $\Sigma$ be an orientable closed surface and $f$ be a $\mathcal{MB}$ function defined on it. Then, the Reeb graph $\mathcal{R}_f$ associated to $f$ is a finite graph, $G$, with labels at the vertices. Suppose that $G$ has vertices of degree 1, 2 or 3. Each vertex of degree 1 is associated with the basin of a maximum or minimum value of $f$. Each vertex of degree 2 is associated with a neighborhood of a singular circle (a cylinder) and each vertex of degree 3, which is not a endpoint, with a neighborhood of a saddle point. Moreover, we endow each edge of the graph with an $f$-orientation given by the direction in which the function $f$ increases (see [**29**]). Thus, each vertex of a saddle point has one (respectively two) incoming edge and two (respectively one) outgoing edges and a vertex of a singular circle has two incoming edges or two outgoing edges.

**5.1. Proof of Theorem 1.2.** We prove here Theorem 1.2 that gives necessary and sufficient conditions for a connected oriented and finite graph $G$ to be associated to Morse Bott function on an orientable closed surface with values in $\mathbb{R}$. To prove this result, we recall the following theorem.

THEOREM 5.1 ([**18**]). *Let $G$ be a finite graph without loops. Then, there exists a smooth function $f : \Sigma \to \mathbb{R}$ on a closed surface $\Sigma$ with finitely many critical values such that its Reeb graph $\mathcal{R}_f$ is homeomorphic to $G$.*

From Theorem 5.1, any finite graph satisfying the conditions (b) and (c) of Theorem 1.2 can be realized as the Reeb graph of a smooth function with finitely many critical values on a compact surface without boundary.

The proof of this theorem is constructive, Masumoto and Saeki [**18**] showed how to construct a smooth function $f$ by means of conditions (a)-(c) of Theorem 1.2. At first, they construct a continuous function $f_1$ from $G$ to $\mathbb{R}$ that is an embedding on each edge. Such a function can be constructed considering first any injective map $V(G) \to \mathbb{R}$ and then by extending it to the edges so that it is linear on each edge, where $V(G)$ denote the set of vertices of $G$. Then, for each vertice $v \in V(G)$, its neighborhood satisfies one of the conditions in (a) of Theorem 1.2 and for each cases in (a) Masumoto and Saeki construct a smooth function $g_v : N_v \to \mathbb{R}$ on a compact surface with boundary where $N_v$ is a neighborhood of $v$.

Finally, it is performed a gluing operation of the smooth function $g_v$ for each vertice $v$ of $G$ in order to get the smooth function $f$ from $\Sigma$ to $\mathbb{R}$, such that $f|_{N_v} = g_v$ for each vertice $v$ of $G$ and $f_1$ can be identified with the function $\bar{f} : \mathcal{R}_f \to \mathbb{R}$.

PROOF. (*Necessity*) Let $\mathcal{R}_f$ be the Reeb graph associated to a $\mathcal{MB}$ function $f$ on $\Sigma$. Condition (a) of Theorem 1.2 is a consequence of the type of singularities in a $\mathcal{MB}$ function and from the definition of the Reeb graph. Condition (b-1) is necessary since the surface is embedded in $\mathbb{R}^3$ so there is an isotopy defined on $\mathbb{R}^3$ that preserves level sets and contracts the surface to the Reeb graph. Moreover, $f$ determines an order on the vertices of $\mathcal{R}_f$ by the following relation: $v_1 < v_2$ if $f(v_1) < f(v_2)$, so we can provide $\mathcal{R}_f$ with an orientation given by the sense of growth of $f$. With this order the conditions (b-2) and (c) are satisfied.

(*Sufficiency*) We must guarantee the existence of a $\mathcal{MB}$ function $f : \Sigma \to \mathbb{R}$ such that the Reeb graph associated to $f$ is isomorphic to the abstract graph $G$ satisfying the conditions (a)-(c) of Theorem 1.2.

We begin by enumerating the vertices of $G$, $v_1, \ldots, v_n$ in such a way that this order will be consistent with the order defined in $G$. Consider a neighborhood $N_i$ for the vertex $v_i$ in $G$ that contains $v_i$ but not any other vertex of $G$. According to the degree of $v_i$ we associate a $\mathcal{MB}$ function to $N_i$ for $i = 1, \ldots, n$. These functions are constructed using the same arguments and ideas by Masumoto and Saeki (see more details in the proof of the Theorem 5.1 in [**18**]). $\square$

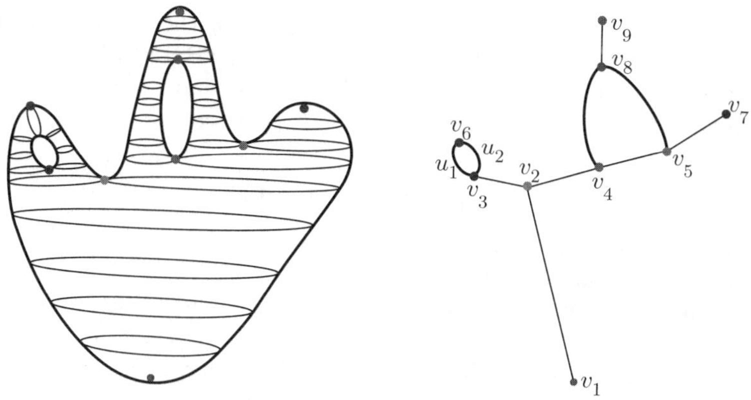


FIGURE 9. On the left a Morse Bott foliation on the 2-torus; on the right the associated Reeb graph.

Given an ordered Reeb graph associated to a Morse function $f$ on a fixed surface $\Sigma$ it is possible to transform the graph into another one corresponding to another Morse function $g$ on $\Sigma$ by means of a set of elementary deformations. These deformations are listed in [**9**] and we call them *Fabio-Landi transformations.* To obtain a similar result in the case of Morse-Bott functions a new elementary deformation must be added:

DEFINITION 5.2. Consider a Reeb graph, $\mathcal{R}_1^\eta(f)$, a saddle vertex $v$ with three adjacent edges $u_1, u_2, u_3$ where $u_3$ connects $v$ with a final center vertex $c$. A direct MB elementary deformation applied to $\mathcal{R}_1^\eta(f)$ yields another Reeb graph without the edge $u_3$ and the center $c$. The vertex $v$ is now associated with a singular circle.

DEFINITION 5.3. Given a Reeb graph $\mathcal{R}_1^\eta(f)$, $f$ a Morse Bott function, $v$ a vertex associated with a singular circle, with $u_1, u_2$ adjacent edges, an inverse MB elementary deformation on a $\mathcal{R}_1^\eta(f)$ consist of adding a new adjacent edge to $v$ connecting $v$ with a new vertex $c$ associated to a center singular point. Moreover, $min(f(x), x \in u_1 \cup u_2) < f(c) < max(f(x), x \in u_1 \cup u_2)$.

In the Figure 9 we consider a Morse Bott foliation on the 2-torus associated to a particular $\mathcal{MB}$ function $g$. In the Reeb graph $\mathcal{R}_1^\eta(g)$ of $g$, the vertex $v_6$ is associated with a singular circle of $g$ with $u_1, u_2$ adjacent edges. Applying an inverse MB elementary deformation on $\mathcal{R}_1^\eta(g)$, we add a new adjacent edge to $v_6$ connecting $v$ with a new vertex $c$ associated to a center singular point. Then the Reeb graph obtained is associated to the foliation of the height function in the Figure 1.

PROPOSITION 5.4. *Every $\mathcal{R}^\eta(f)$ can be transformed into a canonical one through a finite sequence of elementary deformations.*

PROOF. An inverse MB elementary deformation does not change the graph genus or cycles. A sequence of inverse MB elementary deformations transforms a Reeb graph associated with a $\mathcal{MB}$ function in a graph associated with a Morse function in the same surface. By applying Fabio-Landi transformations it is possible to transform the graph into a canonical graph. □

A realization theorem based in transformations of the graph is important to define a topology in the set of Reeb graphs of Morse Bott functions (see [**5**, **9**]). When Reeb graphs are used to recognize images, to have stability of the graph against small perturbations is a very convenient property.

## Acknowledgements

The authors would like thank the referee for his or her comments and substantial suggestions that made our article improved.

The first author is supported by Grant MTM2012-33073 MEC and Fonds Feder. The second author is partially supported by PEC-PG CAPES grant number 470611-0 and by the projet "Singularidades, Geometría Genérica y Morfología Matemática" MTM2012-33073. The third authors is partially supported by the CAPES grant number 88881.030454/2013-01 from the program CSF-PVE, CNPq grant "Projeto Universal 472796/2013-5" and FAPESP grant "Projeto Temático" 2014/00304-2. The second and third authors want to acknowledge the support of the joint projects FP7-PEOPLE-2012-IRSES numbers 316338.

## References

[1] V. I. Arnol′d, *Singularities of smooth mappings* (Russian), Uspehi Mat. Nauk **23** (1968), no. 1, 3–44. MR0226655

[2] Vladimir I. Arnold, *Topological classification of Morse functions and generalisations of Hilbert's 16-th problem*, Math. Phys. Anal. Geom. **10** (2007), no. 3, 227–236, DOI 10.1007/s11040-007-9029-0. MR2368960

[3] V. I. Arnold, V. V. Goryunov, O. V. Lyashko, and V. A. Vasil′ev, *Singularity theory. I*, Springer-Verlag, Berlin, 1998. Translated from the 1988 Russian original by A. Iacob; Reprint of the original English edition from the series Encyclopaedia of Mathematical Sciences [*Dynamical systems. VI*, Encyclopaedia Math. Sci., 6, Springer, Berlin, 1993; MR1230637 (94b:58018)]. MR1660090

[4] Augustin Banyaga and David E. Hurtubise, *A proof of the Morse-Bott lemma*, Expo. Math. **22** (2004), no. 4, 365–373, DOI 10.1016/S0723-0869(04)80014-8. MR2075744

[5] U. Bauer, X. Ge and Y. Wang, *Measuring distance between Reeb graphs*, arXiv:1307.2839 (2013), 1–21.

[6] A. V. Bolsinov and A. T. Fomenko, *Integrable Hamiltonian systems*, Chapman & Hall/CRC, Boca Raton, FL, 2004. Geometry, topology, classification; Translated from the 1999 Russian original. MR2036760

[7] Raoul Bott, *Nondegenerate critical manifolds*, Ann. of Math. (2) **60** (1954), 248–261. MR0064399

[8] Alberto Candel and Lawrence Conlon, *Foliations. II*, Graduate Studies in Mathematics, vol. 60, American Mathematical Society, Providence, RI, 2003. MR1994394

[9] Barbara Di Fabio and Claudia Landi, *Stable shape comparison of surfaces via Reeb graphs*, Discrete geometry for computer imagery, Lecture Notes in Comput. Sci., vol. 8668, Springer, Cham, 2014, pp. 202–213, DOI 10.1007/978-3-319-09955-2_17. MR3279676

[10] Peter Giblin, *Graphs, surfaces and homology*, 3rd ed., Cambridge University Press, Cambridge, 2010. MR2722281

[11] F. Harary, *Graph theory*, Perseus Books, 1994.

[12] Morris W. Hirsch, *Differential topology*, Graduate Texts in Mathematics, vol. 33, Springer-Verlag, New York, 1994. Corrected reprint of the 1976 original. MR1336822

[13] Marek Kaluba, Wacław Marzantowicz, and Nelson Silva, *On representation of the Reeb graph as a sub-complex of manifold*, Topol. Methods Nonlinear Anal. **45** (2015), no. 1, 287–307, DOI 10.12775/TMNA.2015.015. MR3365016

[14] Georgi Khimshiashvili and Dirk Siersma, *Remarks on minimal round functions*, Geometry and topology of caustics—CAUSTICS '02, Banach Center Publ., vol. 62, Polish Acad. Sci. Inst. Math., Warsaw, 2004, pp. 159–172. MR2056438

[15] E. V. Kulinich, *On topologically equivalent Morse functions on surfaces*, Methods Funct. Anal. Topology **4** (1998), no. 1, 59–64. MR1770821

[16] M. A. Marchiolli, *Mecânica quântica no espaço de fase: I. Formulação de Weyl-Wigner*, Revista Brasileira de Ensino de Física **24** (2002), 421–436.

[17] J. Martinez-Alfaro, I. S. Meza-Sarmiento and R. D. S. Oliveira, *Singular levels and topological invariants of Morse Bott systems on surfaces*, Notas da Série Matemática do ICMC-USP, **394** (2014), 1–23, Available in http://www.icmc.usp.br/CMS/Arquivos/arquivos_enviados/ESTAGIO-BIBLIO_171_Serie Mat 394.pdf.

[18] Yasutaka Masumoto and Osamu Saeki, *A smooth function on a manifold with given Reeb graph*, Kyushu J. Math. **65** (2011), no. 1, 75–84, DOI 10.2206/kyushujm.65.75. MR2828379

[19] Yukio Matsumoto, *An introduction to Morse theory*, Translations of Mathematical Monographs, vol. 208, American Mathematical Society, Providence, RI, 2002. Translated from the 1997 Japanese original by Kiki Hudson and Masahico Saito; Iwanami Series in Modern Mathematics. MR1873233

[20] Liviu I. Nicolaescu, *An invitation to Morse theory*, Universitext, Springer, New York, 2007. MR2298610

[21] Liviu I. Nicolaescu, *Counting Morse functions on the 2-sphere*, Compos. Math. **144** (2008), no. 5, 1081–1106, DOI 10.1112/S0010437X08003680. MR2457520

[22] A. O. Prishlyak, *Topological equivalence of smooth functions with isolated critical points on a closed surface*, Topology Appl. **119** (2002), no. 3, 257–267, DOI 10.1016/S0166-8641(01)00077-3. MR1888671

[23] Georges Reeb, *Sur les points singuliers d'une forme de Pfaff complètement intégrable ou d'une fonction numérique* (French), C. R. Acad. Sci. Paris **222** (1946), 847–849. MR0015613

[24] Osamu Saeki, *Topology of singular fibers of differentiable maps*, Lecture Notes in Mathematics, vol. 1854, Springer-Verlag, Berlin, 2004. MR2106689

[25] Bruno Scárdua and José Seade, *Codimension one foliations with Bott-Morse singularities. I*, J. Differential Geom. **83** (2009), no. 1, 189–212. MR2545034

[26] Bruno Scárdua and José Seade, *Codimension 1 foliations with Bott-Morse singularities II*, J. Topol. **4** (2011), no. 2, 343–382, DOI 10.1112/jtopol/jtr004. MR2805995

[27] V. V. Sharko, *On topological equivalence Morse functions on surfaces*, Internat. Conference at Chelyabinsk State Univ., Low-Dimensional Topology and Combinatorial Group Theory (1996), 19–23.

[28] V. V. Sharko, *Smooth and topological equivalence of functions on surfaces* (Russian, with English and Ukrainian summaries), Ukraïn. Mat. Zh. **55** (2003), no. 5, 687–700, DOI 10.1023/B:UKMA.0000010259.21815.d7; English transl., Ukrainian Math. J. **55** (2003), no. 5, 832–846. MR2071708

[29] V. V. Sharko, *About Kronrod-Reeb graph of a function on a manifold*, Methods Funct. Anal. Topology **12** (2006), no. 4, 389–396. MR2279875

[30] *Quantum mechanics in phase space*, World Scientific Series in 20th Century Physics, vol. 34, World Scientific Publishing Co. Pte. Ltd., Hackensack, NJ, 2005. An overview with selected papers; Edited by Cosmas K. Zachos, David B. Fairlie and Thomas L. Curtright. MR2229229

Departamento de Matemática Aplicada, Universitat de València, Burjassot, Spain
*E-mail address*: martinja@uv.es

Departamento de Matemática, ICMC - Universidade de São Paulo, Caixa Postal 668, 13566-590, Sao Carlos, SP, Brazil
*E-mail address*: isofia@icmc.usp.br

Departamento de Matemática, ICMC - Universidade de São Paulo, Caixa Postal 668, 13566-590, Sao Carlos, SP, Brazil
*E-mail address*: regilene@icmc.usp.br

Contemporary Mathematics
Volume **675**, 2016
http://dx.doi.org/10.1090/conm/675/13591

# The link of a frontal surface singularity

R. Martins and J. J. Nuño-Ballesteros

Abstract. We prove that a frontal surface singularity with isolated instability has a cone structure over its link, which is a closed stable frontal curve in the sphere. Then, we define a generalized version of Gauss word of the link, which gives a complete invariant of the singularity. Finally, we also give a complete classification of singularities in the case that the frontal surface is ruled.

## 1. Introduction

The concept of frontal surface was introduced by Kurbatskiĭ and Zakalyukin in [**11**] as a generalization of a front. A front is a singular hypersurface in $\mathbb{R}^{n+1}$ which is obtained as the projection of a Legendrian immersion in the projectivized cotangent bundle $PT^*\mathbb{R}^{n+1}$ with the canonical contact structure. If we allow the Legendrian map to have singularities itself, then the projection is said to be a frontal. Frontal type singularities have nice geometric properties, since they have a well defined limiting tangent hyperplane at the singular points.

The stable frontal singularities are well known in low dimensions. For a frontal curve, the only stable singularities are cusps and nodes. For a frontal surface, the stable singularities are cuspidal edges, swallowtails, cuspidal cross-caps or their transverse self-intersections. Here, we are interested in frontal surfaces with more degenerate singularities, but which are generic in such a way that they have isolated instability. This class contains all the frontal singularities of finite codimension in the sense of [**9, 10**].

Given a frontal surface singularity with isolated instability, we want to describe the local structure of the singularity by means of its link. The link is obtained by taking the intersection of the surface with a small enough sphere centered at the singular point. We show that the link is a stable frontal curve in the sphere and that the surface is locally topologically equivalent to the cone of its link. In fact, the topological equivalence is realized by homeomorphisms in the source and the target which are diffeomorphisms outside the singular point. This point is important, since the cusps are not preserved by homeomorphisms. Our result is similar to the Fukuda's cone theorem for generic non-frontal singularities [**2**].

---

2010 *Mathematics Subject Classification.* Primary 58K15; Secondary 58K40, 58K65.
*Key words and phrases.* Cusps and nodes, frontals, link.
The first author has been supported by CAPES.
The second author has been partially supported by DGICYT Grant MTM2012–33073.

A consequence of the above cone theorem is that the topological classification of generic frontal surface singularities is a discrete problem, i.e., there are no moduli. Thus, it is interesting to have a combinatorial model which codifies the local structure of the singularity. Following the ideas of [**3**], we use a generalized version of the Gauss word, which turns out to be a complete topological invariant. The only difference with the classical Gauss word is that we need to introduce also the information relative to the cusps. We show that two frontal surface singularities with isolated instability are topologically equivalent (by homeomorphisms which are diffeomorphisms outside the singular point) if and only if they have equivalent Gauss words. The use of Gauss words has proven to be also very useful in the topological classification of finitely determined singularities of maps from $\mathbb{R}^n$ to $\mathbb{R}^n$, with $n = 2, 3$ (see [**6, 8**]).

The final section is dedicated to the classification of singularities of frontal ruled surfaces. The classification of all finitely determined singularities of non-frontal ruled surfaces was done by the authors in a previous paper [**4**], so it is natural to ask the same question in the case of frontal surfaces. We also give a complete classification and show that there are only six singularity types which can appear if the frontal surface is ruled. The results of this section are also related to the topological classification of maps from $\mathbb{R}^3$ to $\mathbb{R}^3$ whose discriminant is a ruled surface [**7**].

## 2. The link of a frontal surface

Consider the projectivized cotangent bundle $PT^*\mathbb{R}^{n+1}$ with the canonical contact structure $\alpha$ and denote the projection by $\pi : PT^*\mathbb{R}^{n+1} \to \mathbb{R}^{n+1}$. Given an open subset $U \subset \mathbb{R}^n$, a $C^\infty$ map $\mathcal{L} : U \to PT^*\mathbb{R}^{n+1}$ is called integral if $\mathcal{L}^*\alpha = 0$.

DEFINITION 2.1. A $C^\infty$ map $f : U \to \mathbb{R}^{n+1}$ is called a *frontal* if there exists an integral map $\mathcal{L} : U \to PT^*\mathbb{R}^{n+1}$ such that $\pi \circ \mathcal{L} = f$. If in addition $\mathcal{L}$ is an immersion, then $f$ is called a *front*.

It follows from the definition that $f$ is a frontal if and only if there exists $\nu : U \to T^*\mathbb{R}^{n+1}$ a non-zero $C^\infty$ 1-form along $f$ such that $\nu(df \circ \xi) = 0$ for any tangent vector field $\xi$ in $U$. In local coordinates, if $\nu = \sum_{j=1}^{n+1} \nu_j dx_j$, this means that

$$\sum_{j=1}^{n+1} \nu_j \frac{\partial f_j}{\partial u_i} = 0, \quad \forall i = 1, \dots, n.$$

If $f : U \to \mathbb{R}^{n+1}$ is a frontal, then the multigerm $f : (\mathbb{R}^n, S) \to (\mathbb{R}^{n+1}, y_0)$ at any finite subset $S \subset U$ is also called a frontal multigerm. Since the definition of frontal is local, this is equivalent to the fact that each branch of $f$ is itself a frontal. The next result give us an easy algebraic characterization of frontals in the corank 1 case (the corank is the dimension of the kernel of the differential of $f$).

PROPOSITION 2.2. [**9**] *Let $f : (\mathbb{R}^n, 0) \to (\mathbb{R}^{n+1}, 0)$ be a $C^\infty$ map germ of corank 1. Choose coordinates in such a way that $f(u, v) = (u, p(u, v), q(u, v))$, for some functions $p, q$ with $u \in \mathbb{R}^{n-1}$ and $v \in \mathbb{R}$. Then $f$ is a frontal if and only if either $\partial p/\partial v$ divides $\partial q/\partial v$ or $\partial q/\partial v$ divides $\partial p/\partial v$.*

The stable frontal singularities are well known in low dimensions (see, for instance [**9–11**]). We say that a frontal $f : (\mathbb{R}^n, S) \to (\mathbb{R}^{n+1}, y_0)$ is stable if any unfolding of $f$, which is also frontal, is trivial under Mather's $\mathscr{A}$-equivalence (coordinate changes in the source and the target). For frontal curves and surfaces we have the following classification:

(1) The stable frontal singularities of a frontal curve are cusps (simple cusps) and nodes (transverse double points).
(2) The stable frontal singularities of a frontal surface are either: (semicubic) cuspidal edges, swallowtails, folded Whitney umbrellas or their transverse self-intersections.

DEFINITION 2.3. We say that a frontal $f : (\mathbb{R}^2, 0) \to (\mathbb{R}^3, 0)$ has an *isolated instability* if there is a representative $f : U \to V$, where $U \subset \mathbb{R}^2$ and $V \subset \mathbb{R}^3$ are open neighborhoods of the origin, such that

(1) $f^{-1}(0) = \{0\}$,
(2) $f$ is a finite map (i.e., it is closed and finite-to-one),
(3) $f$ has only semicubic cuspidal edges and transverse double points in $U \setminus \{0\}$.

By using an appropriate version of the Mather-Gaffney finite determinacy criterion, it can be shown that if $f : (\mathbb{R}^2, 0) \to (\mathbb{R}^3, 0)$ is a frontal of finite codimension in the sense of [**9, 10**], then it has an isolated instability. In fact, in the complex analytic case, both concepts are equivalent (see [**9**]).

REMARK 2.4. It follows from Definition 2.3 that the fact that a frontal has isolated instability is preserved under $\mathscr{A}$-equivalence.

From now on, we assume for simplicity that all map germs are real analytic unless otherwise specified. Given $f : (\mathbb{R}^n, 0) \to (\mathbb{R}^p, 0)$ and a representative $f : U \to V$ we denote:

$$D^p_\epsilon = \{y \in \mathbb{R}^p : \|y\|^2 \leq \epsilon\}, \quad S^{p-1}_\epsilon = \{y \in \mathbb{R}^p : \|y\|^2 = \epsilon\}.$$

$$\tilde{D}^n_\epsilon = f^{-1}(D^p_\epsilon), \quad \tilde{S}^{n-1}_\epsilon = f^{-1}(S^{p-1}_\epsilon).$$

The main result is the following cone structure theorem for frontal surfaces, which is an adapted version for frontal surfaces of the Fukuda cone theorem for generic map germs (see [**2**]).

THEOREM 2.5. *Let $f : (\mathbb{R}^2, 0) \to (\mathbb{R}^3, 0)$ be a frontal with an isolated instability. Then there is $\epsilon_0 > 0$ such that for any $\epsilon$, with $0 < \epsilon \leq \epsilon_0$ we have:*

(1) *$\tilde{S}^1_\epsilon$ is diffeomorphic to $S^1$.*
(2) *$f|_{\tilde{S}^1_\epsilon} : \tilde{S}^1_\epsilon \to S^2_\epsilon$ is a stable frontal curve, whose $\mathscr{A}$-class is independent of $\epsilon$.*
(3) *$f|_{\tilde{D}^2_\epsilon \setminus \{0\}} : \tilde{D}^2_\epsilon \setminus \{0\} \to D^3_\epsilon \setminus \{0\}$ is $\mathscr{A}$-equivalent to the product map $\mathrm{id} \times f|_{\tilde{S}^1_\epsilon} : (0, \epsilon] \times \tilde{S}^1_\epsilon \to (0, \epsilon] \times S^2_\epsilon$.*
(4) *By adding the origin, the above $\mathscr{A}$-equivalence can be extended to a $C^0$-$\mathscr{A}$-equivalence between $f|_{\tilde{D}^2_\epsilon} : \tilde{D}^2_\epsilon \to D^3_\epsilon$ and the cone of $f|_{\tilde{S}^1_\epsilon} : \tilde{S}^1_\epsilon \to S^2_\epsilon$.*

PROOF. We take a representative $f : U \to V$, where $U \subset \mathbb{R}^2$ and $V \subset \mathbb{R}^3$ are open neighborhoods of the origin, such that

(1) $f^{-1}(0) = \{0\}$,
(2) $f$ is a finite map (i.e., it is closed and finite-to-one),

(3) $f$ has only semicubic cuspidal edges and transverse double points in $U \setminus \{0\}$.

Let $(\mathcal{A}, \mathcal{B})$ be the stratification by stable types of $f : U \to V$, which is made by a finite number of semianalytic strata. We consider $g : U \to \mathbb{R}$ given by $g = \|f\|^2$ and the restriction $g|_{A_i} : A_i \to \mathbb{R}$, to each stratum $A_i \in \mathcal{A}$. By the Curve Selection Lemma (see [**5**]), $g|_{A_i}$ has a finite number of critical values. There exists $\epsilon_0 > 0$ such that $\forall \epsilon$, $0 < \epsilon \le \epsilon_0$, $\epsilon$ is a regular value of $g|_{A_i}$, $\forall A_i \in \mathcal{A}$.

This implies that $S^2_\epsilon$ is transverse to the stratification $\mathcal{B}$. Then, $\tilde{S}^1_\epsilon = f^{-1}(S^2_\epsilon)$ is a compact 1-dimensional submanifold of $U$ and $f|_{\tilde{S}^1_\epsilon} : \tilde{S}^1_\epsilon \to S^2_\epsilon$ has only simple cusps and transverse double points. We have the first part of (2).

To prove (1), we use Reeb Theorem: since $f^{-1}(0) = \{0\}$, 0 is an isolated minimum of $g$. Then, $\tilde{D}^2_\epsilon$, which is equal to $g^{-1}([0, \epsilon])$, is homeomorphic to the closed disk $D^2$. Then, $\tilde{S}^1_\epsilon = \partial \tilde{D}^2_\epsilon$ is homeomorphic (and thus diffeomorphic) to $S^1$.

Let us see (3). Denote $I = (0, \epsilon]$ and consider the following diffeomorphisms:

$$\Phi : \tilde{D}^2_\epsilon \setminus \{0\} \longrightarrow I \times \tilde{S}^1_\epsilon, \qquad x \longmapsto (g(x), \phi(x)),$$

$$\Psi : D^3_\epsilon \setminus \{0\} \longrightarrow I \times S^2_\epsilon, \qquad y \longmapsto (\|y\|^2, \sqrt{\epsilon}\frac{y}{\|y\|}),$$

where $\phi(x)$ is the point of $\tilde{S}^1_\epsilon$ where the integral curve of the gradient of $g$ passing through $x$ meets $\tilde{S}^1_\epsilon$.

Define $F : I \times \tilde{S}^1_\epsilon \to I \times S^2_\epsilon$ by $F = \Psi \circ f \circ \Phi^{-1}$. By construction, $F(\{t\} \times \tilde{S}^1_\epsilon) \subset \{t\} \times S^2_\epsilon$, for all $t \in I$. Hence, we can write $F$ in the form $F(t, x) = (t, f_t(x))$, with $f_t : \tilde{S}^1_\epsilon \to S^2_\epsilon$ and $t \in I$.

Since $f_t$ is $\mathscr{A}$-equivalent to $f|_{\tilde{S}^1_t}$, $f_t$ is stable. Therefore, $F$ is $\mathscr{A}$-trivial, that is, there exist diffeomorphisms $H$ and $K$ of the form:

$$H : I \times \tilde{S}^1_\epsilon \longrightarrow I \times \tilde{S}^1_\epsilon, \qquad (t, x) \longmapsto (t, h_t(x)),$$

$$K : I \times S^2_\epsilon \longrightarrow I \times S^2_\epsilon, \qquad (t, y) \longmapsto (t, k_t(y)),$$

and such that $K \circ F \circ H^{-1} = \mathrm{id} \times f_\epsilon$. Hence, $f|_{\tilde{D}^2_\epsilon \setminus \{0\}}$ is $\mathscr{A}$-equivalent to $\mathrm{id} \times f_\epsilon$. Finally, (4) and the second part of (2) are immediate consequences of (3). □

DEFINITION 2.6. Let $f : (\mathbb{R}^2, 0) \to (\mathbb{R}^3, 0)$ be a frontal with an isolated instability. We say that $\epsilon_0 > 0$ is a Milnor-Fukuda radius for $f$ if for any $\epsilon$, with $0 < \epsilon \le \epsilon_0$, the above conditions (1), (2), (3) and (4) are satisfied. We call the map $f|_{\tilde{S}^1_\epsilon} : \tilde{S}^1_\epsilon \to S^2_\epsilon$ the *link* of $f$.

We deduce some immediate consequences of the cone structure theorem:

(1) the link is a stable frontal curve $\gamma : S^1 \to S^2$, that is, it has only cusps and nodes,
(2) the link is well defined up to $\mathscr{A}$-equivalence,
(3) $f$ is topologically equivalent to the cone of its link, by homeomorphisms which preserve the origin and which are diffeomorphisms outside the origin.

## 3. Gauss words

In this section we present an extended notion of Gauss word. Usually this concept is defined only for closed curves with nodes, but here we need to extend it to closed curves with cusps (see [**1**]).

Assume that $\gamma : S^1 \to S^2$ is a stable frontal curve which has $r$ nodes and $s$ cusps, which are labeled by $r+s$ letters $\{a_1, a_2, \ldots, a_{r+s}\}$. We choose a base point in $S^1$ and orientations in $S^1, S^2$.

DEFINITION 3.1. The *Gauss word* of $\gamma$ is denoted by $W$ and it is the sequence of cusps and nodes that appear when traveling around $\gamma$ starting from the base point and following the orientation. If we arrive to a point $a_i$, then, we put

- $a_i$, if it corresponds to the positive branch of a node;
- $a_i^{-1}$, if it corresponds to the negative branch of a node;
- $a_i^2$, if it is a cusp and corresponds to the positive branch of a generic deformation of the cusp into a node;
- $a_i^{-2}$, if it is a cusp and corresponds to the negative branch of a generic deformation of the cusp into a node;

(see fig. 1).

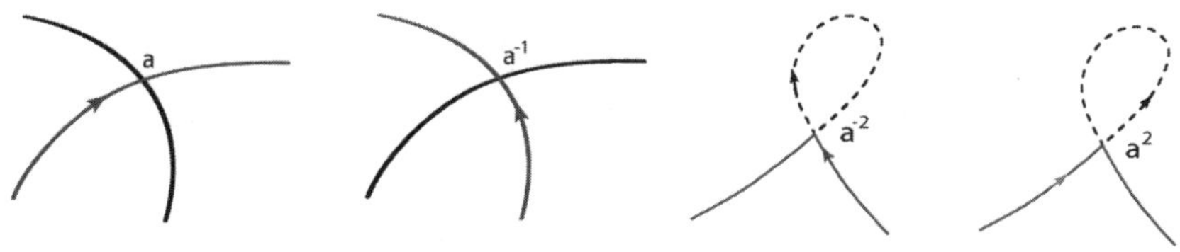


FIGURE 1

It follows that the Gauss word of $\gamma$ is not uniquely defined, since it depends on the choice of the labels of the cusps and nodes, the base point and the orientations in $S^1, S^2$. If we change the labels, the base point or the orientations we get the following changes in the Gauss word:

(1) a permutation in the alphabet $a_1, \ldots, a_{r+s}$,
(2) a cyclic permutation in the sequence,
(3) a reversion of the sequence,
(4) a change of all the exponents from $+1, +2$ to $-1, -2$ respectively and vice-versa.

We say that two Gauss words are equivalent if they are related by means of these four operations. Up to this equivalence, the Gauss word is now well defined.

For curves $\gamma : S^1 \to S^2$ with only node singularities it is well known that the classical Gauss word is a complete topological invariant (see [**1**]). That is, two curves $\gamma, \delta$ are $C^0$-$\mathscr{A}$-equivalent if and only if they have equivalent Gauss words. For stable frontal curves this is not true, since the cusps are not preserved by topological equivalence. However, we will see that the Gauss word is in fact a complete invariant for $\mathscr{A}$-equivalence.

LEMMA 3.2. *Assume that $\gamma, \delta : S^1 \to S^2$ have only node singularities. If $\gamma, \delta$ have equivalent Gauss words then they are $\mathscr{A}$-equivalent.*

PROOF. This is just an adaptation of the classical proof but using diffeomorphisms instead of homeomorphisms. If $\gamma, \delta$ have equivalent Gauss words, then we can always take diffeomorphisms $\alpha, \beta$ such that $\delta$ and $\beta \circ \gamma \circ \alpha^{-1}$ have the same Gauss word. So, it is enough to show the lemma in the particular case that $\gamma, \delta$ have the same Gauss word.

Let us denote $K_1 = \gamma(S^1)$ and $K_2 = \delta(S^1)$. We first construct a diffeomorphism $\beta : S^2 \to S^2$ such that $\beta(K_1) = K_2$. Consider in $S^2$ the stratifications $\mathscr{S}_1$ and $\mathscr{S}_2$ induced by $K_1$ and $K_2$ respectively. The 0-strata of $\mathscr{S}_i$ are labelled by the letters $a_1, \dots, a_r$ of the Gauss word. The 1-strata of $\mathscr{S}_i$ are the oriented edges labelled by the pairs $a_k^{\epsilon_k} a_\ell^{\epsilon_\ell}$ of adjacent letters in the Gauss word (including the oriented edge joining the last with the first letters). The 2-strata of $\mathscr{S}_i$ are the connected components of $S^2 \setminus K_i$ which are labelled by all the closed sequences of oriented edges or their inverses. Thus, the Gauss word induces a bijection $b : \mathscr{S}_1 \to \mathscr{S}_2$ which preserves the types and the adjacencies between the strata. We construct the desired diffeomorphism $\beta : S^2 \to S^2$ just by means of a system of adapted tubular neighbourhoods between corresponding strata as shown in fig. 2. Finally, since $\gamma, \delta$ are local diffeomorphisms onto $K_1, K_2$ respectively, there exists a unique diffeomorphism $\alpha : S^1 \to S^1$ such that $\beta \circ \gamma \circ \alpha^{-1} = \delta$. $\square$

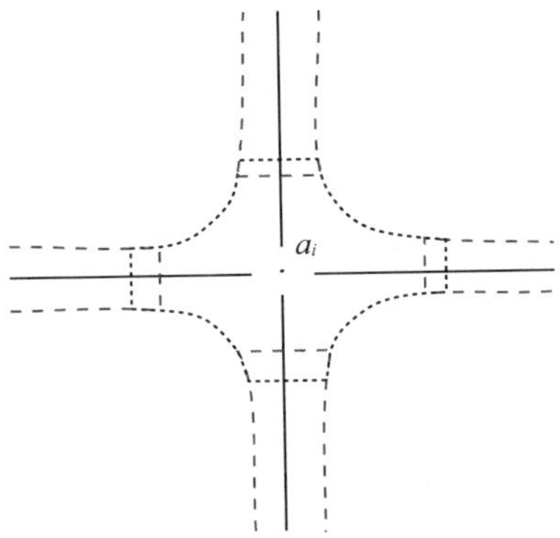


FIGURE 2. Tubular neighbourhoods

THEOREM 3.3. *Two stable frontal curves are $\mathscr{A}$-equivalent if and only if they have equivalent Gauss words.*

PROOF. Assume first that $\gamma, \delta : S^1 \to S^2$ are $\mathscr{A}$-equivalent. There exist diffeomorphisms $h : S^1 \to S^1$ and $k : S^2 \to S^2$ such that $\delta = k \circ \gamma \circ h^{-1}$. We choose letters $a_1, \dots, a_{r+s}$ which label the cusps and nodes of $\gamma$. We also chose a base point in $S^1$ and orientations in $S^1, S^2$ so that we have a well defined Gauss word for $\gamma$. Now, we take in $\delta$ the labels, the base point and the orientations induced by the diffeomorphisms $h, k$. It follows that the Gauss word of $\delta$ will be equal to the Gauss word of $\gamma$. In fact, the cusps and nodes are preserved by $\mathscr{A}$-equivalence and moreover, the sign of the exponents is the same if the diffeomorphisms preserve the orientations.

We see now the converse. Assume that $\gamma, \delta$ have equivalent Gauss words. We take $\tilde{\gamma}, \tilde{\delta}$ small generic deformations of $\gamma, \delta$ respectively, such that $\tilde{\gamma}, \tilde{\delta}$ have only node singularities, such that they coincide with $\gamma, \delta$ respectively, outside a small neighbourhood of each cusp and $\tilde{\gamma}'(t) \neq 0 \neq \tilde{\delta}'(t)$, for all $t$. The Gauss word of $\tilde{\gamma}$ (resp. $\tilde{\delta}$) is obtained from the Gauss word of $\gamma$ (resp. $\delta$) just by substituting each cusp with exponent $a^{-2}$ by $aa^{-1}$ and each cusp with exponent $a^2$ by $a^{-1}a$, as shown in fig. 3.

FIGURE 3. Cusps sign

We have now that $\tilde{\gamma}, \tilde{\delta}$ have equivalent Gauss words and hence, they are $\mathscr{A}$-equivalent by Lemma 3.2. But this implies that $\gamma, \delta$ are also $\mathscr{A}$-equivalent outside a small neighbourhood of each cusp. Since $\gamma, \delta$ are also locally $\mathscr{A}$-equivalent in a neighbourhood of each cusp, we can glue together all the $\mathscr{A}$-equivalences in order to get a global $\mathscr{A}$-equivalence between $\gamma, \delta$. □

DEFINITION 3.4. We say that two map germs $f, g : (\mathbb{R}^2, 0) \to (\mathbb{R}^3, 0)$ are $\mathscr{A}^*$-*equivalent* if there are homeomorphisms $h, k$ which are diffeomorphisms outside the origin and such that $g = k \circ f \circ h^{-1}$.

COROLLARY 3.5. *Let $f, g : (\mathbb{R}^2, 0) \to (\mathbb{R}^3, 0)$ be frontals with isolated instability. They are $\mathscr{A}^*$-equivalent if and only if the Gauss words associated to their respective links are equivalent.*

PROOF. If the Gauss words are equivalent then the links are $\mathscr{A}$-equivalent by Theorem 3.3. Then, by Theorem 2.5, $f$ and $g$ are $\mathscr{A}^*$-equivalent.

To see the converse, we claim that the Gauss word of the link of a map germ $f$ can be obtained by looking at the relative position of the branches of the transverse double point curves and the cuspidal edge curves and the orientation of the image of $f$ with respect to these curves.

We denote by $D(f)$ the set germ of points $z$ in $\mathbb{R}^2$ such that $z$ is either a double point or a singular point. Then $D(f)$ is a real analytic curve in $(\mathbb{R}^2, 0)$ and the restriction of $f$ to $D(f)$ is an immersion outside the origin. We take a small enough representative of $f$ such that $D(f) \setminus \{0\}$ and $f(D(f)) \setminus \{0\}$ are a finite union of open arcs and such that $f$ maps diffeomorphically each component of $D(f) \setminus \{0\}$ into a component of $f(D(f)) \setminus \{0\}$.

We choose labels $a_1, \ldots, a_{r+s}$ for the components of $f(D(f)) \setminus \{0\}$ and moreover we order the components of $D(f) \setminus \{0\}$ by choosing some base arc and some orientation of $\mathbb{R}^2$. We orient all the arcs in $f(D(f)) \setminus \{0\}$ from the origin and also choose an orientation in $\mathbb{R}^3$. We construct a Gauss word as the sequence of letters which appear in the components of $D(f) \setminus \{0\}$ with the following exponents:

- $a_i$, if it correspond to a positive sheet of a transverse double point;
- $a_i^{-1}$, if it correspond to a negative sheet of a transverse double point;
- $a_i^2$, if it is a cuspidal edge and corresponds to the positive sheet of a generic deformation of the cuspidal edge into a transverse double point;
- $a_i^{-2}$, if it is a cuspidal edge and corresponds to the negative sheet of a generic deformation of the cuspidal edge into a transverse double point;

(see fig. 4). The Gauss word constructed in this way coincides with the Gauss word of the link of $f$.

Assume now that $g$ is $\mathscr{A}^*$-equivalent to $f$ by homeomorphisms $h, k$ which are diffeomorphisms outside the origin, such that $g = k \circ f \circ h^{-1}$. Then, $h(D(f)) = D(g)$

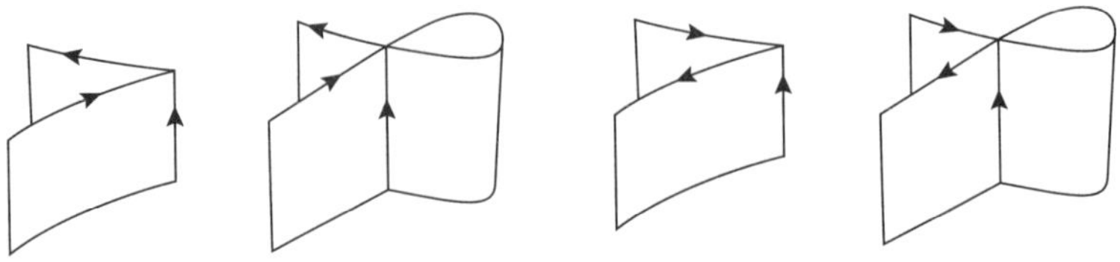

FIGURE 4

and $k(f(D(f))) = g(D(g))$ and we can choose in the components of $g(D(g)) \setminus \{0\}$ the labels induced by $k$. We also choose the base arc and the orientations in $\mathbb{R}^2, \mathbb{R}^3$ induced by $h, k$ respectively. With these choices, it follows that the Gauss word of $g$ is the same as the Gauss word of $f$. □

## 4. Frontal Ruled Surfaces

A ruled surface in $\mathbb{R}^3$ is (locally) the image of the map $F : I \times J \to \mathbb{R}^3$ defined by $F(t, v) = \alpha(t) + v\omega(t)$, where $\alpha : I \to \mathbb{R}^3$ and $\omega : I \to S^2$ are smooth mappings and $I$, $J$ are open intervals. We assume that $I$ is bounded. We call $\alpha$ a base curve and $\omega$ a directrix curve. The straight lines $L_t(v) = \alpha(t) + v\omega(t)$ are called rulings.

EXAMPLE 4.1. We find simple examples of ruled surfaces like the cylinder, the hyperbolic paraboloid, the one sheet hyperboloid, the helicoid and the Möbius strip. From the singularity point of view, we also see that the simplest singular surface, the Whitney umbrella or cross-cap is ruled. Moreover, the generic fronts given by the cuspidal edge and the swallowtail are also ruled (see fig. 5).

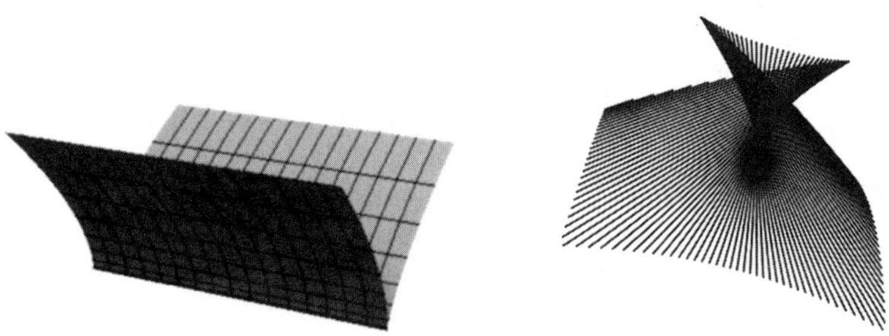

FIGURE 5

By the results of [**4**], given a germ of ruled surface, we can choose affine coordinates in $\mathbb{R}^3$ such that it is parametrized by a map germ $f : (\mathbb{R}^2, 0) \to (\mathbb{R}^3, 0)$ of the form $f(x, t) = \alpha(t) + x\gamma(t)$, where

$$\alpha(t) = (0, \alpha_1(t), \alpha_2(t)), \quad \gamma(t) = (1, \gamma_1(t), \gamma_2(t)). \tag{1}$$

By Proposition 2.2, $f$ is a frontal ruled surface if given $p(x, t) = \alpha_1(t) + x\gamma_1(t)$ and $q(x, t) = \alpha_2(t) + x\gamma_2(t)$, there exists $r : \mathbb{R}^2 \to \mathbb{R}$ such that $q_t(x, t) = r(x, t)p_t(x, t)$, where the subscripts denote the partial derivatives and, in our case, $r(x, t) = a(t) + xb(t)$.

LEMMA 4.2. *A necessary condition for a ruled surface $f(x,t) = \alpha(t) + x\gamma(t)$ to be frontal is that either $\gamma_1(t) \equiv c$ or $b(t) \equiv 0$.*

PROOF. We have that $q_t(x,t) = r(x,t)p_t(x,t)$, so

$$\begin{aligned} q(x,t) &= \int r(x,t)p_t(x,t)dt = \int p_t(x,t)a(t)dt + \int xp_t(x,t)b(t)dt \\ &= \int \alpha_1'(t)a(t)dt + x\left(\int \gamma_1'(t)a(t)dt + \int \alpha_1'(t)b(t)dt\right) + x^2 \int \gamma_1'(t)b(t)dt. \end{aligned}$$

Therefore, $\int \gamma_1'(t)b(t)dt \equiv 0$. □

We first consider $\gamma_1(t) \equiv c$ and, by a change of coordinates, suppose that $\gamma_1(t) \equiv 0$.

PROPOSITION 4.3. *Let $f(x,t) = \alpha(t) + x\gamma(t)$ be a frontal ruled surface with $\gamma_1(t) \equiv 0$ and with isolated instability. Then $f$ is $\mathscr{A}^*$-equivalent either to the cuspidal edge or the cuspidal cross-cap.*

PROOF. Let us denote

$$w_x(t) = (\alpha_1(t), \alpha_2(t) + x\gamma_2(t)).$$

Moreover, we can assume that

$$\alpha_1(t) = t^m + \overline{p}(t), \quad \alpha_2(t) = \overline{a}t^{k_0} + \overline{q}(t), \quad \gamma_2(t) = \overline{b}t^{k_1} + \overline{r}(t),$$

with $\overline{a}, \overline{b} \neq 0$; $k_0, k_1 > m$; $\overline{p} \in m_1^{m+1}$; $\overline{q} \in m_1^{k_0+1}$; $\overline{r} \in m_1^{k_1+1}$. ($m_r$ denotes the maximal ideal of $\mathcal{E}_r$)

The singular locus of $f$ is given by $w_x'(t) = 0$, so it is just the curve $t = 0$. The condition that $f$ has isolated instability implies $f$ has a cuspidal edge along this curve for $x \neq 0$, that is, $|w_x''(0), w_x'''(0)| \neq 0$ if $x \neq 0$. But if $m \geq 3$, we would have:

$$|w_x''(t), w_x'''(t)| = m(m-1)t^{m-2}((\widetilde{a}t^{k_0-3} + \widetilde{p}(t)) + x(\widetilde{b}t^{k_1-3} + \widetilde{q}(t)),$$

for some $\widetilde{a}, \widetilde{b} \neq 0$, $\widetilde{p} \in m_1^{k_0-2}$ and $\widetilde{q} \in m_1^{k_1-2}$. Thus, we must have $m = 2$ and either $k_0 = 3$ or $k_1 = 3$.

If $k_0 = 3$, then $f$ is $\mathscr{A}$-equivalent to the cuspidal edge, so we can assume $k_0 > 3$ and $k_1 = 3$. In this case, the Gauss word of the link of $f$ has exactly two cusps, corresponding to the two half branches of the cuspidal edge $t = 0$. Moreover, since the sign of $|w_x''(0), w_x'''(0)|$ only depends on the sign of $x$, the two cusps will have different exponents.

On the other hand, the equations of the double point curve of $f$ are $(w_x(t) - w_x(v))/(t-v) = 0$, which gives exactly one branch given by

$$\begin{aligned} &t + v + P(t,v) = 0, \\ &\overline{a}(t^{k_0-1} + t^{k_0-2}v + \cdots + v^{k_0-1} + Q(t,v)) + \overline{b}x(t^2 + tv + v^2 + R(t,v)) = 0, \end{aligned}$$

with $P \in m_2^2$, $Q \in m_2^{k_0}$ and $R \in m_2^3$. Thus, the link of $f$ has exactly one node. Note also that after $\mathscr{A}$-equivalence, we can assume that $k_0$ is odd. Then, the projection onto the plane $(x,t)$ gives the curve whose initial part is the parabola $\overline{a}t^{k_0-3} + \overline{b}x = 0$. Such parabola is transverse to the cuspidal edge $t = 0$, which implies that the Gauss word of $f$ must be equal to $a^2b^{-1}c^{-2}b$ and $f$ is $\mathscr{A}^*$-equivalent to the cuspidal cross-cap. □

Now we consider the case $\gamma_1(t) \not\equiv c$, so $b(t) \equiv 0$. We remark that the previous proposition also applies whenever

$$f(x,t) = (x, (t^m + \overline{p}(t)) + x(t^k + \overline{q}(t)), \alpha_2(t) + x\gamma_2(t)),$$

with $m \leq k$, so we will assume from now on that $k < m$. It follows from Lemma 4.2 that $f$ must be of the form

$$f(x,t) = (x, \alpha_1(t) + x\gamma_1(t), \int a(t)\alpha_1'(t)dt + x \int a(t)\gamma_1'(t)dt), \tag{2}$$

for some functions $\alpha_1, \gamma_1, a$.

First, we analyze the possibilities for the cusps in the Gauss word of the link of $f$. Recall that each cusp corresponds to a branch of cuspidal edges in $f$.

PROPOSITION 4.4. *Let $f$ be a frontal ruled surface given as in (2) with isolated instability. Then, $f$ has a cuspidal edge at $(x,t)$ if and only if $\alpha_1'(t) + x\gamma_1'(t) = 0$. Moreover, if $f$ has a cuspidal edge at $(x,t)$, the exponent of the corresponding cusp in the Gauss word only depends on the sign of $a'(t)$.*

PROOF. Let us denote

$$w_x(t) = (\alpha_1(t) + x\gamma_1(t), \int a(t)\alpha_1'(t)dt + x \int a(t)\gamma_1'(t)dt).$$

The cuspidal edges of $f$ correspond to the points where $w_x(t)$ has cusps for $x \in (-\epsilon, \epsilon)$. Therefore, we have a cusp when $w_x'(t) = 0$ and $|w_x''(t), w_x'''(t)| \neq 0$ for $x \neq 0$ close to zero. Moreover, the exponent of the cusp in the Gauss word is determined by the sign of $|w_x''(t_0), w_x'''(t_0)|$.

Note that $w'(t) = 0$ if and only if $\alpha_1'(t) + x\gamma_1'(t) = 0$. Moreover,

$$|w_x''(t_0), w_x'''(t_0)| = 2a'(t)\left(\alpha_1''(t) + x\gamma_1''(t)\right)^2$$
$$+ (\alpha_1'(t) + x\gamma_1'(t))\left(a''(t)\left(\alpha_1''(t) + x\gamma_1''(t)\right) - a'(t)(\alpha_1^{(3)}(t) + x\gamma_1^{(3)}(t))\right).$$

So the sign of $|w_x''(t), w_x'''(t)|$ depends only on the sign of $a'(t)$. $\square$

Let $f$ be a frontal ruled surface given as in (2) with isolated instability. We put $f = f_0 + f_1 : (\mathbb{R}^2, 0) \to (\mathbb{R}^3, 0)$, where

$$\begin{cases} f_0(x,t) = (x, xt^k + t^m, xt^{k+p} + \frac{m(k+p)}{k(m+p)}t^{m+p}), \\ f_1(x,t) = (0, g_1(t) + xg_2(t), g_3(t) + xg_4(t)), \end{cases} \tag{3}$$

with $g_1 \in m_1^{k+1}$, $g_2 \in m_1^{m+1}$, $g_3 \in m_1^{k+p+1}$ and $g_4 \in m_1^{m+p+1}$. By taking the coordinate change in $(\mathbb{R}^2, 0)$ given by $X = x$ and $T = t(1 + \frac{g_2(t)}{t^k})^{1/k}$, we can assume that $g_2(t) \equiv 0$.

COROLLARY 4.5. *Let $f$ be a frontal ruled surface given as in (3) with isolated instability, then either $k = 1$ or $k = 2$ and $p = 1$. Furthermore:*

1. *If $k = 1$, then there is only one branch of cuspidal edges given by $x + mt^{m-1} + g_1'(t) = 0$ and hence two cusps in the link of $f$.*
2. *If $k = 2$, then there are two branches of cuspidal edges given by $t = 0$ and $2x + mt^{m-2} + g_1'(t)/t = 0$ and hence four cusps in the link of $f$.*

*In both cases, the exponents of the cusps in the Gauss word only depend on the parity of $p$.*

PROOF. This is just a direct consequence of Proposition 4.4. We just remark that if either $k \geq 3$ or $k = 2$ and $p \geq 2$, then we find a curve of degenerate cuspidal edges at $t = 0$, which is not possible because of the condition of isolated instability. □

Now we analyze the possibilities for the nodes in the link of $f$. We first do this for the initial part $f_0$.

LEMMA 4.6. *Let $f$ be a frontal ruled surface given as in (3) with isolated instability. We have the following possibilities for the number of branches of transverse double points of $f_0$:*

1. *If $k = 1$ and $m, p$ have different parity, then there are no branches.*
2. *If $k = 1$ and $m, p$ have the same parity, then there is only one branch given by $x + t^{m-1} = 0$ if $m$ is odd or $x = 0$ if $m$ is even.*
3. *If $k = 2$, $p = 1$ and $m$ is odd then there are two branches given by $(x + \mu_1 t^{m-2})(x + \mu_2 t^{m-2}) = 0$, with $\mu_1 > 0 > \mu_2$.*
4. *If $k = 2$, $p = 1$ and $m$ is even, then there are three branches given by $(x + \mu_1 t^{m-2})(x + \mu_2 t^{m-2})(x + \mu_3 t^{m-2}) = 0$, with $\mu_1 > \mu_2 > \mu_3 > 0$. Furthermore, the Gauss word of the link of $f_0$ is equal to abcabc (up the cusps and the exponents).*

PROOF. Let us denote

$$w_x(t) = \left( xt^k + t^m, xt^{k+p} + \frac{m(k+p)}{k(m+p)} t^{m+p} \right).$$

The double point curve of $f_0$ is defined by the the equations $(w_x(t) - w_x(v))/(t-v) = 0$. This gives two equations:

$$F_{k-1}(t,v)x + F_{m-1}(t,v) = 0,$$

$$F_{k+p-1}(t,v)x + \frac{m(k+p)}{k(m+p)} F_{m+p-1}(t,v) = 0,$$

where $F_r(t,v) = \frac{t^r - t^v}{t-v} = t^r + t^{r-1}v + \cdots + v^r$. Since both equations are linear in $x$, we can eliminate the variable $x$ and get just one equation in $(t,v)$:

$$P_{k,m,p}(t,v) := F_{m-1}(t,v)F_{k+p-1}(t,v) - \frac{m(k+p)}{k(m+p)} F_{m+p-1}(t,v)F_{k-1}(t,v) = 0.$$

The polynomial $P_{k,m,p}$ is homogeneous of degree $k + m + p - 2$ and the number of transverse double point branches is determined by the number of real linear factors of $P_{k,m,p}$ of the form $t - \lambda_i v$, with $\lambda_i \neq 1$. Once we have such a solution of $P_{k,m,p}(t,v) = 0$, then we substitute $t = \lambda_i v$ in the two equations above and get the equation for the corresponding branch of transverse double points.

By Corollary 4.5, we have either $k = 1$ or $k = 2$ and $p = 1$. In the case $k = 1$, we get branches of the form

$$x + (\lambda_i^{m-1} + \cdots + 1)t^{m-1} = 0.$$

In the case $k = 2$ and $p = 1$, it depends on the parity of $m$,

$$\begin{cases} x + \dfrac{\lambda_i^{m-1} + \cdots + 1}{\lambda_i + 1} t^{m-2} = 0, & \text{if } m \text{ is odd}, \\ \\ x + (\lambda_i^{m-2} + \lambda_i^{m-4} + \cdots + 1)t^{m-2} = 0, & \text{if } m \text{ is even}. \end{cases}$$

We put $v = 1$ and denote $p(t) = P_{k,m,p}(t, 1)$, so we need to compute the real roots of $p$. It is not difficult to see that $p(1) = p'(1) = 0$, thus $t = 1$ is a double root of $p$. Let

$$\begin{aligned} q(t) = (t-1)^2 p(t) &= (t^m - 1)(t^{k+p} - 1) - \frac{m(k+p)}{k(m+p)}(t^{m+p} - 1)(t^k - 1) \\ &= \left(1 - \frac{m(k+p)}{k(m+p)}\right) t^{k+m+p} + \frac{m(k+p)}{k(m+p)} t^{m+p} - t^m - t^{k+p} + \frac{m(k+p)}{k(m+p)} t^k \\ &\quad + \left(1 - \frac{m(k+p)}{k(m+p)}\right). \end{aligned}$$

Since $1 - \frac{m(k+p)}{k(m+p)} < 0$, by the Descartes rule of signs, $q(t)$ has at most four real roots for $t > 0$ counting multiplicity. But $t = 1$ is a real root of multiplicity four, hence $p(t)$ has no real roots for $t > 0$ distinct from $t = 1$.

We claim that $p(t)$ has at most three real roots for $t < 0$. We split the proof of this fact in several cases. By Corollary 4.5, we have either $k = 1$ or $k = 2$ and $p = 1$. If $k = 1$, the result follows from the proof of [**4**, Theorem 6.1]. It is proved there that in such case, $p(t)$ has at most one negative real root $t = -1$, depending on the parities of $m, p$. This implies (1) and (2).

Let us assume now that $k = 2$ and $p = 1$. We change $t$ by $-t$ in $q(t)$ and apply again the Descartes rule of signs. If $m$ is odd, then $q(t)$ has at most two negative real roots counting multiplicity. We have $q(-1) > 0$ and $\lim_{t\to-\infty} q(t) = -\infty$, so we have one real root $\lambda_1 < -1$. Again by symmetry, there is another real root $\lambda_2 = 1/\lambda_1$. Hence, we have (3).

If $m$ is even, then $q(t)$ has at most three negative real roots counting multiplicity. In this case, $t = -1$ is a simple root with $q'(-1) > 0$. Since $\lim_{t\to-\infty} q(t) = +\infty$, we must have at least one real root $\lambda_1 < -1$. But $q(t) = q(1/t)$, so we have another real root $\lambda_2 = 1/\lambda_1$.

If $m$ is odd, then $q(t)$ has at most two negative real roots counting multiplicity. We have $q(-1) > 0$ and $\lim_{t\to-\infty} q(t) = -\infty$, so we have one real root $\lambda_1 < -1$. Again by symmetry, there is another real root $\lambda_2 = 1/\lambda_1$, which gives the first part of (4).

Finally, we see the last part of (4). We observe that in this case, the branches are given by parabolas of the form $x + \mu_i t^{m-2} = 0$, with $\mu_i = \lambda_i^{m-2} + \lambda_i^{m-4} + \cdots + 1$. By symmetry, given a double point $f(x,t) = f(t,v)$ such that $(x,t)$ is in the branch $x + \mu_1 t^{m-2} = 0$, then $(x,v)$ is in the branch $x + \mu_3 t^{m-2} = 0$ and vice-versa. If $(t,v)$ is in the branch $x + \mu_2 t^{m-2} = 0$, then $(t,v)$ is also in $x + \mu_2 t^{m-2} = 0$. This shows that the relative position of the three branches must be as in fig. 6. □

PROPOSITION 4.7. *Let $f$ be a frontal ruled surface given as in (3) with isolated instability, then the Gauss word of $f$ is the same as that of $f_0$.*

PROOF. By Corollary 4.5, the cuspidal edge curve has either one or two branches. If $k = 1$, we have just one branch $x + mt^{m-1} + g_1'(t) = 0$. If $k = 2$, we have two branches, $t = 0$ and $2x + mt^{m-2} + g_1'(t)/t = 0$. In both cases, the exponents of the cusps in the Gauss word only depend on the parity of $p$.

By Lemma 4.6, there are at most three branches of transverse double points given by equations

$$x + \mu_i t^{m-k} = 0 + H_i(t), \quad i = 1, \ldots, s,$$

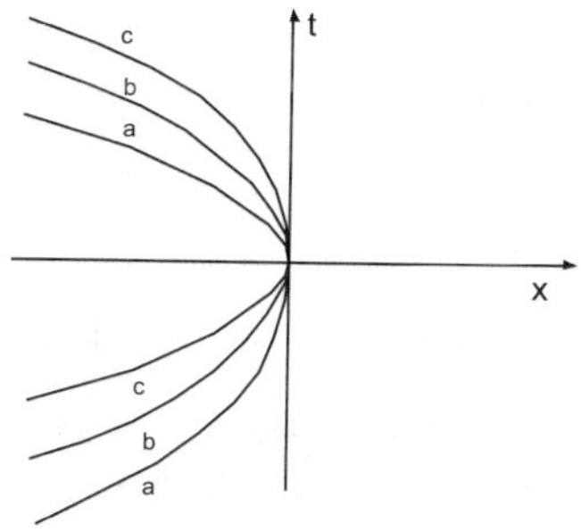


FIGURE 6

with $s \leq 3$, $\mu_i \neq \mu_j$ if $i \neq j$ and $H_i \in m_1^{m-k+1}$. We deduce that the relative position of all the branches of cuspidal edges and transverse double points only depends on the values $\mu_1, \ldots, \mu_s$. Hence $f$ and $f_0$ have the same Gauss word, up to the exponents. To finish the proof, we follow the same arguments as in the proof of [**4**, Proposition 6.10].

The exponent of a branch of transverse double points is determined by the sign of the determinant of some partial derivatives of $f$. By taking $v = \lambda_i t + K_i(t)$, we have

$$\left|\frac{\partial f}{\partial x}(x,t), \frac{\partial f}{\partial x}(x,v), \frac{\partial f}{\partial t}(x,t)\right| =$$
$$\left|\frac{\partial f_0}{\partial x}(x,t), \frac{\partial f_0}{\partial x}(x,\lambda_i t), \frac{\partial f_0}{\partial t}(x,t)\right| + \operatorname{ord}(t),$$

where $\operatorname{ord}(t)$ are higher order terms in $t$ (see the proof of [**4**, Proposition 6.10] for details). □

THEOREM 4.8. *Let $f : (\mathbb{R}^2, 0) \to (\mathbb{R}^3, 0)$ be a frontal ruled surface with an* isolated instability, *then the $\mathscr{A}^*$-class of $f$ is determined by one of the six links given in fig. 7.*

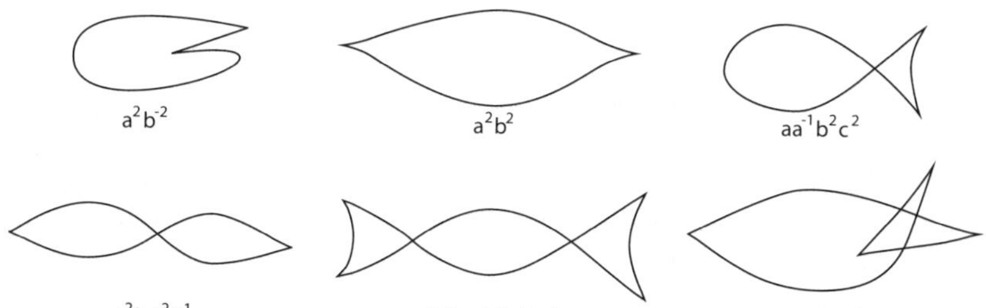


FIGURE 7

PROOF. If $\gamma_1(t) \equiv 0$, we know from Proposition 4.3 that the Gauss word of $f$ is either $a^2b^2$ (the cuspidal edge) or $a^2b^{-1}c^{-2}b$ (the cuspidal cross-cap). Hence, we can assume that $\gamma_1(t) \not\equiv 0$ and that $f$ is given as in (3). By Proposition 4.7, the Gauss word of $f$ is determined by $f_0$. We split the proof according to the number of nodes, according to Lema 4.6.

(1) *Zero nodes.* Since $k = 1$, by Corollary 4.5 there are two cusps corresponding to the half-branches of $x + mt^{m-1} = 0$, whose exponents depend on the parity of $p$. We obtain the two possible Gauss words $a^2b^2$ and $a^2b^{-2}$.

(2) *One node.* Again $k = 1$ and there are two cusps corresponding to the half-branches of $x + mt^{m-1} = 0$, whose exponents depend on the parity of $p$. The node corresponds to the branch of transverse double points, given by $x + t^{m-1} = 0$ if $m$ is odd or $x = 0$ if $m$ is even. Looking at the relative position of the branches, we deduce that the only possible Gauss words are either $aa^{-1}b^2c^2$ if $m$ is odd or $a^2bc^{-2}b^{-1}$ if $m$ is even.

(3) *Two nodes.* In this case, $k = 2$, $p = 1$ and $m$ is odd. By Corollary 4.5 there are four cusps corresponding to the half-branches of $t = 0$ and $2x + mt^{m-2} = 0$, all of them with positive exponent. The nodes corresponds to two branches of transverse double points $(x + \mu_1 t^{m-2})(x + \mu_2 t^{m-2}) = 0$, with $\mu_1 > 0 > \mu_2$. The relative position of the four branches is presented in fig. 8, so the Gauss word is $a^2b^2cde^2f^2dc$, up to the exponents of the double points. But the only possibility for the exponents is $a^2b^2cd^{-1}e^2f^2dc^{-1}$.

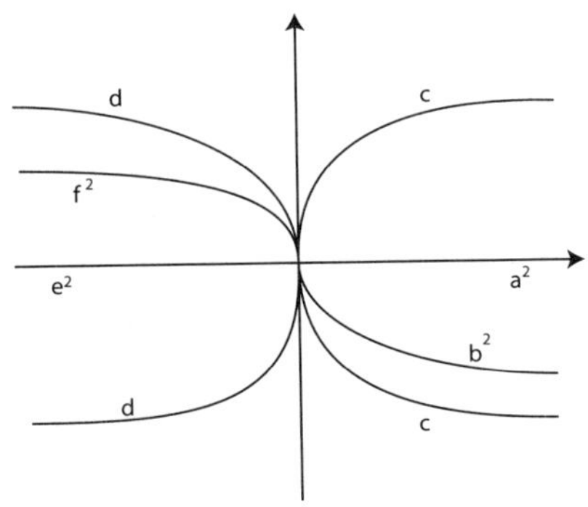


FIGURE 8

(4) *Three nodes.* We have that $k = 2$, $p = 1$ and $m$ is even. Again, there are four cusps with positive exponent corresponding to the half-branches of $t = 0$ and $2x + mt^{m-2} = 0$. The nodes corresponds to three branches of transverse double points $(x + \mu_1 t^{m-2})(x + \mu_2 t^{m-2})(x + \mu_3 t^{m-2}) = 0$, with $\mu_1 > \mu_2 > \mu_3 > 0$. We also know from Lema 4.6 that the Gauss word without exponents and without cusps is $abcabc$. It follows that the relative position of the five branches is given in fig. 9. The Gauss word is now $a^2bc^2def^2bdg^2e$, up to the exponents of the double points. But again there is only one possibility for the exponents which fits into this format, namely $a^2bc^2d^{-1}ef^2b^{-1}dg^2e^{-1}$.

□

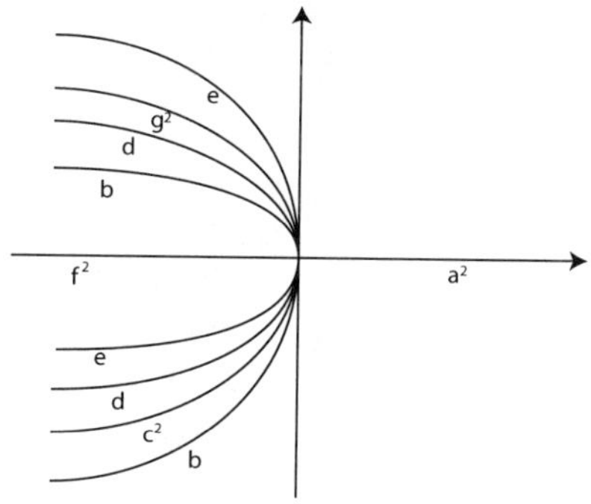


FIGURE 9

## References

[1] J. Scott Carter, *How surfaces intersect in space*, 2nd ed., Series on Knots and Everything, vol. 2, World Scientific Publishing Co., Inc., River Edge, NJ, 1995. An introduction to topology. MR1341588

[2] Takuo Fukuda, *Local topological properties of differentiable mappings. I*, Invent. Math. **65** (1981/82), no. 2, 227–250, DOI 10.1007/BF01389013. MR641129

[3] W. L. Marar and J. J. Nuño-Ballesteros, *The doodle of a finitely determined map germ from $\mathbb{R}^2$ to $\mathbb{R}^3$*, Adv. Math. **221** (2009), no. 4, 1281–1301, DOI 10.1016/j.aim.2009.02.008. MR2518639

[4] R. Martins and J. J. Nuño-Ballesteros, *Finitely determined singularities of ruled surfaces in $\mathbb{R}^3$*, Math. Proc. Cambridge Philos. Soc. **147** (2009), no. 3, 701–733, DOI 10.1017/S0305004109002618. MR2557150

[5] John Milnor, *Singular points of complex hypersurfaces*, Annals of Mathematics Studies, No. 61, Princeton University Press, Princeton, N.J.; University of Tokyo Press, Tokyo, 1968. MR0239612

[6] J. A. Moya-Pérez and J. J. Nuño-Ballesteros, *The link of a finitely determined map germ from $\mathbf{R}^2$ to $\mathbf{R}^2$*, J. Math. Soc. Japan **62** (2010), no. 4, 1069–1092. MR2761914

[7] J. A. Moya-Pérez and J. J. Nuño-Ballesteros, *Topological classification of corank 1 map germs from $\mathbb{R}^3$ to $\mathbb{R}^3$*, Rev. Mat. Complut. **27** (2014), no. 2, 421–445, DOI 10.1007/s13163-013-0137-z. MR3223574

[8] Juan Antonio Moya-Pérez and Juan José Nuño-Ballesteros, *Gauss words and the topology of map germs from $\mathbb{R}^3$ to $\mathbb{R}^3$*, Rev. Mat. Iberoam. **31** (2015), no. 3, 977–988. MR3420482

[9] Juan J. Nuño Ballesteros, *Unfolding plane curves with cusps and nodes*, Proc. Roy. Soc. Edinburgh Sect. A **145** (2015), no. 1, 161–174, DOI 10.1017/S0308210513000632. MR3304580

[10] Go-o Ishikawa, *Infinitesimal deformations and stability of singular Legendre submanifolds*, Asian J. Math. **9** (2005), no. 1, 133–166. MR2150696

[11] V. M. Zakalyukin and A. N. Kurbatskiĭ, *Envelope singularities of families of planes in control theory* (Russian, with Russian summary), Tr. Mat. Inst. Steklova **262** (2008), no. Optim. Upr., 73–86, DOI 10.1134/S0081543808030073; English transl., Proc. Steklov Inst. Math. **262** (2008), no. 1, 66–79. MR2489728

Departament of Mathematics, Universidade Estadual de Maringá, 87020-900, Maringá, Brazil

*E-mail address*: `rmartins@uem.br`

Departament de Geometria i Topologia, Universitat de València, Campus de Burjassot, 46100 Burjassot, Spain

*E-mail address*: `Juan.Nuno@uv.es`

Contemporary Mathematics
Volume **675**, 2016
http://dx.doi.org/10.1090/conm/675/13592

# Non-isolated hypersurface singularities and Lê cycles

David B. Massey

ABSTRACT. In this series of lectures, I will discuss results for complex hypersurfaces with non-isolated singularities.

In Lecture 1, I will review basic definitions and results on complex hypersurfaces, and then present classical material on the Milnor fiber and fibration. In Lecture 2, I will present basic results from Morse theory, and use them to prove some results about complex hypersurfaces, including a proof of Lê's attaching result for Milnor fibers of non-isolated hypersurface singularities. This will include defining the relative polar curve. Lecture 3 will begin with a discussion of intersection cycles for proper intersections inside a complex manifold, and then move on to definitions and basic results on Lê cycles and Lê numbers of non-isolated hypersurface singularities. Lecture 4 will explain the topological importance of Lê cycles and numbers, and then I will explain, informally, the relationship between the Lê cycles and the complex of sheaves of vanishing cycles.

## 1. Lecture 1: Topology of Hypersurfaces and the Milnor fibration

Suppose that $\mathcal{U}$ is an open subset of $\mathbb{C}^{n+1}$; we use $(z_0, \dots, z_n)$ for coordinates.

Consider a complex analytic (i.e., holomorphic) function $f : \mathcal{U} \to \mathbb{C}$ which is not locally constant. Then, the *hypersurface* $V(f)$ defined by $f$ is the purely $n$-dimensional complex analytic space defined by the vanishing of $f$, i.e.,

$$V(f) \;:=\; \{\mathbf{x} \in \mathcal{U} \mid f(\mathbf{x}) = 0\}.$$

To be assured that $V(f)$ is not empty and to have a convenient point in $V(f)$, one frequently assumes that $\mathbf{0} \in V(f)$, i.e., that $f(\mathbf{0}) = 0$. This assumption is frequently included in specifying the function, e.g., we frequently write $f : (\mathcal{U}, \mathbf{0}) \to (\mathbb{C}, 0)$.

Near each point $\mathbf{x} \in V(f)$, we are interested in the local topology of how $V(f)$ is embedded in $\mathcal{U}$. This is question of how to describe the *local, ambient topological-type of* $V(f)$ at each point.

A *critical point of* $f$ is a point $\mathbf{x} \in \mathcal{U}$ at which all of the complex partial derivatives of $f$ vanish. The *critical locus of* $f$ is the set of critical points of $f$, and is denoted by $\Sigma f$, i.e.,

$$\Sigma f \;:=\; V\left(\frac{\partial f}{\partial z_0}, \frac{\partial f}{\partial z_1}, \dots, \frac{\partial f}{\partial z_n}\right).$$

2010 *Mathematics Subject Classification.* Primary 32B15, 32C35, 32C18, 32B10.
*Key words and phrases.* Complex hypersurface, singularities, Morse theory, Lê cycle.

The complex analytic Implicit Function Theorem implies that, if $\mathbf{x} \in V(f)$ and $\mathbf{x} \notin \Sigma f$, then, in an open neighborhood of $\mathbf{x}$, $V(f)$ is a **complex analytic submanifold** of $\mathcal{U}$; thus, we completely understand the ambient topology of $V(f)$ near a non-critical point. However, if $\mathbf{x} \in \Sigma f$, then it is possible that $V(f)$ is not even a topological submanifold of $\mathcal{U}$ near $\mathbf{x}$.

Note that, as sets, $V(f) = V(f^2)$, and that **every** point of $V(f^2)$ is a critical point of $f^2$. In fact, this type of problem occurs near a point $\mathbf{p}$ any time that an irreducible component of $f$ (in its unique factorization in the unique factorization domain $\mathcal{O}_{\mathcal{U},\mathbf{p}}$) is raised to a power greater than one. Hence, when considering the topology of $V(f)$ near a point $\mathbf{p} \in V(f)$, it is standard to assume that $f$ is *reduced*, i.e., has no such repeated factors. This is equivalent to assuming that $\dim_{\mathbf{p}} \Sigma f < n$.

Let's look at a simple, but important, example.

---

EXAMPLE 1.1. Consider $f : (\mathbb{C}^2, \mathbf{0}) \to (\mathbb{C}, 0)$ given by $f(x, y) = y^2 - x^3$.

It is trivial to check that $\Sigma f = \{\mathbf{0}\}$. Thus, at (near) every point of $V(f)$ other than the origin, $V(f)$ is a complex analytic submanifold of $\mathbb{C}^2$.

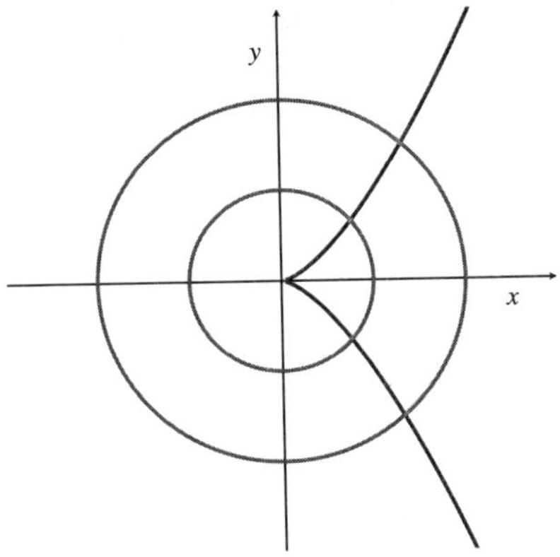


FIGURE 1. A cusp, intersected by two "spheres".

In the figure, ignoring for now the two circles, you see the graph of $V(f)$, but drawn over the real numbers. We draw graphs over the real numbers since we can't draw a picture over the complex numbers, but we hope that the picture over the real numbers gives us some intuition for what happens over the complex numbers.

Note that $f$ has a critical point at the origin, and so the complex analytic Implicit Function Theorem does **not** guarantee that $V(f)$ is a complex submanifold of $\mathbb{C}^2$ near $\mathbf{0}$. If the real picture is not misleading, it appears that $V(f)$ is not even a smooth ($C^\infty$) submanifold of $\mathbb{C}^2$ near $\mathbf{0}$; this is true.

However, the real picture is, in fact, misleading in one important way. Over the real numbers, $V(f)$ is a topological submanifold of $\mathbb{R}^2$ near $\mathbf{0}$, i.e., there exist open neighborhoods $\mathcal{U}$ and $\mathcal{W}$ of the origin in $\mathbb{R}^2$ and a homeomorphism of triples

$$(\mathcal{U}, \mathcal{U} \cap V(f), \mathbf{0}) \cong (\mathcal{W}, \mathcal{W} \cap V(x), \mathbf{0}).$$

However, over the complex numbers $V(f)$ is **not** a topological submanifold of $\mathbb{C}^2$ near $\mathbf{0}$. This takes some work to show.

Why have we drawn the two circles in the figure? Because we want you to observe two things, which correspond to a theorem that we shall state below. First, that the topological-type of the hypersurface seems to stabilize inside open balls of sufficiently small radius, e.g., the hypersurface "looks" the same inside the open disk $\overset{\circ}{B}_2$ bounded by the bigger circle as it does inside the open disk $\overset{\circ}{B}_1$ bounded by the smaller circle; of course, in $\mathbb{C}^2$, "disk" becomes "4-dimensional ball", and "circle" becomes "3-dimensional sphere". Second, it appears that this ambient topological-type can be obtained by taking the (open) cone on the bounding sphere and its intersection with the hypersurface.

We make all of this precise below.

---

Let us first give rigorous definitions of the *local, ambient topological-type of a hypersurface* and of a *singular point.*

DEFINITION 1.2. Suppose that $\mathcal{U}$ is an open subset of $\mathbb{C}^{n+1}$, and that we have a complex analytic function $f : \mathcal{U} \to \mathbb{C}$ which is not locally constant. Let $\mathbf{p} \in V(f)$. Then, the **local, ambient topological-type of** $V(f)$ **at** $\mathbf{p}$ is the homeomorphism-type of the germ at $\mathbf{p}$ of the triple $(\mathcal{U}, V(f), \mathbf{p})$.

In other words, if $g : \mathcal{W} \to \mathbb{C}$ is another such function, and $\mathbf{q} \in V(g)$, then the local, ambient topological-type of $V(f)$ at $\mathbf{p}$ is the same as that of $V(g)$ at $\mathbf{q}$ if and only if there exist open neighborhoods $\mathcal{U}'$ and $\mathcal{W}'$ of $\mathbf{p}$ and $\mathbf{q}$, respectively, and a homeomorphism of triples

$$(\mathcal{U}', \mathcal{U}' \cap V(f), \mathbf{p}) \cong (\mathcal{W}', \mathcal{W}' \cap V(g), \mathbf{q}).$$

The **trivial local, ambient topological-type** is that of $(\mathbb{C}^{n+1}, V(z_0), \mathbf{0})$. To say that $V(f)$ has the trivial topological-type at a point $\mathbf{p}$ is simply to say that $V(f)$ is a topological submanifold of $\mathcal{U}$ near $\mathbf{p}$.

A point on a hypersurface at which it has the trivial local, ambient topological-type is called a **regular point** of the hypersurface. A non-regular point on a hypersurface is called a **singular point** or a **singularity**. The **set of singular points** of $V(f)$ is denoted by $\Sigma V(f)$.

REMARK 1.3. You may question our terminology above. Shouldn't "regular" and "singular" have something to do with smoothness, not just topological data? In fact, it turns out that there is a very strong dichotomy here.

If $f$ is reduced at $\mathbf{p}$, then, in an open neighborhood of $\mathbf{p}$, $\Sigma f = \Sigma V(f)$. This is not trivial to see, and uses the Curve Selection Lemma (see Lemma 5.1 in the Appendix) to show that, near a point in $V(f)$, $\Sigma f \subseteq V(f)$, and then uses results on Milnor fibrations.

But, what it implies is that, at a point on a hypersurface, the hypersurface is either an analytic submanifold or is not even a topological submanifold. Therefore, all conceivable notions of "regular" and "singular" agree for complex hypersurfaces.

This also explains the frequent, mildly bad, habit of using the terms "critical point of $f$" and "singular point of $V(f)$" interchangeably.

The following theorem can be found in the work of Łojasiewicz in [**18**], and is now a part of the general theory of Whitney stratifications. We state the result for hypersurfaces in affine space, but the general result applies to arbitrary analytic sets. We recall the definition of the cone and the cone on a pair in the Appendix; in particular, recall that the cone on a pair is a triple, which includes the cone point.

THEOREM 1.4. (Łojasiewicz, 1965) Suppose that $\mathcal{U}$ is an open subset of $\mathbb{C}^{n+1}$, and that we have a complex analytic function $f : \mathcal{U} \to \mathbb{C}$ which is not locally constant. Let $\mathbf{p} \in V(f)$, and for all $\epsilon > 0$, let $B_\epsilon(\mathbf{p})$ and $S_\epsilon(\mathbf{p})$ denote the closed ball and sphere of radius $\epsilon$, centered at $\mathbf{p}$, in $\mathbb{C}^{n+1}$. Let $B_\epsilon^\circ(\mathbf{p})$ denote the corresponding open ball.

Then, there exists $\epsilon_0 > 0$ such that $B_{\epsilon_0}(\mathbf{p}) \subseteq \mathcal{U}$ and such that, if $0 < \epsilon \leq \epsilon_0$, then:

(1) $\big(B_\epsilon(\mathbf{p}),\ B_\epsilon(\mathbf{p}) \cap V(f), \mathbf{p}\big)$ is homeomorphic to the triple $c\big(S_\epsilon(\mathbf{p}),\ S_\epsilon(\mathbf{p}) \cap V(f)\big)$ by a homeomorphism which is the "identity" on $S_\epsilon(\mathbf{p})$ when it is identified with $S_\epsilon(\mathbf{p}) \times \{0\}$.

In particular,
$$\big(B_\epsilon^\circ(\mathbf{p}),\ B_\epsilon^\circ(\mathbf{p}) \cap V(f), \mathbf{p}\big) \cong c^\circ\big(S_\epsilon(\mathbf{p}),\ S_\epsilon(\mathbf{p}) \cap V(f)\big);$$

(2) the homeomorphism-type of the pair $\big(S_\epsilon(\mathbf{p}),\ S_\epsilon(\mathbf{p}) \cap V(f)\big)$ is independent of the choice of $\epsilon$ (provided $0 < \epsilon \leq \epsilon_0$).

Thus, the local, ambient topological-type of $V(f)$ at $\mathbf{p}$ is determined by the homeomorphism-type of the pair $\big(S_\epsilon(\mathbf{p}),\ S_\epsilon(\mathbf{p}) \cap V(f)\big)$, for sufficiently small $\epsilon > 0$.

DEFINITION 1.5. The space $S_\epsilon(\mathbf{p}) \cap V(f)$ (or its homeomorphism-type) for sufficiently small $\epsilon > 0$ is called the **real link of $V(f)$ at $\mathbf{p}$** and is frequently denoted by $K$.

REMARK 1.6. The letter $K$ is used because, in the first interesting case, of complex curves in $\mathbb{C}^2$, the real link is a knot (or link) in $S^3$, and how this knot is embedded in $S^3$ completely determines the local, ambient topological-type.

EXERCISE 1.7. Consider the following examples:

(1) $f : \mathbb{C}^2 \to \mathbb{C}$ given by $f(x, y) = xy$. Show that $V(f)$ is not a topological manifold at $\mathbf{0}$ (and so, is certainly not a topological submanifold).

(2) $f : \mathbb{C}^2 \to \mathbb{C}$ given by $f(x, y) = y^2 - x^3$. Show that $V(f)$ is homeomorphic to a disk near $\mathbf{0}$, and so is a topological manifold. Now, parameterize $K = S_\epsilon^3 \cap V(f)$ and show that you obtain the trefoil knot in $S_\epsilon^3$. Conclude that $V(f)$ is not a topological submanifold of $\mathbb{C}^2$ near $\mathbf{0}$.

(3) $f : \mathbb{C}^3 \to \mathbb{C}$ given by $f(x, y, z) = 2xy - z^2$. Show that $K$ is homeomorphic to real projective 3-space. Conclude that $V(f)$ is not a topological manifold near $\mathbf{0}$. (Hint: Use $x = s^2$, $y = t^2$, and $z = \sqrt{2}st$, and note that a point of $V(f)$ is not represented by a unique choice of $(s, t)$.)

---

As you can probably tell, the functions used in Exercise 1.7 were chosen very specially, and, in general, it is unreasonable to expect to analyze the topology of a hypersurface at a singular point via such concrete unsophisticated techniques.

So...how does one go about understanding how the real link $K$ embeds in a small sphere?

One large piece of data that one can associate to this situation is the topology of the **complement**. This, of course, is not complete data about the embedding, but it is a significant amount of data.

For ease of notation, assume that we have a complex analytic function $f : (\mathcal{U}, \mathbf{0}) \to (\mathbb{C}, 0)$ which is not locally constant, and that we wish to understand the local, ambient topology of $V(f)$ at $\mathbf{0}$. We will suppress the references to the center $\mathbf{p} = \mathbf{0}$ in our notation for spheres and balls. So, how do you analyze $S_\epsilon - S_\epsilon \cap V(f) = S_\epsilon - K$ for sufficiently small $\epsilon > 0$?

Milnor gave us many tools in his 1968 book [**28**]. He proved that, for sufficiently small $\epsilon > 0$, the map

$$\frac{f}{|f|} : S_\epsilon - K \to S^1 \subseteq \mathbb{C}$$

is a smooth, locally trivial fibration, and then proved many results about the fiber. (To review what a smooth, locally trivial fibration is, see the Appendix.)

We will state some of the results of Milnor and others about the above fibration, which is now known as the *Milnor fibration.* Below, $\mathbb{D}_\delta$ denotes a disk in $\mathbb{C}$, centered at the origin, of radius $\delta$, and so $\partial\mathbb{D}_\delta$ is its boundary circle.

The following theorem is a combination of Theorem 4.8 and Theorem 5.11 of [**28**], together with Theorem 1.1 of [**14**].

THEOREM 1.8. (Milnor, 1968 and Lê, 1976) Suppose that $f : (\mathcal{U}, \mathbf{0}) \to (\mathbb{C}, 0)$ is a complex analytic function. Then, there exists $\epsilon_0 > 0$ such that, for all $\epsilon$ with $0 < \epsilon \le \epsilon_0$, there exists $\delta_\epsilon > 0$, such that, for all $\delta$ with $0 < \delta \le \delta_\epsilon$, the map $f/|f|$ from $S_\epsilon - S_\epsilon \cap V(f) = S_\epsilon - K$ to $S^1$ is a smooth locally trivial fibration.

Furthermore, this smooth locally trivial fibration is diffeomorphic to the restriction

$$f : B^\circ_\epsilon \cap f^{-1}(\partial\mathbb{D}_\delta) \to \partial\mathbb{D}_\delta.$$

Finally, the restriction

$$f : B_\epsilon \cap f^{-1}(\partial\mathbb{D}_\delta) \to \partial\mathbb{D}_\delta$$

(note the closed ball) is a smooth locally trivial fibration, in which the fiber is a smooth manifold with boundary. This fibration is fiber-homotopy-equivalent to the one using the open ball (i.e., is isomorphic up to homotopy).

REMARK 1.9. It will be important to us later that Milnor's proof of the above theorem also shows that $B_\epsilon \cap f^{-1}(\mathbb{D}_\delta)$ is homeomorphic to $B_\epsilon$ and, hence, is contractible. This is sometimes referred to as a **Milnor tube**.

DEFINITION 1.10. Either one of the first two isomorphic fibrations given in the definition above is called the **Milnor fibration** of $f$ at $\mathbf{0}$, and the corresponding fiber is called the **Milnor fiber**.

The third and final fibration from the theorem above is called the **compact Milnor fibration**, and the corresponding fiber is called the **compact Milnor fiber**.

If we are interested in the Milnor fibration and/or Milnor fiber only up to homotopy, then any of the three fibrations and fibers are called the **Milnor fibration** and **Milnor fiber**.

REMARK 1.11. As $\mathcal{U}$ is an open subset of $\mathbb{C}^{n+1}$, the Milnor fiber is a complex $n$-manifold, and so is a real $2n$-manifold. The compact Milnor fiber is thus a compact real $2n$-manifold with boundary.

We should also remark that the Milnor fibration exists at each point $\mathbf{p} \in V(f)$; one simply replaces the ball and spheres centered at $\mathbf{0}$ with balls and spheres centered at $\mathbf{p}$.

As a final remark, we should mention that the phrase "there exists $\epsilon_0 > 0$ such that, for all $\epsilon$ with $0 < \epsilon \leq \epsilon_0$, there exists $\delta_\epsilon > 0$, such that, for all $\delta$ with $0 < \delta \leq \delta_\epsilon$" is usually abbreviated by writing simply "For $0 < \delta \ll \epsilon \ll 1$". This is read aloud as "for all sufficiently small positive $\epsilon$, for all sufficiently small positive $\delta$ (small compared to the choice of $\epsilon$)".

We will now list a number of results on the Milnor fibration and Milnor fiber. Below, we let $\mathcal{U}$ be an open neighborhood of the origin in $\mathbb{C}^{n+1}$, $f : (\mathcal{U}, \mathbf{0}) \to (\mathbb{C}, 0)$ is a complex analytic function, $F_{f,\mathbf{0}}$ denotes the Milnor fiber of $f$ at $\mathbf{0}$, and we let $s := \dim_{\mathbf{0}} \Sigma f$.

(1) If $\mathbf{0} \notin \Sigma f$, then $F_{f,\mathbf{0}}$ is diffeomorphic to a ball and so, in particular, is contractible and has trivial homology (i.e., the homology of a point).

(2) $F_{f,\mathbf{0}}$ has the homotopy-type of a finite $n$-dimensional CW-complex. In particular, if $k > n$, then the homology $H_k(F_{f,\mathbf{0}}; \mathbb{Z}) = 0$, and $H_n(F_{f,\mathbf{0}}; \mathbb{Z})$ is free Abelian. (See [**28**], Theorem 5.1.)

(3) $F_{f,\mathbf{0}}$ is $(n-s-1)$-connected. (For $s = 0$, see [**28**], Lemma 6.4. For general $s$, see [**11**].)

(4) Suppose that $s = 0$. Then Items 1 and 2 imply that $F_{f,\mathbf{0}}$ has the homotopy-type of the one-point union of a finite collection of $n$-spheres; this is usually referred to as a **bouquet of spheres**. The number of spheres in the bouquet, i.e., the rank of $H_n(F_{f,\mathbf{0}}; \mathbb{Z})$, is called the **Milnor number of $f$ at $\mathbf{0}$** and is denoted by either $\mu_f(\mathbf{0})$ or $\mu_{\mathbf{0}}(f)$.

(5) The Milnor number of $f$ at an isolated critical point can be calculated algebraically by taking the complex dimension of the **Jacobian algebra**, i.e.,

$$\mu_f(\mathbf{0}) = \dim_{\mathbb{C}} \frac{\mathbb{C}\{z_0, \ldots, z_n\}}{\left\langle \frac{\partial f}{\partial z_0}, \ldots, \frac{\partial f}{\partial z_n} \right\rangle},$$

where $\mathbb{C}\{z_0, \ldots, z_n\}$ is the ring of convergent power series at the origin. (This follows at once from [**28**], Theorem 7.2, by using a result of V. Palamodov in [**32**].)

In particular, if $s = 0$, then $\mu_f(\mathbf{0}) > 0$ if and only if $\mathbf{0} \in \Sigma f$.

(6) In Lemma 9.4 of [**28**], Milnor proves that, if $f$ is a weighted homogeneous polynomial, then the Milnor fiber of $f$ at $\mathbf{0}$ is diffeomorphic to the global fiber $f^{-1}(1)$ in $\mathbb{C}^{n+1}$.

(7) If $f : (\mathcal{U}, \mathbf{0}) \to (\mathbb{C}, 0)$ and $g : (\mathcal{U}', \mathbf{0}) \to (\mathbb{C}, 0)$ are analytic functions, then the Milnor fibre of the function $h : (\mathcal{U} \times \mathcal{U}', \mathbf{0}) \to (\mathbb{C}, 0)$ defined by $h(\mathbf{w}, \mathbf{z}) := f(\mathbf{w}) + g(\mathbf{z})$ is homotopy-equivalent to the join (see the Appendix), $F_{f,\mathbf{0}} * F_{g,\mathbf{0}}$, of the Milnor fibres of $f$ and $g$.

This determines the homology of $F_{h,\mathbf{0}}$ in a simple way, since the reduced homology of the join of two spaces $X$ and $Y$ is given by

$$\widetilde{H}_{j+1}(X * Y) = \sum_{k+l=j} \widetilde{H}_k(X) \otimes \widetilde{H}_l(Y) \oplus \sum_{k+l=j-1} \operatorname{Tor}\left(\widetilde{H}_k(X), \widetilde{H}_l(Y)\right),$$

where all homology groups are with $\mathbb{Z}$ coefficients.

This is the Sebastiani-Thom Theorem, proved in different cases by many people. See, for instance, [**34**], [**31**], [**33**], [**29**], [**30**], and [**24**].

(8) Let $\mathcal{U}$ and $\mathcal{W}$ be open neighborhoods of $\mathbf{0}$ in $\mathbb{C}^{n+1}$, let $f : (\mathcal{U}, \mathbf{0}) \to (\mathbb{C}, 0)$ and $g : (\mathcal{U}, \mathbf{0}) \to (\mathbb{C}, 0)$ be reduced complex analytic functions which define hypersurfaces with the same ambient topological-type at the origin. Then, there exists a homotopy-equivalence $\alpha : F_{f,\mathbf{0}} \to F_{g,\mathbf{0}}$ such that the induced isomorphism on homology commutes with the respective Milnor monodromy automorphisms.

In particular, the homotopy-type of the Milnor fiber of a reduced complex analytic function $f$ is an invariant of the local, ambient topological-type of $V(f)$, and so, for hypersurfaces defined by a reduced function with an isolated critical point, the Milnor number is an invariant of the local, ambient topological-type.

(For $s = 0$, this result appears in a remark of Teissier in [**37**] in 1972 and in [**38**] in 1973. The general result, with a monodromy statement, is due to Lê in [**13**] and [**12**], which both appeared in 1973.)

(9) Suppose that $\mathbf{0} \in \Sigma f$. Let $T^i_{f,\mathbf{0}} : H_i(F_{f,\mathbf{0}}; \mathbb{Z}) \to H_i(F_{f,\mathbf{0}}; \mathbb{Z})$ denote the monodromy automorphism in degree $i$. Then, the Lefschetz number of the monodromy $T^*_{f,\mathbf{0}}$ is zero, i.e.,

$$\sum_i (-1)^i \operatorname{trace}\left(T^i_{f,\mathbf{0}}\right) = 0.$$

(See [**1**].)

(10) The previous item implies that the converse to Item 1 is true. Thus, the Milnor fiber $F_{f,\mathbf{0}}$ has trivial homology (i.e., has the homology of a point) if and only if $\mathbf{0} \notin \Sigma f$ (and so, in particular, $V(f)$ is a topological submanifold of affine space at $\mathbf{0}$).

EXERCISE 1.12. In some/many cases, the Milnor number can be calculated by hand.

(1) Calculate the Milnor number at $\mathbf{0}$ of $f(x, y) = y^2 - x^3$, which defines a cusp.

(2) Calculate the Milnor number at $\mathbf{0}$ of $f(x, y) = y^2 - x^3 - x^2$, which defines a node. Conclude that the node and cusp have different ambient topological types.

(3) Show that $f = y^2 - x^5$ and $g = y^3 - x^3$ both have Milnor number 4 at the origin, but do not define hypersurfaces with the same ambient topological-type at the origin (actually, these hypersurfaces do not have the same topological-type at the origin, leaving out the term "ambient").

Thus, even for isolated critical points, the converse of Item 8, above, is false.

EXERCISE 1.13. In special cases, one can calculate the homology groups of the Milnor fiber of a non-isolated critical point. Consider $f(x, y, t) = y^2 - x^3 - tx^2$. Show that $\dim_{\mathbf{0}} \Sigma f = 1$ and calculate the homology groups of $F_{f,\mathbf{0}}$. (Hint: Use the Sebastiani-Thom Theorem. Also, use Milnor's result for weighted homogeneous polynomials, and that the homotopy-type of the Milnor fiber is certainly invariant under local analytic coordinate changes.)

The function $f$ can be thought of as a family of hypersurfaces, parameterized by $t$, where each member of the family has an isolated critical point at the origin; so, this is usually described as a **family of nodes which degenerates to a cusp at** $t = 0$.

Despite Item 3 of Exercise 1.12, the stunning conclusion of Lê and Ramanujam is that the converse of Item 7, above, **is** true in the case of isolated critical points if $f$ and $g$ are in the same analytic family (with one dimension restriction):

THEOREM 1.14. (Lê-Ramanujam, [**17**]) Suppose $n \neq 2$, and $f$ and $g$ are part of an analytic family of functions with isolated critical points, all of which have the same Milnor number, then $f$ and $g$ define hypersurfaces with the same local, ambient topological-type.

---

Thus, for hypersurfaces with isolated singularities, the Milnor number is algebraically calculable, determines the homology of the Milnor fiber, and its constancy in a family (with one dimension restriction) controls the local ambient topology in the family.

We would like similar data for hypersurfaces with non-isolated singularities. The Lê numbers succeed at generalizing the Milnor number in many ways, but do not yield such strong results. We shall discuss Lê cycles and Lê numbers in the third lecture.

In the second lecture, we will discuss the basics of Morse Theory, and use it to prove an important result of Lê from [**12**] on the homology of the Milnor fiber for non-isolated hypersurface singularities.

## 2. Lecture 2: Morse Theory, the relative polar curve, and two applications

Many of the results in [**28**] are proved using *Morse Theory*, and so we wish to give a quick introduction to the subject. We will then give some examples of how Morse Theory is used in the study of singular hypersurfaces.

Morse Theory is the study of what happens at the most basic type of critical point of a smooth map. The classic, beautiful references for Morse Theory are [**26**] and [**27**]. We also recommend the excellent, new introductory treatment in [**25**].

In this section, until we explicitly state otherwise, $f : N \to \mathbb{R}$ will be a smooth function from a smooth manifold of dimension $n$ into $\mathbb{R}$. For all $a \in \mathbb{R}$, let $N_{\leq a} := f^{-1}((-\infty, a])$. Note that if $a$ is a regular value of $f$, then $N_{\leq a}$ is a *smooth manifold with boundary* $\partial N_{\leq a} = f^{-1}(a)$ (see, for instance, [**36**]).

The following is essentially Theorem 3.1 of [**26**].

THEOREM 2.1. Suppose that $a, b \in \mathbb{R}$ and $a < b$. Suppose that $f^{-1}([a,b])$ is compact and contains no critical points of $f$.

Then, the restriction $f : f^{-1}([a,b]) \to [a,b]$ is a trivial fibration, and $N_{\leq a}$ is a deformation retract of $N_{\leq b}$ via a smooth isotopy. In particular, $N_{\leq a}$ is diffeomorphic to $N_{\leq b}$.

Now, let $\mathbf{p} \in N$, and let $(x_1, \dots, x_n)$ be a smooth, local coordinate system for $N$ in an open neighborhood of $\mathbf{p}$.

DEFINITION 2.2. The point $\mathbf{p}$ is a **non-degenerate critical point of** $f$ provided that $\mathbf{p}$ is a critical point of $f$, and that the Hessian matrix $\left(\frac{\partial^2 f}{\partial x_i \partial x_j}(\mathbf{p})\right)_{i,j}$ is non-singular.

The **index of** $f$ **at a non-degenerate critical point** $\mathbf{p}$ is the number of negative eigenvalues of $\left(\frac{\partial^2 f}{\partial x_i \partial x_j}(p)\right)_{i,j}$, counted with multiplicity.

Note that since the Hessian matrix is a real symmetric matrix, it is diagonalizable and, hence, the algebraic and geometric multiplicities of eigenvalues are the same.

EXERCISE 2.3. Prove that $\mathbf{p}$ being a non-degenerate critical point of $f$ is independent of the choice of local coordinates on $N$.

The index of $f$ at a non-degenerate critical point $\mathbf{p}$ can also can characterized as the index of the bilinear form $B$ defined by the Hessian matrix; this is defined to be the dimension of a maximal subspace on which $B$ is negative-definite. Using this, prove that the index of $f$ at a non-degenerate critical is also independent of the coordinate choice.

The following is Lemma 2.2 of [**26**], which tells us the basic structure of $f$ near a non-degenerate critical point.

LEMMA 2.4. (The Morse Lemma) Let $\mathbf{p}$ be a non-degenerate critical point of $f$. Then, there is a local coordinate system $(y_1, \dots, y_n)$ in an open neighborhood $\mathcal{U}$ of $\mathbf{p}$, with $y_i(\mathbf{p}) = 0$, for all $i$, and such that, for all $\mathbf{x} \in \mathcal{U}$,

$$f(\mathbf{x}) = f(\mathbf{p}) - (y_1(\mathbf{x}))^2 - (y_2(\mathbf{x}))^2 - \cdots - (y_\lambda(\mathbf{x}))^2 + (y_{\lambda+1}(\mathbf{x}))^2 + \cdots + (y_n(\mathbf{x}))^2,$$

where $\lambda$ is the index of $f$ at $\mathbf{p}$.

In particular, the point $\mathbf{p}$ is an isolated critical point of $f$.

The fundamental result of Morse Theory is a description of how $N_{\leq b}$ is obtained from $N_{\leq a}$, where $a < b$, and where $f^{-1}([a,b])$ is compact and contains a single critical point of $f$, and that critical point is contained in $f^{-1}((a,b))$ and is non-degenerate. See [**26**].

Recall that "attaching a $\lambda$-cell to a space $X$" means taking a closed ball of dimension $\lambda$, and attaching it to $X$ by identifying points on the boundary of the ball with points in $X$.

THEOREM 2.5. Suppose that $a < b$, $f^{-1}([a,b])$ is compact and contains exactly one critical point of $f$, and that this critical point is contained in $f^{-1}((a,b))$ and is non-degenerate of index $\lambda$.

Then, $N_{\leq b}$ has the homotopy-type of $N_{\leq a}$ with a $\lambda$-cell attached, and so $H_i(N_{\leq b}, N_{\leq a};\ \mathbb{Z}) = 0$ if $i \neq \lambda$, and $H_\lambda(N_{\leq b}, N_{\leq a};\ \mathbb{Z}) \cong \mathbb{Z}$.

Thus, functions $f : N \to \mathbb{R}$ that have only non-degenerate critical points are of great interest, and so we make a definition.

DEFINITION 2.6. The smooth function $f : N \to \mathbb{R}$ is a **Morse function** if and only if all of the critical points of $f$ are non-degenerate.

Definition 2.6 would not be terribly useful if there were very few Morse functions. However, there are a number of theorems which tell us that Morse functions are very plentiful. We remind the reader that "almost all" means except for a set of measure zero.

THEOREM 2.7. ([**27**], p. 11) If $g$ is a $C^2$ function from an open subset $\mathcal{U}$ of $\mathbb{R}^n$ to $\mathbb{R}$, then, for almost all linear functions $L : \mathbb{R}^n \to \mathbb{R}$, the function $g + L : \mathcal{U} \to \mathbb{R}$ is a Morse function.

THEOREM 2.8. ([**26**], Theorem 6.6) Let $M$ be a smooth submanifold of $\mathbb{R}^n$, which is a closed subset of $\mathbb{R}^n$. For all $\mathbf{p} \in \mathbb{R}^n$, let $L_\mathbf{p} : M \to \mathbb{R}$ be given by $L_\mathbf{p}(\mathbf{x}) := ||\mathbf{x} - \mathbf{p}||^2$. Then, for almost all $\mathbf{p} \in \mathbb{R}^n$, $L_\mathbf{p}$ is a proper Morse function such that $M_{\leq a}$ is compact for all $a$.

COROLLARY 2.9. ([**26**], p. 36) Every smooth manifold $M$ possesses a Morse function $g : M \to \mathbb{R}$ such that $M_{\leq a}$ is compact for all $a \in \mathbb{R}$. Given such a function $g$, $M$ has the homotopy-type of a CW-complex with one cell of dimension $\lambda$ for each critical point of $g$ of index $\lambda$.

While we stated the above as a corollary to Theorem 2.8, it also strongly uses two other results: Theorem 3.5 of [**26**] and Whitney's Embedding Theorem, which tells us that any smooth manifold can be smoothly embedded as a closed subset of some Euclidean space.

---

We now wish to mention a few complex analytic results which are of importance.

THEOREM 2.10. ([**26**], p. 39-41) Suppose that $M$ is an $m$-dimensional complex analytic submanifold of $\mathbb{C}^n$. For all $\mathbf{p} \in \mathbb{C}^n$, let $L_\mathbf{p} : M \to \mathbb{R}$ be given by $L_\mathbf{p}(\mathbf{x}) := ||\mathbf{x} - \mathbf{p}||^2$. If $\mathbf{x} \in M$ is a non-degenerate critical point of $L_\mathbf{p}$, then the index of $L_\mathbf{p}$ at $\mathbf{x}$ is less than or equal to $m$.

Corollary 2.9 immediately implies:

COROLLARY 2.11. ([**26**], Theorem 7.2) If $M$ is an $m$-dimensional complex analytic submanifold of $\mathbb{C}^n$, which is a closed subset of $\mathbb{C}^n$, then $M$ has the homotopy-type of an $m$-dimensional CW-complex. In particular, $H_i(M;\ \mathbb{Z}) = 0$ for $i > m$.

Note that this result should **not** be considered obvious; $m$ is the **complex** dimension of $M$. Over the real numbers, $M$ is $2m$-dimensional, and so $m$ is frequently referred to as the **middle dimension**. Thus, the above corollary says that the homology of a complex analytic submanifold of $\mathbb{C}^n$, which is closed in $\mathbb{C}^n$, has trivial homology above the middle dimension.

The reader might hope that the corollary above would allow one to obtain nice results about compact complex manifolds; this is **not** the case. The maximum modulus principle, applied to the coordinate functions on $\mathbb{C}^n$, implies that the only compact, connected, complex submanifold of $\mathbb{C}^n$ is a point.

Suppose now that $M$ is a connected complex $m$-manifold, and that $c : M \to \mathbb{C}$ is a complex analytic function. Let $\mathbf{p} \in M$, and let $(z_1, \ldots, z_m)$ be a complex analytic coordinate system for $M$ in an open neighborhood of $\mathbf{p}$.

Analogous to our definition in the smooth case, we have:

DEFINITION 2.12. The point $\mathbf{p}$ is a **complex non-degenerate critical point of** $c$ provided that $\mathbf{p}$ is a critical point of $c$, and that the Hessian matrix $\left(\frac{\partial^2 c}{\partial z_i \partial z_j}(\mathbf{p})\right)_{i,j}$ is non-singular.

There is a complex analytic version of the Morse Lemma, Lemma 2.4:

LEMMA 2.13. *Let $\mathbf{p}$ be a complex non-degenerate critical point of $c$. Then, there is a local complex analytic coordinate system $(y_1, \ldots, y_m)$ in an open neighborhood $\mathcal{U}$ of $\mathbf{p}$, with $y_i(\mathbf{p}) = 0$, for all $i$, and such that, for all $\mathbf{x} \in \mathcal{U}$,*

$$c(\mathbf{x}) = c(\mathbf{p}) + (y_1(\mathbf{x}))^2 + (y_2(\mathbf{x}))^2 + \cdots + (y_m(\mathbf{x}))^2.$$

*In particular, the point $\mathbf{p}$ is an isolated critical point of $c$.*

PROPOSITION 2.14. *The map $c$ has a complex non-degenerate critical point at $\mathbf{p}$ if and only if $c - c(\mathbf{p})$ has an isolated critical point at $\mathbf{p}$ and the Milnor number $\mu_{c-c(\mathbf{p})}(\mathbf{p})$ equals 1.*

The first statement of the following theorem is proved in exactly the same manner as Theorem 2.7; one uses the open mapping principle for complex analytic functions to obtain the second statement.

THEOREM 2.15. *If $c$ is a complex analytic function from an open subset $\mathcal{U}$ of $\mathbb{C}^m$ to $\mathbb{C}$, then, for almost all complex linear functions $L : \mathbb{C}^m \to \mathbb{C}$, the function $c + L : \mathcal{U} \to \mathbb{C}$ has no complex degenerate critical points.*

*In addition, for all $\mathbf{x} \in \mathcal{U}$, there exists an open, dense subset $\mathcal{W}$ in $\mathrm{Hom}_{\mathbb{C}}(\mathbb{C}^m, \mathbb{C}) \cong \mathbb{C}^m$ such that, for all $L \in \mathcal{W}$, there exists an open neighborhood $\mathcal{U}' \subseteq \mathcal{U}$ of $\mathbf{x}$ such that $c + L$ has no complex degenerate critical points in $\mathcal{U}'$.*

Finally, we leave the following result as an exercise for the reader. We denote the real and imaginary parts of $c$ by $\mathrm{Re}\, c$ and $\mathrm{Im}\, c$, respectively.

EXERCISE 2.16. Show that:

(1)
$$\Sigma c \;=\; \Sigma(\mathrm{Re}\, c) \;=\; \Sigma(\mathrm{Im}\, c)$$
and that, if $c(\mathbf{p}) \neq 0$, then $\mathbf{p} \in \Sigma c$ if and only if $\mathbf{p} \in \Sigma\left(|c|^2\right)$.

(2) Suppose that $\mathbf{p}$ is a complex non-degenerate critical point of $c$. Prove that the real functions $\mathrm{Re}\, c : M \to \mathbb{R}$ and $\mathrm{Im}\, c : M \to \mathbb{R}$ each have a (real, smooth) non-degenerate critical point at $\mathbf{p}$ of index precisely equal to $m$, the complex dimension of $M$.

In addition, if $c(\mathbf{p}) \neq 0$, then prove that the real function $|c|^2 : M \to \mathbb{R}$ also has a non-degenerate critical point of index $m$ at $\mathbf{p}$.

---

Now we want to use Morse Theory to sketch the proofs of two important results: one due to Milnor (see [**28**] Theorems 6.5 and 7.2, but our statement and proof are different) and one due to Lê [**12**].

First, it will be convenient to define the **relative polar curve** of Hamm, Lê, and Teissier; see [**9**] and [**38**]. Later, we will give the relative polar curve a cycle structure, but – for now – we give the classical definition as a (reduced) analytic set.

Suppose that $\mathcal{U}$ is an open subset of $\mathbb{C}^{n+1}$ and that $f : (\mathcal{U}, \mathbf{0}) \to (\mathbb{C}, 0)$ is a complex analytic function which is not locally constant. Let $L$ denote a non-zero linear form, and let $\Sigma(f, L)$ denote the critical locus of the map $(f, L) : \mathcal{U} \to \mathbb{C}^2$.

THEOREM 2.17. ([**9**], [**38**]) For a generic choice of $L$:

(1) the analytic set
$$\Gamma_{f,L} := \overline{\Sigma(f, L) - \Sigma f}$$
is purely 1-dimensional at the origin (this allows for the case where $\mathbf{0} \notin \Gamma_{f,L}$);

(2) $\dim_{\mathbf{0}} \Gamma_{f,L} \cap V(L) \leq 0$ and $\dim_{\mathbf{0}} \Gamma_{f,L} \cap V(f) \leq 0$ (the $< 0$ cases allow for $\Gamma_{f,L} = \emptyset$);

(3) for each 1-dimensional irreducible component $C$ of $\Gamma_{f,L}$ which contains the origin, for $\mathbf{p} \in C - \{\mathbf{0}\}$, close enough to the origin, $f_{|_{V(L-L(\mathbf{p}))}}$ has an isolated critical point at $\mathbf{p}$, and
$$\mu_{\mathbf{p}}\left(f_{|_{V(L-L(\mathbf{p}))}}\right) = 1.$$

EXERCISE 2.18. Show that $\dim_{\mathbf{0}} \Gamma_{f,L} \cap V(f) \leq 0$ if and only if $\dim_{\mathbf{0}} \Gamma_{f,L} \cap V(L) \leq 0$, and that these equivalent conditions imply that $\Gamma_{f,L}$ is purely 1-dimensional at $\mathbf{0}$. (Hint: Give yourself a coordinate system on $\mathcal{U}$ that has $L$ as one of its coordinates, and parameterize the components of $\Gamma_{f,L}$.)

REMARK 2.19. Suppose that we choose (re-choose) our coordinate system $(z_0, \ldots, z_n)$ for $\mathbb{C}^{n+1}$ so that $z_0 = L$. Then, we may consider the scheme
$$V\left(\frac{\partial f}{\partial z_1}, \ldots, \frac{\partial f}{\partial z_n}\right).$$
Then $\Gamma_{f,L}$ consists of those irreducible components of this scheme which are **not** contained in $\Sigma f$. The condition that $\mu_{\mathbf{p}}\left(f_{|_{V(L-L(\mathbf{p}))}}\right) = 1$ in Item 3 of Theorem 2.17 is equivalent to saying that these irreducible components of the scheme are reduced at points other than the origin.

In light of Theorem 2.17 and Exercise 2.18, we make the following definition.

DEFINITION 2.20. If $L$ is generic enough so that $\Gamma_{f,L}$ is purely 1-dimensional at $\mathbf{0}$, then we refer to $\Gamma_{f,L}$ as the **relative polar curve** (of $f$ with respect to $L$ at the origin), and denote it by $\Gamma^1_{f,L}$ (note the superscript by the dimension). In this case, we say that **the relative polar curve exists** or, simply, that $\Gamma^1_{f,L}$ exists.

If $\dim_{\mathbf{0}} \Gamma_{f,L} \cap V(f) \leq 0$ (so that, in particular, $\Gamma^1_{f,L}$ exists), then we say that $V(L)$ (or $L$ itself) is a **Thom slice** (for $f$ at $\mathbf{0}$).

If Item 3 of Theorem 2.17 holds, then we say that the relative polar curve is **reduced**.

EXERCISE 2.21. Suppose that $\dim_{\mathbf{0}} \Sigma f = 0$. Conclude that

$$V\left(\frac{\partial f}{\partial z_1}, \dots, \frac{\partial f}{\partial z_n}\right)$$

is a purely 1-dimensional local complete intersection, equal to $\Gamma^1_{f,z_0}$ and that

$$\dim_{\mathbf{0}} \Gamma^1_{f,z_0} \cap V\left(\frac{\partial f}{\partial z_0}\right) = 0.$$

If you are familiar with intersection numbers, conclude also that

$$\mu_f(\mathbf{0}) \;=\; \dim_{\mathbb{C}} \frac{\mathbb{C}\{z_0, \dots, z_n\}}{\left\langle \frac{\partial f}{\partial z_0}, \dots, \frac{\partial f}{\partial z_n} \right\rangle} \;=\; \left(\Gamma^1_{f,z_0} \cdot V\left(\frac{\partial f}{\partial z_0}\right)\right)_{\mathbf{0}}.$$

Now we're ready to prove, modulo many technical details, two important results on Milnor fibers.

First, the classic result of Milnor:

THEOREM 2.22. ([**28**], Theorem 6.5 and 7.2) Suppose that $\dim_{\mathbf{0}} \Sigma f = 0$. Then, the Milnor fiber $F_{f,\mathbf{0}}$ is homotopy-equivalent to a bouquet of $n$-spheres, and the number of spheres in the bouquet is

$$\dim_{\mathbb{C}} \frac{\mathbb{C}\{z_0, \dots, z_n\}}{\left\langle \frac{\partial f}{\partial z_0}, \dots, \frac{\partial f}{\partial z_n} \right\rangle} \;=\; \left(\Gamma^1_{f,z_0} \cdot V\left(\frac{\partial f}{\partial z_0}\right)\right)_{\mathbf{0}}.$$

PROOF. We sketch a proof.

Recall, from Remark 1.9, that the Milnor tube $T := B_\epsilon \cap f^{-1}(\mathbb{D}_\delta)$, for $0 < \delta \ll \epsilon \ll 1$, is contractible. Select a complex number $a$ such that $0 < |a| \ll \delta$. We wish to show that $T$ is obtained from $F_{f,\mathbf{0}} = T \cap f^{-1}(a)$, up to homotopy, by attaching $\left(\Gamma^1_{f,z_0} \cdot V\left(\frac{\partial f}{\partial z_0}\right)\right)_{\mathbf{0}}$ $(n+1)$-cells.

The number $a$ is a regular value of $f$ restricted to the compact manifold with boundary $B_\epsilon$, i.e., a regular value when restricted to the open ball and when restricted to the bounding sphere. Consequently, for $0 < \eta \ll |a|$, the closed disk, $\mathbb{D}_\eta(a)$, of radius $\eta$, centered at $a$, consists of regular values, and so the restriction of $f$ to a map from $T \cap f^{-1}(\mathbb{D}_\eta(a))$ to $\mathbb{D}_\eta(a)$ is a trivial fibration; in particular, $T \cap f^{-1}(\mathbb{D}_\eta(a))$ is homotopy-equivalent to $F_{f,\mathbf{0}}$. Furthermore, $T$ is diffeomorphic to $T' := B_\epsilon \cap f^{-1}(\mathbb{D}_\delta(a))$ and, hence, $T'$ is contractible.

We wish to apply Morse Theory to $|f-a|^2$ as its value grows from $\eta$ to $\delta$. However, there is no reason for the critical points of $f-a$ to be complex non-degenerate. Thus, we assume that the coordinate $z_0$ is chosen to be a generic linear form and, for $0<|t|\ll\eta$, we consider the map $r:=|f-tz_0-a|^2$ as a map from $B_\epsilon$ to $\mathbb{R}$.

Then, $r^{-1}[0,\eta]$ is homotopy-equivalent to $F_{f,0}$, $T'':=r^{-1}[0,\delta]$ is contractible, $r$ has no critical points on $S_\epsilon$, and all of the critical points of $f-tz_0-a$ in $B^\circ_\epsilon$ are complex non-degenerate. Consequently, complex Morse Theory tells us that the contractible set $T''$ is constructed by attaching $(n+1)$-cells to $F_{f,\mathbf{0}}$. Hence, $F_{f,0}$ has the homotopy-type of a finite bouquet of $n$-spheres.

How many $n$-spheres are there? One for each critical point of $f-tz_0-a$ in $B^\circ_\epsilon$. Therefore, the number of $n$-spheres in the homotopy-type is the number of points in

$$B^\circ_\epsilon\cap V\left(\frac{\partial f}{\partial z_0}-t,\frac{\partial f}{\partial z_1},\dots,\frac{\partial f}{\partial z_n}\right) = B^\circ_\epsilon\cap\Gamma^1_{f,z_0}\cap V\left(\frac{\partial f}{\partial z_0}-t\right),$$

where $0<|t|\ll\epsilon\ll 1$. This is precisely the intersection number

$$\left(\Gamma^1_{f,z_0}\cdot V\left(\frac{\partial f}{\partial z_0}\right)\right)_{\mathbf{0}}.$$

□

Now we wish to sketch the proof of Lê's main result of [**12**] for hypersurface singularities of arbitrary dimension.

We continue to assume that $\mathcal{U}$ is an open subset of $\mathbb{C}^{n+1}$ and that $f:(\mathcal{U},\mathbf{0})\to(\mathbb{C},0)$ is a complex analytic function which is not locally constant. In order to appreciate the inductive applications of the theorem, one should note that, if $s:=\dim_0\Sigma f\geq 1$, then, for generic $z_0$, $\dim_{\mathbf{0}}\Sigma(f_{|_{V(z_0)}})=s-1$.

THEOREM 2.23. (Lê, [**12**]) Suppose that $\dim_0\Sigma f$ is arbitrary, Then, for a generic non-zero linear form $z_0$, $F_{f,\mathbf{0}}$ is obtained up to homotopy from $F_{f_{|_{V(z_0)}},\mathbf{0}}$ by attaching $\left(\Gamma^1_{f,z_0}\cdot V(f)\right)_{\mathbf{0}}$ $n$-cells.

PROOF. We once again assume that $z_0$ is a coordinate. The main technical issue, which we will not prove, is that one needs to know that one may use a disk times a ball, rather than a ball itself, when defining the Milnor fiber. More precisely, we shall assume that, up to homotopy, $F_{f,0}$ is given by

$$F'_{f,0}:=\left(\mathbb{D}_\delta\times B^{2n}_\epsilon\right)\cap f^{-1}(a),$$

where $0<|a|\ll\delta\ll\epsilon\ll 1$. Note that $F_{f_{|_{V(z_0)}},\mathbf{0}}=V(z_0)\cap F'_{f,0}$.

The idea of the proof is simple: one considers $r:=|z_0|^2$ on $F'_{f,0}$. As in our previous proof, there is the problem that $r$ has a critical point at each point where $z_0=0$. But, again, as in our previous proof, 0 is a regular value of $z_0$ restricted to $F'_{f,0}$. Hence, for $0<\eta\ll|a|$,

$$r^{-1}[0,\eta]\cap F'_{f,0}\cong F_{f_{|_{V(z_0)}},\mathbf{0}}\times\mathbb{D}_\eta.$$

One also needs to prove a little lemma that, for $a$, $\delta$, and $\epsilon$ as we have chosen them, $z_0$ itself has no critical points on $\left(\mathbb{D}_\delta\times S^{2n-1}_\epsilon\right)\cap f^{-1}(a)$.

Now, one lets the value of $r$ grow from $\eta$ to $\delta$. Note that the critical points of $z_0$ restricted to $F'_{f,0}$ occur precisely at points in

$$V\left(\frac{\partial f}{\partial z_1}, \ldots, \frac{\partial f}{\partial z_n}\right) \cap V(f-a) \;=\; \Gamma^1_{f,z_0} \cap V(f-a),$$

and, by the choice of generic $z_0$ all of these critical points will be complex non-degenerate. The result follows. □

REMARK 2.24. Since attaching $n$-cells does not affect connectivity in dimensions $\leq n-2$, by inductively applying the above attaching theorem, one obtains that the Milnor fiber of a hypersurface in $\mathbb{C}^{n+1}$ with a critical locus of dimension $s$ is $(n-s-1)$-connected. Thus, one recovers the main result of [**11**].

It is also worth noting that Lê's attaching theorem leads to a Lefschetz hyperplane result. It tells one that, for $k \leq n-2$, $H_k(F_{f,\mathbf{0}}) \cong H_k(F_{f_{|V(z_0)},\mathbf{0}})$ and that there is an exact sequence

$$0 \to H_n(F_{f,\mathbf{0}}) \to \mathbb{Z}^\tau \to H_{n-1}(F_{f_{|V(z_0)},\mathbf{0}}) \to H_{n-1}(F_{f,\mathbf{0}}) \to 0,$$

where $\tau = \left(\Gamma^1_{f,z_0} \cdot V(f)\right)_{\mathbf{0}}$.

Theorem 2.23 seems to have been the first theorem about hypersurface singularities of arbitrary dimension that actually allowed for algebraic calculations. By induction, the theorem yields the Euler characteristic of the Milnor fiber and also puts bounds on the Betti numbers, such as $b_n(F_{f,\mathbf{0}}) \leq \left(\Gamma^1_{f,z_0} \cdot V(f)\right)_{\mathbf{0}}$.

However, if $\dim_{\mathbf{0}} \Sigma f = 0$, then $\left(\Gamma^1_{f,z_0} \cdot V(f)\right)_{\mathbf{0}} > \mu_f(\mathbf{0})$ (this is **not** obvious), and so the question is: are there numbers that we can calculate that are "better" than inductive versions of $\left(\Gamma^1_{f,z_0} \cdot V(f)\right)_{\mathbf{0}}$? We want numbers that are actual generalizations of the Milnor number of an isolated critical point.

Our answer to this is: yes – the Lê numbers, as we shall see in the next lecture.

## 3. Lecture 3: Proper intersection theory and Lê numbers

Given a hypersurface $V(f)$ and a point $\mathbf{p} \in V(f)$, if $\dim_{\mathbf{p}} \Sigma f = 0$, then the Milnor number of $f$ at $\mathbf{p}$ provides a great deal of information about the local ambient topology of $V(f)$ at $\mathbf{p}$.

But now, suppose that $s := \dim_{\mathbf{p}} \Sigma f > 0$. What data should replace/generalize a number associated to a point? For instance, suppose $s = 1$. A reasonable hope for "good data" to associate to $f$ at $\mathbf{p}$ would be to assign a number to each irreducible component curve of $\Sigma f$ at $\mathbf{p}$, and also assign a number to $\mathbf{p}$. More generally, if $s$ is arbitrary, one could hope to produce effectively calculable topologically important data which consists of analytic sets of dimensions 0 through $s$, with numbers assigned to each irreducible component.

This is what the **Lê cycles**, $\Lambda^s_{f,\mathbf{z}}, \ldots, \Lambda^1_{f,\mathbf{z}}, \Lambda^0_{f,\mathbf{z}}$ ([**20**], [**21**], [**23**]), give you.

We briefly need to discuss what analytic cycles are, and give a few basic properties. Then we will define the Lê cycles and the associated **Lê numbers**, and calculate some examples.

We need to emphasize that, in this lecture and the last one, when we write $V(\alpha)$, where $\alpha$ is an ideal (actually a coherent sheaf of ideals in $\mathcal{O}_{\mathcal{U}}$) we mean $V(\alpha)$ as a scheme, not merely an analytic set, i.e., we keep in mind what the defining ideal is.

---

We restrict ourselves to the case of analytic cycles in an open subset $\mathcal{U}$ of some affine space $\mathbb{C}^N$. An *analytic cycle* in $\mathcal{U}$ is a formal sum $\sum m_V[V]$, where the $V$'s are (distinct) irreducible analytic subsets of $\mathcal{U}$, the $m_V$'s are integers, and the collection $\{V\}$ is a locally finite collection of subsets of $\mathcal{U}$. As a cycle is a locally finite sum, and as we will normally be concentrating on the germ of an analytic space at a point, usually we can safely assume that a cycle is actually a finite formal sum. If $C = \sum m_V[V]$, we write $C_V$ for the coefficient of $V$ in $C$, i.e., $C_V = m_V$.

For clarification of what structure we are considering, we shall at times enclose cycles in square brackets, $[\ ]$ , and analytic sets in a pair of vertical lines, $||$; with this notation,

$$\left|\sum m_V[V]\right| = \bigcup_{m_V\neq 0} V.$$

Occasionally, when the notation becomes cumbersome, we shall simply state explicitly whether we are considering V as a scheme, a cycle, or a set.

Essentially all of the cycles that we will use will be of the form $\sum m_V[V]$, where all of the $V$'s have the same dimension $d$ (we say that the cycle is of pure dimension $d$) and $m_V \geq 0$ for all $V$ (a non-negative cycle).

We need to consider not necessarily reduced complex analytic spaces (analytic schemes) $(X, \mathcal{O}_X)$ (in the sense of [**8**] and [**7**]), where $X \subseteq \mathcal{U}$. Given an analytic space, $(X, \mathcal{O}_X)$, we wish to define the cycle associated to $(X, \mathcal{O}_X)$. The cycle is defined to be the sum of the irreducible components, $V$, each one with a coefficient $m_V$, which is its geometric multiplicity, i.e., $m_V$ is how many times that component should be thought of as being there.

In the algebraic context, this is given by Fulton in section 1.5 of [**6**] as

$$[X] := \sum m_V[V],$$

where the $V$'s run over all the irreducible components of $X$, and $m_V$ equals the length of the ring $\mathcal{O}_{V,X}$, the Artinian local ring of $X$ along $V$. In the analytic context, we wish to use the same definition, but we must be more careful in defining the $m_V$. Define $m_V$ as follows. Take a point $\mathbf{p}$ in $V$. The germ of $V$ at $\mathbf{p}$ breaks up into irreducible germ components $(V_{\mathbf{p}})_i$. Take any one of the $(V_{\mathbf{p}})_i$ and let $m_V$ equal the Artinian local length of the ring $(\mathcal{O}_{X,\mathbf{p}})_{(V_{\mathbf{p}})_i}$ (that is, the local ring of $X$ at $\mathbf{p}$ localized at the prime corresponding to $(V_{\mathbf{p}})_i$). This number is independent of the point $\mathbf{p}$ in $V$ and the choice of $(V_{\mathbf{p}})_i$.

Note that, in particular, if $\mathbf{p}$ is an isolated point in $V(\alpha)$, then the coefficient of $\mathbf{p}$ (really $\{\mathbf{p}\}$) in $[V(\alpha)]$ is given by

$$[V(\alpha)]_{\mathbf{p}} = \dim_{\mathbb{C}} \frac{\mathbb{C}\{z_0 - p_0, \dots, z_n - p_n\}}{\alpha}.$$

Two cycles $C := \sum m_V[V]$ and $D := \sum m_W[W]$, of pure dimension $a$ and $b$, respectively, in $\mathcal{U}$ are said to **intersect properly** if and only if, for all $V$ and $W$, $\dim(V \cap W) = a + b - N$ (recall that $N$ is the dimension of $\mathcal{U}$).

When $C$ and $D$ intersect properly, there is a well-defined **intersection product** which yields an **intersection cycle** (**not** a rational equivalence class); this intersection cycle is denoted by $(C \cdot D; \mathcal{U})$ or simply $C \cdot D$ if the ambient complex manifold is clear. See Fulton [**6**], Section 8.2 and pages 207-208. Recalling our earlier notation, if $V$ is an irreducible component of $|C| \cap |D|$, then $(C \cdot D)_V$ is the coefficient of $V$ in the intersection cycle. In particular, if $C$ and $D$ intersect properly in an isolated point $\mathbf{p}$, then $(C \cdot D)_{\mathbf{p}}$ is called the **intersection number of $C$ and $D$ at $\mathbf{p}$.**

We will now give some properties of intersection cycles and numbers. All of these can found in, or easily derived from, [**6**]. We assume that all intersections written below are proper.

(1) Suppose that $Y$ and $Z$ are irreducible analytic sets, and that $V$ is an irreducible component of their proper intersection. Then, $([Y]\cdot[Z])_V \geq 1$, with equality holding if and only if, along a generic subset of $V$, $Y$ and $Z$ are smooth and intersect transversely.

(2) If $f, g \in \mathcal{O}_U$, then $[V(fg)] = [V(f)] + [V(g)]$; in particular, $[V(f^m)] = m[V(f)]$.

(3) $C \cdot D = D \cdot C$, $(C \cdot D) \cdot E = C \cdot (D \cdot E)$, and

$$C \cdot \sum_i m_i D_i = \sum_i m_i (C \cdot D_i).$$

(4) **Locality**: Suppose that $Z$ is a component of $|C \cdot D|$ and that $\mathcal{W}$ is an open subset of $\mathcal{U}$ such that $Z \cap \mathcal{W} \neq \emptyset$ and $Z \cap \mathcal{W}$ is irreducible in $\mathcal{W}$. Then,

$$(C \cap \mathcal{W} \ \cdot \ D \cap \mathcal{W};\ \mathcal{W})_{Z\cap\mathcal{W}} = (C \cdot\ D;\ \mathcal{U})_Z$$

(5) If $f$ contains no isolated or embedded components of $V(\alpha)$, then $V(\alpha) \cdot V(f) = V(\alpha + \langle f \rangle)$. In particular, if $f_1, f_2, \dots f_k$ is a regular sequence, then

$$V(f_1) \cdot V(f_2) \cdot \dots \cdot V(f_k) = [V(f_1, f_2, \dots, f_k)].$$

(6) **Reduction to the normal slice**: Let $Z$ be a $d$-dimensional component of $C \cdot D$. Let $\mathbf{p}$ be a smooth point of $Z$. Let $M$ be a **normal slice** to $Z$ at $\mathbf{p}$, i.e., let $M$ be a complex submanifold of $\mathcal{U}$ which transversely intersects $Z$ in the isolated point $\mathbf{p}$. Furthermore, assume that $M$ transversely the smooth parts of $|C|$ and $|D|$ in an open neighborhood of $\mathbf{p}$. Then,

$$(C \cdot\ D;\ \mathcal{U})_Z = ((C \cdot M) \ \cdot\ (D \cdot M);\ M)_{\mathbf{p}}.$$

(7) **Conservation of number**: Let $E$ be a purely $k$-dimensional cycle in $\mathcal{U}$. Let $g_1(\mathbf{z}, t)$, $g_2(\mathbf{z}, t)$, $\dots$, $g_k(\mathbf{z}, t)$ be in $\mathcal{O}_{\mathcal{U}\times\mathbb{D}^\circ}$, for some open disk $\mathbb{D}^\circ$ containing the origin in $\mathbb{C}$. For fixed $t \in \mathbb{D}^\circ$, let $C_t$ be the cycle $[V(g_1(\mathbf{z}, t), g_2(\mathbf{z}, t), \dots, g_k(\mathbf{z}, t))]$ in $\mathcal{U}$. Assume that $E$ and $C_0$ intersect properly in the isolated point $\mathbf{p}$.

Then,

$$(E\cdot C_0)_{\mathbf p} \;=\; \sum_{\mathbf q\in B^\circ_\epsilon(\mathbf p)\cap|E|\cap|C_t|} (E\cdot C_t)_{\mathbf q},$$

for $|t|\ll\epsilon\ll 1$.

(8) Suppose that $Z$ is a curve which is irreducible at $\mathbf p$. Let $\mathbf r(t)$ be a reduced parametrization of the germ of $Z$ at $\mathbf p$ such that $\mathbf r(0)=\mathbf p$. (Here, by reduced, we mean that if $\mathbf r(t)=\mathbf p+\mathbf a_1t+\mathbf a_2t^2+\cdots$, then the exponents of the non-zero powers of $t$ with non-zero coefficients have no common factor, other than 1.) Suppose that $f\in\mathcal O_{\mathcal U}$ is such that that $V(f)$ intersects $Z$ in the isolated point $\mathbf p$. Then,

$$(Z\cdot V(f))_{\mathbf p} \;=\; \operatorname{mult}_t f(\mathbf r(t)),$$

that is, the exponent of the lowest power of $t$ that appears in $f(\mathbf r(t))$.

Exercise 3.1. Use the last property of intersection numbers above to show the following:

(a) Suppose that $C=\sum_W m_W[W]$ is a purely 1-dimensional cycle, and that $C$ properly intersects $V(f)$ and $V(g)$ at a point $\mathbf p$. Suppose that, for all $W$, $(W\cdot V(f))_{\mathbf p}<(W\cdot V(g))_{\mathbf p}$. Then, $C$ properly intersects $V(f+g)$ at $\mathbf p$ and $(C\cdot V(f+g))_{\mathbf p}=(C\cdot V(f))_{\mathbf p}$.

(b) Suppose that $C$ is a purely 1-dimensional cycle, that $|C|\subseteq V(f)$, and that $C$ properly intersects $V(g)$ at a point $\mathbf p$. Then, $C$ properly intersects $V(f+g)$ at $\mathbf p$ and $(C\cdot V(f+g))_{\mathbf p}=(C\cdot V(g))_{\mathbf 0}$.

---

We are now (almost) ready to define the Lê cycles and Lê numbers. However, first, we need a piece of notation and we need to define the **(relative) polar cycles**.

Suppose once again that $\mathcal U$ is an open neighborhood of the origin in $\mathbb C^{n+1}$, and that $f:(\mathcal U,\mathbf 0)\to(\mathbb C,0)$ is a complex analytic function, which is not locally constant. We use coordinates $\mathbf z=(z_0,\dots,z_n)$ on $\mathcal U$. We will at times assume, after possibly a linear change of coordinates, that $\mathbf z$ is generic in some sense with respect to $f$ at $\mathbf 0$. As before, we let $s:=\dim_{\mathbf 0}\Sigma f$.

If $C=\sum_V m_V[V]$ is a cycle in $\mathcal U$ and $Z$ is an analytic subset of $\mathcal U$, then we let

$$C_{\subseteq Z} \;=\; \sum_{V\subseteq Z} m_V[V] \qquad\text{and}\qquad C_{\not\subseteq Z} \;=\; \sum_{V\not\subseteq Z} m_V[V].$$

Definition 3.2. For $0\le k\le n+1$, we define the **$k$-th polar cycle of $f$ with respect to $\mathbf z$** to be

$$\Gamma^k_{f,\mathbf z} \;:=\; \left[V\left(\frac{\partial f}{\partial z_k},\frac{\partial f}{\partial z_{k+1}},\dots,\frac{\partial f}{\partial z_n}\right)\right]_{\not\subseteq\Sigma f}.$$

Here, by $\Gamma^{n+1}_{f,\mathbf z}$, we mean simply $[\mathcal U]$. Also, note that $\Gamma^0_{f,\mathbf z}=0$.

As a set, this definition of $\Gamma^1_{f,\mathbf z}$ agrees with our earlier definition of $\Gamma^1_{f,z_0}$; however, now we give this relative polar curve a cycle structure. If $z_0$ is generic enough, then all of the coefficients of components of the cycle $\Gamma^1_{f,\mathbf z}$ will be 1, but we typically do not want to assume this level of genericity.

Also note that every irreducible component of $V\left(\frac{\partial f}{\partial z_k}, \frac{\partial f}{\partial z_{k+1}}, \dots, \frac{\partial f}{\partial z_n}\right)$ necessarily has dimension at least $k$; hence, for $k \geq s+1$, there can be no components contained in $\Sigma f$ near the origin. Therefore, near the origin, for $k \geq s+1$,

$$\Gamma^k_{f,\mathbf{z}} := \left[V\left(\frac{\partial f}{\partial z_k}, \frac{\partial f}{\partial z_{k+1}}, \dots, \frac{\partial f}{\partial z_n}\right)\right].$$

EXERCISE 3.3. In this exercise, you will be asked to prove what we generally refer to as the *Teissier trick*, since it was first proved by Teissier in [**38**] in the case of isolated critical points, but the proof is the same for arbitrary $s$.

Suppose that $\dim_{\mathbf{0}} \Gamma^1_{f,\mathbf{z}} \cap V(f) \leq 0$. Then, $\dim_{\mathbf{0}} \Gamma^1_{f,\mathbf{z}} \cap V(z_0) \leq 0$, $\dim_{\mathbf{0}} \Gamma^1_{f,\mathbf{z}} \cap V\left(\frac{\partial f}{\partial z_0}\right) \leq 0$, and

$$\left(\Gamma^1_{f,\mathbf{z}} \cdot V(f)\right)_{\mathbf{0}} = \left(\Gamma^1_{f,\mathbf{z}} \cdot V(z_0)\right)_{\mathbf{0}} + \left(\Gamma^1_{f,\mathbf{z}} \cdot V\left(\frac{\partial f}{\partial z_0}\right)\right)_{\mathbf{0}}.$$

(Hint: Parameterize the irreducible components of the polar curve and use the Chain Rule for differentiation.)

In particular, if $\dim_{\mathbf{0}} \Gamma^1_{f,\mathbf{z}} \cap V(f) \leq 0$ and $\Gamma^1_{f,\mathbf{z}} \neq 0$, then

$$\left(\Gamma^1_{f,\mathbf{z}} \cdot V(f)\right)_{\mathbf{0}} > \left(\Gamma^1_{f,\mathbf{z}} \cdot V\left(\frac{\partial f}{\partial z_0}\right)\right)_{\mathbf{0}}.$$

EXERCISE 3.4. Suppose that $k \leq n$, and that $\Gamma^{k+1}_{f,\mathbf{z}}$ is purely $(k+1)$-dimensional and is intersected properly by $V\left(\frac{\partial f}{\partial z_k}\right)$.

Prove that

$$\left(\Gamma^{k+1}_{f,\mathbf{z}} \cdot V\left(\frac{\partial f}{\partial z_k}\right)\right)_{\not\subseteq \Sigma f} = \Gamma^k_{f,\mathbf{z}}.$$

DEFINITION 3.5. Suppose that $k \leq n$, and that $\Gamma^{k+1}_{f,\mathbf{z}}$ is purely $(k+1)$-dimensional and is intersected properly by $V\left(\frac{\partial f}{\partial z_k}\right)$.

Then we say that the **$k$-dimensional Lê cycle exists** and define it to be

$$\Lambda^k_{f,\mathbf{z}} := \left(\Gamma^{k+1}_{f,\mathbf{z}} \cdot V\left(\frac{\partial f}{\partial z_k}\right)\right)_{\subseteq \Sigma f}.$$

Hence,

$$\Gamma^{k+1}_{f,\mathbf{z}} \cdot V\left(\frac{\partial f}{\partial z_k}\right) = \Gamma^k_{f,\mathbf{z}} + \Lambda^k_{f,\mathbf{z}}.$$

REMARK 3.6. Note that, if $\Lambda^s_{f,\mathbf{z}}$ exists, then

$$\Gamma^{s+1}_{f,\mathbf{z}} = V\left(\frac{\partial f}{\partial z_{s+1}}, \dots, \frac{\partial f}{\partial z_n}\right)$$

is purely $(s+1)$-dimensional and, for all $k \geq s$, $\Gamma^{k+1}_{f,\mathbf{z}}$ is purely $(k+1)$-dimensional and is intersected properly by $V\left(\frac{\partial f}{\partial z_k}\right)$.

The point is that saying that $\Lambda^s_{f,\mathbf{z}}$ exists implies that, for $s+1 \leq k \leq n$, $\Lambda^k_{f,\mathbf{z}}$ exists and is 0.

Furthermore, you should note that if $\Lambda^k_{f,\mathbf{z}}$ exists for all $k \leq s$, then each $\Lambda^k_{f,\mathbf{z}}$ is purely $k$-dimensional and $\Sigma f = \bigcup_{k \leq s} \left|\Lambda^k_{f,\mathbf{z}}\right|$.

EXERCISE 3.7. We wish to use intersection cycles to quickly show that the Milnor number is *upper-semicontinuous* in a family.

(1) Suppose that $\dim_{\mathbf{0}} \Sigma(f_{|_{V(z_0)}}) \leq 0$. Show that

$$\mu_{\mathbf{0}}(f_{|_{V(z_0)}}) \ = \ (\Gamma^1_{f,\mathbf{z}} \cdot V(z_0))_{\mathbf{0}} \ + \ (\Lambda^1_{f,\mathbf{z}} \cdot V(z_0))_{\mathbf{0}}.$$

(2) Suppose that we have a complex analytic function $F : (\mathbb{D}^\circ \times \mathcal{U}, \mathbb{D}^\circ \times \{\mathbf{0}\}) \to (\mathbb{C}, 0)$. For each $t \in \mathbb{D}^\circ$, define $f_t : (\mathcal{U}, \mathbf{0}) \to (\mathbb{C}, 0)$ by $f_t(\mathbf{z}) := F(t, \mathbf{z})$, and assume that $\dim_{\mathbf{0}} \Sigma f_t = 0$. Thus, $f_t$ defines a one-parameter family of isolated singularities.

Show, for all $t$ such that $|t|$ is sufficiently small, that $\mu_{\mathbf{0}}(f_0) \geq \mu_{\mathbf{0}}(f_t)$, with equality if and only $\Gamma^1_{f,\mathbf{z}} = 0$ and $\mathbb{D}^\circ \times \mathcal{U}$ is the only component of $\Sigma F$ (near $\mathbf{0}$).

DEFINITION 3.8. Suppose that, for all $k$ such that $0 \leq k \leq n$, $\Lambda^k_{f,\mathbf{z}}$ exists, and

$$\dim_{\mathbf{0}} \Lambda^k_{f,\mathbf{z}} \cap V(z_0, \ldots, z_{k-1}) \leq 0 \qquad \text{and} \qquad \dim_{\mathbf{0}} \Gamma^k_{f,\mathbf{z}} \cap V(z_0, \ldots, z_{k-1}) \leq 0.$$

Then, we say that the **Lê numbers** and **polar numbers** of $f$ with respect to **z exist** in a neighborhood of $\mathbf{0}$ and define them, respectively, for $0 \leq k \leq n$ and for each $\mathbf{p} = (p_0, \ldots, p_n)$ near the origin, to be

$$\lambda^k_{f,\mathbf{z}}(\mathbf{p}) := \left(\Lambda^k_{f,\mathbf{z}} \cdot V(z_0 - p_0, \ldots, z_{k-1} - p_{k-1})\right)_{\mathbf{p}}$$

and

$$\gamma^k_{f,\mathbf{z}}(\mathbf{p}) := \left(\Gamma^k_{f,\mathbf{z}} \cdot V(z_0 - p_0, \ldots, z_{k-1} - p_{k-1})\right)_{\mathbf{p}}.$$

REMARK 3.9. It is the Lê numbers which will serve as our generalization of the Milnor of an isolated critical point.

However, the existence of the polar numbers tells us that our coordinates are generic enough for many of our results to be true. In general, the condition that we will require of our coordinates – which is satisfied generically – will be that the Lê and polar numbers, $\lambda^k_{f,\mathbf{z}}(\mathbf{0})$ and $\gamma^k_{f,\mathbf{z}}(\mathbf{0})$, exist for $1 \leq k \leq s$. (The existence when $k = 0$ is automatic.) Note that when $s = 0$, there is no requirement.

EXERCISE 3.10. Suppose that $s = 1$. Show that the condition that $\lambda^1_{f,\mathbf{z}}(\mathbf{0})$ and $\gamma^1_{f,\mathbf{z}}(\mathbf{0})$ exist is equivalent to requiring $\dim_{\mathbf{0}} \Sigma(f_{|_{V(z_0)}}) = 0$.

Now we wish to look at three examples of Lê cycle and Lê number calculations.

EXAMPLE 3.11. Suppose that $s = 0$. Then, regardless of the coordinate system $\mathbf{z}$, the only possibly non-zero Lê number is $\lambda^0_{f,\mathbf{z}}(\mathbf{0})$. Moreover, as $V\left(\frac{\partial f}{\partial z_0}, \frac{\partial f}{\partial z_1}, \dots, \frac{\partial f}{\partial z_n}\right)$ is 0-dimensional, $\Gamma^1_{f,\mathbf{z}} = V\left(\frac{\partial f}{\partial z_1}, \dots, \frac{\partial f}{\partial z_n}\right)$ is a 1-dimensional complete intersection, and so has no components contained in $\Sigma f$ and has no embedded components.

Therefore,

$$\lambda^0_{f,\mathbf{z}}(\mathbf{0}) = \left(\Gamma^1_{f,\mathbf{z}} \cdot V\left(\frac{\partial f}{\partial z_0}\right)\right)_{\mathbf{0}} = \left(V\left(\frac{\partial f}{\partial z_1}, \dots, \frac{\partial f}{\partial z_n}\right) \cdot V\left(\frac{\partial f}{\partial z_0}\right)\right)_{\mathbf{0}} =$$
$$\left(V\left(\frac{\partial f}{\partial z_0}, \dots, \frac{\partial f}{\partial z_n}\right)\right)_{\mathbf{0}} = \text{the Milnor number of } f \text{ at } \mathbf{0}.$$

EXAMPLE 3.12. Let $f = y^2 - x^a - tx^b$, where $a > b > 1$. We fix the coordinate system $(t, x, y)$ and will suppress any further reference to it.

$$\Sigma f = V(-x^b,\ -ax^{a-1} - btx^{b-1},\ 2y) = V(x, y).$$
$$\Gamma^2_f = V\left(\frac{\partial f}{\partial y}\right) = V(2y) = V(y).$$
$$\Gamma^2_f \cdot V\left(\frac{\partial f}{\partial x}\right) = V(y) \cdot V(-ax^{a-1} - btx^{b-1}) = V(y) \cdot (V(-ax^{a-b} - bt) + V(x^{b-1})) =$$
$$V(-ax^{a-b} - bt, y) + (b-1)V(x, y) = \Gamma^1_f + \Lambda^1_f.$$
$$\Gamma^1_f \cdot V\left(\frac{\partial f}{\partial t}\right) = V(-ax^{a-b} - bt, y) \cdot V(-x^b) = bV(t, x, y) = b[\mathbf{0}] = \Lambda^0_f.$$

Thus, $\lambda^0_f(\mathbf{0}) = b$ and $\lambda^1_f(\mathbf{0}) = b - 1$.

Notice that the exponent $a$ does not appear; this is actually good, for $f = y^2 - x^a - tx^b = y^2 - x^b(x^{a-b} - t)$ which, after an analytic coordinate change at the origin, equals $y^2 - x^b u$.

EXAMPLE 3.13. Let $f = y^2 - x^3 - (u^2 + v^2 + w^2)x^2$ and fix the coordinates $(u, v, w, x, y)$.

$$\Sigma f = V(-2ux^2,\ -2vx^2,\ -2wx^2,\ -3x^2 - 2x(u^2 + v^2 + w^2),\ 2y) = V(x, y).$$

As $\Sigma f$ is three-dimensional, we begin our calculation with $\Gamma^4_f$.

$$\Gamma^4_f = V(-2y) = V(y).$$

$$\Gamma^4_f \cdot V\left(\frac{\partial f}{\partial x}\right) = V(y) \cdot V(-3x^2 - 2x(u^2 + v^2 + w^2)) =$$
$$V(-3x - 2(u^2 + v^2 + w^2), y) + V(x, y) = \Gamma^3_f + \Lambda^3_f.$$

$$\Gamma^3_f \cdot V\left(\frac{\partial f}{\partial w}\right) = V(-3x - 2(u^2 + v^2 + w^2), y) \cdot V(-2wx^2) =$$
$$V(-3x - 2(u^2 + v^2), w, y) + 2V(u^2 + v^2 + w^2, x, y) = \Gamma^2_f + \Lambda^2_f.$$

$$\Gamma^2_f \cdot V\left(\frac{\partial f}{\partial v}\right) = V(-3x - 2(u^2+v^2), w, y) \cdot V(-2vx^2) =$$
$$V(-3x - 2u^2, v, w, y) + 2V(u^2+v^2, w, x, y) = \Gamma^1_f + \Lambda^1_f.$$

$$\Gamma^1_f \cdot V\left(\frac{\partial f}{\partial u}\right) = V(-3x - 2u^2, v, w, y) \cdot V(-2ux^2) =$$
$$V(u, v, w, x, y) + 2V(u^2, v, w, x, y) = 5[\mathbf{0}] = \Lambda^0_f.$$

Hence, $\Lambda^3_f = V(x,y)$, $\Lambda^2_f = 2V(u^2+v^2+w^2, x, y)$ = a cone (as a set), $\Lambda^1_f = 2V(u^2+v^2, w, x, y)$, and $\Lambda^0_f = 5[\mathbf{0}]$. Thus, at the origin, $\lambda^3_f = 1$, $\lambda^2_f = 4$, $\lambda^1_f = 4$, and $\lambda^0_f = 5$.

Note that $\Lambda^1_f$ depends on the choice of coordinates - for, by symmetry, if we re-ordered $u, v$, and $w$, then $\Lambda^1_f$ would change correspondingly. Moreover, one can check that this is a generic problem.

Such "non-fixed" Lê cycles arise from the absolute polar varieties of Lê and Teissier (see [**15**], [**39**], [**40**]) of the higher-dimensional Lê cycles. For instance, in the present case, $\Lambda^2_f$ is a cone, and its one-dimensional polar variety varies with the choice of coordinates, but generically always consists of two lines; this is the case for $\Lambda^1_f$ as well. Though the Lê cycles are not even generically fixed, the Lê numbers are, of course, generically independent of the coordinates.

---

Of course, you should be asking yourself: what does the calculation of the Lê numbers tell us? We shall discuss this in the next lecture.

## 4. Lecture 4: Properties of Lê numbers and vanishing cycles

Now that we know what Lê cycles and Lê numbers are, the question is: what good are they?

Throughout this section, we will be in our usual set-up. We let $\mathcal{U}$ be an open neighborhood of the origin in $\mathbb{C}^{n+1}$, $f : (\mathcal{U}, \mathbf{0}) \to (\mathbb{C}, 0)$ is a complex analytic function which is not locally constant, $s := \dim_0 \Sigma f$, and $\mathbf{z} = (z_0, \ldots, z_n)$ is a coordinate system on $\mathcal{U}$.

**We shall also assume throughout this section, for all $k \leq s$, that $\lambda^k_{f,\mathbf{z}}(\mathbf{0})$ and $\gamma^k_{f,\mathbf{z}}(\mathbf{0})$ exist. We assume that $\mathcal{U}$ is chosen (or re-chosen) that $\dim \Sigma f = s$ ("globally" in $\mathcal{U}$) and that $\lambda^k_{f,\mathbf{z}}(\mathbf{p})$ and $\gamma^k_{f,\mathbf{z}}(\mathbf{p})$ exist for all $k \leq s$ and $\mathbf{p} \in \mathcal{U}$.**

All of the results on Lê numbers given here can be found in [**23**]. However, we have recently replaced our previous assumptions with coordinates being *pre-polar* with the condition that Lê numbers and polar numbers exist (as assumed above).

Let us start with the generalization of Milnor's result that the Milnor fiber at an isolated critical point has the homotopy-type of a bouquet of spheres.

THEOREM 4.1. The Milnor fiber $F_{f,\mathbf{0}}$ has the homotopy-type obtained by beginning with a point and successively attaching $\lambda^{s-k}_{f,\mathbf{z}}(\mathbf{0})$ $(n-s+k)$-cells for $0 \leq k \leq s$.

In particular, there is a chain complex

$$0 \to \mathbb{Z}^{\lambda^s_{f,\mathbf{z}}(\mathbf{0})} \to \mathbb{Z}^{\lambda^{s-1}_{f,\mathbf{z}}(\mathbf{0})} \to \cdots \to \mathbb{Z}^{\lambda^0_{f,\mathbf{z}}(\mathbf{0})} \to 0$$

whose cohomology at the $\lambda^k_{f,\mathbf{z}}(\mathbf{0})$ term is isomorphic to the reduced integral cohomology $\widetilde{H}^{n-k}(F_{f,\mathbf{0}})$.

Thus, the Euler characteristic of the Milnor fiber is given by

$$\chi(F_{f,\mathbf{0}}) = 1 + \sum_{0\leq k\leq s} (-1)^{n-k}\lambda^k_{f,\mathbf{z}}(\mathbf{0}).$$

EXERCISE 4.2. Let us go back and see what this tells us about our previous examples.

(1) Look back at Example 3.12. We calculated that $\lambda^0_{f,\mathbf{z}}(\mathbf{0}) = b$ and $\lambda^1_{f,\mathbf{z}}(\mathbf{0}) = b-1$. Determine precisely the homology/cohomology of $F_{f,\mathbf{0}}$. Compare this with what Theorem 4.1 tells us.

(2) Look back at Example 3.13. What is the Euler characteristic of the Milnor fiber at the origin? What upper-bounds do you obtain for the ranks of $H^1(F_{f,\mathbf{0}})$ and $H^4(F_{f,\mathbf{0}})$?

In a recent paper with Lê [**16**], we showed the upper-bound of $\lambda^s_{f,\mathbf{z}}(\mathbf{0})$ on the rank of $\widetilde{H}^{n-s}(F_{f,\mathbf{0}})$ is obtained only in trivial cases, as given in the following theorem:

THEOREM 4.3. Suppose that the rank of $\widetilde{H}^{n-s}(F_{f,\mathbf{0}})$ is $\lambda^s_{f,\mathbf{z}}(\mathbf{0})$. Then, near $\mathbf{0}$, the critical locus $\Sigma f$ is itself smooth, and $\Lambda^k_{f,\mathbf{z}} = 0$ for $0 \leq k \leq s-1$.

This is equivalent saying that $f$ defines a family, parameterized by $\Sigma f$, of isolated hypersurface singularities with constant Milnor number.

Now we want to look at the extent to which the constancy of the Lê numbers in a family controls the local, ambient topology in the family.

THEOREM 4.4. Suppose that we have a complex analytic function $F : (\mathbb{D}^\circ \times \mathcal{U}, \mathbb{D}^\circ \times \{\mathbf{0}\}) \to (\mathbb{C}, 0)$. For each $t \in \mathbb{D}^\circ$, define $f_t : (\mathcal{U}, \mathbf{0}) \to (\mathbb{C}, 0)$ by $f_t(\mathbf{z}) := F(t, \mathbf{z})$, and let $s = \dim_{\mathbf{0}} \Sigma f_0$. Suppose that the coordinates $\mathbf{z}$ are chosen so that, for all $t \in \mathbb{D}^\circ$, for all $k$ such that $k \leq s$, $\lambda^k_{f_t,\mathbf{z}}(\mathbf{0})$ and $\gamma^k_{f_t,\mathbf{z}}(\mathbf{0})$ exist, and assume that, for each $k$, the value of $\lambda^k_{f_t,\mathbf{z}}(\mathbf{0})$ is constant as a function of $t$.

Then,

(1) The pair $(\mathbb{D}^\circ \times \mathcal{U} - \Sigma f,\ \mathbb{D}^\circ \times \{\mathbf{0}\})$ satisfies Thom's $a_f$ condition at $\mathbf{0}$, i.e., all of the limiting tangent spaces from level sets of $f$ at the origin contain the $t$-axis (or, actually, its tangent line).

(2) (a) The homology of the Milnor fibre of $f_t$ at the origin is constant for all $t$ small.

(b) If $s \leq n-2$, then the fibre-homotopy-type of the Milnor fibrations of $f_t$ at the origin is constant for all $t$ small;

(c) If $s \leq n-3$, then the diffeomorphism-type of the Milnor fibrations of $f_t$ at the origin is constant for all $t$ small.

REMARK 4.5. Note that we do not conclude the constancy of the local, ambient topological-type. However, part of what the theorem says is that, if $s \leq n-3$, then at least the homeomorphism-type (in fact, diffeomorphism-type) of the small sphere minus the real link (i.e., the total space of the Milnor fibration) is constant.

It was an open question until 2004-2005 if the constancy of the Lê numbers implies constancy of the topological-type. In [**3**], Bobadilla proved this is the case when $s = 1$, but, in [**2**], he produced a counterexample when $s = 3$. On the other hand, Bobadilla did show that, for general $s$, in addition to the Milnor fibrations being constant, the homotopy-type of the real link in a family with constant Lê numbers is also constant.

Theorem 4.1 and Theorem 4.4 are the reasons **why** one wants algorithms for calculating Lê numbers.

EXERCISE 4.6. The formulas in this exercise allow one to reduce some problems on a hypersurface to problems on a hypersurface with a singular set of smaller dimension. Suppose that $s \geq 1$.

(1) Near $\mathbf{0}$, $\Sigma(f_{|V(z_0)}) = \Sigma f \cap V(z_0)$, and has dimension $s-1$. Let $\tilde{\mathbf{z}} = (z_1, \ldots, z_n)$ on $V(z_0)$. Then,

$$\lambda^0_{f_{|V(z_0)},\tilde{\mathbf{z}}}(\mathbf{0}) = \gamma^1_{f,\mathbf{z}}(\mathbf{0}) + \lambda^1_{f,\mathbf{z}}(\mathbf{0})$$

and, for all $k \geq 1$,

$$\lambda^k_{f_{|V(z_0)},\tilde{\mathbf{z}}}(\mathbf{0}) = \lambda^{k+1}_{f,\mathbf{z}}(\mathbf{0}).$$

(2) Suppose that $\gamma^1_{f,\mathbf{z}}(\mathbf{0}) = 0$ or that $j > 1 + \left(\lambda^0_{f,\mathbf{z}}(\mathbf{0})/\gamma^1_{f,\mathbf{z}}(\mathbf{0})\right)$.

Then, in a neighborhood of the origin,

$$\Sigma(f + z_0^j) = \Sigma f \cap V(z_0) \quad \text{and} \quad \dim_{\mathbf{0}} \Sigma(f + z_0^j) = s - 1.$$

Furthermore, if we let $\tilde{\mathbf{z}}$ denote the "rotated" coordinate system

$$(z_1, z_2, \ldots, z_n, z_0),$$

then

$$\lambda^0_{f+z_0^j,\tilde{\mathbf{z}}}(\mathbf{0}) = \lambda^0_{f,\mathbf{z}}(\mathbf{0}) + (j-1)\lambda^1_{f,\mathbf{z}}(\mathbf{0})$$

and, for all $k \geq 1$,

$$\lambda^k_{f+z_0^j,\tilde{\mathbf{z}}}(\mathbf{0}) = (j-1)\lambda^{k+1}_{f,\mathbf{z}}(\mathbf{0}).$$

---

We will now discuss the Lê numbers from a very different, very sophisticated, point of view. We refer you once again to [**23**] if you wish to see many results for calculating Lê numbers.

We want to discuss the relationship between the Lê numbers and the vanishing cycles as a bounded, constructible complex of sheaves. As serious references to complex of sheaves, we recommend [**10**] and [**4**]. Here, we will try to present enough material to give you the flavor of the machinery and results.

Suppose that $f$ has a non-isolated critical point at the origin. Then, at each point in $\Sigma f$, we have a Milnor fiber and a Milnor fibration. The question is: are there restrictions on the topology of the Milnor fiber at $\mathbf{0}$ imposed by the Milnor fibers at nearby points in $\Sigma f$?

This is a question of how local data patches together to give global data, which is exactly what sheaves encode. So it is easy to believe that sheaves would be useful in describing the situation. The stalk of a sheaf at a point describes the sheaf, in some sense, at the specified point. In our setting, at a point $\mathbf{p} \in \Sigma f$, we would like for the stalk at $\mathbf{p}$ to give us the cohomology of $F_{f,\mathbf{p}}$. This means that, after we take a stalk, we need a chain complex. Thus, one is lead to consider complexes of sheaves, where the stalks of the individual sheaves are simply $\mathbb{Z}$-modules (we could use other base rings). Note that we are absolutely **not** looking at coherent sheaves of modules over the structure sheaf of an analytic space. We should mention now that, for the most elegant presentation, it is essential that we allow our complex to have non-zero terms in **negative** degrees.

As a quick example, on a complex space $X$, one important complex of sheaves is simply referred to as the *constant sheaf*, $\mathbb{Z}^\bullet_X$. This complex of sheaves is the constant sheaf $\mathbb{Z}_X$ in degree 0 and 0 in all other degrees. Frequently, we will *shift* this complex; the complex of sheaves $\mathbb{Z}^\bullet_X[k]$ is the complex which has $\mathbb{Z}_X$ in degree $-k$ (note the negation) and 0 in all other degrees.

Suppose that we have a complex of sheaves (of $\mathbb{Z}$-modules) $\mathbf{A}^\bullet$ on a space $X$. There are three cohomologies associated to $\mathbf{A}^\bullet$:

(1) The *sheaf cohomology*, that is, the cohomology of the complex: $\mathbf{H}^k(\mathbf{A}^\bullet)$.

(2) The *stalk cohomology* at each point $\mathbf{p} \in X$: $H^k(\mathbf{A}^\bullet)_\mathbf{p}$. This is obtained by either taking stalks first and then cohomology of the resulting complex of $Z$-modules or by first taking the cohomology sheaves and then taking their stalks.

(3) The *hypercohomology* of $X$ with coefficients in $\mathbf{A}^\bullet$: $\mathbb{H}^k(X; \mathbf{A}^\bullet)$. This is a generalization of sheaf cohomology of the space. As an example, $\mathbb{H}^k(X; \mathbb{Z}^\bullet_X)$ simply yields the usual integral cohomology of $X$ in degree $k$.

For all of these cohomologies, the convention on shifting tells us that the index is added to the shift to produce the degree of the unshifted cohomology, e.g., $\mathbf{H}^k(\mathbf{A}^\bullet[m]) \cong \mathbf{H}^{k+m}(\mathbf{A}^\bullet)$.

It is standard that, for a subspace $Y \subseteq X$, one writes $\mathbb{H}^k(Y; \mathbf{A}^\bullet)$ for the hypercohomology of $Y$ with coefficients in the restriction of $\mathbf{A}^\bullet$ to $Y$; the point being that the restriction of the complex is usually not explicitly written in this context.

Our complexes of sheaves will be *constructible*. One thing that this implies is that restriction induces an isomorphism between the hypercohomology in a small ball and the stalk cohomology at the center of the ball, i.e., for all $\mathbf{p} \in X$, there exists $\epsilon > 0$ such that, for all $k$,

$$\mathbb{H}^k(B^\circ_\epsilon(\mathbf{p}); \mathbf{A}^\bullet) \;\cong\; H^k(\mathbf{A}^\bullet)_\mathbf{p}.$$

If $\mathcal{W}$ is an open subset of $X$ (and in other cases), the hypercohomology $\mathbb{H}^k(X, \mathcal{W}; \mathbf{A}^\bullet)$ of the pair $(X, \mathcal{W})$ is defined, and one has the expected long exact sequence

$$\cdots \to \mathbb{H}^{k-1}(\mathcal{W}; \mathbf{A}^\bullet) \to \mathbb{H}^k(X, \mathcal{W}; \mathbf{A}^\bullet) \to \mathbb{H}^k(X; \mathbf{A}^\bullet) \to \mathbb{H}^k(\mathcal{W}; \mathbf{A}^\bullet) \to \dots .$$

The *k-th support of* $\mathbf{A}^\bullet$, $\operatorname{supp}^k \mathbf{A}^\bullet$, is defined to be the closure of the set of points where the stalk cohomology in degree $k$ is not zero, i.e.,

$$\operatorname{supp}^k \mathbf{A}^\bullet \ := \ \overline{\{\mathbf{p} \in X \mid H^k(\mathbf{A}^\bullet)_\mathbf{p} \neq 0\}}.$$

The *support of* $\mathbf{A}^\bullet$, $\operatorname{supp} \mathbf{A}^\bullet$, is given by

$$\operatorname{supp} \mathbf{A}^\bullet \ := \bigcup_k \operatorname{supp}^k \mathbf{A}^\bullet \ = \ \overline{\{\mathbf{p} \in X \mid H^*(\mathbf{A}^\bullet)_\mathbf{p} \neq 0\}}.$$

(We used above that $\mathbf{A}^\bullet$ is bounded, so that the union is finite.)

The *k-cosupport of* $\mathbf{A}^\bullet$, $\operatorname{cosupp}^k \mathbf{A}^\bullet$, is defined to be

$$\operatorname{cosupp}^k \mathbf{A}^\bullet \ := \ \overline{\{\mathbf{p} \in X \mid \mathbb{H}^k\big(B^\circ_\epsilon(\mathbf{p}), B^\circ_\epsilon(\mathbf{p}) - \{\mathbf{p}\};\ \mathbf{A}^\bullet\big)\}}.$$

A **perverse sheaf** (using middle perversity) $\mathbf{P}^\bullet$ is a bounded, constructible complex of sheaves which satisfies two conditions: the *support* and *cosupport* conditions, as given below. We remark that we set the dimension of the empty set to be $-\infty$, so that saying that a set has negative dimension means that the set is empty.

The support and cosupport conditions for a perverse sheaf are that, for all $k$,

(1) (support condition) $\dim \operatorname{supp}^k \mathbf{A}^\bullet \ \leq -k$;

(2) (cosupport condition) $\dim \operatorname{cosupp}^k \mathbf{A}^\bullet \ \leq k$.

Note that the stalk cohomology is required to be 0 in positive degrees.

If $s := \dim \operatorname{supp} \mathbf{P}^\bullet$, then the stalk cohomology is possibly non-zero only in degrees $k$ where $-s \leq k \leq 0$. In particular, if $\mathbf{p}$ is an isolated point in the support of $\mathbf{P}^\bullet$, then the stalk cohomology of $\mathbf{P}^\bullet$ at $\mathbf{p}$ is concentrated in degree 0. Furthermore, the cosupport condition tells us that, for all $\mathbf{p} \in X$, for negative degrees $k$,

$$\mathbb{H}^k\big(B^\circ_\epsilon(\mathbf{p}), B^\circ_\epsilon(\mathbf{p}) - \{\mathbf{p}\};\ \mathbf{P}^\bullet\big) = 0.$$

Note that the restriction of a perverse sheaf to its support remains perverse.

EXERCISE 4.7. Suppose that $M$ is a complex manifold of pure dimension $d$. Show that the shifted constant sheaf $\mathbb{Z}^\bullet_M[d]$ is perverse.

What does any of this have to do with the Milnor fiber and Lê numbers?

Given an analytic $f : (\mathcal{U}, \mathbf{0}) \to (\mathbb{C}, 0)$, there are two functors, $\psi_f[-1]$ and $\phi_f[-1]$, from the category of perverse sheaves on $\mathcal{U}$ to the category of perverse sheaves on $V(f)$ called the (shifted) **nearby cycles along** $f$ and the **vanishing cycles along** $f$, respectively. These functors encode, on a chain level, the (hyper-)cohomology of the Milnor fiber and "reduced" (hyper-)cohomology of the Milnor fiber, and how they patch together. For all $\mathbf{p} \in V(f)$, the stalk cohomology is what we want:

$$H^k\left(\psi_f[-1]\mathbf{P}^\bullet\right)_\mathbf{p} \ \cong \mathbb{H}^{k-1}(F_{f,\mathbf{p}}; \mathbf{P}^\bullet)$$

and

$$H^k\left(\phi_f[-1]\mathbf{P}^\bullet\right)_\mathbf{p} \ \cong \mathbb{H}^k(B^\circ_\epsilon(\mathbf{p}), F_{f,\mathbf{p}}; \mathbf{P}^\bullet).$$

EXERCISE 4.8. Recall from the previous exercise that $\mathbb{Z}^\bullet_\mathcal{U}[n+1]$ is perverse. As $\phi_f[-1]$ takes perverse sheaves to perverse sheaves, it follows that the complex of sheaves $\phi_f[-1]\mathbb{Z}^\bullet_\mathcal{U}[n+1]$ is a perverse sheaf on $V(f)$.

(1) Show that, for all $\mathbf{p} \in V(f)$, for all $k$,

$$H^k(\phi_f[-1]\mathbb{Z}^\bullet_{\mathcal{U}}[n+1])_{\mathbf{p}} \cong \widetilde{H}^{n+k}(F_{f,\mathbf{p}};\mathbb{Z}),$$

and so $\operatorname{supp}\phi_f[-1]\mathbb{Z}^\bullet_{\mathcal{U}}[n+1] = V(f)\cap\Sigma f$.

(2) Explain why this tells you that $\widetilde{H}^j(F_{f,\mathbf{0}};\mathbb{Z})$ is possibly non-zero only for $n-s \le j \le n$.

EXAMPLE 4.9. In this example, we wish to show some of the power of perverse techniques.

Suppose that $s = 1$, so that $\Sigma f$ is a curve at the origin. Let $C_i$ denote the irreducible components of $\Sigma f$ at $\mathbf{0}$. Then, $\Lambda^1_{f,\mathbf{z}} = \sum_i \mu_i^\circ [C_i]$, where $\mu_i^\circ$ is the Milnor number of $f$ restricted to a transverse hyperplane slice to $C_i$ at any point on $C_i - \{\mathbf{0}\}$ near $\mathbf{0}$. Furthermore, in a neighborhood of $\mathbf{0}$, $C_i - \{\mathbf{0}\}$ is a punctured disk, and there is an internal monodromy action $h_i : \mathbb{Z}^{\mu_i^\circ} \to \mathbb{Z}^{\mu_i^\circ}$ induced by following the Milnor fiber once around this puncture.

Let $\mathbf{P}^\bullet := (\phi_f[-1]\mathbb{Z}^\bullet_{\mathcal{U}}[n+1])_{|_{\Sigma f}}$. This is a perverse sheaf on $\Sigma f$. We are going to look at the long exact hypercohomology sequence of the pair

$$\big(B_\epsilon \cap \Sigma f,\ (B_\epsilon - \{\mathbf{0}\})\cap\Sigma f\big),$$

for small $\epsilon$. For convenience of notation, we will assume that we are working in a small enough ball around the origin, and replace $B_\epsilon\cap\Sigma f$ with simply $\Sigma f$, and $(B_\epsilon - \{\mathbf{0}\})\cap\Sigma f$ with $\Sigma f - \{\mathbf{0}\} = \bigcup_i (C_i - \{\mathbf{0}\})$.

Because we are working in a small ball,

$$\mathbb{H}^k(\Sigma f;\mathbf{P}^\bullet) \ \cong\ H^k(\mathbf{P}^\bullet)_{\mathbf{0}} \ \cong\ \widetilde{H}^{n+k}(F_{f,\mathbf{0}};\mathbb{Z}).$$

Furthermore, $\mathbf{P}^\bullet$ restricted to $C_i - \{\mathbf{0}\}$ is a local system (a locally constant sheaf), shifted into degree $-1$ with stalk cohomology $\mathbb{Z}^{\mu_i^\circ}$. Hence,

$$\mathbb{H}^{-1}(C_i - \{\mathbf{0}\};\mathbf{P}^\bullet) \cong \ker\{\mathrm{id} - h_i\},$$

and

$$\mathbb{H}^{-1}\Big(\bigcup_i (C_i - \{\mathbf{0}\});\mathbf{P}^\bullet\Big) \ \cong\ \bigoplus_i \mathbb{H}^{-1}((C_i - \{\mathbf{0}\});\mathbf{P}^\bullet) \ \cong\ \bigoplus_i \ker\{\mathrm{id} - h_i\}.$$

Finally note that the cosupport condition tells us that $\mathbb{H}^{-1}(\Sigma f, \Sigma f - \{\mathbf{0}\};\mathbf{P}^\bullet) = 0$.

Thus, the portion of the long exact sequence on hypercohomology

$$\mathbb{H}^{-1}(\Sigma f, \Sigma f - \{\mathbf{0}\};\mathbf{P}^\bullet) \to \mathbb{H}^{-1}(\Sigma f;\mathbf{P}^\bullet) \to \mathbb{H}^{-1}(\Sigma f - \{\mathbf{0}\};\mathbf{P}^\bullet) \to$$

becomes

$$0 \to \widetilde{H}^{n-1}(F_{f,\mathbf{0}};\mathbb{Z}) \to \bigoplus_i \ker\{\mathrm{id} - h_i\} \to .$$

Even without knowing the $h_i$'s, this tells us that the rank of $\widetilde{H}^{n-1}(F_{f,\mathbf{0}};\mathbb{Z})$ is at most $\sum_i \mu_i^\circ$.

Note that, if all of the components of $\Sigma f$ are smooth, then $\lambda^1_{f,\mathbf{z}}(\mathbf{0}) = \sum_i \mu_i^\circ$, and this bound agrees with what we obtain from Theorem 4.1, but – if one of the $C_i$'s is singular at $\mathbf{0}$ – then this example produces a better bound.

This result is quite complicated to prove without using perverse sheaves; see [**35**].

---

We have yet to tell you what the vanishing cycles have to do with the Lê numbers.

Suppose that $X$ is a complex analytic subspace of $\mathcal{U}$, that $\mathbf{P}^\bullet$ is a perverse sheaf on $X$, and that $\mathbf{p} \in X$. For a generic affine linear form $L$ such that $L(\mathbf{p}) = 0$, the point $\mathbf{p}$ is an isolated point in the support of $\phi_L[-1]\mathbf{P}^\bullet$; for instance, $L$ can be chosen so that $V(L)$ is transverse to all of the strata – near $\mathbf{p}$ but not at $\mathbf{p}$ – of a Whitney stratification with respect to which $\mathbf{P}^\bullet$ is constructible. For such an $L$, since $\mathbf{p}$ is an isolated point in the support of the perverse sheaf $\phi_L[-1]\mathbf{P}^\bullet$, the stalk cohomology of $\phi_L[-1]\mathbf{P}^\bullet$ at $\mathbf{p}$ is concentrated in degree 0.

Our coordinates $(z_0, z_1, \dots, z_n)$ have been chosen so that all of the iterated vanishing and nearby cycles below have $\mathbf{0}$ as an isolated point in their support.

As we showed in [**22**], the connection with the Lê numbers is:

$$\begin{aligned}
\lambda^0_{f,\mathbf{z}}(\mathbf{0}) &= \operatorname{rank} H^0\big(\phi_{z_0}[-1]\phi_f[-1]\mathbb{Z}^\bullet_{\mathcal{U}}[n+1]\big),\\
\lambda^1_{f,\mathbf{z}}(\mathbf{0}) &= \operatorname{rank} H^0\big(\phi_{z_1}[-1]\psi_{z_0}[-1]\phi_f[-1]\mathbb{Z}^\bullet_{\mathcal{U}}[n+1]\big),\\
&\ \ \vdots\\
\lambda^s_{f,\mathbf{z}}(\mathbf{0}) &= \operatorname{rank} H^0\big(\phi_{z_s}[-1]\psi_{z_{s-1}}[-1]\dots\psi_{z_0}[-1]\phi_f[-1]\mathbb{Z}^\bullet_{\mathcal{U}}[n+1]\big).
\end{aligned}$$

The chain complex in Theorem 4.1 can be derived as a formal consequence of the equalities above.

## 5. Appendix

**Curve Selection Lemma**:

The following lemma is an extremely useful tool; see [**28**], §3 and [**19**], §2.1. The complex analytic statement uses Lemma 3.3 of [**28**]. Below, $\overset{\circ}{\mathbb{D}}_\delta$ denotes an open disk of radius $\delta > 0$ centered at the origin in $\mathbb{C}$.

LEMMA 5.1. (Curve Selection Lemma) *Let $\mathbf{p}$ be a point in a real analytic manifold $M$. Let $Z$ be a semianalytic subset of $M$ such that $\mathbf{p} \in \overline{Z}$. Then, there exists a real analytic curve $\gamma : [0, \delta) \to M$ with $\gamma(0) = \mathbf{p}$ and $\gamma(t) \in Z$ for $t \in (0, \delta)$.*

*Let $\mathbf{p}$ be a point in a complex analytic manifold $M$. Let $Z$ be a constructible subset of $M$ such that $\mathbf{p} \in \overline{Z}$. Then, there exists a complex analytic curve $\gamma : \overset{\circ}{\mathbb{D}}_\delta \to M$ with $\gamma(0) = \mathbf{p}$ and $\gamma(t) \in Z$ for $t \in \overset{\circ}{\mathbb{D}}_\delta - \{\mathbf{0}\}$.*

---

**Cones and joins**:

Recall that the abstract cone on a topological space $Y$ is the identification space

$$c(Y) := \frac{Y \times [0,1]}{Y \times \{1\}},$$

where the point (the equivalence class of) $Y \times \{1\}$ is referred to as the *cone point*. Then define the open cone to be $c^\circ(Y) := c(Y)\backslash(Y \times \{0\})$.

If $Z \subseteq Y$, then $c(Z) \subseteq c(Y)$, where we consider the two cone points to be the same, and denote this common cone point simply by $*$.

We define the cones on the pair $(Y, Z)$ to be triples, which include the cone point:

$$c(Y,Z) \ := \ (c(Y),\, c(Z),\, *) \quad \text{ and } \quad c^\circ(Y,Z) \ := \ (c^\circ(Y),\, c^\circ(Z),\, *).$$

The *join* $X * Y$ of topological spaces $X$ and $Y$ is the space $X \times Y \times [0,1]$, where at one end the interval "$X$ is crushed to a point" and at the other end of the interval "$Y$ is crushed to a point". This means that we take the identification space obtained from $X \times Y \times [0,1]$ by identifying $(x_1, y, 0) \sim (x_2, y, 0)$ for all $x_1, x_2 \in X$ and $y \in Y$, and also identifying $(x, y_1, 1) \sim (x, y_2, 1)$ for all $x \in x$ and $y_1, y_2 \in Y$.

The join of a point with a space $X$ is just the cone on $X$. The join of the 0-sphere (two discrete points) with $X$ is the suspension of $X$.

---

**Locally trivial fibrations**:

Suppose that $X$ and $Y$ are smooth manifolds, where $X$ may have boundary, but $Y$ does not. A smooth map $g : X \to Y$ is a *smooth trivial fibration* if and only if there exists a smooth manifold $F$, possibly with boundary, and a diffeomorphism $\alpha : F \times Y \to X$ such that the following diagram commutes:

$$\begin{array}{ccc} F \times Y & \xrightarrow[\cong]{\ \alpha\ } & X \\ & \searrow^{\pi} & \downarrow g \\ & & Y, \end{array}$$

where $\pi$ denotes projection.

A smooth map $g : X \to Y$ is a *smooth locally trivial fibration* if and only if, for all $y \in Y$, there exists an open neighborhood $B$ of $y$ in $Y$ such that the restriction $g_{|_{g^{-1}(B)}} : g^{-1}(B) \to B$ is a trivial fibration.

If the base space $Y$ is connected and $g : X \to Y$ is a smooth locally trivial fibration, it is an easy exercise to show that the diffeomorphism-type of the fibers $g^{-1}(y)$ is independent of $y \in Y$; in this case, any fiber is referred to as simply *the fiber of the fibration.*

It is a theorem that a locally trivial fibration over a contractible base space is, in fact, a trivial fibration.

Now we state the theorem of Ehresmann [**5**], which yields a common for method for concluding that a map is a locally trivial fibration.

THEOREM 5.2. Suppose that $N$ is a smooth manifold, possibly with boundary, $P$ is a smooth manifold, and $f : N \to P$ is a proper submersion. Then, $f$ it is a smooth, locally trivial fibration.

Locally trivial fibrations over a circle are particularly easy to characterize. Begin with a smooth fiber $F$ and a diffeomorphism $\tau : F \to F$, called a *characteristic diffeomorphism.* Then there is a smooth locally trivial fibration

$$p : \frac{F \times [0,1]}{(x,0) \sim (\tau(x), 1)} \to S^1 \subseteq \mathbb{C}$$

given by $p([x,t]) := e^{2\pi i t}$. A characteristic diffeomorphism describes how the fiber is "glued" to itself as one travels counterclockwise once around the base circle.

Every smooth locally trivial fibration $g$ over a circle is diffeomorphic to one obtained as above, but the characteristic diffeomorphism $\tau$ is **not** uniquely determined by $g$; however, the maps on the homology and cohomology of $F$ induced by $\tau$ **are** independent of the choice of $\tau$. These induced maps are called the *monodromy automorphisms* of $g$. Thus, for each degree $i$, there are well-defined monodromy maps $T_i : H_i(F;\mathbb{Z}) \to H_i(F;\mathbb{Z})$ and $T^i : H^i(F;\mathbb{Z}) \to H^i(F;\mathbb{Z})$.

## References


[1] A'Campo, N. Le nombre de Lefschetz d'une monodromie. *Proc. Kon. Ned. Akad. Wet., Series A*, 76:113–118, 1973.

[2] Javier Fernández de Bobadilla, *Answers to some equisingularity questions*, Invent. Math. **161** (2005), no. 3, 657–675, DOI 10.1007/s00222-005-0441-4. MR2181471

[3] Javier Fernández de Bobadilla, *Topological equisingularity of hypersurfaces with 1-dimensional critical set*, Adv. Math. **248** (2013), 1199–1253, DOI 10.1016/j.aim.2013.08.001. MR3107539

[4] Alexandru Dimca, *Sheaves in topology*, Universitext, Springer-Verlag, Berlin, 2004. MR2050072

[5] Charles Ehresmann, *Les connexions infinitésimales dans un espace fibré différentiable* (French), Colloque de topologie (espaces fibrés), Bruxelles, 1950, Georges Thone, Liège; Masson et Cie., Paris, 1951, pp. 29–55. MR0042768

[6] William Fulton, *Intersection theory*, Ergebnisse der Mathematik und ihrer Grenzgebiete (3) [Results in Mathematics and Related Areas (3)], vol. 2, Springer-Verlag, Berlin, 1984. MR732620

[7] Hans Grauert and Reinhold Remmert, *Theory of Stein spaces*, Grundlehren der Mathematischen Wissenschaften [Fundamental Principles of Mathematical Sciences], vol. 236, Springer-Verlag, Berlin-New York, 1979. Translated from the German by Alan Huckleberry. MR580152

[8] Hans Grauert and Reinhold Remmert, *Coherent analytic sheaves*, Grundlehren der Mathematischen Wissenschaften [Fundamental Principles of Mathematical Sciences], vol. 265, Springer-Verlag, Berlin, 1984. MR755331

[9] Helmut A. Hamm and Lê Dũng Tráng, *Un théorème de Zariski du type de Lefschetz*, Ann. Sci. École Norm. Sup. (4) **6** (1973), 317–355. MR0401755

[10] Masaki Kashiwara and Pierre Schapira, *Sheaves on manifolds*, Grundlehren der Mathematischen Wissenschaften [Fundamental Principles of Mathematical Sciences], vol. 292, Springer-Verlag, Berlin, 1990. With a chapter in French by Christian Houzel. MR1074006

[11] Mitsuyoshi Kato and Yukio Matsumoto, *On the connectivity of the Milnor fiber of a holomorphic function at a critical point*, Manifolds—Tokyo 1973 (Proc. Internat. Conf., Tokyo, 1973), Univ. Tokyo Press, Tokyo, 1975, pp. 131–136. MR0372880

[12] Lê Dũng Tráng, *Calcul du nombre de cycles évanouissants d'une hypersurface complexe* (French, with English summary), Ann. Inst. Fourier (Grenoble) **23** (1973), no. 4, 261–270. MR0330501

[13] Lê Dũng Tráng, *Topologie des singularités des hypersurfaces complexes* (French), Singularités à Cargèse (Rencontre Singularités Géom. Anal., Inst. Études Sci., Cargèse, 1972), Soc. Math. France, Paris, 1973, pp. 171–182. Astérisque, Nos. 7 et 8. MR0361147

[14] Lê Dũng Tráng, *Some remarks on relative monodromy*, Real and complex singularities (Proc. Ninth Nordic Summer School/NAVF Sympos. Math., Oslo, 1976), Sijthoff and Noordhoff, Alphen aan den Rijn, 1977, pp. 397–403. MR0476739

[15] Lê Dũng Tráng, *Ensembles analytiques complexes avec lieu singulier de dimension un (d'après I. N. Iomdine)* (French), Seminar on Singularities (Paris, 1976/1977), Publ. Math. Univ. Paris VII, vol. 7, Univ. Paris VII, Paris, 1980, pp. 87–95. MR683620

[16] Lê Dũng Tráng and David B. Massey, *Hypersurface singularities and Milnor equisingularity*, Pure Appl. Math. Q. **2** (2006), no. 3, 893–914, DOI 10.4310/PAMQ.2006.v2.n3.a13. MR2252122

[17] Lê Dũng Tráng and C. P. Ramanujam, *The invariance of Milnor's number implies the invariance of the topological type*, Amer. J. Math. **98** (1976), no. 1, 67–78. MR0399088

[18] Łojasiewicz, S. *Ensemble semi-analytiques.* IHES Lecture notes, 1965.

[19] E. J. N. Looijenga, *Isolated singular points on complete intersections*, London Mathematical Society Lecture Note Series, vol. 77, Cambridge University Press, Cambridge, 1984. MR747303

[20] David B. Massey, *The Lê varieties. I*, Invent. Math. **99** (1990), no. 2, 357–376, DOI 10.1007/BF01234423. MR1031905

[21] David B. Massey, *The Lê varieties. II*, Invent. Math. **104** (1991), no. 1, 113–148, DOI 10.1007/BF01245068. MR1094048

[22] David B. Massey, *Numerical invariants of perverse sheaves*, Duke Math. J. **73** (1994), no. 2, 307–369, DOI 10.1215/S0012-7094-94-07315-8. MR1262210

[23] David B. Massey, *Lê cycles and hypersurface singularities*, Lecture Notes in Mathematics, vol. 1615, Springer-Verlag, Berlin, 1995. MR1441075

[24] David B. Massey, *The Sebastiani-Thom isomorphism in the derived category*, Compositio Math. **125** (2001), no. 3, 353–362, DOI 10.1023/A:1002608716514. MR1818986

[25] Yukio Matsumoto, *An introduction to Morse theory*, Translations of Mathematical Monographs, vol. 208, American Mathematical Society, Providence, RI, 2002. Translated from the 1997 Japanese original by Kiki Hudson and Masahico Saito; Iwanami Series in Modern Mathematics. MR1873233

[26] John Milnor, *Morse theory*, Based on lecture notes by M. Spivak and R. Wells. Annals of Mathematics Studies, No. 51, Princeton University Press, Princeton, N.J., 1963. MR0163331

[27] John Milnor, *Lectures on the h-cobordism theorem*, Notes by L. Siebenmann and J. Sondow, Princeton University Press, Princeton, N.J., 1965. MR0190942

[28] John Milnor, *Singular points of complex hypersurfaces*, Annals of Mathematics Studies, No. 61, Princeton University Press, Princeton, N.J.; University of Tokyo Press, Tokyo, 1968. MR0239612

[29] András Némethi, *Generalized local and global Sebastiani-Thom type theorems*, Compositio Math. **80** (1991), no. 1, 1–14. MR1127057

[30] András Némethi, *Global Sebastiani-Thom theorem for polynomial maps*, J. Math. Soc. Japan **43** (1991), no. 2, 213–218, DOI 10.2969/jmsj/04320213. MR1096431

[31] Mutsuo Oka, *On the homotopy types of hypersurfaces defined by weighted homogeneous polynomials*, Topology **12** (1973), 19–32. MR0309950

[32] V. P. Palamodov, *The multiplicity of a holomorphic transformation* (Russian), Funkcional. Anal. i Priložen **1** (1967), no. 3, 54–65. MR0236424

[33] Koichi Sakamoto, *The Seifert matrices of Milnor fiberings defined by holomorphic functions*, J. Math. Soc. Japan **26** (1974), 714–721. MR0362329

[34] M. Sebastiani and R. Thom, *Un résultat sur la monodromie* (French), Invent. Math. **13** (1971), 90–96. MR0293122

[35] Dirk Siersma, *Variation mappings on singularities with a 1-dimensional critical locus*, Topology **30** (1991), no. 3, 445–469, DOI 10.1016/0040-9383(91)90025-Y. MR1113689

[36] Michael Spivak, *A comprehensive introduction to differential geometry. Vol. I*, 2nd ed., Publish or Perish, Inc., Wilmington, Del., 1979. MR532830

[37] Bernard Teissier, *Déformations à type topologique constant II*. Séminaire Douady-Verdier, Secrétariat E.N.S. Paris, 1972.

[38] Bernard Teissier, *Cycles évanescents, sections planes et conditions de Whitney* (French), Singularités à Cargèse (Rencontre Singularités Géom. Anal., Inst. Études Sci., Cargèse, 1972), Soc. Math. France, Paris, 1973, pp. 285–362. Astérisque, Nos. 7 et 8. MR0374482

[39] Bernard Teissier, *Variétés polaires. I. Invariants polaires des singularités d'hypersurfaces* (French), Invent. Math. **40** (1977), no. 3, 267–292. MR0470246

[40] Bernard Teissier, *Variétés polaires. II. Multiplicités polaires, sections planes, et conditions de Whitney* (French), Algebraic geometry (La Rábida, 1981), Lecture Notes in Math., vol. 961, Springer, Berlin, 1982, pp. 314–491, DOI 10.1007/BFb0071291. MR708342

Contemporary Mathematics
Volume **675**, 2016
http://dx.doi.org/10.1090/conm/675/13593

# Knots and the topology of singular surfaces in $\mathbb{R}^4$

R. Mendes and J. J. Nuño-Ballesteros

ABSTRACT. We consider the link of an isolated surface singularity in $\mathbb{R}^4$ given as the image of an analytic map germ $f : (\mathbb{R}^2, 0) \to (\mathbb{R}^4, 0)$. The link is a knot in a small enough sphere which determines the topological type of the singularity. We prove that the link of a generic linear projection into $\mathbb{R}^3$ is equivalent to a generic projection of the link of $f$. We also give sufficient conditions for a 1-parameter family to have constant knot type in terms of the constancy of the delta invariant and the multiplicity.

## 1. Introduction

The topology of an irreducible complex plane curve singularity $(C, 0) \subset (\mathbb{C}^2, 0)$ is determined by its link $K$. This is a knot obtained by taking the intersection of $C$ with a small enough sphere $S^3_\epsilon$ centered at the origin. The classical results due to Brauner [**2**], Burau [**3**] and Zariski [**13**] say that $K$ is always an iterated torus knot and that two plane curve singularities are topologically equivalent if and only if their links are equivalent as knots.

Since an irreducible complex plane curve singularity can be parametrized as the image of a holomorphic map germ $f : (\mathbb{C}, 0) \to (\mathbb{C}^2, 0)$, it is natural to ask what happens if we consider the more general situation of an isolated surface singularity in $\mathbb{R}^4$, parametrized as the image of a real analytic map germ $f : (\mathbb{R}^2, 0) \to (\mathbb{R}^4, 0)$. The link is again a knot $K$ in $S^3_\epsilon$ and we still have a cone structure on its link. However, we expect that more general knot types can appear, apart from the iterated torus knots. Partial results for this type of knots can be found in [**1, 10**].

In this paper, we consider the general case of an analytic map germ $f : (\mathbb{R}^2, 0) \to (\mathbb{R}^4, 0)$ which parametrizes an isolated surface singularity. We first show that if we take a generic linear projection $p_v : \mathbb{R}^4 \to \mathbb{R}^3$, then the link of $p_v \circ f : (\mathbb{R}^2, 0) \to (\mathbb{R}^3, 0)$ is equivalent to a generic projection of the link of $f$. This allows us to use the results of [**7**] and deduce that any germ with Boardman symbol $\Sigma^{1,0}$ is unknotted.

We also introduce the $\delta$-invariant of $f$, which has the property that if $\delta(f) < \infty$ then $f$ has isolated singularity. The main result in this section is that $f_t$ is a 1-parameter deformation of a corank one germ $f_0$, such that $\delta(f_t)$ and the multiplicity $m(f_t)$ are constant on $t$, then the link of $f_t$ has also constant knot type.

---

2010 *Mathematics Subject Classification.* Primary 58K15; Secondary 57M25, 58K65.
*Key words and phrases.* Link, isolated surface singularity, delta invariant.
The second author has been partially supported by DGICYT Grant MTM2012–33073.

We apply this result to conclude that any map germ $f$ of the form $f(x,y) = (x, xy, q(x,y), r(x,y))$ must be unknotted.

*Acknowledgement.* The authors would like to thank Lev Birbrair for his valuable comments and suggestions.

## 2. The link of a singular surface in $\mathbb{R}^4$

Two $C^\infty$ map germs $f, g : (\mathbb{R}^n, 0) \to (\mathbb{R}^p, 0)$ are said $\mathscr{A}$-equivalent if there are diffeomorphisms $\phi, \psi$ such that $g = \psi \circ f \circ \phi^{-1}$. If $\phi, \psi$ are homeomorphisms instead of diffeomorphisms, then we say that $f, g$ are $C^0$-$\mathscr{A}$-equivalent.

Given $f : (\mathbb{R}^n, 0) \to (\mathbb{R}^p, 0)$, we denote by $j^k f(0)$ its $k$-jet, that is, its Taylor expansion of degree $k$. We say that $f$ is $k$-determined if for any $g$ such that $j^k g(0) = j^k f(0)$, then $g, f$ are $\mathscr{A}$-equivalent. We say that $f$ is finitely determined if it is $k$-determined for some $k$.

Here, we are interested in singular surfaces in $\mathbb{R}^4$ which are locally parametrized as the image of a map germ $f : (\mathbb{R}^2, 0) \to (\mathbb{R}^4, 0)$. We assume that $f$ is analytic and that has isolated singularity in the sense of the following definition.

DEFINITION 2.1. Let $f : (\mathbb{R}^2, 0) \to (\mathbb{R}^4, 0)$ be an analytic map germ. We say that $f$ has *isolated singularity* if there is a small enough representative $f : U \to V$, where $U, V$ are open neighborhoods of the origin in $\mathbb{R}^2, \mathbb{R}^4$ respectively, such that $f$ is a $C^0$-embedding and an immersion on $U \setminus \{0\}$.

REMARK 2.2. If $f : (\mathbb{R}^2, 0) \to (\mathbb{R}^4, 0)$ is $C^\infty$ and finitely determined, then it is $\mathscr{A}$-equivalent to an analytic map germ with isolated singularity. In fact, after $\mathscr{A}$-equivalence, we can assume that $f$ is polynomial. Denote by $\hat{f} : (\mathbb{C}^2, 0) \to (\mathbb{C}^4, 0)$ the complexification of $f$, then $\hat{f}$ is also finitely determined as a complex analytic germ. By the Mather-Gaffney finite determinacy criterion (see [**11**]), there is a small enough representative $\hat{f} : \hat{U} \to \hat{V}$, where $\hat{U}, \hat{V}$ are open neighborhoods of the origin in $\mathbb{C}^2, \mathbb{C}^4$ such that $\hat{f}$ is proper, $\hat{f}^{-1}(0) = \{0\}$ and the restriction to $\hat{U} \setminus \{0\}$ is stable. This implies that $\hat{f}$ is an immersion with a finite number of transverse double points on $\hat{U} \setminus \{0\}$. Finally, by shrinking $\hat{U}$ if necessary, we can assume that $\hat{f}$ has no double points on $\hat{U}$ and we take $f : U \to V$ as the real projection of $\hat{f} : \hat{U} \to \hat{V}$.

Note that the converse in not true in general. For instance, the map germ $f : (\mathbb{R}^2, 0) \to (\mathbb{R}^4, 0)$ given by $f(x,y) = (x, y^2, y(x^2+y^2), 0)$ has isolated singularity but is not finitely determined, since its complexification has a curve of double points along $x^2 + y^2 = 0$.

We introduce the delta invariant, which gives an algebraic tool to know if a surface has isolated singularity. Our definition is based on the Mond's double point ideal [**9**]. Let us denote by $\mathscr{E}_n$ the local ring of $C^\infty$ function germs from $(\mathbb{R}^n, 0) \to \mathbb{R}$. Given a $C^\infty$ map germ $f : (\mathbb{R}^n, 0) \to (\mathbb{R}^p, 0)$, with $n \le p$, we can write

$$f_i(x) - f_i(x') = \sum_{j=1}^{n} \alpha_{ij}(x, x')(x_j - x'_j), \quad i = 1, \dots, p,$$

for some functions $\alpha_{ij} \in \mathscr{E}_{2n}$. The double point ideal of $f$ is the ideal $I^2(f) \subset \mathscr{E}_{2n}$ generated by $f_i(x) - f_i(x')$, $i = 1, \dots, p$ and by the $n \times n$ minors of the matrix $(\alpha_{ij}(x, x'))$. The main property of this ideal is that its zero locus, denoted by

$D^2(f)$, is the set germ in $(\mathbb{R}^n \times \mathbb{R}^n, 0)$ of points $(x, x')$ such that either $x \neq x'$ and $f(x) = f(x')$ or $x = x'$ and $x$ is a critical point of $f$.

In case $f$ has corank 1, after $\mathscr{A}$-equivalence we can write $f$ in the form

$$f(z, y) = (z, f_n(z, y), \ldots, f_p(z, y)), \quad z \in \mathbb{R}^{n-1}, \ y \in \mathbb{R}. \tag{1}$$

Then, $I^2(f)$ can be considered in $\mathscr{E}_{n+1}$ instead of $\mathscr{E}_{2n}$ and is generated by the so-called divided differences of the functions $f_i$:

$$\Delta f_i(z, y, y') := \frac{f_i(z, y) - f_i(z, y')}{y - y'}, \qquad i = n, \ldots, p.$$

(We refer to **[9]** for details about this construction).

DEFINITION 2.3. Given a $C^\infty$ map germ $f : (\mathbb{R}^2, 0) \to (\mathbb{R}^4, 0)$ we define

$$\delta(f) := \dim_{\mathbb{R}} \frac{\mathscr{E}_4}{I^2(f)}.$$

In case $f$ has corank 1 and is given by $f(x, y) = (x, p(x, y), q(x, y), r(x, y))$, with $p, q, r \in \mathscr{E}_2$, we consider the construction of $I^2(f)$ by divided differences, so

$$\delta(f) = \dim_{\mathbb{R}} \frac{\mathscr{E}_3}{\langle \Delta p, \Delta q, \Delta r \rangle}.$$

We remark that the above definitions of $I^2(f)$, $D^2(f)$ and $\delta(f)$ can be extended in a natural way for holomorphic map germs $f : (\mathbb{C}^n, 0) \to (\mathbb{C}^p, 0)$ if we substitute $\mathscr{E}_n$ by $\mathscr{O}_n$, the local ring of holomorphic function germs in $(\mathbb{C}^n, 0)$.

LEMMA 2.4. *Let $f : (\mathbb{R}^2, 0) \to (\mathbb{R}^4, 0)$ be an analytic map germ. We have:*

(1) *$D^2(f) = \{0\}$ if and only if $f$ has isolated singularity;*
(2) *$\delta(f) < \infty$ if and only if $f$ is finitely determined.*

PROOF. The first part is immediate from the definitions. For the second part, we denote by $\hat{f}$ the complexification of $f$. Since $\delta(f) < \infty$, we have $\delta(\hat{f}) = \delta(f) < \infty$ and thus, $D^2(\hat{f}) = \{0\}$ by the Nullstellensatz. By the Mather-Gaffney finite determinacy criterion, $\hat{f}$ is finitely determined and hence, so is $f$. $\square$

The following result about the cone structure of an analytic surface with isolated singularity is well known, but we include the proof in order to be precise about the definition of the link. For a finitely determined $C^\infty$ map germ, it follows from Fukuda's cone theorem **[4]**.

Given $f : (\mathbb{R}^n, 0) \to (\mathbb{R}^p, 0)$ and a representative $f : U \to V$ we denote:

$$D^p_\epsilon = \{y \in \mathbb{R}^p : \|y\|^2 \leq \epsilon\}, \quad S^{p-1}_\epsilon = \{y \in \mathbb{R}^p : \|y\|^2 = \epsilon\},$$
$$\tilde{D}^n_\epsilon = f^{-1}(D^p_\epsilon), \quad \tilde{S}^{n-1}_\epsilon = f^{-1}(S^{p-1}_\epsilon).$$

THEOREM 2.5. *Let $f : (\mathbb{R}^2, 0) \to (\mathbb{R}^4, 0)$ be an analytic map germ with isolated singularity. There exists $\epsilon_0 > 0$ such that for any $\epsilon$, with $0 < \epsilon \leq \epsilon_0$ we have:*

(1) *$\tilde{S}^1_\epsilon$ is diffeomorphic to $S^1$.*
(2) *$f|_{\tilde{S}^1_\epsilon} : \tilde{S}^1_\epsilon \to S^3_\epsilon$ is an embedding, whose $\mathscr{A}$-class is independent of $\epsilon$.*
(3) *$f|_{\tilde{D}^2_\epsilon} : \tilde{D}^2_\epsilon \to D^4_\epsilon$ is $C^0$-$\mathscr{A}$-equivalent to the cone of $f|_{\tilde{S}^1_\epsilon} : \tilde{S}^1_\epsilon \to S^3_\epsilon$.*

PROOF. We take a representative $f : U \to V$, where $U \subset \mathbb{R}^2$ and $V \subset \mathbb{R}^4$ are open neighborhoods of the origin, such that $f$ is a $C^0$-embedding and an immersion on $U \setminus \{0\}$. We consider $g : U \to \mathbb{R}$ given by $g = \|f\|^2$. By the Curve Selection Lemma, $g$ has a finite number of critical values. There exists $\epsilon_0 > 0$ such that $\forall \epsilon$, $0 < \epsilon \leq \epsilon_0$, $\epsilon$ is a regular value of $g$. This implies that $S^3_\epsilon$ is transverse to $f(U)$. Then, $\tilde{S}^1_\epsilon = f^{-1}(S^3_\epsilon)$ is a compact 1-dimensional submanifold of $U$ and $f|_{\tilde{S}^1_\epsilon} : \tilde{S}^1_\epsilon \to S^3_\epsilon$ is an embedding. We have the first part of (2).

To prove (1), we use Reeb Theorem: since $f^{-1}(0) = \{0\}$, 0 is an isolated minimum of $g$. Then, $\tilde{D}^2_\epsilon$, which is equal to $g^{-1}([0, \epsilon])$, is homeomorphic to the closed disk $D^2$. Then, $\tilde{S}^1_\epsilon = \partial \tilde{D}^2_\epsilon$ is homeomorphic (and thus diffeomorphic) to $S^1$.

Denote $I = (0, \epsilon]$ and consider the following diffeomorphisms:

$$\Phi : \tilde{D}^2_\epsilon \setminus \{0\} \longrightarrow I \times \tilde{S}^1_\epsilon, \qquad \Psi : D^4_\epsilon \setminus \{0\} \longrightarrow I \times S^3_\epsilon,$$
$$x \longmapsto (g(x), \phi(x)), \qquad y \longmapsto (\|y\|^2, \sqrt{\epsilon}\frac{y}{\|y\|}),$$

where $\phi(x)$ is the point of $\tilde{S}^1_\epsilon$ where the integral curve of the gradient of $g$ passing through $x$ meets $\tilde{S}^1_\epsilon$.

Define $F : I \times \tilde{S}^1_\epsilon \to I \times S^3_\epsilon$ by $F = \Psi \circ f \circ \Phi^{-1}$. By construction, $F(\{t\} \times \tilde{S}^1_\epsilon) \subset \{t\} \times S^3_\epsilon$, for all $t \in I$. Hence, we can write $F$ in the form $F(t, x) = (t, f_t(x))$, with $f_t : \tilde{S}^1_\epsilon \to S^3_\epsilon$ and $t \in I$.

Since $f_t$ is an embedding, then $f_t$ is stable. Therefore, $F$ is $\mathscr{A}$-trivial, that is, there exist diffeomorphisms $H$ and $K$ of the form:

$$H : I \times \tilde{S}^1_\epsilon \longrightarrow I \times \tilde{S}^1_\epsilon, \qquad K : I \times S^3_\epsilon \longrightarrow I \times S^3_\epsilon,$$
$$(t, x) \longmapsto (t, h_t(x)), \qquad (t, y) \longmapsto (t, k_t(y)),$$

and such that $K \circ F \circ H^{-1} = \mathrm{id} \times f_\epsilon$. This shows the second part of (2).

Finally, we see (3). We have that $f|_{\tilde{D}^2_\epsilon \setminus \{0\}}$ is $\mathscr{A}$-equivalent to $\mathrm{id} \times f_\epsilon$. By adding the origin, this $\mathscr{A}$-equivalence can be extended to a $C^0$-$\mathscr{A}$-equivalence between $f|_{\tilde{D}^2_\epsilon} : \tilde{D}^2_\epsilon \to D^4_\epsilon$ and the cone of $f|_{\tilde{S}^1_\epsilon} : \tilde{S}^1_\epsilon \to S^3_\epsilon$. □

DEFINITION 2.6. Let $f : (\mathbb{R}^2, 0) \to (\mathbb{R}^4, 0)$ be an analytic map germ with isolated singularity. We say that $\epsilon_0 > 0$ is a Milnor-Fukuda radius for $f$ if for any $\epsilon$, with $0 < \epsilon \leq \epsilon_0$, the above conditions (1), (2) and (3) are satisfied. We call the map $f|_{\tilde{S}^1_\epsilon} : \tilde{S}^1_\epsilon \to S^3_\epsilon$ the *link* of $f$ and denote it by $L(f)$. We call the image of the link the *knot* of $f$ and denote it by $K(f)$.

We deduce some immediate consequences of the cone structure theorem:

(1) $L(f)$ is an embedding $S^1 \to S^3$,
(2) $L(f)$ is well defined up to $\mathscr{A}$-equivalence,
(3) $K(f)$ is well defined up to equivalence of knots,
(4) $f$ is topologically equivalent to the cone of $L(f)$.

We also have the following obvious corollary.

COROLLARY 2.7. *Let $f, g : (\mathbb{R}^2, 0) \to (\mathbb{R}^4, 0)$ be analytic map germs with isolated singularity. If $L(f), L(g)$ are $C^0$-$\mathscr{A}$-equivalent, then $f, g$ are $C^0$-$\mathscr{A}$-equivalent.*

In general, we do not know whether or not the converse of the corollary is also true, that is, if the $C^0$-$\mathscr{A}$-equivalence of the germs implies the $C^0$-$\mathscr{A}$-equivalence of their links. In fact, we have the following weaker property.

PROPOSITION 2.8. *Let $f, g : (\mathbb{R}^2, 0) \to (\mathbb{R}^4, 0)$ be analytic map germs with isolated singularity. If $f, g$ are $C^0$-$\mathscr{A}$-equivalent, then $K(f), K(g)$ are equivalent by a knot equivalence and a knot cobordism.*

PROOF. Assume there exist homeomorphisms $\phi, \psi$ such that $g = \psi \circ f \circ \phi^{-1}$. Let $\epsilon > 0$ be a Milnor-Fukuda radius for $f$ and denote $P_\epsilon = \phi(\tilde{D}^2_\epsilon)$ and $Q_\epsilon = \psi(D^4_\epsilon)$. We take now $\delta > 0$ a Milnor-Fukuda radius for $g$ such that $D^4_\delta \subset Q_\epsilon$ and consider

$$V_{\epsilon,\delta} = P_\epsilon \setminus \mathrm{Int}(\tilde{D}^2_\delta), \quad W_{\epsilon,\delta} = Q_\epsilon \setminus \mathrm{Int}(D^4_\delta).$$

By the annulus theorem, $V_{\epsilon,\delta}$ is homeomorphic to $S^1 \times I$ and $W_{\epsilon,\delta}$ is homeomorphic to $S^3 \times I$. Since the restriction $g : V_{\epsilon,\delta} \to W_{\epsilon,\delta}$ is an embedding, it defines a knot cobordism between $(\partial P_\epsilon, \psi(K(f)))$ and $(S^3_\delta, K(g))$. On the other hand, $\psi$ gives a knot equivalence between $(S^3_\epsilon, K(f))$ and $(\partial P_\epsilon, \psi(K(f)))$. □

REMARK 2.9. As it is mentioned in the introduction, in the case that $f, g : (\mathbb{C}, 0) \to (\mathbb{C}^2, 0)$ are holomorphic map germs, the converse of Corollary 2.7 is true. That is, if $f, g$ are $C^0$-$\mathscr{A}$-equivalent then $K(f), K(g)$ are equivalent knots and hence, $L(f), L(g)$ are also $C^0$-$\mathscr{A}$-equivalent. In fact, the knot is always an iterated torus knot, whose type is determined by the exponents of the Newton-Puiseux expansion of the germ. Moreover, the knot type of the link is a topological invariant of the germ (see for instance [**12**] for a proof).

DEFINITION 2.10. We say that $f : (\mathbb{R}^2, 0) \to (\mathbb{R}^4, 0)$ is *unknotted* if it is $C^0$-$\mathscr{A}$-equivalent to the standard embedding $(x, y) \mapsto (x, y, 0, 0)$.

It follows from the definition and from Corollary 2.7 that if $K(f)$ is the trivial knot, then $f$ is unknotted.

## 3. Generic projections

Given a knot $K \subset S^3_\epsilon$, we can obtain a lot of information by means of a generic projection onto $S^2$. In fact, if the projection is generic, the projected knot has only a finite number of transverse crossings. If we know which is the relative position of the branches near each crossing (over or under passing), then we can recover the knot type of $K$ from its projection.

Let $v \in S^3$ be a vector such that $\pm v \notin K$. Denote by $p_v : \mathbb{R}^4 \to \mathbb{R}^3$ the linear projection whose kernel is the line generated by $v$. Then, $\pi_v : S^3 \setminus \{\pm v\} \to S^2$ is the associated spherical projection $\pi_v(x) = p_v(x)/\|p_v(x)\|$. It is well known that for a generic $v$, $\pi_v(K)$ is an immersed curve with only transverse double points. We call this type of curves a *doodle* in $S^2$.

If we have now a doodle $D$ in $S^2$, we can associate the so-called Gauss word: we label the crossings of $D$ with letters $a_1, \dots, a_k$ (one for each crossing) and we construct a sequence by following some chosen orientation of the curve and starting from some chosen base point. Each letter appears twice in the sequence, one with exponent $+1$ and another one with exponent $-1$, according to the relative position of the branches of the curve with respect to some chosen orientation of $S^2$ (see fig. 1).

The following lemma gives a sufficient condition for a knot to be trivial in terms of the Gauss word of a generic projection.

LEMMA 3.1. *Let $K \subset S^3$ be a knot and assume there is a generic projection $\pi_v(K) \subset S^2$ whose Gauss word is the identity when considered as an element of the free group generated by the letters of the Gauss word. Then $K$ is the trivial knot.*

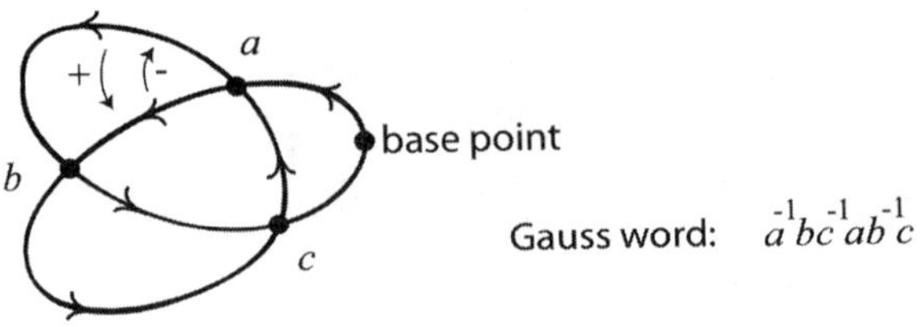


FIGURE 1

PROOF. We proceed by induction on the number $k$ of crossings of $\pi_v(K)$. If $k=1$, the result is trivial. We suppose that it is true for any knot whose generic projection has $k-1$ crossings and assume that $\pi_v(K)$ has $k$ crossings. By assumption, the Gauss word of $\pi_v(K)$ must contain at least a pair of consecutive letters of the form $a_i a_i^{-1}$ or $a_i^{-1} a_i$. Then, the crossing labeled by $a_i$ forms a loop in $\pi_v(K)$ which can be eliminated by means of the first Reidemeister move (see fig. 2). Thus, we obtain another presentation of the same knot $K$ with a generic projection whose Gauss word has $k-1$ crossings and has again the same property (it is the identity when considered as an element of the free group generated by the letters). By induction hypothesis, $K$ is the trivial knot. □

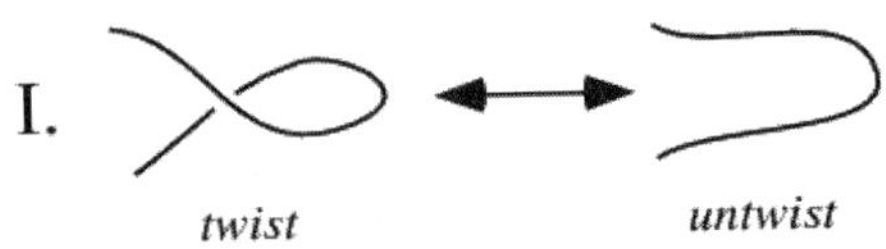


FIGURE 2

Let $f:(\mathbb{R}^2,0)\to(\mathbb{R}^4,0)$ be an analytic map germ with isolated singularity and consider the projection $p_v\circ f:(\mathbb{R}^2,0)\to(\mathbb{R}^3,0)$. We fix a representative $f:U\to V$ such that $f$ is a $C^0$-embedding and an immersion on $U\setminus\{0\}$. It follows from the Thom Transversality Theorem that for a generic $v$, the map $\tilde{f}:=p_v\circ f$ is also generic on $U\setminus\{0\}$. Hence, $\tilde{f}$ is an immersion with normal crossings except at isolated points, where $\tilde{f}$ has singularities of Whitney umbrella type. By shrinking $U$ if necessary, we can avoid the triple points and the Whitney umbrellas, so that $\tilde{f}$ has only transverse double points on $U\setminus\{0\}$. It follows that $\tilde{f}$ has a well defined link $L(\tilde{f})$ in the sense of [**7**]. This link is an immersed closed curve in $S^2_\epsilon$ with only transverse double points and is called in [**7**] the doodle of $\tilde{f}$.

THEOREM 3.2. *Let $f:(\mathbb{R}^2,0)\to(\mathbb{R}^4,0)$ be an analytic map germ with isolated singularity. Then the knot of $f$ has a generic projection equivalent to the doodle of $p_v\circ f$, for some generic vector $v$.*

PROOF. We assume for simplicity that $v=(0,0,0,1)$. We fix a small enough representative $f:U\to V$ and for each $t\in[0,1]$, we define

$$f_t(x,y)=(f_1(x,y),f_2(x,y),f_3(x,y),tf_4(x,y)).$$

We have that $f_1=f$ and $f_0$ can be identified with $p_v\circ f$.

We choose $\epsilon > 0$ small enough such that $K_1 = f_1(U) \cap S^3_\epsilon$ is the knot of $f_1$ and $K_0 = f_0(U) \cap S^3_\epsilon$ is the doodle of $f_0$. We take now $\delta > 0$ small enough such that for all $t \in (0, \delta)$, $K_t = f_t(U) \cap S^3_\epsilon$ is a knot whose projection $\pi_v(K_t)$ is generic. On one hand, $\pi_v(K_t)$ is isotopic, and hence equivalent, to $\pi_v(K_0) = K_0$. On the other hand, $K_1$ is obviously equivalent to $K_t$ just by taking a linear change of coordinates. Thus, $K_1$ has a generic projection equivalent to $\pi_v(K_t)$. □

We denote by $J^2(2,4)$ the space of 2-jets $j^2f(0)$ of smooth map germs $f : (\mathbb{R}^2, 0) \to (\mathbb{R}^4, 0)$ and by $\Sigma^1 J^2(2,4)$ the subset of 2-jets of corank 1. Moreover, $\mathscr{A}^2$ denotes the group of 2-jets of diffeomorphisms in the source and target. The following proposition gives the classification of 2-jets, which is analogous to the case $(\mathbb{R}^2, 0) \to (\mathbb{R}^3, 0)$ (see [**8**]).

PROPOSITION 3.3. *There exist four orbits in $\Sigma^1 J^2(2,4)$ under the action of $\mathscr{A}^2$, which are*

$$(x, y^2, xy, 0), \quad (x, y^2, 0, 0), \quad (x, xy, 0, 0), \quad (x, 0, 0, 0).$$

PROOF. Since $f$ has corank 1, we can assume that

$$j^2f(0) = (x, a_{20}x^2 + a_{11}xy + a_{02}y^2, b_{20}x^2 + b_{11}xy + b_{02}y^2, c_{20}x^2 + c_{11}xy + c_{02}y^2).$$

Moreover, by using a coordinate change in the target, we can eliminate the terms in $x^2$, so we can take $a_{20} = b_{20} = c_{20} = 0$. We denote by $A$ the matrix

$$A = \begin{pmatrix} a_{11} & a_{02} \\ b_{11} & b_{02} \\ c_{11} & c_{02} \end{pmatrix}.$$

If $A$ has rank 2, then $j^2f(0)$ is equivalent to $(x, y^2, xy, 0)$ by a linear coordinate change in the target. Analogously, if $A$ has rank 1, then $j^2f(0)$ is equivalent to either $(x, y^2, 0, 0)$ or $(x, xy, 0, 0)$. Finally, if $A$ has rank 0, then $j^2f(0) = (x, 0, 0, 0)$. □

It is convenient to use the Thom-Boardman symbol to discriminate the two first types in the list above. In fact, it is easy to see that $f : (\mathbb{R}^2, 0) \to (\mathbb{R}^4, 0)$ has type $\Sigma^{1,0}$ if and only if its 2-jet is equivalent to either $(x, y^2, xy, 0)$ or $(x, y^2, 0, 0)$.

COROLLARY 3.4. *Let $f : (\mathbb{R}^2, 0) \to (\mathbb{R}^4, 0)$ be an analytic map germ with isolated singularity of type $\Sigma^{1,0}$. Then $K(f)$ is the trivial knot.*

PROOF. It follows from Proposition 3.3 that if $f$ has type $\Sigma^{1,0}$, then its generic projection $p_v \circ f : (\mathbb{R}^2, 0) \to (\mathbb{R}^3, 0)$ has also type $\Sigma^{1,0}$. By [**7**, Theorem 4.2], $p_v \circ f$ has a doodle whose Gauss word is of the form

$$a_1 a_2^{-1} a_3 \dots a_k^{\epsilon} a_k^{-\epsilon} \dots a_3^{-1} a_2 a_1^{-1},$$

with $\epsilon = (-1)^{k+1}$. Moreover, we have from Theorem 3.2 that $K(f)$ has a generic projection equivalent to the doodle of $p_v \circ f$. But this implies that $K(f)$ is the trivial knot by Lemma 3.1. □

## 4. $\delta$-constant families

By a 1-parameter family we mean an analytic map germ $F : (\mathbb{R} \times \mathbb{R}^n, 0) \to (\mathbb{R}^p, 0)$ such that $F(t, 0) = 0$ for all $t$. If we denote $f_t(x) = F(t, x)$, then we can consider for each $t$, the germ $f_t : (\mathbb{R}^n, 0) \to (\mathbb{R}^p, 0)$.

DEFINITION 4.1. We say that a family $F : (\mathbb{R} \times \mathbb{R}^2, 0) \to (\mathbb{R}^4, 0)$ is *$\delta$-constant* if $\delta(f_0) < \infty$ and $\delta(f_t) = \delta(f_0)$ for all $t$ in a neighbourhood of 0.

The first interesting property is that a $\delta$-constant family has isolated singularity "uniformly", that is, in a neighbourhood which is independent of $t$. For simplicity, we will restrict ourselves to the corank 1 case.

LEMMA 4.2. *Let $F : (\mathbb{R} \times \mathbb{R}^2, 0) \to (\mathbb{R}^4, 0)$ be a $\delta$-constant family such that $f_0$ has corank 1. There is a representative $F : (-\delta, \delta) \times U \to V$ such that for all $t \in (-\delta, \delta)$, $f_t : U \to V$ is a $C^0$-embedding and an immersion on $U \setminus \{0\}$.*

PROOF. Let $\hat{F} : (\mathbb{C} \times \mathbb{C}^2, 0) \to (\mathbb{C}^4, 0)$ be the complexification of $F$ and denote $\hat{f}_t(x) = \hat{F}(t, x)$. We have $\delta(\hat{f}_0) < \infty$ and $\delta(\hat{f}_t) = \delta(\hat{f}_0)$ for all $t$ in a neighbourhood of 0. Moreover, since $\hat{f}_0$ has corank 1, we can choose coordinates in such a way that

$$\hat{f}_t(x, y) = (x, p_t(x, y), q_t(x, y), r_t(x, y)),$$

and thus,

$$\delta(\hat{f}_t) = \dim_{\mathbb{C}} \frac{\mathscr{O}_3}{\langle \Delta p_t, \Delta q_t, \Delta r_t \rangle}.$$

By the Nullstelensatz, we fix the representatives in an open neighbourhood $\hat{U}$ of 0 in $\mathbb{C}^3$ such that $D^2(\hat{f}_0) = \{0\}$. We use now the principle of conservation of number (see for instance [**6**, p. 250]): there is $\delta > 0$ such that for all $t$ with $|t| < \delta$,

$$\delta(\hat{f}_0) = \sum_{(x,y,y') \in D^2(\hat{f}_t)} \dim_{\mathbb{C}} \frac{\mathscr{O}_{\mathbb{C}^3,(x,y,y')}}{\langle \Delta p_t, \Delta q_t, \Delta r_t \rangle}.$$

But the right hand side must contain at least one term equal to $\delta(\hat{f}_t)$, corresponding to the origin, so that $D^2(\hat{f}_t) = \{0\}$. Finally, if $U$ is the real projection of $\hat{U}$, we have $D^2(f_t) = \{0\}$ on $U$. □

In order to state our main result, we also require our family to have constant multiplicity. We recall that the multiplicity of a map germ $f : (\mathbb{R}^n, 0) \to (\mathbb{R}^p, 0)$ is the number

$$m(f) = \dim_{\mathbb{R}} \frac{\mathscr{E}_n}{\langle f_1, \dots, f_p \rangle}.$$

In case $f$ has corank 1 an is written as in (1), we have

$$m(f) = \dim_{\mathbb{R}} \frac{\mathscr{E}_1}{\langle f_1(0, y), \dots, f_p(0, y) \rangle}.$$

Again, this definition also makes sense for holomorphic map germs if we consider $\mathscr{O}_n$ instead of $\mathscr{E}_n$. In analogy with Definition 4.1, we say that a family $F$ is $m$-constant if $m(f_0) < \infty$ and $m(f_t) = m(f_0)$ for all $t$ small enough.

LEMMA 4.3. *Let $F : (\mathbb{R} \times \mathbb{R}^2, 0) \to (\mathbb{R}^4, 0)$ be an $m$-constant family such that $f_0$ has corank 1. There exist a representative $F : (-\delta, \delta) \times U \to V$ and $\epsilon_0 > 0$ such that for all $t \in (-\delta, \delta)$ and for all $\epsilon$ with $0 < \epsilon \leq \epsilon_0$, $\epsilon$ is a regular value of $g_t = \|f_t\|^2$.*

PROOF. We follow the same notation as in the proof of Lemma 4.2, but in addition, we denote by $\hat{g}_t$ the complexification of $g_t$:

$$\hat{g}_t(x, y) = x^2 + p_t(x, y)^2 + q_t(x, y)^2 + r_t(x, y)^2.$$

We also denote by $\hat{G} : (\mathbb{C} \times \mathbb{C}^2) \to (\mathbb{C} \times \mathbb{C})$ the unfolding associated to the family $\hat{g}_t$, that is, $\hat{G}(t, x, y) = (t, \hat{g}_t(x, y))$.

Assume $m(f_t) = m$ and consider the weights $(m, 1)$ for the variables $(x, y)$. After a linear coordinate change in the target, we can suppose that the monomial $y^m$ only appears in one of the functions $p_t, q_t$ or $r_t$. Then $\hat{g}_t$ has a principal part $x^2 + a_m^2 y^{2m}$, for some $a_m \neq 0$, and all the other monomials in $\hat{g}_t$ have higher weighted degree. In fact, if the term $y^m$ appears in the function $p_t$, we have

$$\begin{aligned} p_t(x,y)^2 &= (a_m y^m + a_{m+1}y^{m+1} + \cdots + x(b_\ell y^\ell + b_{\ell+1}y^{\ell+1} + \dots) + \dots)^2 \\ &= a_m^2 y^{2m} + 2b_\ell x y^{m+\ell} + \dots \end{aligned}$$

Hence all the monomials apart from $y^{2m}$ have weighted degree $\geq 2m + \ell > 2m$. The same argument can be used with the other functions $q_t, r_t$.

This implies that the family $\hat{g}_t$ has constant Milnor number $\mu(\hat{g}_t) = 2m - 1$. By a theorem of Greuel [**5**], the singular locus of $\hat{G}$ is the germ of the $t$ axis $(\mathbb{C} \times \{0\}, 0)$. Then, the discriminant (the image of the singular locus) is again the germ of the $t$ axis in the target. There exist $\hat{U}$ an open neighbourhood of the origin $\mathbb{C}^2$ and $\delta, \epsilon_0 > 0$ such that for all $t$ with $|t| < \delta$ and for all $z \in \mathbb{C}$ with $0 < |z| \leq \epsilon_0$, $z$ is a regular value of $\hat{g}_t$. Finally, we just take $U$ as the real projection of $\hat{U}$. □

THEOREM 4.4. *Let $F : (\mathbb{R} \times \mathbb{R}^2, 0) \to (\mathbb{R}^4, 0)$ be a family which is a $\delta$-constant, $m$-constant and such that $f_0$ has corank 1. Then $K(f_t)$ is knot equivalent to $K(f_0)$ for all $t$ in a neighbourhood of* 0.

PROOF. By lemmas 4.2 and 4.3, there exist a representative $F : (-\delta, \delta) \times U \to V$ and $\epsilon_0 > 0$ such that for all $t \in (-\delta, \delta)$, $f_t : U \to V$ is a $C^0$-embedding and an immersion on $U \setminus \{0\}$ and for all $\epsilon$ with $0 < \epsilon \leq \epsilon_0$, $\epsilon$ is a regular value of $g_t = \|f_t\|^2$. It follows from the proof of Theorem 2.5 that for all $t \in (-\delta, \delta)$, $\epsilon_0$ is a Milnor-Fukuda radius for $f_t$ and $K(f_t) = f_t(U) \cap S^3_{\epsilon_0}$ is the knot of $f_t$.

Moreover, by using the same arguments as in the proof of Theorem 2.5, we can construct a diffeomorphism $\Phi : F^{-1}(S^3_{\epsilon_0}) \to (-\delta, \delta) \times S^1$ of the form $\Phi(t, x, y) = (t, \phi_t(x, y))$ and such that $\phi_t(f_t^{-1}(S^3_{\epsilon_0})) = S^1$. Then $F \circ \Phi^{-1} : (-\delta, \delta) \times S^1 \to S^3_{\epsilon_0}$ give us an isotopy between the knots $K(f_t)$, which will be all of them equivalent. □

We give an application of this theorem to an analyze germs $f : (\mathbb{R}^2, 0) \to (\mathbb{R}^4, 0)$ whose 2-jet has type $(x, xy, 0, 0)$. In general, such a germ can be written in the form:

$$f(x, y) = (x, xy + p(x, y), q(x, y), r(x, y)),$$

for some functions $p, q, r \in \mathfrak{m}_2^3$, where $\mathfrak{m}_2$ denotes the maximal ideal of $\mathscr{E}_2$. We will restrict to the simplest case, namely, when $p = 0$.

THEOREM 4.5. *Let $f : (\mathbb{R}^2, 0) \to (\mathbb{R}^4, 0)$ be a finitely determined map germ of the form*

$$f(x, y) = (x, xy, q(x, y), r(x, y)),$$

*for some functions $q, r \in \mathfrak{m}_2^3$. Then $f$ is unknotted.*

PROOF. We write the functions $q, r$ as $q(x, y) = q_0(y) + xq_1(x, y)$ and $r(x, y) = r_0(y) + xr_1(x, y)$ and consider the 1-parameter family

$$f_t(x, y) = (x, xy, q_0(y) + txq_1(x, y), r_0(y) + txr_1(x, y)).$$

We have that

$$\delta(f_t) = \dim_{\mathbb{R}} \frac{\mathscr{E}_3}{\langle x, \Delta q_0, \Delta r_0 \rangle}, \quad m(f_t) = \dim_{\mathbb{R}} \frac{\mathscr{E}_2}{\langle x, q_0, r_0 \rangle},$$

thus $f_t$ is $\delta$-constant and $m$-constant. By Theorem 4.4, $f_1 = f$ is $C^0$-$\mathscr{A}$-equivalent to $f_0$.

Since $\delta(f_0) < \infty$, we also have $\delta(\gamma_0) < \infty$, where $\gamma_0$ is the plane curve $\gamma_0(y) = (q_0(y), r_0(y))$. Then, $\gamma_0$ has isolated singularity and is $C^0$-$\mathscr{A}$-equivalent to the standard embedding $y \mapsto (y, 0)$. This $C^0$-$\mathscr{A}$-equivalence can be extended to a $C^0$-$\mathscr{A}$-equivalence between $f_0$ and $(x, y) \mapsto (x, xy, y, 0)$. Finally, by using a coordinate change in $\mathbb{R}^4$, we arrive to the standard embedding $(x, y) \mapsto (x, y, 0, 0)$. $\square$

We finish with an example which shows that we may find nontrivial knots in the class of map germs with 2-jet of type $(x, xy, 0, 0)$, if we do not assume $p = 0$.

EXAMPLE 4.6. Let $f : (\mathbb{R}^2, 0) \to (\mathbb{R}^4, 0)$ be the map germ given by

$$f(x, y) = (x, y^3 - xy, y^4 - \frac{3}{2}xy^2, y^5 + \frac{3}{5}x^2y - \frac{5}{3}xy^3).$$

We consider $f$ as $f(x, y) = (x, \gamma_x(y))$, so that $\gamma_x$ is a 1-parameter family of space curves. For $x = 0$, $\gamma_0(y) = (y^3, y^4, y^5)$ has isolated singularity at the origin, whilst for $x \neq 0$, $\gamma_x$ is a stabilization of $\gamma_0$. We show in fig. 3 the pictures of $\gamma_x$ for $x < 0$ (left hand side) and for $x > 0$ (right hand side). The knot of $f$ can be obtained (up to knot equivalence) by taking the union of these two curves, which gives in this case the trefoil knot.

FIGURE 3

## References

[1] Grazielle Feliciani Barbosa, Rodrigo Martins, and Marcelo José Saia, *Topology of simple singularities of ruled surfaces in* $\mathbb{R}^p$, Topology Appl. **159** (2012), no. 2, 397–404, DOI 10.1016/j.topol.2011.09.013. MR2868898

[2] Karl Brauner, *Das verhalten der funktionen in der umgebung ihrer verzweigungsstellen* (German), Abh. Math. Sem. Univ. Hamburg **6** (1928), no. 1, 1–55, DOI 10.1007/BF02940600. MR3069487

[3] Werner Burau, *Kennzeichnung der Schlauchknoten* (German), Abh. Math. Sem. Univ. Hamburg **9** (1933), no. 1, 125–133, DOI 10.1007/BF02940635. MR3069587

[4] Takuo Fukuda, *Local topological properties of differentiable mappings. I*, Invent. Math. **65** (1981/82), no. 2, 227–250, DOI 10.1007/BF01389013. MR641129

[5] Gert-Martin Greuel, *Constant Milnor number implies constant multiplicity for quasihomogeneous singularities*, Manuscripta Math. **56** (1986), no. 2, 159–166, DOI 10.1007/BF01172153. MR850367

[6] Theo de Jong and Gerhard Pfister, *Local analytic geometry*, Advanced Lectures in Mathematics, Friedr. Vieweg & Sohn, Braunschweig, 2000. Basic theory and applications. MR1760953

[7] W. L. Marar and J. J. Nuño-Ballesteros, *The doodle of a finitely determined map germ from* $\mathbb{R}^2$ *to* $\mathbb{R}^3$, Adv. Math. **221** (2009), no. 4, 1281–1301, DOI 10.1016/j.aim.2009.02.008. MR2518639

[8] David Mond, *On the classification of germs of maps from* $\mathbf{R}^2$ *to* $\mathbf{R}^3$, Proc. London Math. Soc. (3) **50** (1985), no. 2, 333–369, DOI 10.1112/plms/s3-50.2.333. MR772717

[9] David Mond, *Some remarks on the geometry and classification of germs of maps from surfaces to 3-space*, Topology **26** (1987), no. 3, 361–383, DOI 10.1016/0040-9383(87)90007-3. MR899055

[10] Marc Soret and Marina Ville, *Singularity knots of minimal surfaces in* $\mathbb{R}^4$, J. Knot Theory Ramifications **20** (2011), no. 4, 513–546, DOI 10.1142/S0218216511009406. MR2796225

[11] C. T. C. Wall, *Finite determinacy of smooth map-germs*, Bull. London Math. Soc. **13** (1981), no. 6, 481–539, DOI 10.1112/blms/13.6.481. MR634595

[12] C. T. C. Wall, *Singular points of plane curves*, London Mathematical Society Student Texts, vol. 63, Cambridge University Press, Cambridge, 2004. MR2107253

[13] Oscar Zariski, *On the Topology of Algebroid Singularities*, Amer. J. Math. **54** (1932), no. 3, 453–465, DOI 10.2307/2370887. MR1507926

Departamento of Matemática, Universidade Federal do Ceará, Campus do Pici Bloco 914, CEP 60455–760, Fortaleza CE, Brazil

*E-mail address*: rodrigoxolidio@gmail.com

Departament de Geometria i Topologia, Universitat de València, Campus de Burjassot, 46100 Burjassot, Spain

*E-mail address*: Juan.Nuno@uv.es

Contemporary Mathematics
Volume **675**, 2016
http://dx.doi.org/10.1090/conm/675/13594

# A presentation matrix associated to the discriminant of a co-rank one map-germ from $\mathbb{C}^n$ to $\mathbb{C}^n$

Aldicio José Miranda and Marcelo José Saia

ABSTRACT. In this work we show how to obtain a presentation matrix of the critical locus of a finite co-rank one map germ $f$ in $\mathcal{O}_{n,n}$, as an $\mathcal{O}_n$-module via $f^*$. We apply a result given by Mond and Pellikaan which guarantees the existence of the presentation matrix, but does not show how to obtain it and even in simple cases this is not easy. Here we show how to obtain the entries of this presentation matrix, moreover we implemented this algorithm of Mond and Pellikaan using the software Maple and Singular. Then we show how to use this presentation matrix to obtain invariants associated to the isolated stable singularities of such germs, called 0-stable singularities. We also compare our method with other and show that in this set-up it is much faster to obtain the presentation.

## 1. Introduction

A presentation of an $R$-module $M$ ($R$, a commutative ring with unit) is an exact sequence

$$R^p \xrightarrow{\lambda} R^q \xrightarrow{\alpha} M \longrightarrow 0 \tag{1.1}$$

of $R$-modules, when $M$ is finitely presented such a presentation always exists and $\lambda$ is called a *presentation matrix* of relations among the generators of the module.

If $(X, x)$ is the multi-germ of a Cohen-Macaulay variety of dimension $n$, Mond and Pellikaan in [**MP**, 2.2] showed an algorithm to obtain a presentation of the $\mathcal{O}_{n+1}$-module $\mathcal{O}_{(X,x)}$ for finite analytic map germs $f : (X, x) \to (\mathbb{C}^{n+1}, 0)$. However this procedure only guarantees the existence of the presentation matrix, but does not show how to obtain it and in general this is not easy. In this paper we give an explicit construction of a presentation matrix when $(X, x)$ is the singular locus of a holomorphic co-rank one map germ $f : (\mathbb{C}^n, 0) \to (\mathbb{C}^n, 0)$.

For such germs the presentation matrix is strongly related to the study of the analytic invariants of the map germ $f$, which are given in terms of the multiple point schemes of its restriction to the critical locus of $f$, and these spaces can be computed by means of the Fitting ideals of the presentation matrix.

2010 *Mathematics Subject Classification.* Primary 58K60, 58K65, 32S05, 32S10.

*Key words and phrases.* Presentation matrix of discriminants, fitting ideals, invariants of singularities.

The first named author was partially supported by FAPEMIG Grant #APQ-00781-13.

This work was partially supported by CAPES, CNPq and FAPESP.

Moreover we implemented this construction using the software *Maple*, [**Mp**] and *Singular*, [**Si**] showing explicitly how to compute the entries of the matrix $\lambda$ for such maps. Then we show how to apply this construction to obtain some invariants associated to the singularities of stable map germs from $(\mathbb{C}^n, 0)$ to $(\mathbb{C}^n, 0)$.

## 2. The presentation

We describe here the algorithm given by Mond and Pellikaan in [**MP**, Section 2.2] to construct the presentation of any $\mathcal{O}_{n+1}$-module $f_*\mathcal{O}_X$.

Denote by $\mathcal{O}_n$ the local ring of holomorphic function germs $h : (\mathbb{C}^n, 0) \to \mathbb{C}$ and $\mathcal{O}_{(n,n)}$ denotes the module of holomorphic map germs $f : (\mathbb{C}^n, 0) \to \mathbb{C}^n$.

Let $(X, x)$ be the multi-germ of a Cohen-Macaulay variety of dimension $n$ and $f : (X, x) \to (\mathbb{C}^{n+1}, 0)$ be the germ of a finite analytic map, by the Weierstrass preparation theorem it follows that $\mathcal{O}_{(X,x)}$ is a finite $\mathcal{O}_{n+1}$-module via the function $f^*$.

To compute the matrix $\lambda$, of relations among the generators of the module $\frac{\mathcal{O}_{(X,x)}}{f^*\mathfrak{m}_0}$ we remark that if the classes of $g_1, g_2, \dots, g_h$ in $\frac{\mathcal{O}_{(X,x)}}{f^*\mathfrak{m}_0}$ generate it as a vector space over $\mathbb{C}$, then $g_1, g_2, \dots, g_h$ generate $\mathcal{O}_{(X,x)}$ as an $\mathcal{O}_{n+1}$-module, therefore it is enough to obtain the presentation of the $\mathcal{O}_n$-module $\mathcal{O}_{(X,x)}$. the relations among the $g_i$, $i = 1, \dots, h$ over $\mathcal{O}_{n+1}$.

**Procedure to obtain the matrix $\lambda$:** Choose a projection $\pi : (\mathbb{C}^{n+1}, 0) \to (\mathbb{C}^n, 0)$ such that $\widetilde{f} = \pi \circ f$ is also finite. After a change of coordinates we suppose that $f(x) = (\widetilde{f}(x), f_{n+1}(x))$ and as $\mathcal{O}_{X,x}$ is Cohen-Macaulay, then $\mathcal{O}_{X,x}$ is a free $\mathcal{O}_n$-module via $\widetilde{f}^*$.

Then for all $i, j$ with $1 \le i, j \le h$ there exist unique $\alpha^i_j \in \mathcal{O}_n$, such that

$$g_j \cdot f_{n+1} = \sum (\alpha^i_j \circ \widetilde{f}) \cdot g_i. \tag{2.1}$$

As the germs $g_i$ generate $\mathcal{O}_{X,x}$ over $\mathcal{O}_{n+1}$ via $f^*$, then

$$\lambda^i_j = \alpha^i_j \circ \pi, \quad i \neq j$$

$$\lambda^i_i = \alpha^i_i \circ \pi - X_{n+1},$$

since $f_{n+1} = X_{n+1} \circ f$ and $(X_1, \dots, X_{n+1})$ denotes the coordinates of $\mathbb{C}^{n+1}$ in the target.

Therefore one has the matrix $(\lambda) = \left(\lambda^i_j\right)$ for the exact sequence of the $\mathcal{O}_{n+1}$-module $\mathcal{O}_{X,x}$.

$$\mathcal{O}^p_{n+1} \xrightarrow{\lambda} \mathcal{O}^q_{n+1} \xrightarrow{\alpha} \mathcal{O}_{X,x} \longrightarrow 0 \tag{2.2}$$

We note that the algorithm of Mond-Pellikaan only guarantees the existence of the entries $\alpha^i_j$ in the matrix, but does not show how to obtain these. In Section 3 we show explicitly how to get the $\alpha^i_j$ in the case dealt with in this article.

If $g_1, g_2, \dots, g_h$ generate $\mathcal{O}_{X,x}$, we can considere $g_1 = 1$, $q = p = h$ and the matrix $\alpha$ is given in such a way that $\alpha(e_i) = g_i$, where $e_i$ is the $i^{th}$ element of the usual basis.

## 3. The presentation for co-rank one map germs in $\mathcal{O}(n,n)$

For any co-rank one finite map germ $f$ in $\mathcal{O}(n,n)$, the set $(\Sigma(f),0)$ is an $(n-1)$-dimensional Cohen-Macaulay variety, therefore we can use the procedure of Mond-Pellikaan to compute its presentation. Here we show explicitly this construction.

**The presentation matrix:**

Let $f \in \mathcal{O}(n,n)$ be a finite map germ of co-rank one, after choosing linearly adapted coordinates one can write $f(x,z) = (x_1,\ldots,x_{n-1},g(x,z))$, where $x = (x_1,\ldots,x_{n-1}) \in \mathbb{C}^{n-1}$, $z \in \mathbb{C}$ and $g : (\mathbb{C}^n,0) \to (\mathbb{C},0)$ is a polynomial that can be written in the form $g(x,z) = z^{k+1} + h(x,z)$, with

$$h(x,z) = h_{k-1}(x)z^{k-1} + h_{k-2}(x)z^{k-2} + \cdots h_1(x)z + h_0(x)$$

and $h_i(0) = 0$ for all $h_i : (\mathbb{C}^n,0) \to (\mathbb{C},0)$, $i = 0,\ldots,k-1$.

To describe the presentation matrix explicitly we remember that the Jacobian determinant of the matrix of the derivatives of $f$ at any point $(x,z)$ is $J(f) = (k+1)z^k + h_z(x,z)$, where $h_z(x,z)$ denotes the derivative of $h(x,z)$ with respect to the variable $z$.

From the fact that $f$ is of co-rank one it follows that the local algebra $\frac{\mathcal{O}_{(\Sigma(f),0)}}{f^*\mathfrak{m}_{\mathbb{C}^n,0}}$ is isomorphic to $\frac{\mathbb{C}\{z\}}{\langle z^k\rangle}$, or in other words, $\mathcal{O}_{(\Sigma(f),0)}$ is minimally generated as $\mathcal{O}_n$-module (target) via $f|^*_{(\Sigma(f),0)}$ and in this case a system of generators $\{g_0,g_1,\ldots,g_{k-1}\}$ is given by $\{1,z,z^2,\ldots,z^{k-1}\}$.

The main difficulty to obtain the matrix $\lambda$ as in (1.1) is to find $k$ relations among the target variables $X_i = f_i$, $i = \{1,\ldots,n\}$ and the set of generators $\{1,z,z^2,\ldots,z^{k-1}\}$, module the jacobian ideal $J(f)$. These relations are shown below.

For the first relation, one has

$$g(x,z)\cdot 1 = z^{k+1} + h_{k-1}(x_1,\ldots,x_{n-1})\cdot z^{k-1} + \cdots + h_0(x_1,\ldots,x_{n-1})\cdot 1$$

Let $X_i := f_i(x,z) = x_i$, $i = \{1,\ldots,n-1\}$ and $X_n := g(x,z)$, then

$$X_n\cdot 1 = z^{k+1} + H_{k-1}(X_1,\ldots,X_{n-1})\cdot z^{k-1} + \cdots + H_0(X_1,\ldots,X_{n-1})\cdot 1$$

$$X_n\cdot 1 = z^{k+1} + H_{k-1}(\overline{X})\cdot z^{k-1} + \cdots + H_1(\overline{X})\cdot z + H_0(\overline{X})\cdot 1$$

$$z^{k+1} = (X_n - H_0(\overline{X}))\cdot 1 - H_1(\overline{X})\cdot z - \cdots - H_{k-1}(\overline{X})\cdot z^{k-1} \tag{3.1}$$

where $H_j$ is just $h_j$, $j = 0,\ldots,k-1$, substituting $x_i$ by $X_i$, $i = 1,\ldots,n-1$.

On the other hand, $(k+1)\cdot z^k + h_z(x,z) = 0$ (in $\mathcal{O}_{(\Sigma(f),0)}$). Then

$$(k+1)\cdot z^k = -(k-1)h_{k-1}(x)\cdot z^{k-2} - \cdots - 2h_2(x)\cdot z - h_1(x)\cdot 1$$

$$z^k = -\left(\frac{k-1}{k+1}\right)H_{k-1}(\overline{X})\cdot z^{k-2} - \cdots - \left(\frac{1}{k+1}\right)H_1(\overline{X})\cdot 1 \tag{3.2}$$

From equality (3.1) and equality (3.2) multiplied by $z$, we obtain

$$(X_n - H_0(\overline{X}))\cdot 1 - \left(\frac{k}{k+1}\right)H_1(\overline{X})\cdot z - \cdots - \left(\frac{2}{k+1}\right)H_{k-1}(\overline{X})\cdot z^{k-1} = 0. \tag{3.3}$$

Now, denote $H_{1,1}(\overline{X}) = -H_0(\overline{X})$, $H_{1,j+1}(\overline{X}) = -\left(\frac{k+1-j}{k+1}\right) H_j(\overline{X})$, $j = 1, \ldots, k-1$ and equation (3.3) shows that

$$(X_n + H_{1,1}(\overline{X})) \cdot g_0 + \sum_{i=1}^{k-1} H_{1,i+1}(\overline{X}) g_i = 0. \tag{3.4}$$

Therefore the first row of the matrix $\lambda$ is

$$\begin{bmatrix} X_n + H_{1,1}(\overline{X}) & H_{1,2}(\overline{X}) & \cdots & H_{1,k}(\overline{X}) \end{bmatrix}_{1\times k}.$$

To obtain the second row of the matrix $\lambda$, we multiply the equation (3.4) by $g_1 := z$, then

$$X_n \cdot g_1 + H_{1,1}(\overline{X}) \cdot g_1 + \sum_{i=1}^{k-1} H_{1,i+1}(\overline{X}) g_i \cdot g_1 = 0.$$

As $g_i \cdot g_j = g_{i+j}$, $i + j < k$,

$$X_n \cdot g_1 + H_{1,1}(\overline{X}) \cdot g_1 + H_{1,2}(\overline{X}) \cdot g_2 + \cdots + H_{1,k-1}(\overline{X}) \cdot g_{k-1} + H_{1,k}(\overline{X}) \cdot z^k = 0. \tag{3.5}$$

Substituting the right hand of the equation (3.2) in the equation (3.5), regrouping and renaming the terms

$$H_{2,1}(\overline{X}) \cdot g_0 + X_n \cdot g_1 + H_{2,2}(\overline{X}) \cdot g_1 + H_{2,3}(\overline{X}) \cdot g_2 + \cdots + H_{2,k}(\overline{X}) \cdot g_{k-1} = 0.$$

And the second row of the matrix is given by

$$\begin{bmatrix} H_{2,1}(\overline{X}) & X_n + H_{2,2}(\overline{X}) & \cdots & H_{2,k}(\overline{X}) \end{bmatrix}_{1\times k}.$$

Observe that the equation (3.2) does not depend of the variable $X_n$.

Proceeding in this way, to obtain the $r^{th}$ row multiply the $(r-1)^{th}$ row by $z$ or the first row by $z^{r-1}$ and use the equations (3.1) and (3.2). The process is algorithmic and finishes after $k$ rows, where $k$ is the number of generators. To conclude, by [**MP**, Section 2.2] , the relations among the $g_i$ obtained above generate the module Ker $\alpha$.

Therefore we obtain for any co-rank one finitely determined map germ $f : (\mathbb{C}^n, 0) \to (\mathbb{C}^n, 0)$ written as $f(x, z) = (x_1, \ldots, x_{n-1}, g(x, z))$ the following:

THEOREM 3.1. *The presentation matrix resulting from our algorithm of $\mathcal{O}_{(\Sigma(f),0)}$ over $\mathcal{O}_n$ is given by:*

$$\lambda = \begin{bmatrix} X_n + H_{1,1}(\overline{X}) & H_{1,2}(\overline{X}) & \cdots & H_{1,k}(\overline{X}) \\ H_{2,1}(\overline{X}) & X_n + H_{2,2}(\overline{X}) & & H_{2,k}(\overline{X}) \\ \vdots & & \ddots & \vdots \\ H_{k,1}(\overline{X}) & \cdots & & X_n + H_{k,k}(\overline{X}) \end{bmatrix}_{k\times k}$$

*where, $H_{i,j} : (\mathbb{C}^{n-1}, 0) \longrightarrow (\mathbb{C}, 0)$ are the polynomials constructed above and $(\overline{X}, X_n) = (X_1, \ldots, X_{n-1}, X_n)$ denote the target variables.*

REMARK 3.2. From the results of Mond-Pellikaan we can say that the elements of this matrix $\lambda$ are in the ring $\mathcal{O}_{n+1}$, but in this case the entries are polynomials and the determinant of the matrix $\lambda$ is of the form $X_n^k + B$, where $B$ is a polynomial in the variables $X_1, \ldots, X_n$, with $B(0, \ldots 0, X_n) = 0$.

We implemented this procedure, called **Presentation_CnCn_Cor1** using the software Maple, see section 5 for this implementation. For the implementation of this procedure using the software Singular, see [**Mi**].

## 4. Applications to Singularity Theory

**4.1. Fitting ideals and invariants.** The determinacy of numerical invariants associated to map germs is a powerful tool in the study of problems of its singularities, in general these invariants appear as schemes associated to the discriminant of stable maps, called stable singularities.

For the particular singularities which are isolated, called 0-stable singularities, the type and also the number of such singularities is very relevant because they carry information about the local geometric behavior of such maps, as we can see in [**GM1**], [**M2**] , [**M3**], [**JLS**], [**JS**] and [**JRS**]. In general the computation of the 0-stable singularities is not easy, but when the germ is quasi homogeneous, there are several works that show how to compute such numbers in terms of the weights and degrees of quasi homogeneity, see for instance [**GM2**], [**MMR**] , [**M1**] and [**RS**].

From the presentation matrix one has the key tools to compute the stable singularities of the map germ, its Fitting ideals, as shown by Mond and Pellikaan in [**MP**].

DEFINITION 4.1. Let $M$ be a finitely presented $R$-module (where $R$ is a commutative ring with unit) and let

$$R^p \xrightarrow{\lambda} R^q \xrightarrow{\alpha} M \longrightarrow 0$$

be a presentation. The $k^{th}$ Fitting ideal of $M$, $\mathcal{F}_k(M)$, is defined to be the ideal in $R$ generated by all $(q-k)\times(q-k)$ minors of the matrix $\lambda$, for $q > k \geq q-p$. $\mathcal{F}_k(M) = R$, for $k \geq q$, and $\mathcal{F}_k(M) = 0$ for $k < q-p$.

By [**Lo**, p. 59], the Fitting ideals are well defined, that is, they do not depend upon the choice of a presentation.

According to [**MP**, Proposition 1.5], $V(\mathcal{F}_k(\mathcal{O}_{(\Sigma(f),0)}))$ is the space of points in the target whose pre-image consists of $(k+1)$ or more points in $\Sigma(f)$, counting multiplicities. These spaces are the so called multiple points spaces of singularities of $f$ and their computation is the key tool to study invariants of the singularities of the germ $f$. To calculate the $k^{th}$ Fitting ideals of the matrix $\lambda$, we use the command $fitting(\lambda, k)$ in Singular.

We consider here the usual notation for the $A_k$ singularities introduced by Arnol'd. For a fixed partition $\mathcal{P} = (k_1, k_2, \ldots, k_s)$ of $m \leq n$, we call $A_{\mathcal{P}}$ the singularity in the discriminant of a stable co-rank one germ $f : (\mathbb{C}^n, 0) \to (\mathbb{C}^n, 0)$ formed by the multiple points given as the normal crossing of the singularities $A_{k_\ell}$ with $\ell = 1, \ldots, s$.

In the case of co-rank one map germs from $(\mathbb{C}^3, 0)$ to $(\mathbb{C}^3, 0)$, the multiple points spaces in the discriminant are formed by the isolated singularities: $A_3$ (swallowtail), $A_{2,1}$ (transversal crossing between a cuspidal edge, $A_2$ singularity, and the singularity $A_1$) and $A_{1,1,1}$ (the transversal crossing of 3 $A_1$ singularities). The one dimensional singularities are cuspidal edge and the double points curve $A_{1,1}$ (transversal crossing between two singularities of type $A_1$). For such germs the defining equation of the discriminant of $f$ is the equation defined by the ideal $\mathcal{F}_0(f)$. The union of ordinary double points curve with the cuspidal edge curve is the zero space of the ideal $\mathcal{F}_1(f)$ and the union of all 0-stable singularities appears as the zero space of the ideal $\mathcal{F}_2(f)$, hence, as we see in [**JMS**], the total number of these 0-stable singularities is given as the complex dimension (as vector space) of the algebra $\frac{\mathcal{O}_3}{\mathcal{F}_2(f)}$, or

$$\sharp A_3 + \sharp A_{2,1} + \sharp A_{1,1,1} = \dim_{\mathbb{C}} \frac{\mathcal{O}_3}{\mathcal{F}_2(f)},$$

where $\sharp A_{\mathcal{P}}$ denotes the number of $A_{\mathcal{P}}$-singularities.

**4.2. The defining equation of the discriminant.** Using this algorithm we can compute the *defining equation of the discriminant* in several cases that the classical method of elimination of variables does not terminate because of lack of memory. This method gives the defining equation using the command *eliminate* in the software Maple or Singular.

We see a description of this method in the proof of the Lemma 1.3. of [**GM2**] for the case of map germs $g : \mathbb{C}^2 \to \mathbb{C}^2$. Here we show it for map germs $f : \mathbb{C}^4 \to \mathbb{C}^4$.

Given a map germ $f = (f_1, f_2, f_3, f_4)$ do:

(1) In the variables $\{x, y, z, w, X, Y, Z, W\}$, define the ideal $I$ generated by $\{X - f_1, Y - f_2, Z - f_3, W - f_4, Jf\}$, where $Jf$ denotes the determinant of matrix of the partial derivatives of $f$, and compute its standard basis.
(2) In the standard basis of this ideal, find the unique function which depends of the variables $(X, Y, Z, W)$, denoted $G(X, Y, Z, W)$, this is the defining equation of $\Delta(f)$ in $\mathbb{C}^4$.

The main problem of this algorithm is that it uses a huge amount of memory to compute the standard basis of the ideal $I$ and even for simple cases, using very powerful computers, the process does not terminate.

For instance, consider the map germ $f$ from $\mathbb{C}^4$ to $\mathbb{C}^4$ given by

$$f(x, y, z, w) = (x, y, z, w^{33} + z^{13}w^3 + y^8 w + x^3 w^8),$$

by using a computer equipped with 2.8 *Ghz* Intel Core *I7* processor with 8 *Gb* of RAM memory, the total time to find the $32 \times 32$-matrix $\lambda_f$ of its presentation using the command Presentation_CnCn_Cor1([f]) is less than 0.1 second. To compute the defining equation the total time was 2.82 seconds, while using the command *eliminate* in the software Maple, the total time spent is 460 seconds. Using the software Singular in this computer it is not possible to obtain this defining equation because there is not enough memory.

In the examples $\lambda_f$ denotes the presentation matrix of $\mathcal{O}_{(\Sigma(f),0)}$ as $\mathcal{O}_{\mathbb{C}^n,0}$-module.

**4.3. Examples.**

EXAMPLE 4.2. Let $f : (\mathbb{C}^3, 0) \to (\mathbb{C}^3, 0)$ be defined by $f(x, y, z) = (x, y, z^6 + xz + yz^2)$. As $f$ is of co-rank one, then $\frac{\mathcal{O}_{(\Sigma(f),0)}}{f^*\mathfrak{m}_{(\mathbb{C}^3,0)}} \cong \frac{\mathbb{C}\{z\}}{<z^5>}$ and $\mathcal{O}_{(\Sigma(f),0)}$ is finitely generated as $\mathcal{O}_3$-module via $f^*$.

A presentation matrix of $f$ is given by

$$\lambda_f = \begin{bmatrix} X_3 & -\frac{5}{6}X_1 & -\frac{2}{3}X_2 & 0 & 0 \\ 0 & X_3 & -\frac{5}{6}X_1 & -\frac{2}{3}X_2 & 0 \\ 0 & 0 & X_3 & -\frac{5}{6}X_1 & -\frac{2}{3}X_2 \\ \frac{1}{9}X_1X_2 & \frac{2}{9}X_2^2 & 0 & X_3 & -\frac{5}{6}X_1 \\ \frac{5}{36}X_1^2 & \frac{7}{18}X_1X_2 & \frac{2}{9}X_2^2 & 0 & X_3 \end{bmatrix}_{5\times 5}.$$

To obtain this matrix using our implementation, in Maple software just enter with the command:

$$\text{Presentation\_CnCn\_Cor1}([X_1, X_2, z^6 + X_1z + X_2z^2]).$$

We remember that $X_1 = x$, $X_2 = y$ and $X_3 = z^6 + xz + yz^2$ since $f$ is a co-rank one map germ.

It is of great interest in Singularity Theory to describe the multiple points spaces (in the target) of any germ $f_t$ given in a deformation of the germ $f$. We show here how to use this algorithm for the particular case that the deformation is given by $f_t(x, y, z) = (x, y, z^6 + xz + yz^2 + tz^4)$.

To solve this problem we consider the one parameter unfolding $F$ of $f$, $F : (\mathbb{C} \times \mathbb{C}^3, 0) \to (\mathbb{C} \times \mathbb{C}^3, 0)$ written as

$$F(t, x, y, z) = (t, x, y, z^6 + xz + yz^2 + tz^4)$$

and use the command Presentation_CnCn_Cor1($[t, X_1, X_2, z^6 + X_1z + X_2z^2 + tz^4]$); fixing the correspondence $x = X_1$, $y = X_2$ and $z^6 + xz + yz^2 + tz^4 = X_3$.

Using the command Presentation_CnCn_Cor1([F]) we obtain directly the $5 \times 5$ matrix $\lambda_{f_t}$ for the presentation

$$\mathcal{O}_3^5 \xrightarrow{\lambda_{f_t}} \mathcal{O}_3^5 \xrightarrow{\alpha} \mathcal{O}_{\Sigma(f_t),0} \to 0$$

where $\lambda_{f_t}$ is given by

$$\begin{bmatrix} X_3 & -\frac{5}{6}X_1 & -\frac{2}{3}X_2 & 0 & -\frac{1}{3}t \\ \frac{1}{18}tX_1 & X_3 + \frac{1}{9}tX_2 & -\frac{5}{6}X_1 & -\frac{2}{3}X_2 + \frac{2}{9}t^2 & 0 \\ 0 & \frac{1}{18}tX_1 & X_3 + \frac{1}{9}tX_2 & -\frac{5}{6}X_1 & -\frac{2}{3}X_2 + \frac{2}{9}t^2 \\ \frac{1}{9}X_1(X_2 - \frac{1}{3}t^2) & \frac{2}{9}X_2^2 - \frac{2}{27}t^2X_2 & \frac{1}{18}tX_1 & X_3 + \frac{5}{9}tX_2 - \frac{4}{27}t^3 & -\frac{5}{6}X_1 \\ \frac{5}{36}X_1^2 & \frac{7}{18}X_1X_2 - \frac{1}{27}t^2X_1 & \frac{2}{9}X_2^2 - \frac{2}{27}t^2X_2 & \frac{11}{18}tX_1 & X_3 + \frac{5}{9}tX_2 - \frac{4}{27}t^3 \end{bmatrix}.$$

Using the command $fitting(\lambda, k)$ in the software Singular we compute the corresponding Fitting ideals.

$\mathcal{F}_0(f_t) = 46656X_3^5 + 3125X_1^6 + 22500X_1^4X_2X_3 + 43200X_1^2X_2^2X_3^2 + 32400tX_1^2X_3^3 + 13824X_2^3X_3^3 + 62208tX_2X_3^4 + 2000tX_1^4X_2^2 + 256X_1^2X_2^5 - 1500t^2X_1^43 + 10560tX_1^2X_2^3X_3 + 1024X_2^6X_3 - 6480t^2X_1^2X_2X_3^2 + 9216tX_2^4X_3^2 + 17280t^2X_2^2X_3^3 - 13824t^3X_3^4 - 900t^3X_1^4X_2 - 128t^2X_1^2X_2^4 - 4816t^3X_1^2X_2^2X_3 - 512t^2X_2^5X_3 - 192t^4X_1^2X_3^2 - 4352t^3X_2^3X_3^2 - 9216t^4X_2X_3^3 + 108t^5X_1^4 + 16t^4X_1^2X_2^3 + 576t^5X_1^2X_2X_3 + 64t^4X_2^4X_3 + 512t^5X_2^2X_3^2 + 1024t^6X_3^3.$

$\mathcal{F}_1(f_t) = 1875X_1^4 + 5400X_1^2X_2X_3 + 1728X_2^2X_3^2 + 1296tX_3^3 + 1440X_1^2X_2^2 + 256X_2^5 - 1080t^2X_1^2X_3 + 1344tX_2^3X_3 + 432t^2X_2X_3^2 - 616t^3X_1^2X_2 - 128t^2X_2^4 - 592t^3X_2^2X_3 - 192t^4X_3^2 + 72t^5X_1^2 + 16t^4X_2^3 + 64t^5X_2X_3, 1125X_1^3X_3 + 2160X_1X_2X_3^2 - 100tX_1^3X_2 - 64X_1X_2^4 + 576tX_1X_2^2X_3 - 756t^2X_1X_3^2 + 15t^3X_1^3 + 32t^2X_1X_2^3 - 344t^3X_1X_2X_3 - 4t^4X_1X_2^2 + 48t^5X_1X_3,$

$2700X_1^2X_3^2 + 2592X_2X_3^3 + 125tX_1^4 + 160X_1^2X_2^3 + 180tX_1^2X_2X_3 + 384X_2^4X_3 + 1728tX_2^2X_3^2 - 864t^2X_3^3 - 88t^2X_1^2X_2^2 + 44t^3X_1^2X_3 - 224t^2X_2^3X_3 - 960t^3X_2X_3^2 + 12t^4X_1^2X_2 + 32t^4X_2^2X_3 + 128t^5X_3^2, 1620X_1X_3^3 - 100X_1^3X_2^2 + 150tX_1^3X_3 - 336X_1X_2^3X_3 + 216tX_1X_2X_3^2 + 65t^2X_1^3X_2 + 228t^2X_1X_2^2X_3 + 48t^3X_1X_3^2 - 9t^4X_1^3 - 32t^4X_1X_2X_3, 3888X_3^4 + 250X_1^4X_2 + 1080X_1^2X_2^2X_3 + 540tX_1^2X_3^2 + 576X_2^3X_3^2 + 3024tX_2X_3^3 - 75t^2X_1^4 + 16tX_1^2X_2^3 - 216t^2X_1^2X_2X_3 + 64tX_2^4X_3 + 144t^2X_2^2X_3^2 - 576t^3X_3^3 - 4t^3X_1^2X_2^2 - 24t^4X_1^2X_3 - 16t^3X_2^3X_3 - 64t^4X_2X_3^2.$

$\mathcal{F}_2(f_t) = 75tX_1^2 + 96X_2^3 + 72tX_2X_3 - 56t^2X_2^2 - 24t^3X_3 + 8t^4X_2, 375X_1^3 + 360X_1X_2X_3 + 80tX_1X_2^2 + 12t^2X_1X_3 - 12t^3X_1X_2, 75X_1^2X_2 + 54tX_3^2 - 25t^2X_1^2 + 8tX_2^3 + 36t^2X_2X_3 - 2t^3X_2^2 - 8t^4X_3, 60X_1X_2^2 - 45tX_1X_3 - 43t^2X_1X_2 + 6t^4X_1, 225X_1^2X_3 - 20tX_1^2X_2 + 3t^3X_1^2, 90X_1X_2X_3 - 4tX_1X_2^2 - 33t^2X_1X_3 + t^3X_1X_2, 144X_2^2X_3 + 5t^2X_1^2 - 96t^2X_2X_3 + 16t^4X_3, 540X_1X_3^2 + 25tX_1^3 + 36tX_1X_2X_3 + 8t^3X_1X_3, 216X_2X_3^2 + 135tX_1^2X_2 + 32X_2^4 + 144tX_2^2X_3 - 33t^3X_1^2 - 8t^2X_2^3 - 32t^3X_2X_3, 324X_3^3 + 20X_1^2X_2^2 + 15tX_1^2X_3 + 48X_2^3X_3 + 216tX_2X_3^2 - 5t^2X_1^2X_2 - 12t^2X_2^2X_3 - 48t^3X_3^2, 192X_2^4 - 108t^2X_3^2 + 45t^3X_1^2 - 128t^2X_2^3 - 24t^3X_2X_3 + 20t^4X_2^2.$

With these ideals and using for example the Surfex program, [**Su**], we can visualize the multiple points set in the target in real time moving the parameter $t$. The Figure 4.3 shows the discriminant of $f_t$ for a fixed real value $t \neq 0$.

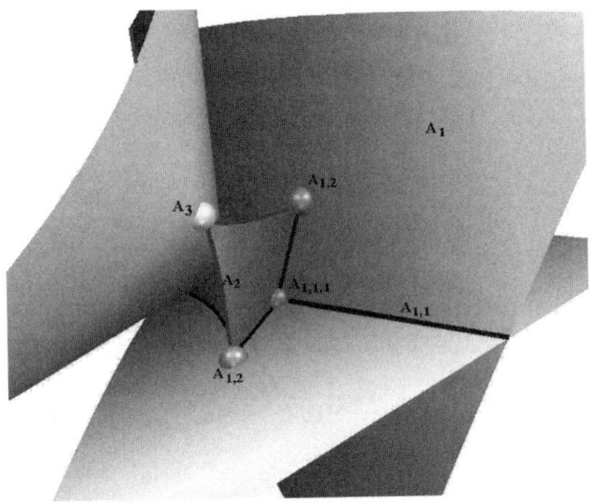


Real part of the discriminant of $f_t(x, y, z) = (x, y, z^6 + xz + yz^2 + tz^4)$.

EXAMPLE 4.3. Let $f : (\mathbb{C}^5, 0) \longmapsto (\mathbb{C}^5, 0)$ be given by:

$$f(a, b, c, d, w) = (a, b, c, d, w^{10} + aw^4 + bw^3 + c^2w^2 + d^2w).$$

The presentation for $\mathcal{O}_{(\Sigma(f),0)}$ is the $9 \times 9$-matrix:

$$\begin{bmatrix}
z & \frac{-9}{10}d^2 & \frac{-4}{5}c^2 & \frac{-7}{10}b & \frac{-3}{5}a & 0 & 0 & 0 & 0 \\
0 & z & \frac{-9}{10}d^2 & \frac{-4}{5}c^2 & \frac{-7}{10}b & \frac{-3}{5}a & 0 & 0 & 0 \\
0 & 0 & z & \frac{-9}{10}d^2 & \frac{-4}{5}c^2 & -\frac{7}{10}b & \frac{-3}{5}a & 0 & 0 \\
0 & 0 & 0 & z & \frac{-9}{10}d^2 & \frac{-4}{5}c^2 & \frac{-7}{10}b & \frac{-3}{5}a & 0 \\
0 & 0 & 0 & 0 & z & \frac{-9}{10}d^2 & \frac{-4}{5}c^2 & \frac{-7}{10}b & \frac{-3}{5}a \\
\frac{3}{50}ad^2 & \frac{3}{25}ac^2 & \frac{9}{50}ab & \frac{6}{25}a^2 & 0 & z & \frac{-9}{10}d^2 & \frac{-4}{5}c^2 & \frac{-7}{10}b \\
\frac{7}{100}bd^2 & \frac{3}{50}ad^2+\frac{7}{50}bc^2 & \frac{3}{25}ac^2+\frac{21}{100}b^2 & \frac{23}{50}ab & \frac{6}{25}a^2 & 0 & z & -\frac{9}{10}d^2 & \frac{-4}{5}c^2 \\
\frac{2}{25}c^2d^2 & \frac{7}{100}bd^2+\frac{4}{25}c^4 & \frac{3}{50}ad^2+\frac{19}{50}bc^2 & \frac{11}{25}ac^2+\frac{21}{100}b^2 & \frac{23}{50}ab & \frac{6}{25}a^2 & 0 & z & \frac{-9}{10}d^2 \\
\frac{9}{100}d^4 & \frac{13}{50}c^2d^2 & \frac{17}{50}bd^2+\frac{4}{25}c^4 & \frac{21}{50}ad^2+\frac{19}{50}bc^2 & \frac{11}{25}ac^2+\frac{21}{100}b^2 & \frac{23}{50}ab & \frac{6}{25}a^2 & 0 & z
\end{bmatrix}$$

We are fixing the variables in the target of this example as $(a, b, c, d, Z)$

Here, by [**JMS**, Corollary 4.5], we obtain that

$$\sharp A_5 + \sharp A_{(4,1)} + \sharp A_{(3,2)} + \sharp A_{(3,1,1)} + \sharp A_{(2,2,1)} + \sharp A_{(2,1,1,1)} + \sharp A_{(1,1,1,1,1)} = \dim_{\mathbb{C}} \frac{\mathcal{O}_5}{\mathcal{F}_4(f)} = 504.$$

We do not show the generators of the Fitting ideal $\mathcal{F}_4(f)$ because it has 126 generators.

On the other hand, as the germ $f$ is weighted homogeneous with weights 12, 14, 8, 9, 2 and weighted degree 20, we can apply the results of [**MMR**] to count each one of the 0-stable singularities associated to this germ. Then we have:

$$\sharp A_5 = 20, \quad \sharp A_{(4,1)} = 80, \quad \sharp A_{(3,2)} = 80, \quad \sharp A_{(3,1,1)} = 120,$$
$$\sharp A_{(2,2,1)} = 120, \quad \sharp A_{(2,1,1,1)} = 80, \ \sharp A_{(1,1,1,1,1)} = 4.$$

EXAMPLE 4.4. Let $f : (\mathbb{C}^5, 0) \longmapsto (\mathbb{C}^5, 0)$ be given by:

$$f(a, b, c, d, w) = (a, b, c, d, w^9 + aw^5 + b^2w^2 + bw^4 + cw^3 + dw^2).$$

The number of generators of the $4^{th}$-fitting ideal is 70, and by [**JMS**, Corollary 4.5], we obtain that

$$\sharp A_5 + \sharp A_{(4,1)} + \sharp A_{(3,2)} + \sharp A_{(3,1,1)} + \sharp A_{(2,2,1)} + \sharp A_{(2,1,1,1)} + \sharp A_{(1,1,1,1,1)} = \dim_{\mathbb{C}} \frac{\mathcal{O}_5}{\mathcal{F}_4(f)} = 112.$$

The total time to get the presentation matrix to this example using the implementation below and to get the ideal $\mathcal{F}_4(f)$ and to calculate the $\dim_{\mathbb{C}} \frac{\mathcal{O}_5}{\mathcal{F}_4(f)}$, was less then 0.5 seconds.

In this case we can not apply the results of [**MMR**] to count each one of the 0-stable singularities because the germ $f$ is not weighted homogeneous.

An alternative way to compute such points is the algorithm given in [**MMR**], see also [**JLS**], for the determination of the multiple points associated to each singularity.

In this example we applied the algorithm given in [**JLS**] to obtain: 8 points $A_{2,2,1}$, 24 points $A_{3,1,1}$, 24 points $A_{2,1,1,1}$, 24 points $A_{3,2}$, 24 points $A_{4,1}$ and 8 points $A_5$.

We can not compute the ideal that defines the set $A_{1,1,1,1,1}$ because lack of memory of the computer.

But as the total number of these singularities is 112, we obtain $\sharp A_{1,1,1,1,1} = 0$.

## 5. Implementation

We present in this section the source code to obtain the matrix of the presentation given in the Theorem 3.1 by using computational methods. The implementation is obtained using the software Maple. To write the source code below we follow directly the proof of the Theorem 3.1.

```
Presentation_CnCn_Cor1:=proc(f)

local Iz,Izz,J,G,t,i,b,j,V,V1,nf,vf,vec,zz,F,vec2,nGer,varf;
global lambda;
print (f=f);
nf:=nops(f); varf:=indets(f); vec:=varf[]; vf:=nops(varf); F:=f;

if nf <> vf then
   ERROR("Dimensions of source and target are different");
fi:

for i to (nf-1) do
  F[nf]:=subs(f[i]=0,F[nf]):
   if F[nf]=0 then
     ERROR("f is not finite,verify input!");
   fi;
od;

  zz:=indets(F[nf])[];
   J:=diff(f[nf],zz):
vec2:=(varf minus {zz})[];
   b:=tcoeff(J,{vec2}):
   G:=[seq(zz^i,i=1..degree(b))];
nGer:=nops(G);

  Iz:=(J-lcoeff(tcoeff(J,{vec2}),{zz}, 'n')*n)*
      (-1/lcoeff(tcoeff(J,{vec2}),{zz}, 'n')):
 Izz:=expand(Iz*zz):
   t:=X[nf]-f[nf];
   t:=simplify(subs(zz^(degree(F[nf]))=Izz,t)):

 if degree(t,zz) > (degree(b)-0) then
    ERROR("No result! Verify input.");
 fi;

 for j from 1 to nGer do
  V[j]:=t;
   for i from 1 to nops(t) do
      if divide(op(i,t),zz^(degree(b))) then
         t:=simplify(subs(zz^(degree(b))=Iz,t)):
         V[j]:=t;
         break:
      fi:
```

```
  od:
 t:=expand(t*zz);
od:

lambda:=Matrix(1..nGer,1..nGer):
 for j from 1 to nGer do
   V1:=expand(zz*V[j]);
      for i from 1 to nGer do
         lambda[j,i]:=coeff(V1,G[i]):
      od:
 od:
print('lambda'=lambda);
end:
```

REMARK 5.1. (1) If dimensions of the source and target are different, a message is displayed to user.

(2) If $\mathcal{O}_{(\Sigma(f),0)}$ is not finitely generated as $\mathcal{O}_n$-module via $f^*$, a message is displayed to user.

(3) This implementation of this algorithm fixes the first $n-1$ variables in the target as $X_1 = x_1$, $X_2 = x_2$, $\ldots, X_{n-1} = x_{n-1}$ and $X_n = g$ denotes the last entry of the germ.

## References


[Mi] https: //sites.google.com/site/aldicio/publicacoes/presenmatrixcor1.

[GM1] T. Gaffney and D. M. Q. Mond, *Cusps and double folds of germs of analytic maps* $\mathbf{C}^2 \to \mathbf{C}^2$, J. London Math. Soc. (2) **43** (1991), no. 1, 185–192, DOI 10.1112/jlms/s2-43.1.185. MR1099097

[GM2] T. Gaffney and D. M. Q. Mond, *Weighted homogeneous maps from the plane to the plane*, Math. Proc. Cambridge Philos. Soc. **109** (1991), no. 3, 451–470, DOI 10.1017/S0305004100069905. MR1094745

[Si] G.-M. Greuel, G. Pfister, H. Schönemann, SINGULAR: A Computer Algebra System for Polynomial Computations. Centre for Computer Algebra, University of Kaiserslautern, free software under the GNU General Public Licence (1990–2007), `http://www.singular.uni-kl.de`.

[Su] S. Holzer, O. Labs, SURFEX 0.90, University of Mainz and University of Saarbrücken (2008), `www.surfex.AlgebraicSurface.net`.

[JS] V. H. Jorge Pérez and M. J. Saia, *Euler obstruction, polar multiplicities and equisingularity of map germs in* $\mathcal{O}(n,p)$, $n < p$, Internat. J. Math. **17** (2006), no. 8, 887–903, DOI 10.1142/S0129167X06003783. MR2261639

[JLS] V. H. Jorge Pérez, D. Levcovitz, and M. J. Saia, *Invariants, equisingularity and Euler obstruction of map germs from* $\mathbb{C}^n$ *to* $\mathbb{C}^n$, J. Reine Angew. Math. **587** (2005), 145–167, DOI 10.1515/crll.2005.2005.587.145. MR2186977

[JRS] V. H. Jorge Pérez, E. C. Rizziolli, M. J. Saia, *Whitney equisingularity, Euler obstruction and invariants of map germs from* $\mathbb{C}^n$ *to* $\mathbb{C}^3, n > 3$. Real and complex singularities, 263–287, Trends Math., Birkhäuser, Basel, (2007).

[JMS] V. H. Jorge-Pérez, A. J. Miranda, and M. J. Saia, *Counting singularities via Fitting ideals*, Internat. J. Math. **23** (2012), no. 6, 1250062, 18, DOI 10.1142/S0129167X12500620. MR2925478

[MMR] W. L. Marar, J. A. Montaldi, and M. A. S. Ruas, *Multiplicities of zero-schemes in quasihomogeneous corank-1 singularities* $\mathbf{C}^n \to \mathbf{C}^n$, Singularity theory (Liverpool, 1996), London Math. Soc. Lecture Note Ser., vol. 263, Cambridge Univ. Press, Cambridge, 1999, pp. 353–367. MR1709362

[M1] David Mond, *The number of vanishing cycles for a quasihomogeneous mapping from* $\mathbf{C}^2$ *to* $\mathbf{C}^3$, Quart. J. Math. Oxford Ser. (2) **42** (1991), no. 167, 335–345, DOI 10.1093/qmath/42.1.335. MR1120994

[MP] David Mond and Ruud Pellikaan, *Fitting ideals and multiple points of analytic mappings*, Algebraic geometry and complex analysis (Pátzcuaro, 1987), Lecture Notes in Math., vol. 1414, Springer, Berlin, 1989, pp. 107–161, DOI 10.1007/BFb0090254. MR1042359

[M2] David Mond, *Some remarks on the geometry and classification of germs of maps from surfaces to* 3*-space*, Topology **26** (1987), no. 3, 361–383, DOI 10.1016/0040-9383(87)90007-3. MR899055

[M3] David Mond, *Vanishing cycles for analytic maps*, Singularity theory and its applications, Part I (Coventry, 1988/1989), Lecture Notes in Math., vol. 1462, Springer, Berlin, 1991, pp. 221–234, DOI 10.1007/BFb0086385. MR1129035

[Lo] E. J. N. Looijenga, *Isolated singular points on complete intersections*, London Mathematical Society Lecture Note Series, vol. 77, Cambridge University Press, Cambridge, 1984. MR747303

[RS] E. C. Rizziolli and M. J. Saia, *Polar multiplicities and Euler obstruction of the stable types in weighted homogeneous map germs from* $\mathbb{C}^n$ *to* $\mathbb{C}^3$, $n \geq 3$, Singularities in geometry and topology, World Sci. Publ., Hackensack, NJ, 2007, pp. 723–748, DOI 10.1142/9789812706812_0025. MR2311507

[Mp] Waterloo Maple Inc., MAPLE, `http://www.maplesoft.com`.

FAMAT-UFU, UNIVERSIDADE FEDERAL DE UBERLÂNDIA, UBERLÂNDIA, MINAS GERAIS, BRASIL

*E-mail address*: `aldicio@ufu.br`

DEPARTAMENTO DE MATEMÁTICA, ICMC-USP, SÃO CARLOS, BRAZIL

*E-mail address*: `mjsaia@icmc.usp.br`

Contemporary Mathematics
Volume **675**, 2016
http://dx.doi.org/10.1090/conm/675/13595

# Critical points of the Gauss map and the exponential tangent map

M. G. Monera and E. Sanabria-Codesal

ABSTRACT. Given $M$ an immersed $k$-dimensional submanifold of $\mathbb{R}^{k+n}$ and $CM$ its unit normal bundle, we study the critical points of the generalized Gauss map $\psi : CM \to \mathbb{R}^{k+n}$, $\psi(m, U_m) = U_m$ with $U_m \in C_m M$ and $m \in M$. Usually, the image of these points is called the parabolic set. The case $k = 1, 2$ is analyzed in detail. Finally, we analyze the relation between the parabolic set and the critical points of the exponential tangent map $\Omega : TM \to \mathbb{R}^{k+n}$, $\psi(m, Y_m) = m + Y_m$ with $Y_m \in T_m M$, in the case $k = n$.

## 1. Introduction

Let $\lambda : M \times S^{k+n-1} \to \mathbb{R}$ be the height function given by $\lambda(m, U_m) = m \cdot U_m$, where $U_m \in S^{k+n-1}$, $M$ is a $k$-dimensional manifold immersed in $\mathbb{R}^{k+n}$ and $S^{k+n-1}$ is the unit sphere in $\mathbb{R}^{k+n}$ centered at the origin. The analysis of the singularities of this function determines that the singular set of $\lambda$ can clearly be identified with the canal hypersurface $CM$ of $M$ in $\mathbb{R}^{k+n}$. Moreover, the restriction of the natural projection $\pi : M \times S^{k+n-1} \to S^{k+n-1}$ to this singular set is defined as the *generalized Gauss map* of $M$ $\psi : CM \to \mathbb{R}^{k+n}$ given by $\psi(m, U_m) = U_m$, for $U_m \in C_m M$ where $C_m M$ is the unit normal subspace of $M$ at $m$, see [**5**]. It is not difficult to verify that $m \in M$ is a degenerate singularity of $\lambda_{U_m}$ if and only if $(m, U_m)$ is a singular point of $\psi$. This occurs when $K(m, U_m) = 0$, where $K$ denotes the gaussian curvature function on $CM$, i.e. $K = det(d\psi)$. The set of these points is called the parabolic set and can also be interpreted as the image of the critical points of the generalized Gauss map.

On the other hand, the concept of curvature ellipse at a point of a surface $M$ immersed in $\mathbb{R}^4$ was treated with full details in [**4**]. It is defined as the locus of all the end points of the curvature vectors of the normal sections along all the tangent directions to $M$ at a point in it. This ellipse lies in the normal subspace of that point and it is completely determined by the second fundamental form. We call *Veronese of curvature* to the natural generalization of the curvature ellipse for higher dimensions of $M$.

---

2010 *Mathematics Subject Classification.* Primary 54C40, 14E20; Secondary 46E25, 20C20.
*Key words and phrases.* Asymptotic direction, binormal direction, critical points, exponential tangent map, generalized Gauss map.
Work partially supported by DGCYT grants no. MTM2012-33073.

In this work, we describe first the parabolic set as the image of the critical points for the generalized Gauss map. In the particular case that $M$ is an immersed surface in $\mathbb{R}^{2+n}$, we see that the number of critical points of the generalized Gauss map at $m \in M$ may depend on the degeneration of the curvature ellipse. We shall use the same techniques that in [**8**], where the authors study the critical points of the exponential normal map $\nu : NM \to \mathbb{R}^{k+n}$ given by $\nu(m, u) = m + u$, where $NM$ is the normal bundle of $M$ and $u$ belongs to the normal subspace at $m$, $N_mM$.

Finally, we study the critical points of the exponential tangent map $\Omega : TM \to \mathbb{R}^{k+n}$, $\psi(m, Y_m) = m + Y_m$ with $Y_m \in T_mM$. We analyze the equivalence between these critical points and the critical points of the generalized Gauss map when $M$ is a $k$-dimensional manifold immersed in $\mathbb{R}^{2k}$.

Our goal in future works is to obtain some information about degenerated critical points of the generalized Gauss map and its relation with asymptotic lines of surfaces in higher codimensions (see [**7**], [**9**]).

## 2. Preliminaries

Let $M$ be a $k$-dimensional submanifold immersed in $\mathbb{R}^{k+n}$. Since all of our study will be local, we gain in brevity by assuming that it is a regular submanifold. For each $m \in M$ we consider the decomposition $T_m\mathbb{R}^{k+n} = T_mM \oplus N_mM$, where $T_mM$ denotes the tangent space to $M$ at $m \in M$, and $N_mM$ denotes the normal subspace to $M$ at $m$, i.e. $N_mM = (T_mM)^{\perp}$. Given $X \in T_m\mathbb{R}^{k+n}$, that decomposition will be written as $X_m = X_m^{\top} + X_m^{\perp}$ where $X^{\top} \in T_mM$, $X^{\perp} \in N_mM$.

Let $\pi : TM \to M$ and $\pi_N : NM \to M$ denote the tangent and normal bundles respectively. If $E$ is the total space of a smooth bundle we will denote by $\Gamma(E)$ the space of its smooth sections. If $s \in \Gamma(E)$ then $s_m$ shall denote usually the value of $s$ at $m$. For the particular case of $TM$, we will put $\mathfrak{X}(M)$ instead of $\Gamma(TM)$. We define the linear connection $\nabla^{\top}$ for $\pi$ by $\nabla_X^{\top}Y = (D_XY)^{\top}$, $X, Y \in \mathfrak{X}(M)$, where $D$ is the Levi-Civita connection in $\mathbb{R}^{k+n}$ which in fact is equal to the directional derivative. For $\pi_N$ we define the connection $\nabla^{\perp}$ by $\nabla_X^{\perp}u = (D_Xu)^{\perp}$, $u \in \Gamma(NM)$. These connections define a new connection $\nabla$ in $\Gamma(T^{(r,s)}M \otimes N^{(p,q)}M)$ such that if, for example, we have $w = u \otimes Y \otimes \gamma$, where $u \in \Gamma(NM)$, $Y \in \mathfrak{X}(M)$, $\gamma \in \Gamma(T^*M)$, then:

$$\nabla_X w = (\nabla_X^{\perp}u) \otimes Y \otimes \gamma + u \otimes (\nabla_X^{\top}Y) \otimes \gamma + u \otimes Y \otimes (\nabla_X^{\top}\gamma).$$

This connection preserves the inner product. Sometimes we shall use here a notation more or less usual, with which from an 1-form $\mu \in T_m^*M$ we build the vector $\mu^{\sharp} \in T_mM$ such that $\mu^{\sharp} \cdot x = \mu(x)$, $\forall x \in T_mM$.

Suppose now that we have a linear connection with covariant derivative $\nabla$ in a manifold $M$. Then we can define the operator $\nabla^r$ which acts over any $r$ differentiable vector fields $X_1, \ldots, X_r \in \mathfrak{X}(M)$ and a tensor field $K$ of type $(r, s)$. This operator is $C^{\infty}(M)$-linear in the first $r$ arguments and gives another tensor field of the same type $(r, s)$ :

$$\begin{aligned} \nabla^r(X_1, \ldots, X_r; K) &= \nabla_{X_1}(\nabla^{r-1}(X_2, \ldots, X_r; K)) \\ &\quad - \sum_{i=2}^{r} \nabla^{r-1}(X_2, \ldots, \nabla_{X_1}X_i, \ldots, X_r; K). \end{aligned}$$

The *first fundamental form* over $M$ is the field $g$ of positive definite symmetric bilinear forms induced over $M$ by the Euclidean inner product in $\mathbb{R}^{k+n}$. Thus, if

$X_m, Y_m \in T_mM$, then

$$g_m(X_m, Y_m) = X_m \cdot Y_m.$$

The *second fundamental form* $\alpha$ in $M$ is defined as the section $\alpha \in \Gamma(NM \otimes T^*M \otimes T^*M)$ such that, if $X, Y \in \mathfrak{X}(M)$ then

$$\alpha(X, Y) = (D_X Y)^\perp.$$

This is well defined, because $\alpha$ is evidently bilinear for the sum, and if $f, l : M \to \mathbb{R}$ are arbitrary smooth functions we have

$$\alpha(fX, lY) = (flD_X Y + fX(l)Y)^\perp = fl(D_X Y)^\perp.$$

Also, it is symmetric because $[X, Y] \in \mathfrak{X}(M)$ whence

$$\alpha(X, Y) - \alpha(Y, X) = (D_X Y - D_Y X)^\perp = [X, Y]^\perp = 0.$$

Another manner of looking at the second fundamental form is to consider for each $u \in N_mM$ the endomorphism

$$\mathcal{A}_m(u, .) : T_mM \to T_mM$$

that sends any vector $X_m \in T_mM$ to the vector $\mathcal{A}_m(u, X) \in T_mM$ whose inner product with any vector $Y_m \in T_mM$ is:

$$Y_m \cdot \mathcal{A}_m(u, X_m) = Y_m \cdot D_X \tilde{u} = -u \cdot \alpha_m(X_m, Y_m) = X_m \cdot \mathcal{A}_m(u, Y_m),$$

where $\tilde{u}$ is any normal vector field whose value at $m$ be $u$. This defines a smooth tensor field $\mathcal{A}$ on $M$ called *Weingarten operator.*

The second fundamental form at a point $m \in M$ describes the geometry of the immersion $M$ at $m$ to the second order of approximation.

**Veronese of curvature.**

DEFINITION 2.1. Let $PT_mM$ denote the projective space of directions in $T_mM$. The *Veronese of curvature* of $M$ at $m$ is the map (or its image) $\eta_m : PT_mM \to N_mM$ given by

$$\eta_m([t]) = \frac{\alpha_m(t, t)}{t \cdot t}, \quad \forall t \in T_mM \backslash \{0\}.$$

If $0 \neq t \in T_mM$, we shall put $\eta_m(t) = \eta_m([t])$ by a small abuse of notation. We note that the Veronese of curvature may also be defined as the restriction of $\eta_m$ to the unit sphere in $T_mM$.

In the particular case that $M$ is a surface, we observe that the Veronese of curvature is an ellipse. Let $m \in M$, and let $(t_1, t_2)$ be an orthonormal basis of $T_mM$. We denote by $S(T_mM)$ the circle of unit vectors of $T_mM$ that is

$$S(T_mM) = \{t \in T_mM : t \cdot t = 1\}.$$

We can parameterize $S(T_mM)$ by the angle $\theta$ with respect to $t_1$. Thus any point of $S(T_mM)$ may be written in the form $t(\theta) = t_1 \cos\theta + t_2 \sin\theta$. Therefore, the ellipse of curvature at $m$ may be written as

$$\eta_m(t(\theta)) = H_m + B_m \cos 2\theta + C_m \sin 2\theta,$$

where

$$H_m = \frac{1}{2}(\alpha_m(t_1, t_1) + \alpha_m(t_2, t_2)),$$

$$B_m = \frac{1}{2}(\alpha_m(t_1, t_1) - \alpha_m(t_2, t_2)),$$

and

$$C_m = \alpha_m(t_1, t_2).$$

Putting $b_1 = \alpha_m(t_1, t_1)$, $b_2 = \alpha_m(t_2, t_2)$ and $b_3 = \alpha_m(t_1, t_2)$ we have that:

$$H_m = \frac{1}{2}(b_1 + b_2), \quad B_m = \frac{1}{2}(b_1 - b_2), \quad C_m = b_3,$$

Consider the affine map $\xi : TM \to NM$ given by $\xi(a_1t_1 + a_2t_2) = H + a_1B + a_2C$, where $a_1, a_2 \in \mathbb{R}$, and the map $\delta : PTM \to S(TM)$ given by $\delta([t(\theta)]) = t(2\theta)$. The map $\delta$ is well defined because

$$\delta([t(\theta + p\pi)]) = t(2\theta + 2p\pi) = t(2\theta), \quad \forall p \in \mathbb{Z}.$$

Since $\eta_m = \xi \circ \delta$, we see that the image of $\eta_m$ is the image, by the affine map $\xi$, of a circle of $T_mM$ and therefore, it is an ellipse, that may degenerate to a segment or to a point. It is called the **curvature ellipse** of $M$ at $m$.

The center of that ellipse, denoted $H_m$, is called the *mean curvature vector*. The vectors $B_m$ and $C_m$ generate an affine subspace $\mathcal{S}_m$ of $N_mM$, passing by $H_m$. It is the affine span of the curvature ellipse. In general, it is an affine plane in $N_mM$. The vector subspace of $N_mM$ of same dimension as $\mathcal{S}_m$ and parallel to it, that is the plane generated by $B_m$ and $C_m$ shall be denoted by $\epsilon_m$. While $B_m$ and $C_m$ depend on the choice of the orthonormal basis $(t_1, t_2)$, $H_m, \mathcal{S}_m$ and $\epsilon_m$ do not.

The following definitions were introduced by Little, see [**4**]. We say that $m$ is:

- *Pseudo-umbilic*, when $\mathcal{S}_m$ is orthogonal to $H_m$.
- *Semiumbilic*, when $B_m$ and $C_m$ are linearly dependent. In this case the curvature ellipse degenerates into a segment or into a point.
- *Inflection point*, when there is a vector line that contains the curvature ellipse. In other words, $m$ is semumbilic and there is an affine line by the origin that contains the ellipse.
- *Umbilic*, when the ellipse degenerates to a point.
- *Plane or flat umbilic*, when the ellipse degenerates to the origen.

In the particular case that $M$ is a surface immersed in $\mathbb{R}^{2+n}$, $n = 1, 2$, we say that $m$ is:

- *Hyperbolic*, when the origin is outside the ellipse.
- *Elliptic*, when the origin is inside the ellipse.
- *Parabolic*, when the origin belongs to the ellipse.

## 3. Critical points of the generalized Gauss map

Let

$$CM = \{U \in PNM\,;\, U_m \cdot U_m = 1\} \subset NM$$

and let $\psi : CM \to \mathbb{R}^{k+n}$ be the map given by $\psi(m, [U_m]) = U_m$, $U_m \in C_mM$, where $C_mM = CM \cap PN_mM$ and by a small abuse of notation $\psi(m, [U_m]) = \psi(m, U_m)$. We observe that this map may also be defined as the restriction of $\psi$ to the unit sphere in $N_mM$. Then we consider the *generalized Gauss map* of $M$ as the positive directions in this unit sphere.

DEFINITION 3.1. A vector $X \in T(CM)$ is a **critical point of $\psi$** iff it is non-zero and $d\psi(X) = 0$. And we say that $U \in CM$ admits a critical point of $\psi$ if there is a critical point $X \in T_U(CM)$.

DEFINITION 3.2. Let $\psi$ be the generalized Gauss map of M. The **parabolic set** of M, denoted here by $\mathcal{P}(M)$ is defined as the set of critical points of $\psi$. Since there will be little risk of confusion, the same name will be used for the image of $\mathcal{P}(M)$ by $\psi$.

PROPOSITION 3.3. The point $U_m \in C_mM$ admits a critical point of $\psi$ denoted by $X \in T_{(m;U_m)}(CM)$ iff $\det(U_m \cdot \alpha_m(\cdot,\cdot)) = 0$, and in such case $X = (v;0)$, where $0 \neq v = d\pi_N(X) \in T_mM$ satisfies $U_m \cdot \alpha_m(v,\cdot) = 0$.

**Proof.** Let $X \in T_{(m;U_m)}(CM)$ be a critical point of $\psi$. Consider a smooth curve $\gamma : I \to NM$ given by $\gamma(t) = (\beta(t); u(t))$, where $I$ is an open neighborhood of $0 \in \mathbb{R}$, $\beta := \pi_N \circ \gamma$ and $u(t) \in N_{\beta(t)}M$, $\forall t \in I$. Suppose that $(m; U_m) = \gamma(0) = (\beta(0); u(0))$ and $X = \gamma'(0) = (\beta'(0); u'(0))$. Let us put $v = \beta'(0)$. Since $X$ is a critical point of $\psi$, we have:

$$d\psi(X) = (\psi \circ \gamma)'(0) = u'(0) = 0.$$

Thus, we must find in what conditions upon $v$ and $U_m$ we may have $u'(0) = 0$.

The fact that $\beta'(0) \neq 0$ implies that $\beta$ is an immersion in a neighborhood of 0. Then there exists a neighborhood $\mathcal{U}$ of $m$ and a section $\tilde{u} : \mathcal{U} \subset M \to NM$ such that $\tilde{u} \circ \beta = \gamma$ and therefore $u(0) = \tilde{u}_m$ and $u'(0) = D_v\tilde{u}$. Hence, taking the normal and tangent components, we conclude that $d\psi(X) = D_v\tilde{u} = 0$ iff $(D_v\tilde{u})^\perp = \nabla_v^\perp \tilde{u} = 0$ and $\mathcal{A}_m(U_m, v) = 0$, or, equivalently $\nabla_v^\perp \tilde{u} = 0$ and $-\tilde{u}_m \cdot \alpha_m(v,x) = 0$, $\forall x \in T_mM$. The last equation says that if $U_m$ admits a critical point, then $\det(-U_m \cdot \alpha_m) = 0$ and the nonzero vector $v \in T_mM$ satisfies $U_m \cdot \alpha_m(v,\cdot) = 0$, where $d\pi_N(X) = v$.

Suppose that these conditions are satisfied. Then, let $\beta : I \to M$ be such that $\beta(0) = m$, $\beta'(0) = v$, and let $u : I \to \mathbb{R}^{k+n}$ be such that $u(t) \in N_{\beta(t)}M$ be the parallel transport of $U_m$ along $\beta$. Then

$$u'(0) = 0.$$

Therefore the tangent of the curve $(\beta(t); u(t))$ at $t = 0$ is equal to $X = (v; 0)$, so that $X \neq 0$ and $d\psi(X) = 0$. Hence $X$ is a critical point of $\psi$.

□

Let $\{t_i\}_{i=1}^k$ be an orthonormal reference of $T_mM$ at $m \in \mathcal{U} \subset M$ and $\{w_i\}_{i=1}^n$ be an orthonormal reference of $C_mM$ in $\mathcal{U}$. Then, by using the previous notation, the matrix associated to $d(\psi \circ \tilde{u})_m$ is given by:

$$\left(\begin{array}{cccc|c} -U_m\cdot\alpha_m(t_1,t_1) & \cdot & \cdot & -U_m\cdot\alpha_m(t_1,t_k) & \\ \cdot & \cdot & & \cdot & \\ \cdot & & \cdot & \cdot & 0_{k\times n} \\ \cdot & & \cdot & \cdot & \\ -U_m\cdot\alpha_m(t_k,t_1) & \cdot & \cdot & -U_m\cdot\alpha_m(t_k,t_k) & \\ \hline & & 0_{n\times k} & & I_{n\times n} \end{array}\right),$$

where $U_m = \tilde{u}_m$. Then $U_m \in C_mM$ admits a critical point of $\psi$ iff the following determinant vanishes:

$$\left|\begin{array}{cccc} U_m\cdot\alpha_m(t_1,t_1) & \cdot & \cdot & U_m\cdot\alpha_m(t_1,t_k) \\ \cdot & \cdot & & \cdot \\ \cdot & & \cdot & \cdot \\ \cdot & & \cdot & \cdot \\ U_m\cdot\alpha_m(t_k,t_1) & \cdot & \cdot & U_m\cdot\alpha_m(t_k,t_k) \end{array}\right| = 0.$$

PROPOSITION 3.4. Let $\mathcal{P}_m(M) = \mathcal{P}(M) \cap C_m M$. The following properties are satisfied:

(1) If $\tilde{u} : \mathcal{U} \to CM$ is a local section in a neighborhood of $m \in M$, then $\det(U_m \cdot \alpha_m) = \det(d(\psi \circ \tilde{u})_m^\top)$, where $U_m = \tilde{u}_m$.
(2) If $U_m \in \mathcal{P}_m(M)$, then there exists $t \in T_m M \backslash \{0\}$, such that $0 = U_m \cdot \alpha_m(t) \in T_m^* M$. In the following item, $t$ and $U_m$ satisfy that property.
(3) $U_m \cdot d\eta_m(X) = 0$, for all $X \in T_t(T_m M \backslash \{0\})$ such that $X \cdot t = 0$.

**Proof.** 1) and 2) are clear by the above proposition.

3) Note that here $\eta_m$ is taken as a map from $T_m M \backslash \{0\}$ to $N_m M$. For making the calculations easier we can assume that $t \cdot t = 1$, i.e. $t \in S(T_m M)$. Then, if $X \in T_t(T_m M \backslash \{0\})$, $X \cdot t = 0$ and

$$d\eta_m(X) = 2\alpha_m(t, X),$$

by inner multiplication of this with $U_m$ we get that $U_m \cdot d\eta_m(X) = 0$. Notice that $d\eta_m(T_t S(T_m M))$ is the tangent space to the Veronese of curvature at $\eta_m(t)$, whose dimension is in general $min(k-1, n)$. □

Notice that the first condition implies that the singularities of this map $\psi$ coincides with the degenerate singularities of the height function $\lambda$, see [**6**].

In general, $\mathcal{P}(M)$ will be a hypersurface of $CM$, with singularities, whose intersection with each fiber $C_m M$ will be an algebraic hypersurface of degree $k$.

Let us see the parabolic set of two particular examples: curves immersed $\mathbb{R}^{1+n}$ and surfaces immersed $\mathbb{R}^3$.

EXAMPLE 3.5. Let $\beta : I \to \mathbb{R}^{1+n}$ be a curve where $\beta(0) = m$. Since dim $T_m \beta = 1$, then $\mathcal{P}_m(\beta)$ is generated by the unit vectors $U_m \in C_m M$ such that $U_m \cdot \alpha_m(t,t) = 0$. Notice that if $\alpha_m(t,t) = 0$, i.e. $m$ is plane, then $\mathcal{P}_m(\beta)$ is $C_m M$. In the particular case that $\beta : I \to \mathbb{R}^3$ then $\mathcal{P}_m(\beta)$ is the binormal vector and if $\beta$ is a plane curve, then $\mathcal{P}_m(\beta)$ is empty except on the case that $m$ is plane.

Let $M$ be a surface immersed in $\mathbb{R}^3$, $m \in M$. Then there exists $t \in T$ such that $\alpha(t, \cdot) = 0$ when $t$ is an asymptotic and principal direction, i. e. $m$ is a parabolic point in the classical sense ( when one of the principal curvatures of the surface in this point vanishes). Then $\mathcal{P}_m(M)$ is the unit normal vector $U_m$. If $m$ is umbilic plane, then $\alpha_m(\cdot,\cdot) = 0$. In this case, $\mathcal{P}_m(M)$ is the unit normal vector $U_m$. Otherwise, $\mathcal{P}_m(M)$ is empty.

Let us see the case of a surface in detail. We shall distinguish different cases depending on the degeneration of the curvature ellipse at the point $m \in M$. We are here interested solely in the study of $\mathcal{P}_m(M)$. This justifies the use of the following simplified notation in these propositions:

$$T = T_m M \backslash \{0\},\ C = C_m M,\ N = N_m M,\ \alpha = \alpha_m,\ \eta = \eta_m$$

and

$$\mathcal{P} = \mathcal{P}_m(M),\ (d\eta)_t = (d\eta_m)_t(T_t(T_m M \backslash \{0\})).$$

Suppose now that $M$ is a surface immersed in $\mathbb{R}^4$. In this case, the conditions 2 and 3 of Proposition 3.4 imply that $\eta(t)$ and $(d\eta)_t$ are parallel, where $t \in T$, and perpendicular to $U_m \in \mathcal{P}$. The tangent directions that satisfy this condition are the asymptotic directions $t_1, t_2 \in T$ and in this case $U_1, U_2$ are the binormal vectors defined in [**5**].

The following propositions are a consequence of the proposition 3.4. The first one was also proved in [**5**].

PROPOSITION 3.6. Let $M$ be a surface immersed in $\mathbb{R}^4$. Depending on the degeneration of the curvature ellipse at the point $m \in M$, we have:

1) If $m$ is elliptic, then $\mathcal{P}$ is empty.
2) If $m$ is hyperbolic, there exists $t_1, t_2 \in T$ such that $\eta(t_i) \, ||(d\eta)_{t_i}$, $i = 1, 2$. In this case $\mathcal{P}$ is the union of two vectors $U_1, U_2 \in C$ such that $U_i$ is perpendicular to the line which contains the origin and the extreme of the segment $\eta(t_i)$, $i = 1, 2$.
3) If $m$ is parabolic, there exists only one $t_1 \in T$ such that $\eta(t_1) = 0$. Then $\mathcal{P}$ is a unit vector $U_m \in C$ perpendicular to the tangent line at $\eta(t_1)$.
4) If $m$ semiumbilic but not inflection point, $\mathcal{P}$ is the union of two vectors $U_1, U_2 \in C$ such that $U_i$ is perpendicular to the line which contains the origin and the extreme of the ellipse $\eta(t_i)$, $i = 1, 2$.
5) If $m$ is an inflection point, $\mathcal{P}$ is a unit vector $U_m \in C$ orthogonal to the line which contains the ellipse.
6) If $m$ is umbilic not plane, $\eta(t)||(d\eta)_t$, $\forall t \in T$. In this case $\mathcal{P}$ is a vector $U_m \in C$ perpendicular to the line which contain the origin and $\eta(t)$.
7) If $m$ is plane then $\mathcal{P}$ is $C$.

In the following proposition we analyze the general case.

PROPOSITION 3.7. Let $M$ be a surface immersed in $\mathbb{R}^{2+n}$, $n > 2$ and $m \in M$. In this case, depending on the degeneration of the curvature ellipse at the point $m \in M$ we have:

1) If the ellipse is not degenerated and the origin is inside the plane containing the ellipse, then $\mathcal{P}$ is the orthogonal vector subspace in $N$ to the plane that contains the ellipse. Otherwise, $\mathcal{P}$ is the union of the unit vectors $U_m$ such that $U_m \cdot \alpha(t, T) = 0$, $\forall t \in T$.
2) If $m$ is semiumbilic but not an inflection point, $\mathcal{P}$ is the orthogonal vector subspace in $N$ to the plane that contains the segment and the origin.
3) If $m$ is an inflection point or umbilic not plane, $\mathcal{P}$ is the orthogonal vector subspace in $N$ to the line that contains the origin and $\eta(t)$.
4) If $m$ is umbilic plane, $\mathcal{P}$ is $C$.

In the case of a surface $M \subset \mathbb{R}^{2+n}$, $\mathcal{P}_m(M)$ will be, in general, an algebraic hypersurface of degree 2 given by:

$$\begin{vmatrix} U_m \cdot \alpha_m(t_1,t_1) & U_m \cdot \alpha_m(t_1,t_2) \\ U_m \cdot \alpha_m(t_2,t_1) & U_m \cdot \alpha_m(t_2,t_2) \end{vmatrix} = 0.$$

This gives an equation of the form:

$$\sum_{i,j=1}^{n} a_{ij} u_i u_j = 0, \tag{3.1}$$

where $U_m = \sum_{i=1}^{n} u_i w_i$ and $a_{ij} \in \mathbb{R}$. In this case $\mathcal{P}_m(M)$ is a cone centered at the origin.

## 4. Critical points of the exponential tangent map

Let $\Omega : TM \to \mathbb{R}^{k+n}$ be the exponential tangent map given by $\Omega(m; Y_m) = m + Y_m$, where $m \in M$ and $Y_m \in T_m M$.

DEFINITION 4.1. A vector $X \in T(TM)$ is a **critical point of $\Omega$** iff it is non-zero and $d\Omega(X) = 0$. We say that $Y \in TM$ admits a critical point of $\Omega$ if there is a critical point $X \in T_Y(TM)$.

DEFINITION 4.2. If $Y \in TM$ is such that admits a critical point of $\Omega$, then we say that it belongs to the **tangent set** of $M$. We shall denote by $\mathcal{T}(M)$ the set of these points.

Let us see the conditions that must satisfy a point $Y_m \in T_m M$ in order to admit a critical point of $\Omega$.

PROPOSITION 4.3. The point $Y_m \in T_m M$ admits a critical point of $\Omega$ denoted by $X \in T_{(m;Y_m)}(TM)$ iff there exists $0 \neq v = d\pi_T(X) \in T_m M$ and a smooth vector field $\tilde{y} : M \to \mathbb{R}^{k+n}$ tangent to $M$ such that $Y_m = y(0)$ and:

a) $\alpha_m(Y_m, v) = 0$,
b) $v + \nabla_v^\top \tilde{y} = 0$.

In such case, $X = (v; -v)$.

**Proof.** Let $X \in T_{(m;Y_m)}(TM)$ be a critical point of $\Omega$. Consider a smooth curve $\gamma : I \to SM$ given by $\gamma(t) = (\beta(t); y(t))$, where $I$ is an open neighborhood of $0 \in \mathbb{R}$ and $y(t) \in T_{\beta(t)}M$, $\forall t \in \mathbb{R}$. Suppose that $(m; Y_m) = \gamma(0) = (\beta(0); y(0))$ and $X = \gamma'(0) = (\beta'(0); y'(0))$. Let us put $v = \beta'(0)$. Since $X$ is a critical point of $\Omega$, we have:

$$d\Omega(X) = (\Omega \circ \gamma)'(0) = \beta'(0) + y'(0) = v + y'(0) = 0.$$

If $v = 0$, then we should have $y'(0) = 0$, that is $X = 0$, against the hypothesis. Therefore $v \neq 0$ and $y'(0) = -v \in T_m M$. Thus, we must find in what conditions upon $v$ and $Y_m$ we may have $y'(0) = -v$.

The fact that $\beta'(0) \neq 0$ implies that $\beta$ is an immersion in a neighborhood of 0. Then there exists a neighborhood $\mathcal{U}$ of $m$ and a vector tangent field $\tilde{y} : \mathcal{U} \subset M \to SM$ such that $\tilde{y} \circ \beta = \gamma$ and therefore $y'(0) = D_v \tilde{y}$.

Hence, taking the tangent and normal components, we conclude that $d\Omega(X) = v + D_v \tilde{y} = 0$ iff $v + \nabla_v^\top y = 0$, where $(D_v \tilde{y})^\top = \nabla_v^\top \tilde{y}$, and $\alpha_m(Y_m, v) = 0$.

Suppose that these conditions are satisfied, where $d\pi(X) = v$. Then, let $\beta : I \to M$ be such that $\beta(0) = m$, $\beta'(0) = v$, and let $y : I \to \mathbb{R}^{k+n}$ a smooth tangent vector field along $\beta$. Then

$$y'(0) = (D_v \tilde{y})^\perp + (D_v \tilde{y})^\top = -v.$$

Therefore the tangent of the curve $(\beta(t); y(t))$ at $t = 0$ is equal to $X = (v; -v)$, so that $X \neq 0$ and $d\Omega(X) = 0$. Hence $X$ is a critical point of $\Omega$.

□

In [**1**] the author gives the following definition.

DEFINITION 4.4. Let $M$ be a k-dimensional submanifold immersed in $\mathbb{R}^{k+n}$. Two directions $X_m, Y_m \in T_m M$ are called conjugate directions if $u \cdot \alpha_m(X_m, Y_m) = 0, \forall u \in N_m M$.

Notice that in this case the directions $Y_m \in T_mM$ that admit a 1-critical point of $\Omega$ are conjugated directions.

Let us see some examples.

EXAMPLE 4.5. Let $\beta : I \to \mathbb{R}^{1+n}$ be a curve where $\beta(0) = m$. Since $\dim T_m\beta = 1$, then $Y_m = t_1 \in T_m\beta$. Hence $Y_m$ admits a 1-critical point of the map $\Omega$ iff $\alpha_m(t_1, t_1) = 0$. In other words, when $m$ is plane.

Suppose now that $M$ is a surface immersed in $\mathbb{R}^3$. In this case, the principal directions are conjugated directions and the asymptotic directions are self conjugated directions at a point $m \in M$.

In general, if $n < k$, then every tangent vector is conjugate to at least one other vector. If $n > k$, the set of conjugate vectors becomes more sparse, until we get to the point where conjugate vectors are isolated in general ($n = 3k - 2$), there are no conjugate directions in general ($n > 3k - 2$) or every tangent vector is conjugate to at least one other vector ($n < 3k - 2$), see [**1**]. Then, the most interesting structure for the conjugate vectors occurs when $k = n$, as we see in the following Proposition proved in [**1**].

PROPOSITION 4.6. ([**1**]) Let $m$ be a point on a $k$-dimensional manifold immersed in $\mathbb{R}^{2k}$. A vector $Y_m \in T_mM$ is a conjugate vector at $m$ iff there is a 1-critical point $U_m \in C_mM$ of $\psi$ such that $U_m \cdot \alpha_m(Y_m, \cdot) = 0$.

By using the above result and the Proposition 4.3 we obtain the relation between the critical points of $\Omega$ and $\psi$ in a $n$-dimensional manifold immersed in $\mathbb{R}^{2k}$.

PROPOSITION 4.7. A vector $Y_m \in T_mM\backslash\{0\}$ is a 1-critical point of $\Omega$ iff there is a 1-critical point $U_m \in C_mM$ of $\psi$ such that $U_m \cdot \alpha_m(Y_m, \cdot) = 0$.

This means that, in the particular case of a surface immersed in $\mathbb{R}^4$ the directions of $T_mM$ that admits a critical point of $\Omega$ are the asymptotic directions and the $U_m \in C_mM$ are the binormal directions defined in [**5**].

Finally, by using the Propositions 4.7 and 3.4 we arrive to the following theorem, where the first 4 conditions where proved in [**2**].

THEOREM 4.8. Let $M$ be a n-dimensional submanifold immersed in $\mathbb{R}^{2n}$ and let $Y_m \in T_mM\backslash\{0\}$ at a point $m \in M$. The following conditions are equivalent:

(1) There exists a unit tangent vector $v$ such that $U_m \cdot \alpha(Y_m, v) = 0$ for all $U_m \in N_mM$.
(2) There exists a unit normal vector $U_m$ such that $U_m \cdot \alpha(Y_m, \cdot) = 0$.
(3) There exists a height function $\lambda(m, U_m) = U_m \cdot m$ such that $\lambda$ has a degenerate (non-Morse) singularity.
(4) Let $t \in S(T_mM)$ be the unit vector associated to the direction $Y_m \in T_mM\backslash\{0\}$, then the vector $d\eta_m(t)$ is tangent to the Veronese of curvature at $\eta_m(t)$ or the Veronese of curvature has a singularity at $t$.
(5) The exponential tangent map $\Omega$ has a singularity at $(m, Y_m)$.

## References

[1] Daniel Dreibelbis, *Conjugate vectors of immersed manifolds*, Real and complex singularities, Contemp. Math., vol. 459, Amer. Math. Soc., Providence, RI, 2008, pp. 1–12, DOI 10.1090/conm/459/08958. MR2444389

[2] Daniel Dreibelbis, *Self-conjugate vectors of immersed 3-manifolds in* $\mathbb{R}^6$, Topology Appl. **159** (2012), no. 2, 450–456, DOI 10.1016/j.topol.2011.09.019. MR2868904

[3] C. G. Gibson, *Singular points of smooth mappings*, Research Notes in Mathematics, vol. 25, Pitman (Advanced Publishing Program), Boston, Mass.-London, 1979. MR533668

[4] John A. Little, *On singularities of submanifolds of higher dimensional Euclidean spaces*, Ann. Mat. Pura Appl. (4) **83** (1969), 261–335. MR0271970

[5] Dirce Kiyomi Hayashida Mochida, Maria Del Carmen Romero Fuster, and Maria Aparecida Soares Ruas, *The geometry of surfaces in 4-space from a contact viewpoint*, Geom. Dedicata **54** (1995), no. 3, 323–332, DOI 10.1007/BF01265348. MR1326737

[6] D. K. H. Mochida, M. C. Romero-Fuster, and M. A. S. Ruas, *Osculating hyperplanes and asymptotic directions of codimension two submanifolds of Euclidean spaces*, Geom. Dedicata **77** (1999), no. 3, 305–315, DOI 10.1023/A:1005145104603. MR1716680

[7] D. K. H. Mochida, M. C. Romero-Fuster, and M. A. S. Ruas, *Inflection points and nonsingular embeddings of surfaces in* $\mathbf{R}^5$, Rocky Mountain J. Math. **33** (2003), no. 3, 995–1009, DOI 10.1216/rmjm/1181069939. MR2038535

[8] M. G. Monera, A. Montesinos-Amilibia, S. M. Moraes, and E. Sanabria-Codesal, *Critical points of higher order for the normal map of immersions in* $\mathbb{R}^d$, Topology Appl. **159** (2012), no. 2, 537–544, DOI 10.1016/j.topol.2011.09.029. MR2868914

[9] M. C. Romero-Fuster, M. A. S. Ruas, and F. Tari, *Asymptotic curves on surfaces in* $\mathbb{R}^5$, Commun. Contemp. Math. **10** (2008), no. 3, 309–335, DOI 10.1142/S0219199708002806. MR2417919

Departamento de Matemática Aplicada, Universitat Politècnica de València C/ Camino de Vera s/n, 46022, Valencia

*E-mail address*: magar21@upv.es

Departamento de Matemática Aplicada, Universitat Politècnica de València C/ Camino de Vera s/n, 46022, Valencia

*E-mail address*: esanabri@mat.upv.es

Contemporary Mathematics
Volume **675**, 2016
http://dx.doi.org/10.1090/conm/675/13596

# Minkowski medial axes and shocks of plane curves

Graham Mark Reeve and Farid Tari

ABSTRACT. In this paper a Minkowski analogue of the Euclidean medial axis of a closed and smooth plane curve is introduced. Its generic local configurations are studied and the types of shocks that can occur on it are determined.

## 1. Introduction

The concept of the *medial axis* of Euclidean plane curves was first introduced by Blum in [**1**]. For a closed and smooth curve $\gamma$ the medial axis is defined to be the locus of the centres of maximal circles that are tangent to $\gamma$ in two or more points. Here a circle is said to be maximal if its radius equals the absolute minimum distance from its centre to $\gamma$: such a circle is either contained in the interior or the exterior of $\gamma$ and cannot be expanded about its centre without crossing $\gamma$. Many applications of medial axes are given in Blum's original paper [**1**] and other applications relating to computer vision can be found in [**10**].

The *symmetry set* of the curve $\gamma$ is the same as that of the medial axis except that the constraint that the circles must be maximal is dropped (see [**4, 5**]). For this reason the medial axis forms a subset of the symmetry set.

In [**6**] the generic shocks that can occur on the Euclidean medial axis are classified. Motivated by applications in fluid mechanics, Bogaevsky uses a different approach in [**3**] to obtain similar results to those in [**6**].

In this paper a Minkowski analogue of the Euclidean medial axis for a closed and smooth plane curve, called the Minkowski medial axis ($MMA$) is introduced. An analogue of the symmetry set for curves in the Minkowski plane, called the *Minkowski Symmetry Set ($MSS$)*, is introduced and studied in [**11**]. For a curve $\gamma$ in the Minkowski plane, the $MSS$ is defined as the locus of the centres of pseudo-circles that are tangent to the curve $\gamma$ in two or more points.

Similarly to the Euclidean version, a point on the $MSS$ is said to belong to the Minkowski medial axis if the radius $r$ of the bi-tangent pseudo-circle equals the absolute maximum (if $r$ is positive) or the absolute minimum (if $r$ is negative) distance from its centre to $\gamma$. It is shown in Theorem 4.8 that this definition implies that the bi-tangent points lie on just one branch of the pseudo-circle. This fact leads on to a new generalised type of $MMA$, called the *1-branch* $MMA$. This is defined to

2010 *Mathematics Subject Classification.* Primary 53A35; Secondary 58K05 53D12.
The first author was supported by the FAPESP grant 2012/0536-9.
The second author was partially supported by the grants FAPESP 2014/00304-2, CNPq 301589/2012-7 and CNPq 472796/2013-5.

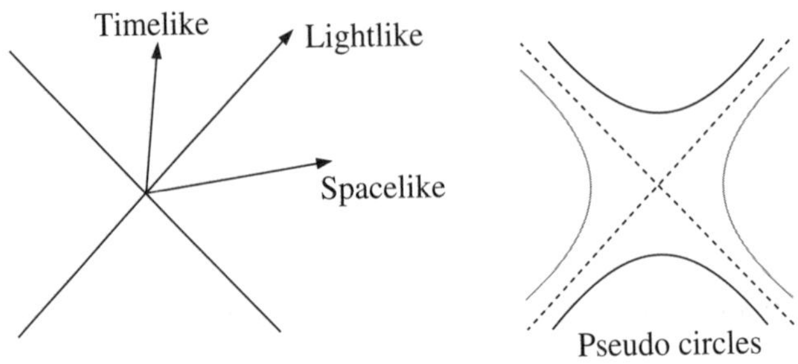


FIGURE 1. The three types of vectors (left) and pseudo-circles (right) in $\mathbb{R}^2_1$.

be the centres of pseudo-circles that have a bi-tangency with just one branch of the pseudo-circles. It follows that the $MMA$ forms a subset of the 1-branch $MMA$. The Minkowski symmetry set together with a radius function, like its Euclidean counterpart ([**10**]), can be used to reconstruct the original curve $\gamma$. Unlike with the Euclidean medial axis, neither the $MMA$ nor the 1-branch $MMA$ can be used to reconstruct non-convex curves. However, the 1-branch MMA together with a radius function can be used to reconstruct convex curves.

The *shock set* of $\gamma$ is obtained by adding an arrow to the $MMA$ indicating the direction of increasing radii of the relevant bi-tangent pseudo-circles. In this paper, the generic local configuration of the $MMA$ and shocks of curves in the Minkowski plane are obtained.

## 2. Preliminaries

The *Minkowski plane* $(\mathbb{R}^2_1, \langle , \rangle)$ is the vector space $\mathbb{R}^2$ endowed with the pseudo-scalar product $\langle \boldsymbol{u}, \boldsymbol{v} \rangle = -u_0v_0 + u_1v_1$, for any $\boldsymbol{u} = (u_0, u_1)$ and $\boldsymbol{v} = (v_0, v_1)$. A vector $\boldsymbol{u} \in \mathbb{R}^2_1 \setminus \{0\}$ is called

*spacelike* if $\langle \boldsymbol{u}, \boldsymbol{u} \rangle > 0$,
*timelike* if $\langle \boldsymbol{u}, \boldsymbol{u} \rangle < 0$ or
*lightlike* if $\langle \boldsymbol{u}, \boldsymbol{u} \rangle = 0$.

The norm of a vector is defined by $||\boldsymbol{u}|| = \sqrt{|\langle \boldsymbol{u}, \boldsymbol{u} \rangle|}$.

The pseudo-circles in $\mathbb{R}^2_1$ with centre $c \in \mathbb{R}^2_1$ and radius $r$ are defined as follows:

$$
\begin{aligned}
H^1(c, r) &= \{p \in \mathbb{R}^2_1 \,|\, \langle p - c, p - c \rangle = -r^2\} \text{ if } r < 0,\\
S^1(c, r) &= \{p \in \mathbb{R}^2_1 \,|\, \langle p - c, p - c \rangle = r^2\} \text{ if } r > 0,\\
LC^*(c) &= \{p \in \mathbb{R}^2_1 \setminus 0 \,|\, \langle p - c, p - c \rangle = 0\} \text{ if } r = 0.
\end{aligned}
$$

Observe that $LC^*(c)$, which is the set of points at zero distance from $c$, is the union of the two lines through $c$ with tangent directions $(1, 1)$ and $(1, -1)$, with the point $c$ removed. The pseudo-circle $H^1(c, -r)$ has two branches which can be parameterised by $c + (\pm r \cosh(t), r \sinh(t))$, $t \in \mathbb{R}$. The pseudo-circle $S^1(c, r)$ also has two branches which can be parameterised by $c + (r \sinh(t), \pm r \cosh(t))$, $t \in \mathbb{R}$. See Figure 1.

Let $\gamma : J \to \mathbb{R}^2_1$ be a smooth curve, where $J$ is an open interval of $\mathbb{R}$ or the unit circle $S^1$ if the curve is closed. The curve $\gamma$ is spacelike if $\gamma'(t)$ is a spacelike vector for all $t \in J$ and is timelike if $\gamma'(t)$ is a timelike vector for all $t \in J$. A point $\gamma(t)$ is called a lightlike point if $\gamma'(t)$ is a lightlike vector.

If $\gamma$ is spacelike or timelike, then it can be reparameterised by arc-length and its curvature is well defined at each point (see, for example, [**11**]). One can also have the notion of vertices (points where the derivative of the curvature vanishes). The curvature of $\gamma$ is not defined at lightlike points (so we have no notion of vertices at such points). An inflection can be defined in terms of the contact of the curve with lines, so the curve can have an inflection at a lightlike point. For a generic curve, the lightlike points are not inflection points.

It is shown in ([**11**], Proposition 2.1) that the set of lightlike points of a closed curve $\gamma$ is the union of at least four disjoint non-empty and closed subsets of $\gamma$ (Figure 2). The complement of these sets are disjoint connected spacelike or timelike pieces of the curve $\gamma$.

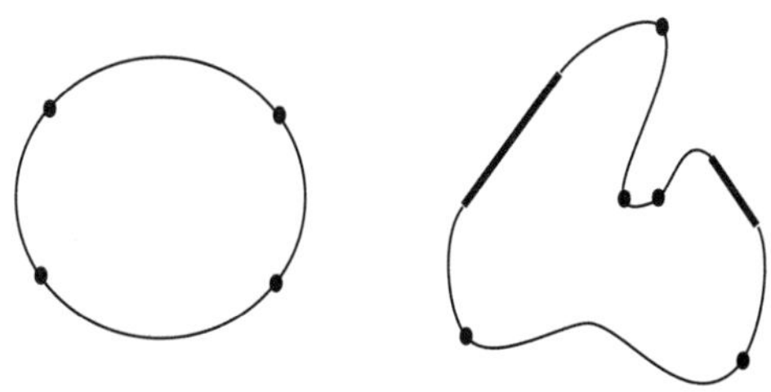

FIGURE 2. Lightlike points in thick on smooth closed curves in $\mathbb{R}^2_1$.

The Minkowski symmetry set is defined in [**11**] as follows.

DEFINITION 2.1. The Minkowski Symmetry Set ($MSS$) of a curve $\gamma$ in the Minkowski plane is the closure of the locus of centres of bi-tangent pseudo-circles to the curve.

In the Minkowski plane $\mathbb{R}^2_1$, the fact that vectors can have negative length has to be considered. For this reason, a pseudo-circle is said to be *maximal* if its radius equals the absolute minimum *modulus* distance from its centre to $\gamma$. The radius of such a circle cannot be increased if it is of type $S^1(p,r)$ or decreased if it is type $H^1(p,r)$ without it crossing $\gamma$.

DEFINITION 2.2. The *Minkowski medial axis* ($MMA$) of a curve $\gamma$ in $\mathbb{R}^2_1$ is the subset of the Minkowski symmetry set formed by the centres of bi-tangent pseudo-circles which are maximal (see Figure 3).

REMARK 2.3. In the Euclidean plane, for a closed curve $\gamma$, maximality implies that the bi-tangent circles are either entirely inside or entirely outside the curve ([**7**]). In the Minkowski plane however, since pseudo-circles are not compact, maximality ensures that the centres are entirely outside the curve, see Lemma 4.3.

*The family of distance-squared functions* $f : J \times \mathbb{R}^2_1 \to \mathbb{R}$ on $\gamma$ is given by

$$f(t,c) = \langle \gamma(t) - c, \gamma(t) - c\rangle,$$

and the *extended family of distance-squared functions* $\tilde{f} : J \times \mathbb{R}^2_1 \times \mathbb{R} \to \mathbb{R}$ is given by

$$\tilde{f}(t,c,r) = \langle \gamma(t) - c, \gamma(t) - c\rangle - r^2.$$

(To simplify notation the extended family of distance-squared functions will also be denoted by $f$.)

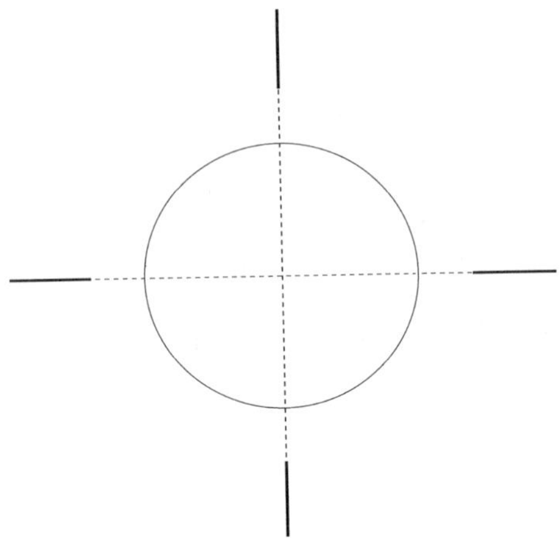

FIGURE 3. A circle and its $MSS$ (the two transverse line segments). The $MMA$ is the subset of the $MSS$ represented by a thick line.

Denote by $f_c : J \to \mathbb{R}$ the function given by $f_c(t) = f(t, c)$. The function $f_c$ is said to have an $A_k$-singularity at $t_0$ if $f'_c(t_0) = f''_c(t_0) = \dots = f_c^{(k)}(t_0) = 0$ and $f_c^{(k+1)}(t_0) \neq 0$. This is equivalent to the existence of a local re-parametrisation $h$ of $\gamma$ at $t_0$ such that $(f \circ h)(t) = \pm(t - t_0)^{k+1}$.

If $f_c$ has a singularity at $t_1$ of type $A_k$ and at $t_2$ of type $A_l$, then $f$ is said to have a multi-local singularity of type $A_kA_l$.

Geometrically, $f_c$ has an $A_k$-singularity if and only if the curve $\gamma$ has contact of order $k+1$ at $\gamma(t_0)$ with the pseudo-circle $C(c, r)$ of centre $c$ and radius $r$, with $|r| = ||\gamma(t_0) - c||$. The distance squared function $f_c$ has $A_kA_l$-singularity if and only if the pseudo circle $C(c, r)$ is tangent to $\gamma$ at two distinct points and has order of contact $k+1$ at one of them and $l+1$ at the other.

It follows from Thom's transversality theorem (see for example [**2**, **8**]) that for an open and dense set of immersions $\gamma : S^1 \to \mathbb{R}^2_1$ the function $f_c$ has only local singularities of type $A_1, A_2, A_3$ and multi-local singularities of type $A_1^2$, $A_1A_2$, $A_1^3$.

The $MSS$ is the multi-local component of the bifurcation set of the family $f$, that is,

$$MSS = \{c \in \mathbb{R}^2_1 \mid \exists t_1, t_2 \text{ such that } t_1 \neq t_2,\ f_c(t_1) = f_c(t_2),\ f'_c(t_1) = f'_c(t_2) = 0\}.$$

It follows from Theorem 3.2 in [**9**] that the family $f$ is always a versal unfolding of its generic singularities, so the $MSS$ is locally diffeomorphic to the bifurcation set of the models of such singularities (Corollary 3.3 in [**9**]). Thus the configuration of the $MSS$ at the generic multi-local singularities of $f_c$ are as in Figure 4.

For a point on the $MSS$ to also belong to the $MMA$ the relevant bi-tangent pseudo-circle must be maximal (Definition 2.2). This means that the $MMA$ forms a subset of the $MSS$ and in particular this condition ensures that only $A_1^2, A_3$ and $A_1^3$ can belong to $MMA$. This is because for the generic singularity types $A_2$ and $A_1A_2$ the pseudo-circle locally crosses the curve and therefore it cannot be maximal.

In [**11**] it was shown that the $MSS$ is a regular curve at $c_0$ if and only if the bi-tangent pseudo-circle to $\gamma$ at $\gamma(t_1)$ and $\gamma(t_2)$ is not osculating at $\gamma(t_1)$ or at $\gamma(t_2)$.

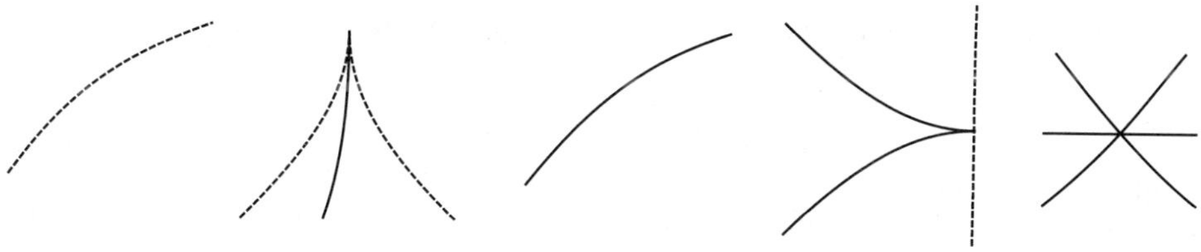


FIGURE 4. Generic local models of the $MSS$ in continuous line (the dashed curve is the caustic of the curve). Only the $A_1^2, A_1^3$ and $A_3$ singularities occur generically on the $MMA$.

If this is the case, the tangent line to the $MSS$ at $p$ is the perpendicular bisector to the chord joining $\gamma(t_1)$ and $\gamma(t_2)$. (This is also true in the Euclidean case, see [**5**].)

## 3. Local reconstruction of the curve from the $MSS$

Suppose we are given the $MSS$ (or $MMA$) of a spacelike or timelike smooth curve $\gamma$. Then the $MSS$ is either a spacelike or a timelike curve [**11**]. If the $MSS$ is not singular, it can be parameterised by arc length $c(s) = (x(s), y(s))$ and denote by $r(s)$ the radius of the bi-tangent circle to $\gamma$ centred at $c(s)$. Then it is possible to reconstruct local parametrisations $\gamma_1$ and $\gamma_2$ of the two corresponding arcs of $\gamma$ as an envelope of the bi-tangent circles $C(c(s), r(s))$.

PROPOSITION 3.1. *If the curve $c(s)$ is timelike and the bi-tangent pseudo-circle is of type $S^1(p,r)$ or if the curve $c(s)$ is spacelike and the bi-tangent pseudo-circle is of type $H^1(p,r)$, then the points of tangency are given by*

$$\gamma_i = c + r\left(\frac{\partial r}{\partial s}\right)T + (-1)^i\left(r\sqrt{\left(\frac{\partial r}{\partial s}\right)^2 + 1}\right)N,\ i = 1,2, \tag{3.1}$$

*where $T$ and $N$ are the unit tangent and unit Minkowski normal to the $MSS$ and $r$ is the radius of the bi-tangent pseudo-circle, all evaluated at $s$.*

*If the curve $c(s)$ is spacelike and the bi-tangent pseudo-circle is of type $S^1(p,r)$ or if the curve $c(s)$ is timelike and the bi-tangent pseudo-circle is of type $H^1(p,r)$, then the points of tangency are given by*

$$\gamma_i = c - r\left(\frac{\partial r}{\partial s}\right)T + (-1)^i\left(r\sqrt{\left(\frac{\partial r}{\partial s}\right)^2 - 1}\right)N,\ i = 1,2. \tag{3.2}$$

PROOF. Suppose that $c$ is spacelike and the pseudo-circles are of type $H^1(p,r)$. Then the pseudo-circle of radius $r(s)$ centred at $c(s)$ is the set of points $w \in \mathbb{R}^2_1$ such that

$$F(s,w) = \langle c(s) - w, c(s) - w\rangle + r(s)^2 = 0.$$

The envelope of this family of pseudo-circles is given by

$$D(F) = \{w \in \mathbb{R}^2 : \exists s \in \mathbb{R} \text{ such that } F(s,w) = \frac{\partial F}{\partial s}(s,w) = 0\}.$$

Differentiating $F$ with respect to $s$ and dropping the arguments yields

$$\frac{\partial F}{\partial s} = 2\langle (c-w, T\rangle + 2r\frac{\partial r}{\partial s}.$$

Since $\gamma$ is spacelike, $\langle T,T\rangle = 1$ and $\langle N,N\rangle = -1$. Writing $c-w = \lambda T + \mu N$, with $\lambda, \mu \in \mathbb{R}$, and substituting into $\partial F/\partial s = 0$ yields $\lambda + r\partial r/\partial s = 0$ so that $\lambda = -r(\partial r/\partial s)$. Substituting into $F=0$ yields $\lambda^2 - \mu^2 + r^2 = 0$ so that $\mu = \pm r\sqrt{(\partial r/\partial s)^2+1}$. It follows that the locii of the envelope points are as given (3.1). The same method can be applied to find the formula for the envelope in the remaining cases. □

REMARK 3.2. In the Euclidean case the rate of change of the radius function is restricted to be less than or equal to 1 in order for the envelope to be real, see [**7**]. In the Minkowski setting this restriction only applies when the pseudo-circles and the curve $c$ are of opposite type (spacelike/timelike), otherwise any function $r$ gives a real envelope.

REMARK 3.3. Theorem 4.1 of [**9**] states that for any point $p$ of a closed smooth curve $\gamma$ there exists another point $q \in \gamma$ and a pseudo-circle that is tangent to $\gamma$ at both $p$ and $q$. From this, and the fact that it is possible to reconstruct the curve locally, it follows that it is possible to reconstruct any smooth closed curve from its Minkowski symmetry set.

## 4. The Minkowski medial axis

DEFINITION 4.1. The *Minkowski hull* $MH(\gamma)$ of a smooth closed curve $\gamma$ in the Minkowski plane is the region of the plane such that for any point $p \in MH(\gamma)$ there exists a point $q \in \gamma$ such that the Minkowski distance between $p$ and $q$ is zero.

PROPOSITION 4.2. *The complement of the Minkowski hull of a closed smooth curve $\gamma$ in the Minkowski plane consists of four disjoint open regions.*

PROOF. Consider a smooth closed curve $\gamma : S^1 \to \mathbb{R}^2_1$ and use coordinates in $\mathbb{R}^2_1$ such that the axes are parallel to the lightlike directions. We write $\gamma(t) = (x(t), y(t))$. Since the functions $x(t)$ and $y(t)$ are bounded, they must both attain an absolute maximum and an absolute minimum. Since the curve is smooth, this gives exactly four extremal sets (two of $x(t)$ and two of $y(t)$) on the curve. The curve $\gamma$ is now contained in the compact region determined by the four tangent lines to points in these sets. The tangent lines divide the plane into the Minkowski hull and four disjoint regions of its complement, see Figure 5. □

LEMMA 4.3. *The points on the $MMA$ of a closed plane curve which are the centres of bi-tangent circles of type $H^1(p,r)$ or $S^1(p,r)$ lie in the complement of the Minkowski hull of the curve.*

PROOF. From the definition, a point $p$ belonging to the Minkowski hull has Minkowski distance zero to some point on the curve. It follows that pseudo-circles of types $H^1(p,r)$ and $S^1(p,r)$ cannot be maximal because the light cone $LC^*(p)$ would intersect the curve. Therefore, pseudo-circles of types $H^1(p,r)$ and $S^1(p,r)$ corresponding to the $MMA$ lie in the complement of the Minkowski hull. □

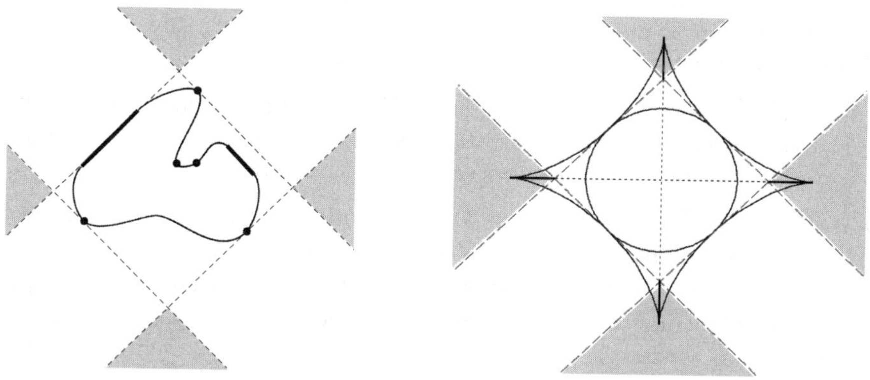

FIGURE 5. Left: the Minkowski hull of a closed curve in the Minkowski plane and its complement (shaded in grey). Right: a circle, its Minkowski caustic (continuous curve with four cusps), Minkowski hull (shaded in grey), $MSS$ and $MMA$ (in thick).

PROPOSITION 4.4. *The $MMA$ of a closed smooth convex plane curve lies in the closure of the complement of the Minkowski hull.*

PROOF. Closed smooth convex plane curves have exactly four closed lightlike regions (see [**11**]). The four centres of the pseudo-circles that are bi-tangent to these four regions are on the boundary of the Minkowski hull. This, together with Lemma 4.3, prove that the centres of all three types of maximal bi-tangent pseudo-circles lie in the complement of the Minkowski hull or on its boundary. Therefore, the $MMA$ of a closed smooth convex plane curve lies in the closure of the complement of the Minkowski hull. See Figure 5, right. □

COROLLARY 4.5. (*of* Lemma 4.3.) *The $MMA$ of a closed smooth (not necessarily convex) plane curve lies in the complement of the Minkowski hull of the curve except for the centres of bi-tangent pseudo-circles of type $LC(p)$ which lie on the boundary of the complement of the Minkowski hull (note that these centres form subsets of lightlike lines).*

REMARK 4.6. It is worth observing the difference between the medial axis of a convex curve in the Euclidean plane and of the $MMA$ of a convex curve in the Minkowski plane. The former lies strictly inside the curve while the latter lies strictly outside the curve.

DEFINITION 4.7. A pseudo-circle of type $H^1(p,r)$ or $S^1(p,r)$ is said to be *1-branch bi-tangent* to a curve $\gamma$ if one of its branches is tangent to $\gamma$ in at least two distinct points.

Since the $MMA$ lies in the closure of the complement of the Minkowski hull we have the following result.

THEOREM 4.8. *The pseudo-circles of types $H^1(p,r)$ and $S^1(p,r)$ corresponding to the $MMA$ of a closed plane curve are all 1-branch bi-tangent.*

PROOF. Lemma 4.3 states that the centres of the bi-tangent pseudo-circles of type $H^1(p,r)$ and $S^1(p,r)$ lie in the complement of the Minkowski hull of the curve. Since the bi-tangent pseudo-circle necessarily has one of its branches completely contained in the complement of the Minkowski hull it follows that the tangencies must occur on only one of the pseudo-circle's two branches. □

REMARK 4.9. Note that the converse of Theorem 4.8 is true for convex curves but is not true for non-convex curves. That is, not all 1-branch bi-tangent pseudo-circles are maximal for non-convex curves. See for example Figure 7 (right).

One of the useful properties of the Euclidean medial axis is that it can be used to reconstruct the original curve. It is shown in Proposition 4.5 of [**9**] that for the piece of curve $\alpha(t) = (t, t^3)$, $-\frac{\sqrt{3}}{3} < t < \frac{\sqrt{3}}{3}$, there are no 1-branch bi-tangent pseudo-circles. The curve $\alpha(t)$ can be extended smoothly to obtain a closed curve $\gamma$. It is possible to construct $\gamma$ so that there do not exist pseudo-circles which are 1-branch tangent to a point of $\alpha(t)$ and are also tangent to some other point of $\gamma(t)$. Consider for example the limaçon whose radius $r$ is given by $r = \frac{3}{2} + \cos(\theta)$ where $-\pi < \theta < \pi$. Splitting the limaçon into timelike and spacelike components, only pairs of points from the regions $\arctan(\frac{\sqrt{35}}{17}) - \pi < \theta < -\arctan(\frac{\sqrt{35}}{17}) + \pi$ have corresponding Minkowski medial axis points, see Figure 6. Therefore, the $MMA$ together with a radius function ($MMA$ transform), unlike its Euclidean counterpart, is not a complete shape describer.

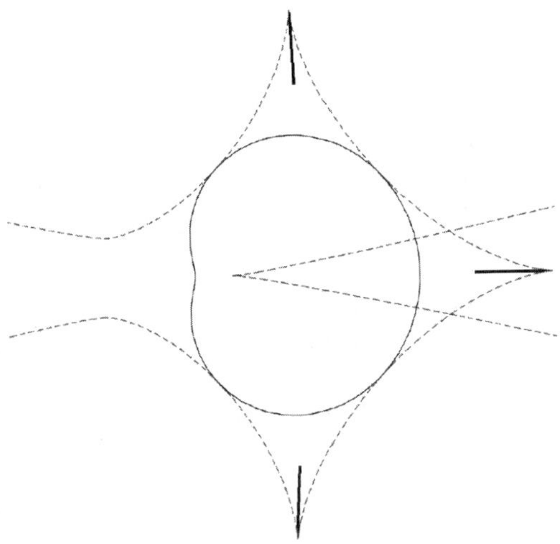

FIGURE 6. The limaçon given by $r = \frac{3}{2} + \cos(\theta)$, $-\pi < \theta < \pi$, its Minkowski caustic (dashed) and its $MMA$ (thick curves).

The fact that the $MMA$ consists of centres of only 1-branch bi-tangent pseudo-circles motivates the following new type of medial axis, which shall be called the *1-branch Minkowski medial axis*.

DEFINITION 4.10. The 1-branch Minkowski medial axis is the locus of the centres of pseudo-circles that are tangent to the curve $\gamma$ in two or more points such that the tangencies occur on just one of the branches of the pseudo-circle.

Since the $MMA$ is made up of *only* 1-branch bi-tangent pseudo-circle centres, it is a subset of the 1-branch medial axis.

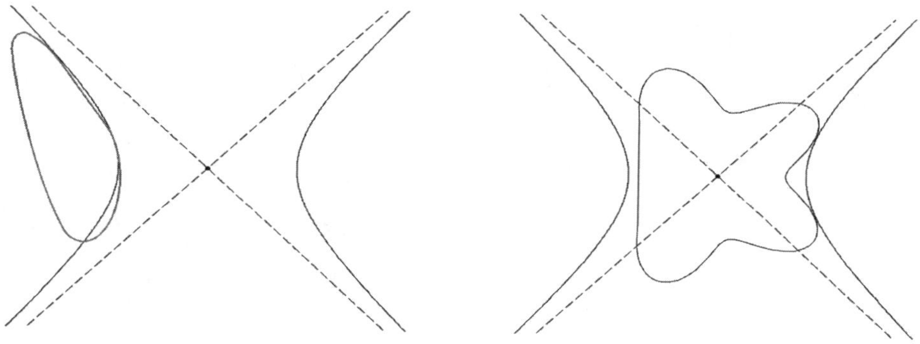

FIGURE 7. Left: an illustration of a 1-branch bi-tangent pseudo circle which is not maximal ($A_1A_2$-singularity). Right: a 1-branch bi-tangent pseudo-circle whose centre lies in the Minkowski hull of the (non-convex) curve.

The two sets are not equal because not all 1-branch bi-tangent pseudo-circles are maximal, see for example Figure 7, left. The 1-branch Minkowski medial axis does not lie in the complement of the Minkowski hull for non-convex curves; see for example Figure 7, right.

REMARK 4.11. Theorem 4.5 of [**9**] states that for any point $p$ on a spacelike or timelike curve $\gamma$ without inflections there exists another point $q$ on $\gamma$ and a pseudo-circle that is tangent to $\gamma$ at both $p$ and $q$ with both points being on a single branch of the pseudo-circle. From this it follows that any closed convex curve can be reconstructed from its 1-branch medial axis. The four lightlike components are either isolated points or lightlike line segments (in the generic case only isolated points are possible). To complete $\gamma$, these components can be added by taking the closure of the curve if they are just isolated points, or by joining up the remaining components with lightlike lines.

## 5. Shocks on the Minkowski medial axis

At each point on the medial axis of a curve $\gamma$ in the Euclidean plane there is an associated radius function $r$ corresponding to the radius of the bi-tangent circle. The direction of the increasing radius function on the medial axis, that is the direction for which $\partial r/\partial s(s) > 0$, can be indicated by an arrow and this gives the shock set (see for example [**1, 6**]).

The shock set is a dynamic view of the medial axis. If there is a propagation of waves from the curve $\gamma$, then this leads to the formation of singularities (the medial axis). The shock set gives the direction along which this formation of singularities propagates.

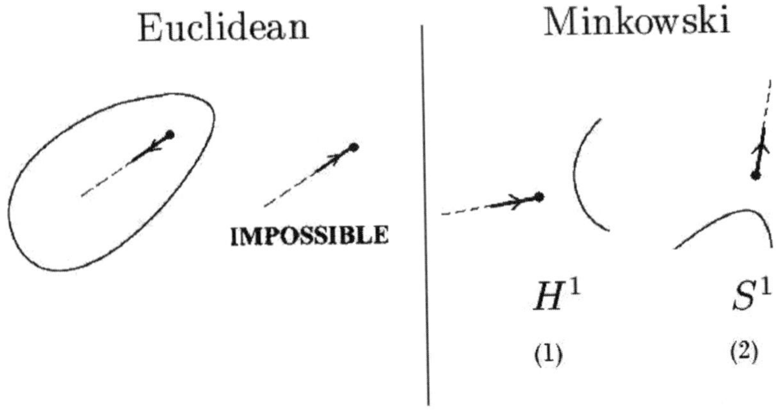


FIGURE 8. Comparing $A_3$ shocks that can occur on the Euclidean and Minkowski medial axes. In the Minkowski plane, both cases can occur and are distinguished by the type of the bi-tangent pseudo-circle.

In [**6**] the local generic forms of shocks that can occur on the medial axis in the Euclidean plane are classified. It is shown that some types of shocks cannot occur generically on the Euclidean medial axis. For example, it is proven that the only form of shock that can occur at an $A_3$-singularity of a given distance squared function on $\gamma$ is that with outward velocity, see Figure 8.

We define the analogue of shocks in the Minkowski plane. This is the Minkowski medial axis together with an arrow in the direction of increasing radius of the corresponding bi-tangent pseudo-circle.

In this section it is shown that for the $MMA$ both types of shock can occur (outwards and inwards) at an $A_3$-singularity of a given distance squared function on $\gamma$, depending on whether the $MMA$ is spacelike or timelike. It will also be shown that the generic shocks that can occur at an $A_1^3$-singularity of a given distance squared function on $\gamma$ are different to those on the Euclidean medial axis.

In what follows, the singularities refer to those of a given distance squared function.

**5.1. Shocks at an $A_3$-singularity.** The $A_3$-singularity occurs at a vertex of the curve. These occur where the two $A_1$-contact points for nearby $A_1^2$-bi-tangent pseudo-circles come into coincidence. The $A_1$-points must therefore lie on the same branch of the pseudo-circle. This means that given a $MMA$ near an $A_3$ point and its associated radius function, formula (3.2) can be used to find the corresponding points on $\gamma$.

THEOREM 5.1. *If the curve $\gamma$ is timelike/spacelike at an ordinary vertex, then the shock on the $MMA$ is of outwards/inwards type respectively.*

PROOF. Consider a neighbourhood of an $A_3$ point on a spacelike $MMA$ (which necessarily corresponds to a timelike piece of $\gamma$, see Theorem 5.2 in [**9**]). Orient the

$MMA$ so that its tangent line points towards the branch of the bi-tangent pseudo-circle that contains the tangent points. In the coordinate system with origin at the $A_3$ point on $MMA$ and with basis $T$ and $N$, the envelope points of the bi-tangent pseudo-circles have positive $T$ coordinate. It follows from formula (3.2) that $-r\partial r/\partial s > 0$. As $r > 0$, $\partial r/\partial s < 0$ so the radius function $r$ must decrease in the direction of the $A_3$ point. Therefore, for a timelike $\gamma$ near a vertex, the shocks are of outwards type; see Figure 8.

If the $MMA$ is timelike (the curve $\gamma$ must be spacelike, Theorem 5.2 in [**9**]). Following the same arguments as above and using formula (3.2) at an $A_3$-point, yields $-r\partial r/\partial s > 0$. Here $r < 0$ so it follows $\partial r/\partial s > 0$. This implies that the shocks are of inwards type; see Figure 8. □

REMARK 5.2. Theorem 5.1 also holds when the $MMA$ is replaced by the 1-branch $MMA$.

**5.2. Shocks at an $A_1^3$-singularity.** As with the shocks at $A_3$-singularity, the shocks at $A_1^3$-singularity turn out to be different from those of the Euclidean medial axis. The type of shocks that can occur depends on whether the bi-tangent pseudo-circle is of type $H^1(p, r)$ or $S^1(p, r)$.

For closed curves, the $MMA$ only consists of the centres of pseudo-circles whose tangencies occur on only one of its branches (Theorem 4.8). Theorem 5.4 gives a classification of shocks that can occur for closed curves. For two disjoint pieces of curves, it is possible that the pseudo-circle centred on the medial axis can be tangent to each piece of curve (so the centre is not on the 1-branch $MMA$). For completeness, Theorem 5.6 gives the classification of shocks at an $A_1^3$-singularity when the relevant pseudo-circle has at least one tangency on each of its branches.

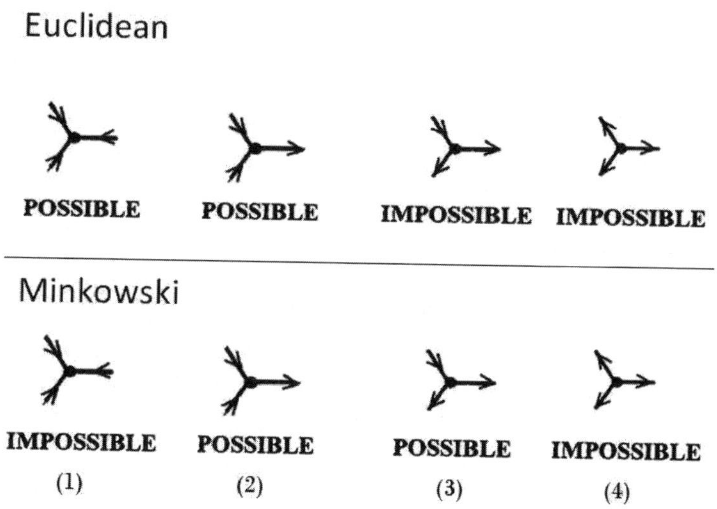


FIGURE 9. $A_1^3$ shocks that can occur on the Euclidean and Minkowski medial axes for closed curves. Case (2) occurs when the pseudo-circle is of type $S^1$. Case (3) occurs when the pseudo-circle is of type $H^1$.

Suppose that $\gamma$ is a closed plane curve and the three tri-tangent points, say $q_1, q_2$ and $q_3$ on $\gamma$ to a pseudo-circle all lie on one branch of the pseudo-circle. Each branch of the MMA corresponds to the centres of bi-tangent pseudo-circles whose points of tangency are near two of the points $q_i, i = 1, 2, 3$. For each branch, the two corresponding tangency points are called the characteristic points and are denoted $X^-$ and $X^+$. If the branch corresponds to points near $q_i$ and $q_j$ then the point $q_k$ with $k \neq i, j$ is denoted by $P$.

LEMMA 5.3. *If the arc that contains $X^-$ and $X^+$ does not contain the point $P$, then the medial axis goes in the direction of entering the arc. If the arc contains the point $P$ then the medial axis goes away from the arc.*

PROOF. Consider the case of a tri-tangent pseudo-circle of type $S^1(p,r)$, the proof is similar for a tri-tangent pseudo-circle of type $H^1(p,r)$ and is omitted.

Consider the function $f(s) = \langle c(s) - P, c(s) - P\rangle - (r(s))^2$, where $s$ is the arc-length parameter of the $MSS$ branch $c(s)$ and $r(s)$ is the radius of the bi-tangent pseudo-circle. Let $s_0$ correspond to the $A_1^3$ point. Note that $f(s_0) = 0$. We have tangency of type $A_1$, so $f'(s_0) \neq 0$. If $f'(s_0) < 0$, then $f(s) < f(s_0)$ for small $s > s_0$. For such $s$, $c(s)$ cannot be on the $MMA$ since the point $P$ will have come 'inside' the pseudo-circle centre $c(s)$ radius $r(s)$. Here, 'inside' means that its absolute distance from the centre is less than $|r|$.

Now $f'(s) = 2\langle c(s) - P, T(s)\rangle - 2r(s)r'(s)$, so that $f'(s_0) < 0$ is equivalent to $\langle c(s_0) - P, T(s_0)\rangle < r(s_0)r'(s_0)$. We have $r(s_0)r'(s_0) = \langle c(s_0) - X^\pm, T(s_0)\rangle$. This in turn implies $\langle X^\pm - P, T\rangle < 0$.

Parameterise the tri-tangent pseudo-circle by $g(t) = (r_0 \sinh(t), r_0 \cosh(t))$. Suppose that $X^+ = g(\theta)$, $X^- = g(\varphi)$ and $P = g(\rho)$ for some $\theta, \varphi$ and $\rho$.

The tangent to the branch of the $MSS$ corresponding to the characteristic points $X^+$ and $X^-$ has direction $T = (\sinh(\frac{\theta+\varphi}{2}), \cosh(\frac{\theta+\varphi}{2}))$ and the vector $X^+ - p$ has direction $(\cosh(\frac{\theta+\rho}{2}), \sinh(\frac{\theta+\rho}{2}))$ if $\theta > \rho$ and $(-\cosh(\frac{\theta+\rho}{2}), -\sinh(\frac{\theta+\rho}{2}))$ if $\theta < \rho$. Assume that $\theta > \rho$. The Minkowski product $\langle X^\pm - P, T\rangle$, up to a positive factor, is given by

$$-\sinh\left(\frac{\theta+\varphi}{2}\right)\cosh\left(\frac{\theta+\rho}{2}\right) + \sinh\left(\frac{\theta+\rho}{2}\right)\cosh\left(\frac{\theta+\varphi}{2}\right) = 2(e^\rho - e^\varphi)e^{-(\theta+\varphi+\rho)}$$

which is negative if and only if $\rho < \varphi$.

Similarly, assuming $\theta < \rho$ gives that $\langle X^\pm - P, T\rangle$ is negative if and only $\varphi < \rho$.

Thus the condition $\langle X^\pm - P, T\rangle < 0$ is equivalent to either $\rho < \varphi, \theta$ or $\rho > \varphi, \theta$ which is equivalent to the statement in Lemma 5.3. □

THEOREM 5.4. *For a smooth closed curve, the Minkowski medial axis at an $A_1^3$-singularity has shock type* (2) *in* Figure 9 *if the tri-tangent pseudo-circle is of type $S^1(p,r)$. It is of type* (3) *in* Figure 9 *if the tri-tangent pseudo-circle is of type $H^1(p,r)$.*

PROOF. Consider tangent pseudo-circles of type $S^1(p,r)$, the other case follows similarly. For three points on the same branch of the pseudo-circle and for $\rho, \varphi$ and $\theta$ as in the proof of Lemma 5.3, the conditions $\rho < \varphi, \theta$ or $\rho > \varphi, \theta$ must be true for two of the three medial axis branches. The arrows, indicating directions of increasing radius, can now be added to the medial axes. Formula (3.2) implies that the radius must be increasing in the direction of the branch of the pseudo-circle that contains the three tri-tangent points. □

REMARK 5.5. Theorem 5.4 gives a complete classification of shocks for closed curves at an $A_1^3$-singularity. It also holds when the $MMA$ is replaced by the 1-branch $MMA$. In this case the condition that the curve be closed can also be dropped.

Blum viewed the (Euclidean) medial axis as a quench point for grass-fire flow initiated from the boundary of the shape [**1**]. He considered the medial axes of curve segments as well as closed curves. In this spirit, and for the purpose of applications, we now consider the $MMA$ and classify its shocks for when the tri-tangency occurs on both branches of the pseudo-circle.

THEOREM 5.6. *The shocks that can occur at an $A_1^3$-singularity when the tangent points occur on both branches of a pseudo-circle of type $S^1(p,r)$ are as shown in* Figure 10. *For pseudo-circles of type $H^1(p,r)$ the shocks are as in* Figure 10 *with the directions of the arrows reversed.*

PROOF. Consider pseudo circles of type $S^1(p,r)$; the proof is similar for the pseudo-circles of type $H^1(p,r)$ and is omitted. Take the centre of the tri-tangent pseudo-circle to be the origin. The pseudo-circle has two branches, parameterise them by $g_1(t) = (r_0 \sinh(t), r_0 \cosh(t))$ and $g_2(t) = (r_0 \sinh(t), -r_0 \cosh(t))$ and assume that the branch containing two tangent points to be the branch in the lower half of the plane. Denote by $X_\theta = g_1(\theta)$ the tangent point on the upper branch and by $X_\varphi = g_2(\varphi)$ and $X_\rho = g_2(\rho)$ the two tangent points on the lower branch.

Denote by $T_{\theta,\varphi}$ the tangent line to the $MMA$ branch that corresponds to the characteristic points $\theta$ and $\varphi$ and similarly for $T_{\theta,\rho}$ and $T_{\rho,\varphi}$. Consider first the branch of the $MMA$ with tangent line $T_{\rho,\varphi}$. This tangent line has direction

$$T_{\rho,\varphi} = (\sinh(\frac{\rho+\varphi}{2}), \cosh(\frac{\rho+\varphi}{2}))$$

and the vector joining the point $X_\rho$ to $X_\theta$ is given by

$$X_\varphi - X_\theta = r_0(\sinh(\varphi) - \sinh(\theta), \cosh(\varphi) + \cosh(\theta)).$$

Taking their Minkowski product yields

$$\langle X_\varphi - X_\theta, T_{\rho,\varphi}\rangle = \langle X_\rho - X_\theta, T_{\rho,\varphi}\rangle = r_0 \cosh\left(\frac{\varphi-\rho}{2}\right) + r_0 \cosh\left(\frac{\varphi+\rho}{2} + \theta\right)$$

which is always positive. Therefore, the corresponding branch of the $MMA$ starts at the $A_1^3$ point and goes in the direction of the branch containing the characteristic points. Formula (3.2) implies that the shock travels along this $MMA$ branch in the direction towards the centre.

Consider now the two branches with tangents $T_{\theta,\rho}$ and $T_{\theta,\varphi}$. For each branch of the $MMA$ there are four possibilities: The $MMA$ can be to the left or to the right of the centre (since the branches are both timelike), and in both cases the direction of the shock can be either towards or away from the centre.

The tangent line $T_{\theta,\varphi}$ to the medial axis at the $A_1^3$, up to a nonzero factor, can be written

$$T_{\theta,\varphi} = \left(\frac{\sinh(\theta)+\sinh(\varphi)}{2}, \frac{\cosh(\theta)-\cosh(\varphi)}{2}\right)$$

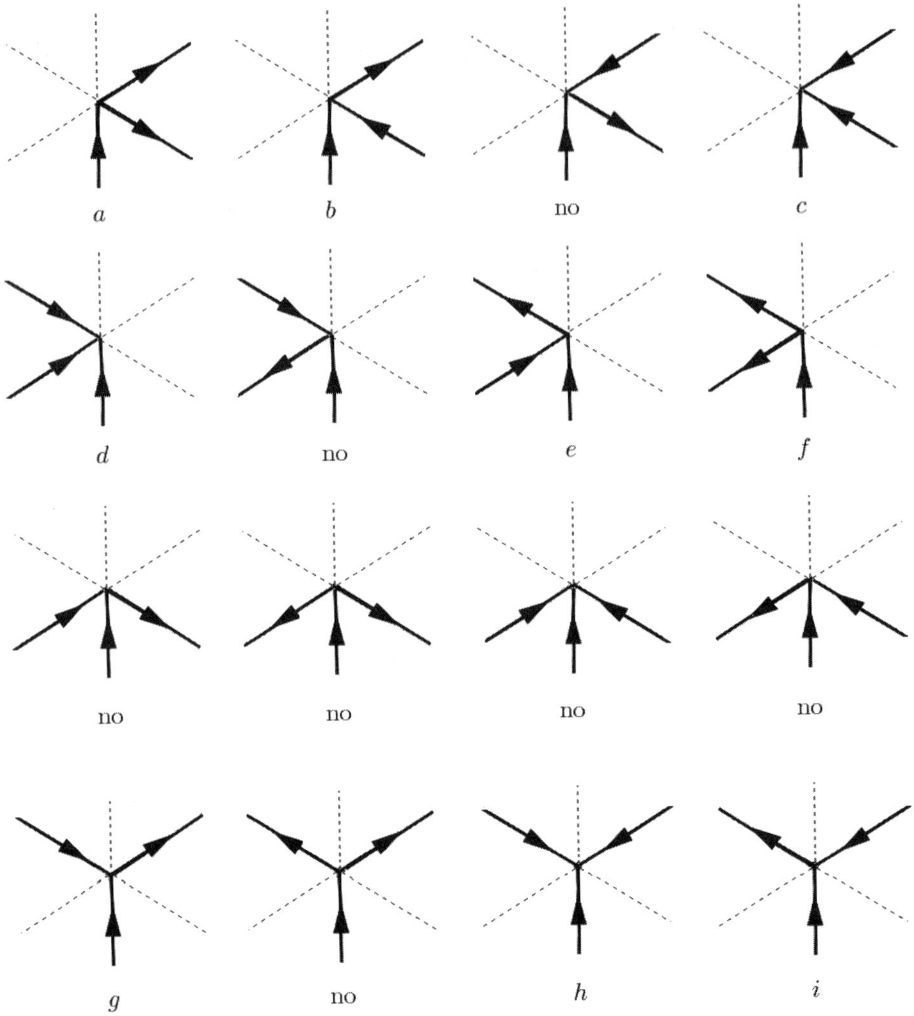


FIGURE 10. The types of shocks that can occur for timelike $A_1^3$ points.

and the vector joining the point $X_\rho$ to $X_\theta$, is given by $X_\theta - X_\rho = r_0(\sinh(\theta) - \sinh(\rho), \cosh(\theta) + \cosh(\rho))$. Taking their Minkowski product yields

$$(5.1) \qquad \langle X_\theta - X_\rho, T_{\theta,\varphi}\rangle = \frac{r_0}{2}(1 - \cosh(\rho - \theta) + \cosh(\rho + \varphi) - \cosh(\theta + \varphi))$$

which is positive when $-\varphi < \theta < \rho$ or $\rho < \theta < -\varphi$ and negative otherwise. This can be shown by considering all the possible orderings of $\theta, -\varphi$ and $\rho$. For example, suppose that $\theta < -\phi < \rho$ so that $\theta = -\phi - \delta$ for some positive $\delta$. Substituting in (5.1) yields

$$\langle X_\theta - X_\rho, T_{\theta,\varphi}\rangle = \frac{r_0}{2}(1 - \cosh(\rho + \phi + \delta) + \cosh(\rho + \varphi) - \cosh(-\delta)).$$

The above expression is always negative because $\cosh(\rho + \varphi + \delta) > \cosh(\rho + \varphi)$ as both $\rho + \varphi$ and $\delta$ are positive, and $1 - \cosh(-\delta)$ is also negative.

Recall that when $\langle X_\theta - X_\rho, T_{\theta,\varphi}\rangle$ is positive, small positive $s > s_0$ belongs to the $MMA$, and when it is negative, it is small $s < s_0$ that belongs to the $MMA$. Now, the direction of increasing radius can be added to the $MMA$. It follows directly from formula (3.1) that if $\theta + \varphi > 0$ the radius function on the branch corresponding to $T_{\theta,\varphi}$ increases from left to right, whereas if $\theta + \varphi < 0$ the radius increases from right to left, and similarly for $T_{\theta,\rho}$.

Considering the two branches together now, it must be determined which branch goes 'over' the other. Comparing the gradients of the two tangent lines it can be seen that $T_{\theta,\varphi}$ is steeper than $T_{\theta,\rho}$ if and only if

$$\theta < \rho < \varphi,\ \rho < \varphi < \theta \text{ or } \varphi < \theta < \rho.$$

Otherwise, $T_{\theta,\rho}$ is the steeper of the two.

Considering these conditions, along with the above conditions for which side the branches lie on, gives the complete list of possible shocks that can occur (see Table 1 and Figure 10).

□

TABLE 1. Conditions for shocks in Figure 10 to occur.

| | | | |
|---|---|---|---|
| a) | $-\rho < -\varphi < \theta < \varphi < \rho$<br>$-\varphi < -\rho < \theta < \rho < \varphi$ | b) | $-\rho < \theta < \varphi < -\varphi < \rho$<br>$-\rho < \theta < -\varphi < \varphi < \rho$<br>$-\varphi < \theta < -\rho < \rho < \varphi$<br>$-\varphi < \theta < -\rho < \rho < \varphi$ |
| c) | $\theta < -\varphi, \varphi, -\rho, \rho$ | d) | $-\varphi, \varphi, -\rho, \rho < \theta$ |
| e) | $\varphi < -\rho < \rho < \theta < -\varphi$<br>$\varphi < \rho < -\rho < \theta < -\varphi$<br>$\rho < -\varphi < \varphi < \theta < -\rho$<br>$\rho < \varphi < -\varphi < \theta < -\rho$ | f) | $\rho < \varphi < \theta < -\varphi < -\rho$<br>$\varphi < \rho < \theta < -\rho < -\varphi$ |
| g) | $-\varphi < -\rho < \rho < \theta < \varphi$<br>$-\varphi < \rho < -\rho < \theta < \varphi$<br>$-\rho < -\varphi < \varphi < \theta < \rho$<br>$-\rho < \varphi < -\varphi < \theta < \rho$ | h) | $-\varphi < \rho < \theta < -\rho < \varphi$<br>$\rho < -\varphi < \theta < \varphi < -\rho$<br>$-\rho < \varphi < \theta < -\varphi < \rho$<br>$\varphi < -\rho < \theta < \rho < -\varphi$ |
| i) | $\rho < \theta < \varphi < -\varphi < -\rho$<br>$\rho < \theta < -\varphi < \varphi < -\rho$<br>$\varphi < \theta < -\rho < \rho < -\varphi$<br>$\varphi < \theta < -\rho < \rho < -\varphi$ | | |

## References


[1] H. Blum, *A transformation for extracting new descriptors of shape.* W. Whaten-Dunn (Ed.), Models for the Perception of Speech and Visual Forms. MIT Press, Cambridge, MA, 1967, 362–380.

[2] J. W. Bruce and P. J. Giblin, *Curves and singularities*, 2nd ed., Cambridge University Press, Cambridge, 1992. A geometrical introduction to singularity theory. MR1206472

[3] Ilya A. Bogaevsky, *Perestroikas of shocks and singularities of minimum functions*, Phys. D **173** (2002), no. 1-2, 1–28, DOI 10.1016/S0167-2789(02)00652-8. MR1945478

[4] J. W. Bruce, P. J. Giblin, and C. G. Gibson, *Symmetry sets*, Proc. Roy. Soc. Edinburgh Sect. A **101** (1985), no. 1-2, 163–186, DOI 10.1017/S0308210500026263. MR824218

[5] P. J. Giblin and S. A. Brassett, *Local symmetry of plane curves*, Amer. Math. Monthly **92** (1985), no. 10, 689–707, DOI 10.2307/2323220. MR820052

[6] P. J. Giblin and B .B. Kimia, *On the local form and transitions of symmetry sets, medial axes, and shocks.* Int. J. Computer Vision **54** (2003), 143–157.

[7] P. J. Giblin and J. P. Warder, *Reconstruction from medial representations*, Amer. Math. Monthly **118** (2011), no. 8, 712–725, DOI 10.4169/amer.math.monthly.118.08.712. MR2843991

[8] M. Golubitsky and V. Guillemin, *Stable mappings and their singularities*, Springer-Verlag, New York-Heidelberg, 1973. Graduate Texts in Mathematics, Vol. 14. MR0341518

[9] G. M. Reeve and F. Tari, *Mikowski symmetry sets of plane curves.* To appear in Proc. Edinburgh Math. Soc.

[10] *Medial representations*, Computational Imaging and Vision, vol. 37, Springer, New York, 2008. Mathematics, algorithms and applications; Edited by Kaleem Siddiqi and Stephen M. Pizer. MR2547467

[11] Amani Saloom and Farid Tari, *Curves in the Minkowski plane and their contact with pseudo-circles*, Geom. Dedicata **159** (2012), 109–124, DOI 10.1007/s10711-011-9649-1. MR2944524

Instituto de Ciências Matemáticas e de Computação - USP, Avenida Trabalhador são-carlense, 400 - Centro, CEP: 13566-590 - São Carlos - SP, Brazil

*E-mail address*: grahamreeve86@gmail.com

Instituto de Ciências Matemáticas e de Computação - USP, Avenida Trabalhador são-carlense, 400 - Centro, CEP: 13566-590 - São Carlos - SP, Brazil

*E-mail address*: faridtari@icmc.usp.br

Contemporary Mathematics
Volume **675**, 2016
http://dx.doi.org/10.1090/conm/675/13597

# Cobordism group of Morse functions on surfaces with boundary

Osamu Saeki and Takahiro Yamamoto

*Dedicated to Professors Taizo Kanenobu, Yasutaka Nakanishi and Makoto Sakuma on their sixtieth birthdays*

ABSTRACT. We consider Morse functions on compact manifolds possibly with boundary, and define their admissible cobordism group, based on generic maps into the plane that are submersions near the boundary. Then, we show that the cobordism group of Morse functions on surfaces with boundary is isomorphic to the cyclic group of order two. Our approach is based on the Stein factorizations: the novelty lies in the challenge that we consider Morse functions on manifolds with boundary and their cobordisms.

## 1. Introduction

Let $N$ be a compact $C^\infty$ manifold possibly with boundary. A $C^\infty$ real-valued function $f: N \to \mathbb{R}$ is called a *Morse function* if it is a submersion on a neighborhood of the boundary, and the critical points of $f$ and $f|_{\partial N}$ are all non-degenerate and have distinct values (see [**1**], for example). Note that such a class of functions is generic in the sense that it forms an open and dense subset in the space of $C^\infty$ functions $C^\infty(N, \mathbb{R})$ endowed with the Whitney $C^\infty$ topology. More precisely, it coincides with the class of $C^\infty$ stable functions. Here, a $C^\infty$ map $f: N \to P$ of a manifold with boundary $N$ into a manifold without boundary $P$ is *stable* (or more precisely, $C^\infty$ *stable*) if there is a neighborhood of $f$ in the space $C^\infty(N, P)$ of $C^\infty$ maps of $N$ into $P$ endowed with the Whitney $C^\infty$ topology such that each map in the neighborhood is $C^\infty$ right-left equivalent to $f$, where two $C^\infty$ maps $f$ and $g: N \to P$ are $C^\infty$ *right-left equivalent* if there exist diffeomorphisms $\Phi: N \to N$ and $\Psi: P \to P$ such that it $\Psi \circ f = g \circ \Phi$ (see, for example, [**4**]).

For a non-negative integer $n$, denote by $b\mathcal{N}(n)$ the set of all Morse functions on compact $n$-dimensional manifolds possibly with boundary.

In the following, a $C^\infty$ map defined on a manifold with boundary is said to be *admissible* if it is a submersion on a neighborhood of the boundary. In this sense, every Morse function is admissible.

Let us introduce a certain cobordism relation for Morse functions on manifolds with boundary as follows.

2010 *Mathematics Subject Classification.* Primary 57R45; Secondary 57R35, 57R90,58K15, 58K65 .

*Key words and phrases.* Morse function, cobordism, stable map, admissible map, Reeb graph, singular fiber, manifold with boundary.

DEFINITION 1.1. Let $N_0$ and $N_1$ be compact $n$-dimensional manifolds possibly with boundary. Two Morse functions $f_i\colon N_i \to \mathbb{R}$, $i = 0, 1$, are said to be *admissibly cobordant* if there exist a compact $(n+1)$-dimensional manifold $X$ with corners and a $C^\infty$ map $F\colon X \to \mathbb{R} \times [0, 1]$ with the following properties.

(1) We have $\partial X = N_0 \cup Q \cup N_1$, where $N_0$, $N_1$ and $Q$ are codimension 0 submanifolds of $\partial X$, $N_0 \cap N_1 = \emptyset$, and $\partial Q = (N_0 \cap Q) \cup (N_1 \cap Q)$.
(2) The manifold $X$ has corners exactly along $\partial Q$.
(3) We have $F|_{N_0 \times [0,\varepsilon)} = f_0 \times \mathrm{id}_{[0,\varepsilon)}$ and $F|_{N_1 \times (1-\varepsilon,1]} = f_1 \times \mathrm{id}_{(1-\varepsilon,1]}$, where $N_0 \times [0, \varepsilon)$ and $N_1 \times (1-\varepsilon, 1]$ denote the collar neighborhoods (with corners) of $N_0$ and $N_1$ in $X$, respectively, and $\mathrm{id}_{[0,\varepsilon)}$ (resp. $\mathrm{id}_{(1-\varepsilon,1]}$) denotes the identity map of $[0, \varepsilon)$ (resp. $(1 - \varepsilon, 1]$).
(4) We have $F^{-1}(\mathbb{R} \times \{i\}) = N_i$, $i = 0, 1$, and the restriction $F|_{X \setminus (N_0 \cup N_1)}\colon X \setminus (N_0 \cup N_1) \to \mathbb{R} \times (0, 1)$ is a proper admissible $C^\infty$ stable map.

In this case we call the map $F\colon X \to \mathbb{R} \times [0, 1]$ an *admissible cobordism* between $f_0$ and $f_1$.

If a Morse function in $b\mathcal{N}(n)$ is admissibly cobordant to the function on the empty set, then we say that it is *null-cobordant*.

Note that in the above situation, the map $F|_{X \setminus (N_0 \cup N_1)}$ has fold points and cusp points as its singularities (for example, see [**12**]). The above definition slightly differs from that of a fold cobordism as defined in [**6, 7, 10**], since we allow cusp points as the singularities.

The above relation clearly defines an equivalence relation on $b\mathcal{N}(n)$ for each $n$. Furthermore, the set of all equivalence classes forms an additive group under the disjoint union: the neutral element is the class corresponding to null-cobordant Morse functions, and the inverse of a class represented by a Morse function $f\colon N \to \mathbb{R}$ is given by the class of $-f\colon N \to \mathbb{R}$, where $(-f)(x) = -f(x)$ for all $x \in N$. We denote by $b\mathfrak{N}_n$ the additive group of all admissible cobordism classes of elements of $b\mathcal{N}(n)$ and call it the *admissible cobordism group of Morse functions on manifolds with boundary of dimension $n$*, or the *$n$-dimensional admissible cobordism group of Morse functions*.

The cobordism theory of $C^\infty$ maps originally dates back to the work of Thom [**25**], who showed that the cobordism group of embeddings is isomorphic to a homotopy group of a certain Thom complex by using the Pontrjagin-Thom construction. Then, the cobordism group of immersions was studied by Wells [**26**] using a similar method. Rimányi and Szűcs [**16**] extended these results to the cobordism group of maps with a certain class of singularities by using the notion of a $\tau$-map. (See also the work of Eliashberg [**3**].) In the literature, such cobordism groups are usually computed by using the method of algebraic topology: they are isomorphic to homotopy groups of certain Thom complexes. This kind of an approach is thoroughly studied by Szűcs [**23, 24**] using sophisticated algebraic topology techniques. Note that in these works they consider the case where the dimension of the target is greater than or equal to that of the source.

The first author [**17**] defined the cobordism group of special generic functions on smooth closed manifolds and showed that it is isomorphic to the $h$-cobordism group of homotopy spheres for high dimensions, by using geometric methods, namely, their Stein factorizations. Ikegami and the first author of the present paper [**6, 7, 19**], and Kalmar [**10**] studied certain cobordism groups of Morse functions on smooth closed manifolds by using geometric methods; the Stein factorizations or Levine's cusp elimination technique [**12**]. In this paper, our approach is based on the Stein factorizations: the novelty lies in the challenge that we consider Morse functions on manifolds with boundary.

The authors [**20**] established the notion of an admissible cobordism for Morse functions defined on compact manifolds with boundary. We studied the case of Morse functions on surfaces, and found a non-trivial $\mathbb{Z}_2$-valued admissible cobordism invariant, as a result of the computation of the cohomology group of a universal complex associated with singular fibers (see [**18**]).

In this paper, we study the admissible cobordism group of Morse functions on surfaces with boundary using geometric methods as developed in [**6, 10, 19**]. More precisely, we use their Stein factorizations or the so-called Reeb graphs (for details, see Definition 2.1).

Our main theorem of this paper is the following.

THEOREM 1.2. *The 2-dimensional admissible cobordism group of Morse functions* $b\mathfrak{N}_2$ *is cyclic of order two.*

This answers the question posed by Terry Gaffney at the 13th International Workshop on Real and Complex Singularities in 2014. It also gives a positive answer to the conjecture proposed in [**21**, Conjecture 6.4].

The paper is organized as follows. In §2, we study the Stein factorizations of Morse functions on surfaces with boundary and those of stable maps of 3-dimensional manifolds possibly with boundary into surfaces without boundary. These will play an essential role in the proof of our main theorem. In §3, we introduce the notion of a Reeb-like function defined on a labeled Reeb-like graph, and we define their cobordism group. Then, we show that this group is isomorphic to the admissible cobordism group of Morse functions on surfaces with boundary. In §4, we prove Theorem 1.2, by using certain moves for Reeb-like functions defined on graphs, which correspond to certain singular fibers studied in [**20**]. In §5, we study the admissible cobordism groups of Morse functions on 0- and 1-dimensional manifolds possibly with boundary. Finally in §6, we propose several open problems concerning cobordism groups of Morse functions on compact manifolds with boundary.

Throughout the paper, all manifolds and maps between them are smooth of class $C^\infty$ unless otherwise stated. For a $C^\infty$ map $f$ between manifolds, $S(f)$ denotes the set of singular points of $f$ in the source manifold. For two continuous maps $f_i\colon X_i \to Y_i$, $i = 0, 1$, between topological spaces, and for points $p_i \in X_i$, $i = 0, 1$, we say that $f_0$ and $f_1$ are *locally equivalent* around $p_0$ and $p_1$ if there exist open neighborhoods $U_i$ of $p_i$ in $X_i$ and $V_i$ of $f_i(p_i)$ in $Y_i$ with $f_i(U_i) \subset V_i$, $i = 0, 1$, and homeomorphisms $\varphi\colon U_0 \to U_1$ and $\psi\colon V_0 \to V_1$ with $\varphi(p_0) = p_1$ and $\psi(f_0(p_0)) = f_1(p_1)$ such that the diagram

$$\begin{array}{ccc} U_0 & \xrightarrow{\varphi} & U_1 \\ {\scriptstyle f_0}\downarrow & & \downarrow{\scriptstyle f_1} \\ V_0 & \xrightarrow{\psi} & V_1 \end{array}$$

is commutative.

## 2. Preliminaries

Let us begin by recalling the notion of a Stein factorization, which will play an essential role in this paper.

DEFINITION 2.1. Let $N$ and $P$ be manifolds possibly with boundary such that $n = \dim N > \dim P = p$, and $f\colon N \to P$ a $C^\infty$ map. Two points in $N$ are equivalent if they lie in the same connected component of the pre-image of a point in $P$ by $f$. Let $W_f$ denote the quotient space of $N$ with respect to this equivalence relation and $q_f\colon N \to W_f$ the quotient map. Then, we see easily that there exists a unique continuous map $\overline{f}\colon W_f \to P$ such that $f = \overline{f} \circ q_f$. The space $W_f$ or the commutative diagram

$$\begin{array}{ccc} N & \xrightarrow{\quad f \quad} & P \\ & {\scriptstyle q_f}\searrow \quad \nearrow {\scriptstyle \overline{f}} & \\ & W_f & \end{array}$$

is called the *Stein factorization* of $f$. Note that $W_f$ can be regarded as the space of connected components of $f$-fibers. The space $W_f$ is also called the *Reeb space* of $f$, and the continuous map $\overline{f}$ is called the *Reeb map* of $f$ (see also [**2**]).

In the above situation, a point in $P$ is a *regular value* of $f$ if it is a regular value of both $f|_{\mathrm{Int}\,N}$ and $f|_{\partial N}$. In particular, if $N$ is compact and $\dim N = \dim P + 1$, then the pre-image of a regular value is a finite disjoint union of circles and arcs.

Let $f\colon N \to \mathbb{R}$ be a Morse function on a compact $n$-dimensional manifold with boundary. It is known that for each critical point $p$ of $f|_{\partial N}$, $f$ is locally given by

$$x_1^2 \pm x_2^2 \pm \cdots \pm x_{n-1}^2 \pm y,$$

where the local coordinates $(x_1, x_2, \ldots, x_{n-1}, y)$ around $p$ satisfy that $\mathrm{Int}\,N$ and $\partial N$ correspond to the sets $\{y > 0\}$ and $\{y = 0\}$, respectively. Using this fact together with arguments as in [**5, 8, 9, 11**], we can prove the following.

LEMMA 2.2. *Let $f\colon N \to \mathbb{R}$ be a Morse function on a compact surface possibly with boundary. Then, its Reeb space $W_f$ is a finite graph whose vertices are the $q_f$-images of the critical points of $f$ and $f|_{\partial N}$. Furthermore, each edge corresponds to a circle regular fiber or an arc regular fiber.*

In the above situation, the Reeb space $W_f$ is also called a *Reeb graph*, and the continuous map $\overline{f}\colon W_f \to \mathbb{R}$ is called a *Reeb function*. In the following, we label each edge of a Reeb graph by 0 or 1, where 0 (resp. 1) means that the edge corresponds to a circle regular fiber (resp. an arc regular fiber). Such a Reeb graph with each edge being labeled by an element of $\{0, 1\}$ is called a *labeled Reeb graph*. In the following, when a labeled Reeb graph is depicted, we adopt the convention that an edge with label 0 is drawn by a thick line, while an edge with label 1 is drawn by a dotted line.

By the above lemma, each vertex corresponds to a codimension one connected singular fiber of $f$ in the sense of [**20**]. Recall that we have nine such singular fibers: $\widetilde{\mathrm{bI}}^2, \widetilde{\mathrm{bI}}^3, \ldots, \widetilde{\mathrm{bI}}^{10}$ (see Figure 1). See also [**1**].

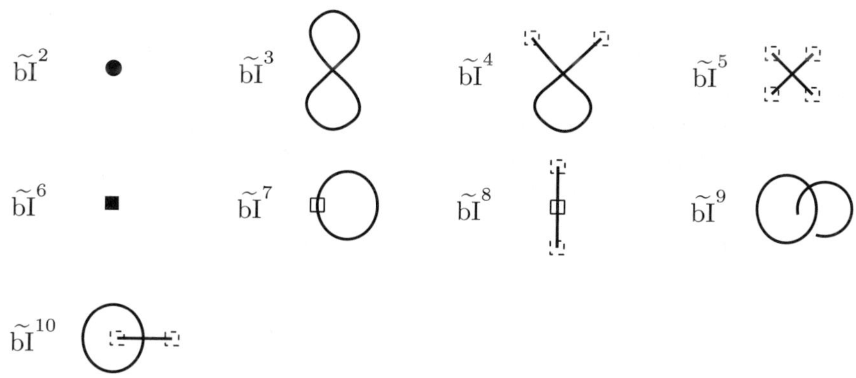


FIGURE 1. List of singular fibers of Morse functions on compact surfaces with boundary

Then, we have the following.

(1) The vertices corresponding to singular fibers of types $\widetilde{\mathrm{bI}}^2$ and $\widetilde{\mathrm{bI}}^6$ have degree 1, those corresponding to $\widetilde{\mathrm{bI}}^7$, $\widetilde{\mathrm{bI}}^9$ and $\widetilde{\mathrm{bI}}^{10}$ have degree 2, those corresponding to $\widetilde{\mathrm{bI}}^3$, $\widetilde{\mathrm{bI}}^4$ and $\widetilde{\mathrm{bI}}^8$ have degree 3, and those corresponding to $\widetilde{\mathrm{bI}}^5$ have degree 4.
(2) Around each vertex, $\overline{f}\colon W_f \to \mathbb{R}$ is locally equivalent to one of the functions as depicted in Figure 2 (or their negatives), where the relevant functions are depicted as the height functions.
(3) The continuous map $\overline{f}$ is an embedding on each edge.

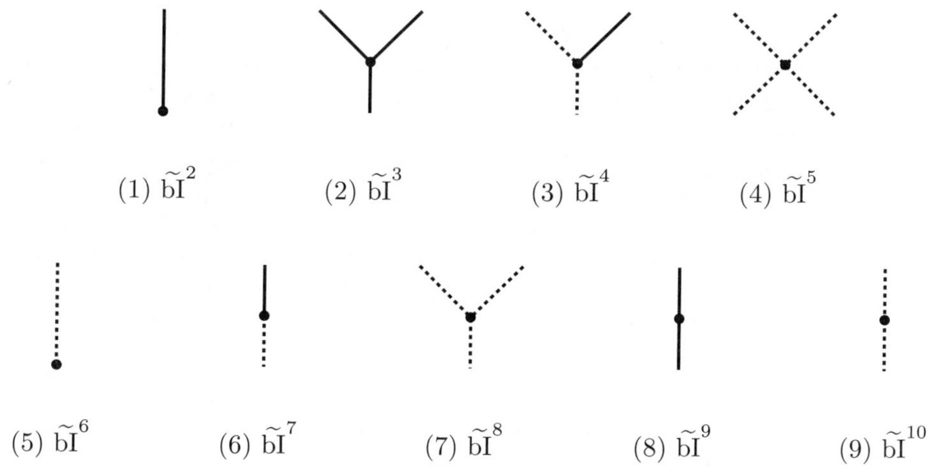


FIGURE 2. Local models of $\overline{f}\colon W_f \to \mathbb{R}$ around vertices

Now, let us proceed to the corresponding descriptions for $C^\infty$ stable maps of 3-dimensional manifolds with boundary into surfaces. Let us first recall the following characterization (see Shibata [**22**] and Martins and Nabarro [**14**]).

PROPOSITION 2.3. *Let $N$ be a 3-dimensional manifold possibly with boundary and $P$ a surface without boundary. A proper $C^\infty$ map $f\colon N \to P$ is $C^\infty$ stable if and only if it satisfies the following conditions:*

(1) (*Local conditions*) *In the following, for $p \in \partial N$, we use local coordinates $(x, y, z)$ around $p$ such that $\operatorname{Int} N$ and $\partial N$ correspond to the sets $\{z > 0\}$ and $\{z = 0\}$, respectively.*

(1a) *For $p \in \operatorname{Int} N$, the germ of $f$ at $p$ is right-left equivalent to one of the following:*

$$(x,y,z) \mapsto \begin{cases} (x,y), & p\text{: regular point,} \\ (x, y^2+z^2), & p\text{: definite fold point,} \\ (x, y^2-z^2), & p\text{: indefinite fold point,} \\ (x, y^3+xy-z^2), & p\text{: cusp point.} \end{cases}$$

(1b) *For $p \in \partial N \setminus S(f)$, the germ of $f$ at $p$ is right-left equivalent to one of the following:*

$$(x,y,z) \mapsto \begin{cases} (x,y), & p\text{: regular point of } f|_{\partial N}, \\ (x, y^2+z), & p\text{: boundary definite fold point,} \\ (x, y^2-z), & p\text{: boundary indefinite fold point,} \\ (x, y^3+xy+z), & p\text{: boundary cusp point.} \end{cases}$$

(1c) *For $p \in \partial N \cap S(f)$, the germ of $f$ at $p$ is right-left equivalent to the map germ*

$$(x,y,z) \mapsto (x, y^2 + xz \pm z^2).$$

(2) (*Global conditions*) *For each $q \in f(S(f)) \cup f(S(f|_{\partial N}))$, the multi-germ*

$$(f|_{S(f)\cup S(f|_{\partial N})}, f^{-1}(q) \cap (S(f) \cup S(f|_{\partial N})))$$

*is right-left equivalent to one of the eight multi-germs whose singular value sets are depicted in Figure 3, where the ordinary curves correspond to $f(S(f))$ and the dotted curves to $f(S(f|_{\partial M}))$: (1) and (4) represent immersion mono-germs $(\mathbb{R}, 0) \ni t \mapsto (t, 0) \in (\mathbb{R}^2, 0)$ which correspond to a single fold point and a single boundary fold point, respectively, (3), (6) and (7) represent normal crossings of two immersion germs, each of which corresponds to a fold point or a boundary fold point, (2) and (5) represent cusp mono-germs $(\mathbb{R}, 0) \ni t \mapsto (t^2, t^3) \in (\mathbb{R}^2, 0)$ which correspond to a cusp point and a boundary cusp point, respectively, and (8) represents the restriction of the mono-germ (1c), corresponding to a single point in $\partial N \cap S(f)$, to the singular point set.*

As can be easily seen, a $C^\infty$ stable map $f\colon N \to P$ of a 3-dimensional manifold possibly with boundary into a surface without boundary is admissible if and only if it has no singular point of type (1c) above.

Then, as before, we can prove the following.

LEMMA 2.4. *Let $f\colon X \to P$ be an admissible $C^\infty$ stable map of a compact 3-dimensional manifold possibly with boundary into a surface without boundary. Then, the Reeb space $W_f$ is a compact 2-dimensional polyhedron, which is labeled: in other words, each component of $W_f \setminus q_f(S(f))$ is labeled with an element of $\{0, 1\}$, where 0 (resp. 1) corresponds to a regular circle fiber (resp. a regular arc fiber).*

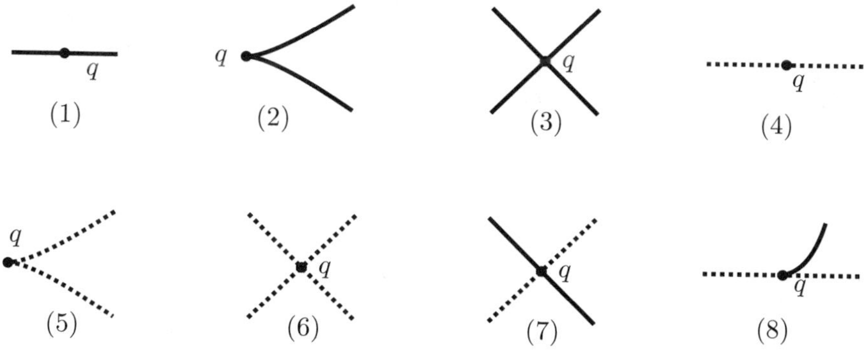


FIGURE 3. Multi-germs of $f|_{S(f)\cup S(f|_{\partial N})}$

In the above situation, each vertex of $W_f$ corresponds to a codimension two connected singular fiber in the sense of [**20**], and each component of the complement in $q_f(S(f))$ of the vertices corresponds to a codimension one connected singular fiber. Furthermore, around each point of $W_f$, $\overline{f}\colon W_f \to P$ is locally equivalent to one of the maps as depicted in Figures 4, 5 and 6, where the relevant map is the vertical projection to a plane. Here, each point of $W_f$ corresponds to a connected fiber and we use the notation for singular fibers as introduced in [**20**]. Note that $\overline{f}$ is a local embedding on $q_f(S(f))$ and on $W_f \setminus q_f(S(f))$.

## 3. Cobordism group of Reeb-like functions

In this section, we introduce the notion of cobordism for functions defined on Reeb-like graphs.

DEFINITION 3.1. Let $G$ be a finite graph which may possibly be disconnected and whose edges are labeled by an element of $\{0, 1\}$. We assume that around each vertex of $G$, it is locally homeomorphic to one of the labeled graphs as depicted in Figure 2 in such a way that the labels are consistent. In this case, we call $G$ a *labeled Reeb-like graph.*

Now, let $r\colon G \to \mathbb{R}$ be a continuous function such that

(1) around each vertex of $G$, $r$ is locally equivalent to one of the functions as depicted in Figure 2 (or their negatives), and
(2) $r$ is an embedding on each edge.

Then, we call $r\colon G \to \mathbb{R}$ a *Reeb-like function.*

Note that for a Morse function $f\colon N \to \mathbb{R}$ on a compact surface possibly with boundary, its labeled Reeb graph is a labeled Reeb-like graph, and its Reeb function $\overline{f}\colon W_f \to \mathbb{R}$ is a Reeb-like function, by the very definition.

DEFINITION 3.2. Let $r_i\colon G_i \to \mathbb{R}$, $i = 0, 1$, be Reeb-like functions on labeled Reeb-like graphs. We say that $r_0$ and $r_1$ are *cobordant* if there exist a compact 2-dimensional polyhedron $W$, a 1-dimensional subpolyhedron $\Sigma(W)$, and a continuous map $R\colon W \to \mathbb{R} \times [0, 1]$ with the following properties.

(1) Each connected component of $W \setminus \Sigma(W)$ is labeled with an element of $\{0, 1\}$.

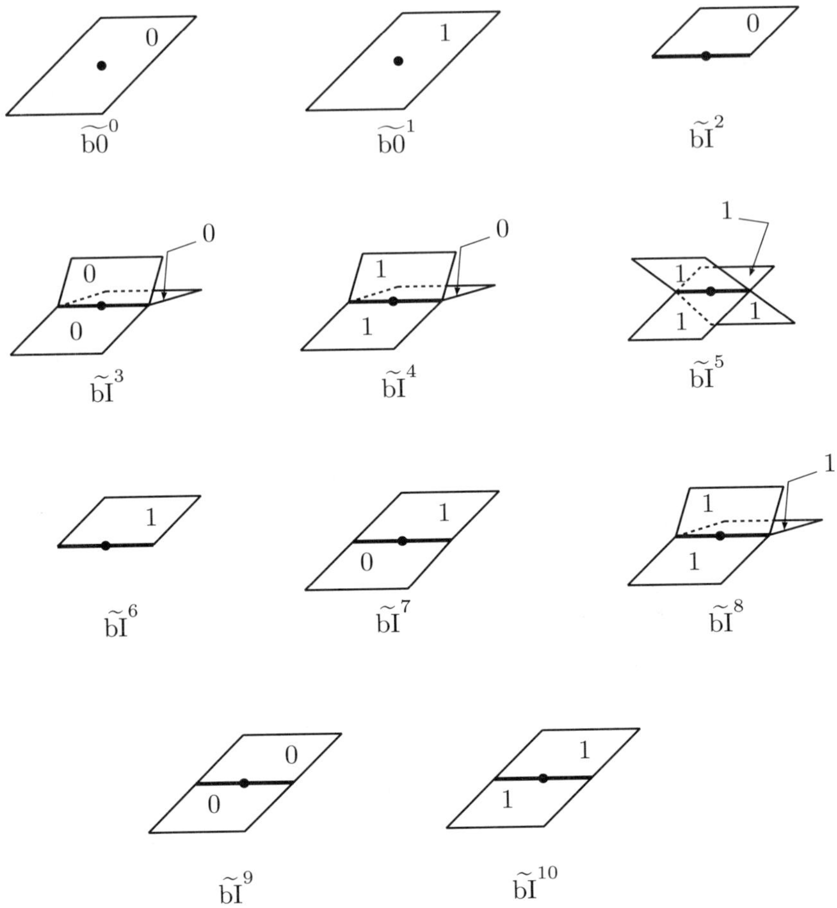


FIGURE 4. Local models of the Reeb map (I)

(2) The graphs $G_0$ and $G_1$ are identified with 1-dimensional subcomplexes of $W$, and we have $G_i = R^{-1}(\mathbb{R} \times \{i\})$, $i = 0, 1$.
(3) The intersection $G_i \cap \Sigma(W)$ coincides with the set of vertices of $G_i$, $i = 0, 1$.
(4) Each edge of $G_i$ is embedded in a component of $W \setminus \Sigma(W)$ with the same label, $i = 0, 1$.
(5) The subcomplexes $G_0$ and $G_1$ have regular neighborhoods of the forms $G_0 \times [0, \varepsilon]$ and $G_1 \times [1-\varepsilon, 1]$, respectively, in $W$, and we have $R|_{G_0 \times [0,\varepsilon)} = r_0 \times \mathrm{id}_{[0,\varepsilon)}$ and $R|_{G_1 \times (1-\varepsilon,1]} = r_1 \times \mathrm{id}_{(1-\varepsilon,1]}$.
(6) Around each point of $W \setminus (G_0 \cup G_1)$, $R\colon (W, \Sigma(W)) \to \mathbb{R} \times [0, 1]$ is locally equivalent to the Reeb map $\overline{f}\colon (W_f, q_f(S(f))) \to \mathbb{R} \times [0, 1]$ of a proper admissible stable map $f$ of a 3-dimensional manifold possibly with boundary into a surface in such a way that $q_f(S(f))$ locally corresponds to $\Sigma(W)$ and that the labels of $W_f \setminus q_f(S(f))$ are consistent with those for $W \setminus \Sigma(W)$. (In other words, $R\colon (W, \Sigma(W)) \to \mathbb{R} \times [0, 1]$ is locally equivalent to one of the local models as depicted in Figures 4, 5 and 6.)

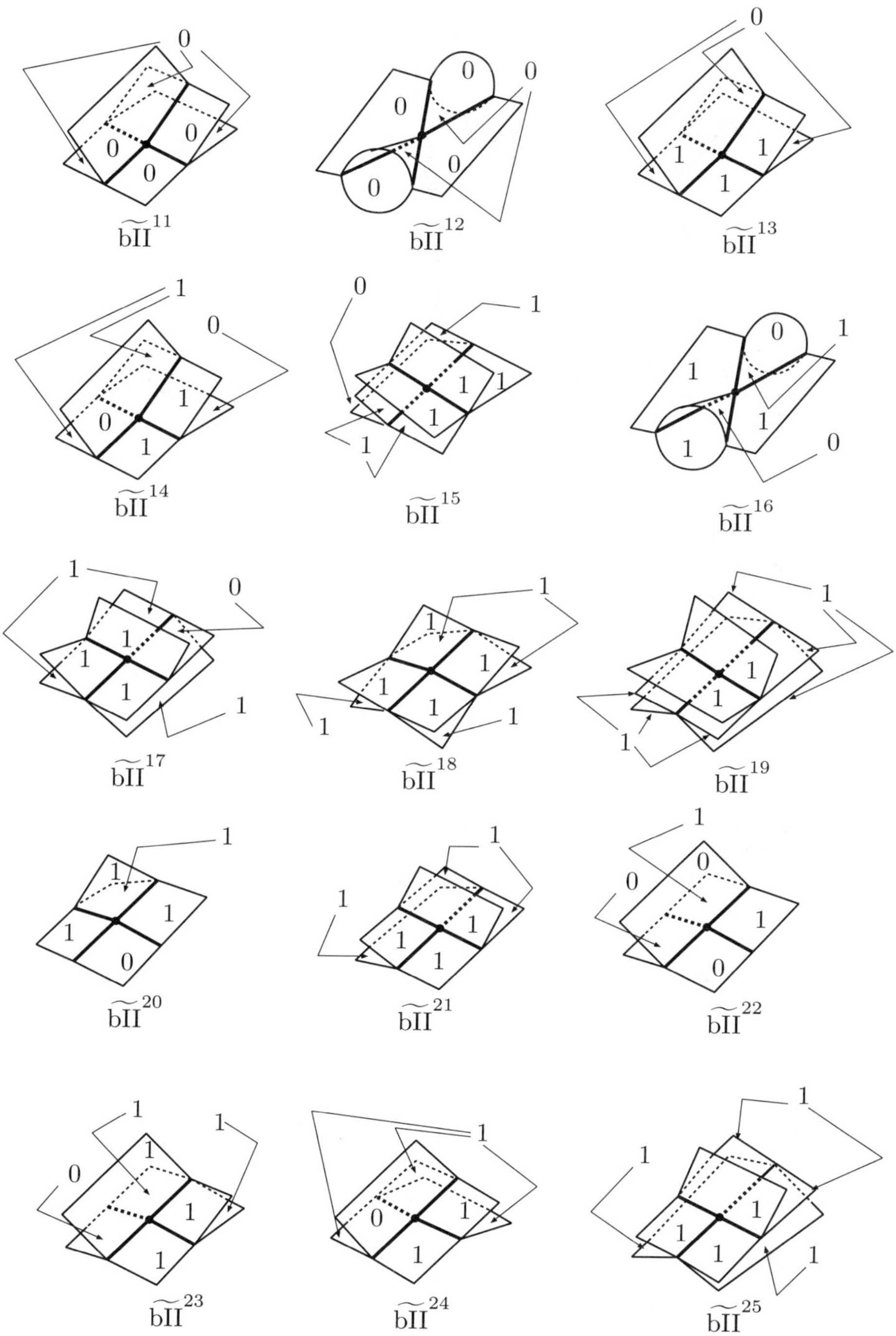


FIGURE 5. Local models of the Reeb map (II)

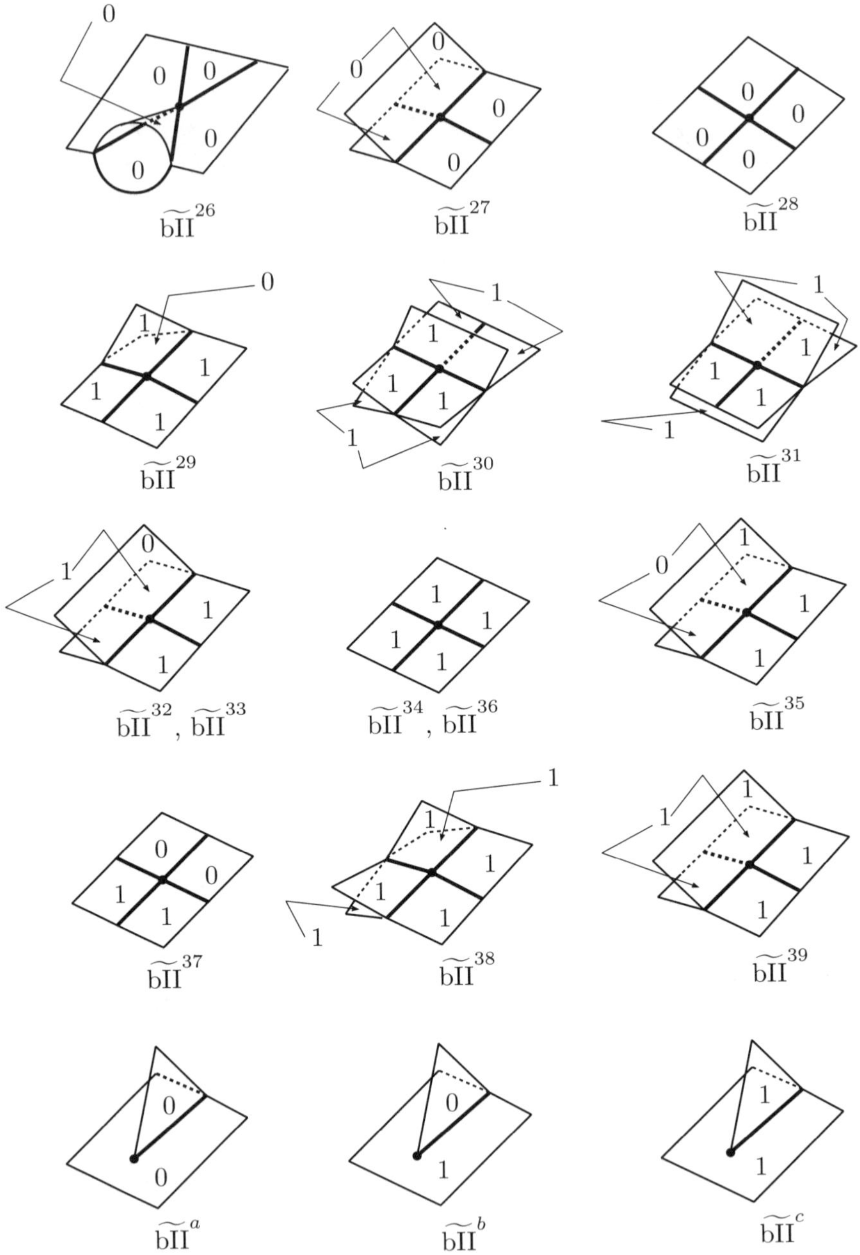


FIGURE 6. Local models of the Reeb map (III)

If a Reeb-like function is cobordant to the function on the empty graph, then we say that it is *null-cobordant*.

Note that the above relation defines an equivalence relation for Reeb-like functions. Furthermore, the set of all cobordism classes forms an additive group under the disjoint union. We denote by $b\mathfrak{R}$ the additive group of all cobordism classes of

Reeb-like functions on labeled Reeb-like graphs and call it the *cobordism group of Reeb-like functions.*

By the very definition, if two Morse functions $f_i\colon N_i \to \mathbb{R}$, $i = 0, 1$, on compact surfaces possibly with boundary are admissibly cobordant, then their Reeb functions $\overline{f}_i\colon W_{f_i} \to \mathbb{R}$, $i = 0, 1$, are cobordant.

Thus, we obtain the natural map

$$\rho\colon b\mathfrak{N}_2 \to b\mathfrak{R},$$

which associates to each admissible cobordism class of a Morse function $f$ on a compact surface possibly with boundary the cobordism class of the Reeb function $\overline{f}\colon W_f \to \mathbb{R}$. It is straightforward to see that this defines a homomorphism of additive groups.

In fact, we have the following.

PROPOSITION 3.3. *The homomorphism $\rho\colon b\mathfrak{N}_2 \to b\mathfrak{R}$ is an isomorphism.*

PROOF. For a given Reeb-like function $r\colon G \to \mathbb{R}$, we can easily construct a compact surface with boundary $N$ and a Morse function $f\colon N \to \mathbb{R}$ such that $\overline{f}\colon W_f \to \mathbb{R}$ is identified with $r$ in the sense that there is a homeomorphism $\psi\colon W_f \to G$ which makes the following diagram commutative.

$$\begin{array}{ccccc} N & & \xrightarrow{\quad f \quad} & & \mathbb{R} \\ & {}_{q_f}\searrow & & \overline{f}\nearrow & & \nwarrow_r \\ & & W_f & \xrightarrow{\quad \psi \quad} & G \end{array}$$

In fact, we can first construct such a Morse function for the restriction of $r$ to the union of small neighborhoods of the vertices of $G$. This is achieved by considering standard Morse functions. Then, we paste them according to the edges of $G$, with the help of the function $r$.

Therefore, the map $\rho$ is surjective.

To show that $\rho$ is injective, suppose that $f_0\colon N_0 \to \mathbb{R}$ and $f_1\colon N_1 \to \mathbb{R}$ are Morse functions whose associated Reeb functions $\overline{f}_0\colon W_{f_0} \to \mathbb{R}$ and $\overline{f}_1\colon W_{f_1} \to \mathbb{R}$ are cobordant, where $N_0$ and $N_1$ are compact surfaces possibly with boundary. Then, there is a continuous map $R\colon W \to \mathbb{R} \times [0, 1]$ of a compact 2-dimensional polyhedron $W$ as in Definition 3.2, where $r_i = \overline{f}_i$ and $G_i = W_{f_i}$, $i = 0, 1$. Such a map $R$ can be regarded as an "immersed-W with boundary" in the sense of Mata-Lorenzo (see [**15**, Definition 3.7]).

In order to realize $R$ by an admissible cobordism between $f_0$ and $f_1$, we imitate the construction due to Mata-Lorenzo, which is sketched in [**15**]: in other words, we try to construct an admissible cobordism in such a way that its Reeb space is identified with $W$, and its Reeb map is identified with $R$.

In Step 1, we construct a part of an admissible cobordism realizing the restrictions of $R$ to $G_0 \times [0, \varepsilon)$ and $G_1 \times (1 - \varepsilon, 1]$ and those of $R$ to neighborhoods of the vertices. This is straightforward.

In Step 2, we extend the map over a regular neighborhood of the subpolyhedron $\Sigma(W)$. This is possible, since each automorphism of the Reeb functions as depicted in Figure 2 can be lifted to an automorphism of the relevant Morse function on the surface.

Finally, in Step 3, we try to extend the map over the components of $W \setminus (\Sigma(W) \cup G_0 \cup G_1)$. Let $Z$ be such a component and let $Z'$ be a compact surface in $Z$, slightly smaller than $Z$, such that $Z \setminus Z'$ is homeomorphic to a finite disjoint

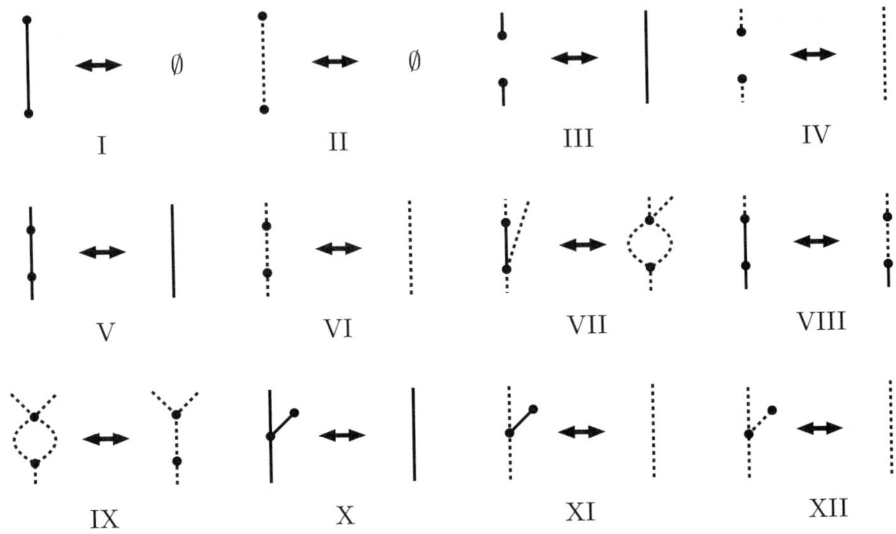


FIGURE 7. Moves for Reeb-like functions

union of thin open annuli. The map constructed so far is the projection of a fiber bundle over $\partial Z'$, where the fiber is a circle or an arc according as the label of $Z$ is 0 or 1. Let $C$ be a closed collar neighborhood of $\partial Z'$ in $Z'$. Then, we see easily that the map constructed so far extends to the quotient map of a stable map over $C$ in such a way that $\partial C \setminus \partial Z'$ corresponds to the image of the definite fold (or boundary definite fold) if the fiber is a circle (resp. if the fiber is an arc). In this way, we can construct the desired admissible cobordism between the Morse functions $f_0$ and $f_1$ by modifying $W$ and $R$. (In fact, a similar argument already appears in [**10**].)

Thus, the homomorphism $\rho$ is injective. □

## 4. Proof of the main theorem

Let us proceed to the proof of Theorem 1.2, by introducing some moves applied to Reeb-like functions.

By an argument similar to that in [**6**], we get the following.

LEMMA 4.1. *Let* $r_i\colon G_i \to \mathbb{R}$, $i = 0, 1$, *be Reeb-like functions on labeled Reeb-like graphs. If* $r_1$ *is obtained from* $r_0$ *by one of the local moves as depicted in Figure 7 or their negatives, then* $r_0$ *and* $r_1$ *are cobordant.*

Note that the moves VII, VIII, IX, X, XI and XII correspond to codimension two singular fibers of types $\widetilde{\mathrm{bII}}^{23}$, $\widetilde{\mathrm{bII}}^{37}$, $\widetilde{\mathrm{bII}}^{38}$, $\widetilde{\mathrm{bII}}^{a}$, $\widetilde{\mathrm{bII}}^{b}$ and $\widetilde{\mathrm{bII}}^{c}$, respectively.

REMARK 4.2. The moves mentioned in Lemma 4.1 form a part of the full list of allowed moves. We can list all the allowed moves in such a way that two Reeb-like functions are cobordant if and only if they are transformed to each other by a finite sequence of the allowed moves. However, the complete list gets tedious and we will need only a part of them for our main purpose: therefore, we do not give such a complete list here.

In order to prove Theorem 1.2, we have only to prove the following proposition by virtue of Proposition 3.3.

PROPOSITION 4.3. *The cobordism group $b\mathfrak{R}$ of Reeb-like functions is a cyclic group of order two generated by the cobordism class of the Reeb function of the Morse function as depicted in Figure* 8.

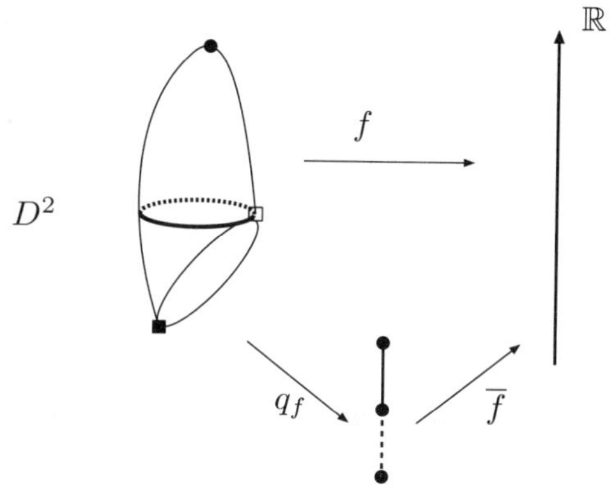


FIGURE 8. Morse function such that the cobordism class of its Reeb function generates $b\mathfrak{R}$

PROOF. Define the map

$$\sigma\colon\, b\mathfrak{R} \to \mathbb{Z}_2$$

by sending the cobordism class of a Reeb-like function $r\colon\, G \to \mathbb{R}$ to the modulo two number of the vertices of type as in Figure 2 (1), (2), (3), (4), (8) and (9). By the definition of cobordism for Reeb-like functions, we see that if two Reeb-like functions $r_i\colon\, G_i \to \mathbb{R}$ on labeled Reeb-like graphs, $i = 0, 1$, are cobordant, then we have $\sigma(r_0) = \sigma(r_1)$, since the allowed moves as mentioned in Remark 4.2 do not change the $\sigma$-value. Thus, the map $\sigma\colon\, b\mathfrak{R} \to \mathbb{Z}_2$ is well-defined and is clearly a homomorphism of abelian groups. In fact, the homomorphism corresponds to nothing but the cohomology class $s_{1*}\alpha$ mentioned in [**20**, Corollary 4.9]. The well-definedness of $\sigma$ is a direct consequence of the fact that the representative of the cohomology class $\alpha$ is a cocycle.

We will show that $\sigma$ is in fact an isomorphism.

For this, let $r\colon\, G \to \mathbb{R}$ be an arbitrary Reeb-like function on a labeled Reeb-like graph. We will show that $r$ is null-cobordant or is cobordant to the Reeb function of the Morse function as depicted in Figure 8.

First, by moves I–IV, we can cut each edge of $G$ so that $r$ is cobordant to the disjoint union of several copies of the Reeb-like functions (or their negatives) as depicted in Figure 9. We call each such function an *elementary Reeb-like function.* (The same procedure already appears in [**19**, Figure 2].)

By moves I and II, we see that the elementary Reeb-like functions $r_1$ and $r_5$ are null-cobordant. Furthermore, by using moves X, XI and XII, we see easily that the elementary Reeb-like functions $r_2, r_3$ and $r_7$ are also null-cobordant.

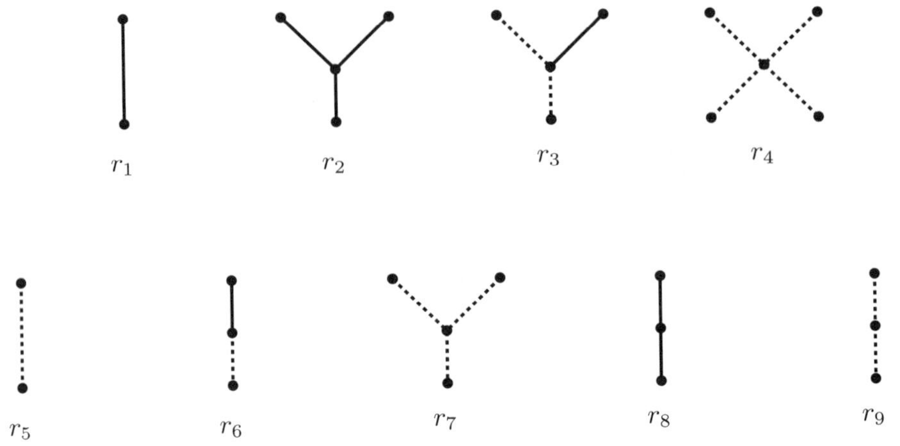


FIGURE 9. Elementary Reeb-like functions

As to $r_4$, we see that it is cobordant to $r_6$ by the sequence of moves as depicted in Figure 10.

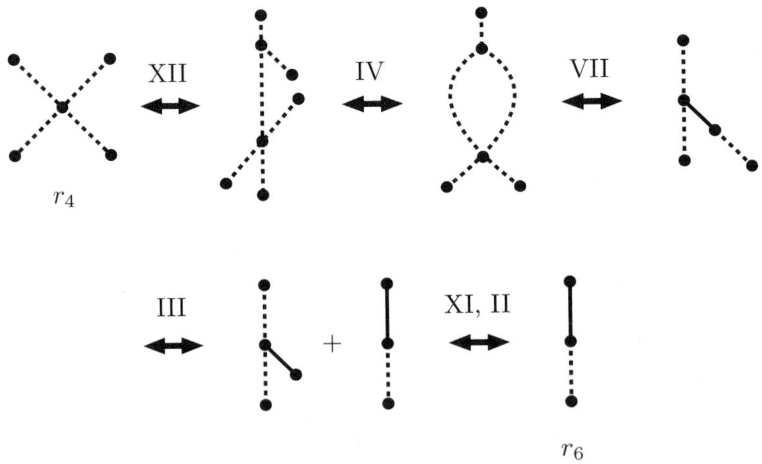


FIGURE 10. Elementary Reeb-like functions $r_4$ and $r_6$ are cobordant

Furthermore, as depicted in Figure 11, the disjoint union of the Reeb-like functions $r_6$ and $r_8$ is cobordant to the disjoint union of $r_6$ and $r_9$. As a consequence, we see that $r_8$ and $r_9$ are cobordant.

Moreover, as depicted in Figure 12, we see that the elementary Reeb-like functions $r_6$ and $r_9$ are cobordant.

As a consequence, we see that the cobordism group $b\mathfrak{R}$ is a cyclic group generated by the cobordism class of $r_6$: moreover, $r_4$, $r_6$, $r_8$ and $r_9$ are all cobordant, and any one of them can be chosen as a representative of the generator.

Finally, we can show that the disjoint union of two copies of $r_8$ is null-cobordant as depicted in Figure 13.

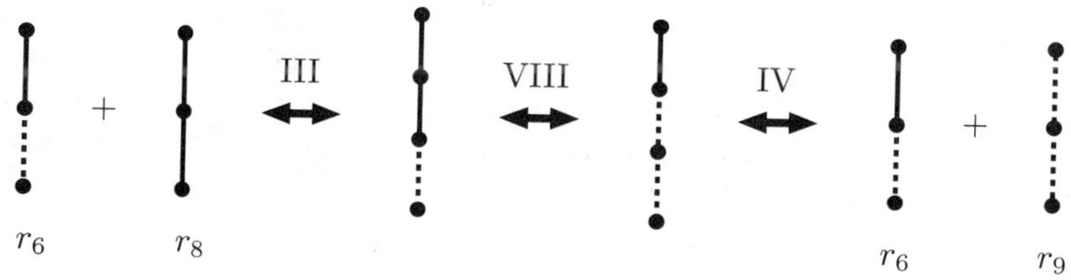


FIGURE 11. The disjoint union of $r_6$ and $r_8$ is cobordant to that of $r_6$ and $r_9$

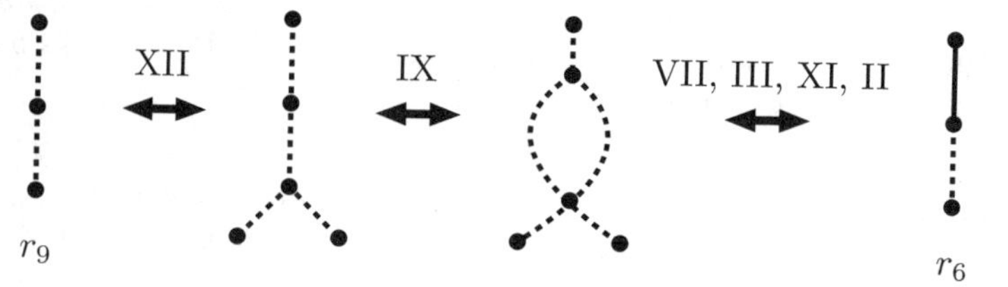


FIGURE 12. Elementary Reeb-like functions $r_6$ and $r_9$ are cobordant

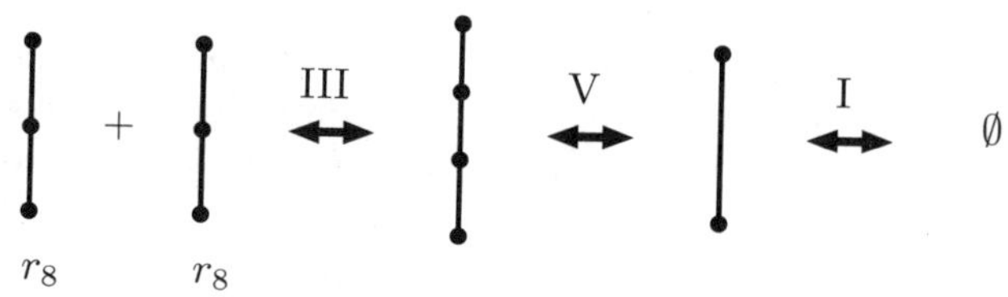


FIGURE 13. The cobordism class of $r_8$ has order two

On the other hand, we see easily that the $\sigma$-value of the cobordism class of $r_6$ is not zero. Therefore, the homomorphism $\sigma$ is an isomorphism and the cobordism group $b\mathfrak{R}$ is cyclic of order two. This completes the proof of Proposition 4.3. □

Combining Propositions 3.3 and 4.3, we get Theorem 1.2. We also see that the admissible cobordism class of the Morse function as depicted in Figure 8 can be chosen to be the generator.

## 5. Lower dimensional cobordism groups

In this section, we determine the group structures of the cobordism groups $b\mathfrak{N}_0$ and $b\mathfrak{N}_1$. In the following, we adopt the definition as follows. A function $f\colon N \to \mathbb{R}$ on a compact 0-dimensional manifold $N$ is a *Morse function* if it is injective. A $C^\infty$ function $f\colon N \to \mathbb{R}$ on a compact 1-dimensional manifold $N$ is a *Morse function* if it is a submersion near the boundary, the critical points are all non-degenerate, and the values of the critical points together with the values at the boundary points are all distinct. We adopt the above convention so that the class of Morse functions coincides with the class of $C^\infty$ stable functions.

We first prove the following.

PROPOSITION 5.1. *The 0-dimensional admissible cobordism group of Morse functions $b\mathfrak{N}_0$ is trivial.*

PROOF. Let $N$ be a compact 0-dimensional manifold, i.e., $N$ is a finite point set, and $f\colon N \to \mathbb{R}$ a Morse function. Then, the embedding $F\colon N \times [0,1/2] \to \mathbb{R} \times [0,1]$ given by $F(x,t) = (f(x),t)$, $(x,t) \in N \times [0,1/2]$, gives an admissible cobordism between $f$ and the function on the empty set, which can be verified in a straightforward way. This completes the proof. □

For the 1-dimensional case, we have the following.

PROPOSITION 5.2. *The 1-dimensional admissible cobordism group of Morse functions $b\mathfrak{N}_1$ is an infinite cyclic group generated by the admissible cobordism class of $f_0\colon [-1,2] \to \mathbb{R}$ given by $f_0(x) = x^2$, $x \in [-1,2]$.*

PROOF. Let $N$ be a compact 1-dimensional manifold and $f\colon N \to \mathbb{R}$ an arbitrary Morse function. Note that $N$ is a finite disjoint union of circles and arcs, and it has an even number of boundary points. Let $x$ be a point in $\partial N$. For an outward normal vector $v \in T_xN \setminus \{0\}$, $df_x(v) \in T_{f(x)}\mathbb{R} = \mathbb{R}$ is not zero, and its sign is independent of the choice of $v$. We say that $x$ is a *positive end point* if $df_x(v)$ is positive, and a *negative end point* otherwise.

Let $f_i\colon N_i \to \mathbb{R}$, $i = 0,1$, be Morse functions on compact 1-dimensional manifolds and suppose that they are admissibly cobordant by the admissible cobordism $F\colon X \to \mathbb{R} \times [0,1]$ as in Definition 1.1. Note that $Q$ is orientable and so is its collar neighborhood (with corners) $C$ in $X$. Furthermore, $F|_C\colon C \to \mathbb{R} \times [0,1]$ is an immersion. Let $Q_0$ be an arc component of $Q$, and $C_0$ be its collar neighborhood in $X$. Then, since $F|_{C_0}$ is an immersion, if both of the two end points of $Q_0$ are mapped to $\mathbb{R} \times \{0\}$ (or to $\mathbb{R} \times \{1\}$) by $F$, then their signs are opposite to each other. If one of the two end points is mapped to $\mathbb{R} \times \{0\}$ and the other to $\mathbb{R} \times \{1\}$, then their signs are the same. Therefore, the number of positive end points of $N_0$ minus the number of negative end points of $N_0$ coincides with that for $N_1$.

This means that we have the homomorphism $\tau\colon b\mathfrak{N}_1 \to \mathbb{Z}$, which sends the admissible cobordism class of a Morse function $f\colon N \to \mathbb{R}$ to one half of the number of positive end points of $N$ minus the number of negative end points of $N$. Note that since the total number of end points of $N$ is even, the resulting number is an integer. Furthermore, by the argument in the previous paragraph, this is a well-defined homomorphism.

This homomorphism $\tau$ is clearly surjective, since the $\tau$-value of the cobordism class of $f_0$ as in the statement of Proposition 5.2 is equal to 1.

Suppose that for a Morse function $f\colon N \to \mathbb{R}$, its $\tau$-value vanishes. Then, the number of positive end points of $N$ is equal to that of negative end points. Let $Q$ be a finite disjoint union of arcs such that each of the arcs connects two end points of $N$ with opposite signs. Then, there exists a compact surface with corners $X$ with the following properties: $\partial X = N \cup Q$, $N$ and $Q$ are codimension 0 submanifolds of $\partial X$, and $X$ has corners exactly along $\partial Q$. Let $C_0$ be a small collar neighborhood of $\partial N$ in $N$ and $C$ a collar neighborhood (with corners) of $Q$ in $X$ such that $C \cap (N \times [0,\varepsilon)) = C_0 \times [0,\varepsilon)$.

Since each component of $Q$ connects end points of $N$ of opposite signs, the immersion $f \times \mathrm{id}_{[0,\varepsilon)}\colon C_0 \times [0,\varepsilon) \to \mathbb{R} \times [0,\varepsilon)$ can be extended to an immersion of $C$ into $\mathbb{R} \times [0,1)$. Then, we can extend this immersion together with the generic map $f \times \mathrm{id}_{[0,\varepsilon)} \colon N \times [0, \varepsilon) \to \mathbb{R} \times [0, \varepsilon)$ to an admissible cobordism $F \colon X \to \mathbb{R} \times [0, 1]$ between $f$ and the function on the empty set by using a generic extension. Therefore, $\tau$ is injective.

This completes the proof. □

## 6. Problems

In this section, we propose several open problems concerning the admissible cobordism groups of Morse functions on compact manifolds possibly with boundary.

The first natural question is the following.

PROBLEM 6.1. Study the group structure of $b\mathfrak{N}_n$, $n \geq 3$.

For a non-negative integer $n$, denote by $b\mathcal{M}(n)$ the set of all Morse functions on compact oriented $n$-dimensional manifolds with boundary. Then, the notion of oriented admissible cobordism between two Morse functions $f_i \colon N_i \to \mathbb{R}$, $i = 0, 1$, on compact oriented $n$-dimensional manifolds possibly with boundary is defined as in Definition 1.1, where the manifold $X$ is now oriented and the orientation of $\partial X$ is consistent with the orientations of $-N_0$ and $N_1$, where $-N_0$ denotes the manifold $N_0$ with the opposite orientation. Then, the corresponding cobordism group is denoted by $b\mathfrak{M}_n$, which is called the *oriented admissible cobordism group of Morse functions on manifolds with boundary of dimension* $n$, or the $n$-*dimensional oriented admissible cobordism group of Morse functions*.

PROBLEM 6.2. Study the group structure of $b\mathfrak{M}_n$.

Furthermore, the notion of a fold cobordism (or oriented fold cobordism) for Morse functions on compact (resp. compact and oriented) $n$-dimensional manifolds possibly with boundary is defined: in Definition 1.1 (4), we impose the condition that the map $F|_{X \setminus (N_0 \cup N_1)} \colon X \setminus (N_0 \cup N_1) \to \mathbb{R} \times (0, 1)$ has only fold points and boundary fold points as its singularities. Then, the corresponding cobordism group is denoted by $b\mathfrak{F}_n$, which is called the *admissible fold cobordism group of Morse functions on manifolds with boundary of dimension* $n$, or the $n$-*dimensional admissible fold cobordism group of Morse functions*.

PROBLEM 6.3. Study the group structures of $b\mathfrak{F}_n$ and its oriented version.

For the above problems, we can consider the corresponding subgroups of the unoriented versions generated by Morse functions on *orientable* manifolds as well; or as in [**21**, §7], we can also consider the respective cobordism groups of Morse functions on *orientable* (but not oriented) manifolds with respect to *orientable cobordisms*.

## Acknowledgment

The authors would like to express their sincere gratitude to Professor Terry Gaffney for posing an interesting question at the 13th International Workshop on Real and Complex Singularities held at São Carlos in 2014, which motivated the work presented in this paper. They would also like to thank the referee for important comments.

The first author has been supported in part by JSPS KAKENHI Grant Number 23244008, 23654028 and 15K13438. The second author has been supported in part by JSPS KAKENHI Grant Number 23654028 and 15K13438.

## References


[1] Kree Cole-McLaughlin, Herbert Edelsbrunner, John Harer, Vijay Natarajan, and Valerio Pascucci, *Loops in Reeb graphs of 2-manifolds*, Discrete Comput. Geom. **32** (2004), no. 2, 231–244, DOI 10.1007/s00454-004-1122-6. MR2074339

[2] Herbert Edelsbrunner, John Harer, and Amit K. Patel, *Reeb spaces of piecewise linear mappings*, Computational geometry (SCG'08), ACM, New York, 2008, pp. 242–250, DOI 10.1145/1377676.1377720. MR2504290

[3] J. Eliashberg, *Cobordisme des solutions de relations différentielles* (French), South Rhone seminar on geometry, I (Lyon, 1983), Travaux en Cours, Hermann, Paris, 1984, pp. 17–31. MR753850

[4] M. Golubitsky and V. Guillemin, *Stable mappings and their singularities*, Springer-Verlag, New York-Heidelberg, 1973. Graduate Texts in Mathematics, Vol. 14. MR0341518

[5] Jorge T. Hiratuka and Osamu Saeki, *Triangulating Stein factorizations of generic maps and Euler characteristic formulas*, Singularity theory, geometry and topology, RIMS Kôkyûroku Bessatsu, B38, Res. Inst. Math. Sci. (RIMS), Kyoto, 2013, pp. 61–89. MR3156903

[6] Kazuichi Ikegami and Osamu Saeki, *Cobordism group of Morse functions on surfaces*, J. Math. Soc. Japan **55** (2003), no. 4, 1081–1094, DOI 10.2969/jmsj/1191418765. MR2003761

[7] Kazuichi Ikegami, *Cobordism group of Morse functions on manifolds*, Hiroshima Math. J. **34** (2004), no. 2, 211–230. MR2086843

[8] S.A. Izar, *Funções de Morse e topologia das superfícies. I) O grafo de Reeb de* $f : M \to \mathbb{R}$ (in Portuguese), Métrica no. 31, 1988, IBILCE, UNESP, available at `http://www.ibilce.unesp.br/Home/Departamentos/Matematica/metrica-31.pdf`

[9] S.A. Izar, *Funções de Morse e topologia das superfícies. II) Classificação das funções de Morse estáveis sobre superfícies* (in Portuguese), Métrica no. 35, 1989, IBILCE, UNESP, available at `http://www.ibilce.unesp.br/Home/Departamentos/Matematica/metrica-35.pdf`

[10] Boldizsár Kalmár, *Cobordism group of Morse functions on unoriented surfaces*, Kyushu J. Math. **59** (2005), no. 2, 351–363, DOI 10.2206/kyushujm.59.351. MR2188598

[11] Mahito Kobayashi and Osamu Saeki, *Simplifying stable mappings into the plane from a global viewpoint*, Trans. Amer. Math. Soc. **348** (1996), no. 7, 2607–2636, DOI 10.1090/S0002-9947-96-01576-0. MR1344209

[12] Harold I. Levine, *Elimination of cusps*, Topology **3** (1965), no. suppl. 2, 263–296. MR0176484

[13] Harold Levine, *Classifying immersions into* $\mathbf{R}^4$ *over stable maps of 3-manifolds into* $\mathbf{R}^2$, Lecture Notes in Mathematics, vol. 1157, Springer-Verlag, Berlin, 1985. MR814689

[14] Luciana F. Martins and Ana Claudia Nabarro, *Projections of hypersurfaces in* $\mathbb{R}^4$ *with boundary to planes*, Glasg. Math. J. **56** (2014), no. 1, 149–167, DOI 10.1017/S001708951300013X. MR3137856

[15] W. Motta, P. Porto and O Saeki, *Stable maps of 3-manifolds into the plane and their quotient spaces*, Proc. London Math. Soc. (3) **71** (1995), 158–174.

[16] R. Rimányi and A. Szűcs, *Pontrjagin-Thom-type construction for maps with singularities*, Topology **37** (1998), no. 6, 1177–1191, DOI 10.1016/S0040-9383(97)00093-1. MR1632908

[17] Osamu Saeki, *Cobordism groups of special generic functions and groups of homotopy spheres*, Japan. J. Math. (N.S.) **28** (2002), no. 2, 287–297. MR1947905

[18] Osamu Saeki, *Topology of singular fibers of differentiable maps*, Lecture Notes in Mathematics, vol. 1854, Springer-Verlag, Berlin, 2004. MR2106689

[19] Osamu Saeki, *Cobordism of Morse functions on surfaces, the universal complex of singular fibers and their application to map germs*, Algebr. Geom. Topol. **6** (2006), 539–572 (electronic), DOI 10.2140/agt.2006.6.539. MR2220688

[20] Osamu Saeki and Takahiro Yamamoto, *Singular fibers of stable maps of 3–manifolds with boundary into surfaces and their applications*, Algebr. Geom. Topol. **16** (2016), no. 3, 1379–1402, DOI 10.2140/agt.2016.16.1379. MR3523043

[21] Osamu Saeki and Takahiro Yamamoto, *Co-orientable singular fibers of stable maps of 3-manifolds with boundary into surfaces*, RIMS Kôkyûroku, Research Institute for Mathematical Sciences, Kyoto University, No. 1948 (2015), 137–152.

[22] Naoki Shibata, *On non-singular stable maps of 3-manifolds with boundary into the plane*, Hiroshima Math. J. **30** (2000), no. 3, 415–435. MR1799298

[23] András Szűcs, *On the cobordism group of Morin maps*, Acta Math. Hungar. **80** (1998), no. 3, 191–209, DOI 10.1023/A:1006598609305. MR1645258
[24] András Szűcs, *Cobordism of singular maps*, Geom. Topol. **12** (2008), no. 4, 2379–2452, DOI 10.2140/gt.2008.12.2379. MR2443969
[25] René Thom, *Quelques propriétés globales des variétés différentiables* (French), Comment. Math. Helv. **28** (1954), 17–86. MR0061823
[26] Robert Wells, *Cobordism groups of immersions*, Topology **5** (1966), 281–294. MR0196760

Institute of Mathematics for Industry, Kyushu University, Motooka 744, Nishi-ku, Fukuoka 819-0395, Japan
*E-mail address*: saeki@imi.kyushu-u.ac.jp

Faculty of Engineering, Kyushu Sangyo University, 3-1 Matsukadai 2-chome, Higashi-ku, Fukuoka, 813-8503, Japan
*E-mail address*: yama.t@ip.kyusan-u.ac.jp

Contemporary Mathematics
Volume **675**, 2016
http://dx.doi.org/10.1090/conm/675/13598

# Affine metric for locally strictly convex manifolds of codimension 2

Marcelo J. Saia and Luis F. Sánchez

ABSTRACT. We introduce a family of metrics, metrics of the transversal vector fields, on a locally strictly convex $m$-surface $M$ in affine $(m+2)$-space, which is an affine invariant. By using the metric of the transversal vector field on $M$ we introduce the affine normal plane and the families of affine distance and height functions on $M$. We show that the singularities of the affine height functions appear at directions on the affine normal plane and the singularities of the affine distance functions appear at points on the affine normal plane.

## 1. Introduction

The main interest in affine differential geometry is the study of properties of $m$-surfaces $M$ in affine $n$-space $(n > m)$ that are invariants under the group of unimodular affine transformations. The theory of affine hypersurfaces is classical, as we can see [**1, 6, 8**].

Concerning submanifolds of codimension 2, there exists an extensive work from the Euclidean viewpoint see [**5, 7, 10, 13**], while in the affine context there are few results. Nomizu and Vrancken in [**9**] developed an affine theory for surfaces in $\mathbb{R}^4$, they used the affine metric of Burstin and Mayer [**2**], which is affine invariant, to construct among other results, the affine normal plane.

On the other side, it is of great interest the study of locally strictly convex surfaces $M$ (i.e., at each point $p \in M$ there is a tangent hyperplane with a non-degenerate contact which locally supports $M$) in $\mathbb{R}^4$, however the affine metric of Burstin and Mayer and the corresponding affine normal plane present several problems in the study these surfaces. To ride out this problem, in [**12**] it was developed the study of locally strictly convex surfaces in $\mathbb{R}^4$ from the affine differential geometry viewpoint, using an other family of metrics which is an affine invariant. Any metric of this family is called a metric of the transversal vector field and for each metric of this family it is defined the corresponding symetric and antissymetric equiaffine plane bundles.

In this work we generalize some tools developed in [**12**] for locally strictly convex $m$-surfaces in $\mathbb{R}^{m+2}$. First we introduce a family of affine metrics, denoted $g_\xi$, on a locally strictly convex $m$-surface $M \subset \mathbb{R}^{m+2}$ which are positive definite. For this

2010 *Mathematics Subject Classification.* Primary 53A15; Secondary 53A07, 58K05.
*Key words and phrases.* xx.
This work is partially supported by CAPES, CNPq and FAPESP.

we consider a transversal vector field $\xi$ such that $\xi$ and the tangent space $T_pM$ span a local support hyperplane with non-degenerate contact at $p$ and then we define the corresponding metric $g_\xi$. Moreover, for each affine metric $g_\xi$ we introduce the affine normal plane $\mathbf{A}$ and the families of affine distance and height functions on $M$. We show in the Theorem 4.6 that the singularities of the affine height functions appear at directions on the affine normal plane and in the Theorem 4.3 we show that the singularities of the affine distance functions appear at points on the affine normal plane.

An interesting class of $m$-surfaces consists of those surfaces that are in a hypersurface contained in affine $(m+2)$-space, the main difficulty in this case is to be sure about the consistence among the metrics related with the $m$-surface and the hypersurface. Here we show that when the $m$-surface $M$ is immersed in a locally strictly convex hypersurface $N$ in affine $(m+2)$-space, then there is a natural choice of $\xi$ in such a way that $g_\xi$ coincides with the Blaschke metric of $N$ restricted to $M$. Moreover, we show in Theorem 5.4 that the affine distance on the $m$-surface $M$ coincides with the affine distance on the hypersurface, and that the affine normal vector field of the hypersurface belong to the affine normal plane of the $m$-surface. Finally, in Corollary 5.5 it is shown that the affine normal plane is the orthogonal plane to the tangent space with respect to the extended Blaschke metric.

## 2. The metric of the transversal vector field

Let $\mathbb{R}^{m+2}$ be the affine $(m+2)$-space and $D$ the usual flat connection on $\mathbb{R}^{m+2}$. Let $M \subset \mathbb{R}^{m+2}$ be an immersed $m$-surface and let $\sigma$ be a transversal plane bundle on $M$. Then, for all $p \in M$, $\sigma_p \subset T_p\mathbb{R}^{m+2}$ is a plane such that

$$T_p\mathbb{R}^{m+2} = T_pM \oplus \sigma_p,$$

and for all tangent vector fields $X, Y$ on $M$,

$$(D_XY)_p = (\nabla_XY)_p + h(X,Y)_p,$$

where $(\nabla_XY)_p \in T_pM$ and $h(X,Y)_p \in \sigma_p$, for all $p \in M$.

We note that for $p \in M$, there are $\xi_1, \xi_2$ transversal vector fields defined on some neighborhood $U_p$ such that: $\sigma_q = \operatorname{span}\{\xi_1(q), \xi_2(q)\}$, $\forall q \in U_p$.

Then for tangent vector fields $X, Y$ on $M$ we have:

$$\begin{aligned} D_XY &= \nabla_XY + h^1(X,Y)\xi_1 + h^2(X,Y)\xi_2, \\ D_X\xi_1 &= -S_1X + \tau_1^1(X)\xi_1 + \tau_1^2(X)\xi_2, \\ D_X\xi_2 &= -S_2X + \tau_2^1(X)\xi_1 + \tau_2^2(X)\xi_2, \end{aligned}$$

where $\nabla = \nabla(\sigma)$ is a torsion free affine connection, $h^1, h^2$ are bilinear symmetric forms, $S_1, S_2$ are $(1,1)$ tensor fields, and $\tau_i^j$ are 1-forms on $M$. We call $\nabla$ the affine connection induced by the transversal plane bundle $\sigma$.

Let $M \subset \mathbb{R}^{m+2}$ be a locally strictly convex $m$-surface and $\mathfrak{u} = \{X_1, X_2, \dots, X_m\}$ be a local tangent frame of a point $p \in M$ and let $\xi$ be a transversal vector field on $M$.

We define the symmetric bilinear form $G_\mathfrak{u}$ on $M$ to be

$$G_\mathfrak{u}(Y,Z) = [X_1, X_2, \dots, X_m, D_ZY, \xi].$$

DEFINITION 2.1. We call $\xi$ a *metric field* when $\xi$ is such that $G_\mathfrak{u}$ is positive definite.

We remark that the metric fields exist because $M$ is locally strictly convex. A metric field $\xi$ is defined only locally, but since all our results are local, we can assume without loss of generality that $\xi$ is globally defined and that $M$ is globally oriented.

DEFINITION 2.2. We define the *metric of the transversal vector field*, denoted by $g_\xi$, by

$$g_\xi(Y,Z) = \frac{G_{\mathfrak{u}}(Y,Z)}{(\det_{\mathfrak{u}} G_{\mathfrak{u}})^{\frac{1}{m+2}}},$$

where $\det_{\mathfrak{u}} G_{\mathfrak{u}} = \det(G_{\mathfrak{u}}(X_i, X_j))$.

LEMMA 2.3. *The symmetric bilinear form $g_\xi$ does not depend on the choice of the local tangent frame $\mathfrak{u}$, provided it is positively oriented.*

PROOF. Let $\mathfrak{v} = \{Y_1, Y_2, \ldots, Y_m\}$ be another local tangent frame on a neighborhood $U$ of $p \in M$, then there exists a matrix $a = (a_{ij})$ with $\det a > 0$, defined on $U$ such that $Y_i = \sum_{j=1}^m a_{ij} X_j$, $(i = 1, \ldots, m)$. Note that

$$G_{\mathfrak{v}}(Y,Z) = [Y_1, Y_2, \ldots, Y_m, D_Z Y, \xi] = \det(a) G_{\mathfrak{u}}(Y,Z).$$

By properties of the determinant, it follows that $\det_{\mathfrak{v}} G_{\mathfrak{v}} = \det(a)^m \det_{\mathfrak{v}} G_{\mathfrak{u}}$. On the other hand, from a simple computation $\det_{\mathfrak{v}} G_{\mathfrak{u}} = \det(a)^2 \det_{\mathfrak{u}} G_{\mathfrak{u}}$, therefore

$$\det_{\mathfrak{v}} G_{\mathfrak{v}} = \det(a)^{m+2} \det_{\mathfrak{u}} G_{\mathfrak{u}}.$$

Finally,

$$\frac{G_{\mathfrak{v}}(Y,Z)}{(\det_{\mathfrak{v}} G_{\mathfrak{v}})^{1/(m+2)}} = \frac{\det(a) G_{\mathfrak{u}}(Y,Z)}{(\det(a)^{(m+2)} \det_{\mathfrak{u}} G_{\mathfrak{u}})^{1/(m+2)}} = \frac{G_{\mathfrak{u}}(Y,Z)}{(\det_{\mathfrak{u}} G_{\mathfrak{u}})^{1/(m+2)}}.$$

□

From now on, we shall restrict ourselves to orthonormal frames $\{X_1, \ldots, X_m\}$ relative to $g = g_\xi$, this is, $g(X_i, X_j) = \delta_{ij}, \forall i,j = 1, \ldots, m$.

THEOREM 2.4. *Let $M \subset \mathbb{R}^{m+2}$ be a locally strictly convex $m$-surface and $\xi$ a metric field. Let $\mathfrak{u} = \{X_1, \ldots, X_m\}$ be a local orthonormal tangent frame of $g_\xi$ and let $\sigma$ be an arbitrary transversal plane bundle. Then there exists a unique local frame $\{\xi_1, \xi_2\}$ of $\sigma$, such that: $[X_1, \ldots, X_m, \xi_1, \xi_2] = 1$, $-\xi_1 \in [\xi]$, $h^1(X_1, X_1) = 0$ and $h^2(X_i, X_j) = \delta_{ij}$.*

PROOF. Let $p$ be a point in $M$ and let $\{\nu_1, \nu_2\}$ be any local frame of $\sigma$ in a neighborhood $U$ of $p$. We can assume that $X_1, \ldots, X_m$ are defined on $U$. Now, we write

$$[\xi] = \lambda_1 \nu_1 + \lambda_2 \nu_2 + T_p M.$$

We denote: $h^1(X_i, X_j) = a_{ij}$, $h^2(X_i, X_j) = b_{ij}$ and $K = [X_1, \ldots, X_m, \nu_1, \nu_2]$. Then the symmetric bilinear form $G_{\mathfrak{u}}$ is given by:

$$G_{\mathfrak{u}}(X_i, X_j) = (\lambda_2 a_{ij} - \lambda_1 b_{ij}) K. \tag{2.1}$$

By using the change: $\nu_1 = \alpha \xi_1 + \beta \xi_2$, $\nu_2 = \varphi \xi_1 + \psi \xi_2$, we obtain the affine fundamental forms in the new frame $\{\xi_1, \xi_2\}$: $\overline{h}^1(X_i, X_j) = \alpha a_{ij} + \varphi b_{ij}$ and $\overline{h}^2(X_i, X_j) = \beta a_{ij} + \psi b_{ij}$. Since $0 \neq G_{\mathfrak{u}}(X_1, X_1) = a_{11}\lambda_2 - b_{11}\lambda_1$ the following system:

$$\begin{aligned} 1 &= \beta a_{11} + \psi b_{11} \\ 0 &= \beta \lambda_1 + \psi \lambda_2, \end{aligned}$$

has solution $(\beta, \psi)$ given by

$$\beta = \frac{\lambda_2}{a_{11}\lambda_2 - b_{11}\lambda_1} \qquad \psi = \frac{-\lambda_1}{a_{11}\lambda_2 - b_{11}\lambda_1}.$$

We substitute $\beta$ and $\psi$ in $\overline{h}^2(X_i, X_j)$ and we prove that $\overline{h}^2(X_i, X_j) = \delta_{ij}$. In fact,

$$\begin{aligned}
\overline{h}^2(X_i, X_j) &= \beta a_{ij} + \psi b_{ij} = (\frac{\lambda_2}{a_{11}\lambda_2 - b_{11}\lambda_1})a_{ij} + (\frac{-\lambda_1}{a_{11}\lambda_2 - b_{11}\lambda_1})b_{ij} \\
&= \frac{K(\lambda_2 a_{ij} - \lambda_1 b_{ij})}{K(\lambda_2 a_{11} - \lambda_1 b_{11})} = \frac{G_{\mathfrak{u}}(X_i, X_j)}{G_{\mathfrak{u}}(X_1, X_1)} = \frac{G_{\mathfrak{u}}(X_i, X_j)/(det_{\mathfrak{u}} G_{\mathfrak{u}})^{\frac{1}{m+2}}}{G_{\mathfrak{u}}(X_1, X_1)/(det_{\mathfrak{u}} G_{\mathfrak{u}})^{\frac{1}{m+2}}} \\
&= \frac{g_\xi(X_i, X_j)}{g_\xi(X_1, X_1)} = g_\xi(X_i, X_j).
\end{aligned}$$

From the equation $0 = \overline{h}^1(X_1, X_1) = \alpha a_{11} + \varphi b_{11}$ we can write $\alpha = Rb_{11}$ and $\varphi = -Ra_{11}$, therefore

$$\begin{aligned}
[X_1, \dots, X_m, \nu_1, \nu_2] &= [X_1, \dots, X_m, \xi_1, \xi_2](\alpha\psi - \beta\varphi) = (\alpha\psi - \beta\varphi) \\
&= ((Rb_{11})\psi - \beta(-Ra_{11})) = R,
\end{aligned}$$

we conclude $R = K$, $\alpha = Kb_{11}$ and $\varphi = -Ka_{11}$.

It only remains to prove that $[\xi] = -[\xi_1]$. First we note that $G_{\mathfrak{u}}(X_i, X_j) = 0$ $(i \neq j)$ because $\{X_1, \dots, X_m\}$ is a orthonormal tangent frame relative to $g_\xi$. Moreover,

$$\frac{\det_{\mathfrak{u}} G_{\mathfrak{u}}}{(\det_{\mathfrak{u}} G_{\mathfrak{u}})^{m/(m+2)}} = \frac{G_{\mathfrak{u}}(X_1,X_1)}{(\det_{\mathfrak{u}} G_{\mathfrak{u}})^{1/(m+2)}} \cdots \frac{G_{\mathfrak{u}}(X_m,X_m)}{(\det_{\mathfrak{u}} G_{\mathfrak{u}})^{1/(m+2)}} = 1.$$

Therefore $\det_{\mathfrak{u}} G_{\mathfrak{u}} = 1$, it follows that

$$\lambda_1\alpha + \lambda_2\varphi = \lambda_1 K b_{11} - \lambda_2 K a_{11} = K(\lambda_1 b_{11} - \lambda_2 a_{11}) = -G_{\mathfrak{u}}(X_1, X_1) = -1.$$

Finally, we compute $[\xi]$:

$$\begin{aligned}
[\xi] &= \lambda_1\nu_1 + \lambda_2\nu_2 + T_pM \\
&= \lambda_1(\alpha\xi_1 + \beta\xi_2) + \lambda_2(\varphi\xi_1 + \psi\xi_2) + T_pM \\
&= \underbrace{(\lambda_1\alpha + \lambda_2\varphi)}_{-1}\xi_1 + \underbrace{(\lambda_1\beta + \lambda_2\psi)}_{0}\xi_2 + T_pM.
\end{aligned}$$

□

LEMMA 2.5. *Let $M \subset \mathbb{R}^{m+2}$ be a locally strictly convex $m$-surface and $\xi$ a metric field. Let $\mathfrak{u} = \{X_1, \dots, X_m\}$ and $\mathfrak{v} = \{Y_1, \dots, Y_m\}$ be two orthonormal frames and let $\sigma$ be a transversal plane bundle. So we can write for $i = 1, \dots, m$*

$$Y_i = \sum_{j=1}^{m} a_{ij} X_j \tag{2.2}$$

*where $\mathbf{a} = (a_{ij})$ is an orthogonal matrix and $\det(\mathbf{a}) = 1$. If we denote by $\{\xi_1, \xi_2\}$ (resp. $\{\overline{\xi}_1, \overline{\xi}_2\}$) the frame of Theorem 2.4 corresponding to $\mathfrak{u}$ (resp. $\mathfrak{v}$), then*

$$\begin{aligned}
\xi_1 &= \overline{\xi}_1, \\
\xi_2 &= \lambda\overline{\xi}_1 + \overline{\xi}_2,
\end{aligned}$$

*for some function $\lambda$.*

PROOF. From Theorem 2.4, we have $[\xi_1] = -[\xi] = [\overline{\xi}_1]$. Since $\xi_1$ and $\overline{\xi}_1$ belong to the same transversal plane we conclude that $\xi_1 = \overline{\xi}_1$. We compute now the affine connection in these two references $\{\xi_1, \xi_2\}$ and $\{\overline{\xi}_1, \overline{\xi}_2\}$. By using the frame $\{\overline{\xi}_1, \overline{\xi}_2\}$, it follows from Theorem 2.4 that

$$D_{Y_1}Y_1 = \nabla_{Y_1}Y_1 + \overline{\xi}_2,$$

and by using the reference $\{\xi_1, \xi_2\}$ we have,

$$D_{Y_1}Y_1 = \nabla_{Y_1}Y_1 + h^1(Y_1, Y_1)\xi_1 + h^2(Y_1, Y_1)\xi_2.$$

Hence, $\overline{\xi}_2 = h^1(Y_1, Y_1)\xi_1 + h^2(Y_1, Y_1)\xi_2$. Now since the frame $\mathfrak{u}$ is orthonormal with $g_\xi$ and $h^2 = g_\xi$

$$h^2(Y_1, Y_1) = h^2(\sum_{j=1}^{m} a_{1j}X_j, \sum_{j=1}^{m} a_{1j}X_j) = \sum_{i}^{m} a_{1i}^2 = 1.$$

Therefore

$$\overline{\xi}_2 = \lambda\xi_1 + \xi_2.$$

□

LEMMA 2.6. *Let $M \subset \mathbb{R}^{m+2}$ be a locally strictly convex $m$-surface. Let $\xi$ be a metric field and $\mathfrak{u} = \{X_1, \dots, X_m\}$ a local orthonormal tangent frame. Let $\sigma$ and $\overline{\sigma}$ be two transversal plane bundles. We denote by $\{\xi_1, \xi_2\}$ and $\{\overline{\xi}_1, \overline{\xi}_2\}$ the transversal frames obtained from Theorem 2.4 for $\sigma$ and $\overline{\sigma}$, respectively. Then there are $Z_1$ and $Z_2$ tangent vector fields on $M$ such that*

$$\overline{\xi}_1 = \xi_1 + Z_1,$$
$$\overline{\xi}_2 = \xi_2 + Z_2.$$

PROOF. We suppose that $\xi_1 = \phi\overline{\xi}_1 + \psi\overline{\xi}_2 + Z_1$ and $\xi_2 = \rho\overline{\xi}_1 + \beta\overline{\xi}_2 + Z_2$. Since $[\xi_1] = [\overline{\xi}_1]$ we have $\psi = 0$ and $\phi = 1$. By Theorem 2.4 $[X_1, \dots, X_m, \xi_1, \xi_2] = 1$, which implies $\phi\beta - \psi\rho = 1$ and it follows that $\beta = 1$. We denote by $\overline{h}^1$ and $\overline{h}^2$ the affine fundamental forms of the frame $\{\overline{\xi}_1, \overline{\xi}_2\}$. We note that

$$0 = \overline{h}^1(X_1, X_1) = \phi h^1(X_1, X_1) + \rho h^2(X_1, X_1) = \rho.$$

□

## 3. The equiaffine and normal plane bundle

Let $M \subset \mathbb{R}^{m+2}$ be a locally strictly convex $m$-surface and $\xi$ a metric field and $g = g_\xi$ the metric of the transversal field $\xi$.

We denote by $\omega_g$ the metric volume form for the metric $g = g_\xi$

$$\omega_g(Z_1, Z_2, \dots, Z_m) = \sqrt{|\det(g(Z_i, Z_j))|},$$

where $\{Z_1, \dots, Z_m\}$ is any positively oriented basis of $T_pM$.

DEFINITION 3.1. We say a transversal plane bundle $\sigma$ is *equiaffine* if the connection $\nabla = \nabla(\sigma)$ induced by $\sigma$ satisfies $\nabla\omega_g = 0$.

If $\mathfrak{u} = \{X_1, \ldots, X_m\}$ is a local orthonormal tangent frame and $\{\xi_1, \xi_2\}$ is the transversal frame given by Theorem 2.4, then $\omega_g = \theta$, where $\theta$ is the volume form induced by the determinant:

$$\theta(Z_1, \ldots, Z_m) = [Z_1, \ldots, Z_m, \xi_1, \xi_2], \ \forall Z_1, \ldots, Z_m \in T_pM.$$

This is because $\omega_g(X_1, \ldots, X_m) = \theta(X_1, \ldots, X_m)$.

LEMMA 3.2. *Let $M \subset \mathbb{R}^{m+2}$ be a locally strictly convex $m$-surface and $\xi$ be a metric field. Let $\sigma$ be a transversal plane bundle and $\mathfrak{u} = \{X_1, \ldots, X_m\}$ a local orthonormal tangent frame. Then the plane bundle $\sigma$ is equiaffine if and only if $B_j = 0$ for all $j = 1, \ldots, m$ where*

$$B_j = (\nabla g)(X_j, X_1, X_1) + (\nabla g)(X_j, X_2, X_2) + \ldots + (\nabla g)(X_j, X_m, X_m). \tag{3.1}$$

PROOF. First, we claim that

$$B_j = -2 \sum_{i=1}^{m} \Gamma^i_{ji}, \tag{3.2}$$

where

$$\nabla_{X_i} X_j = \sum_{k}^{m} \Gamma^k_{ij} X_k. \tag{3.3}$$

In fact, by definition of $B_j$ and since the frame $\mathfrak{u}$ is orthonormal with $g$,

$$\begin{aligned} B_j &= \sum_{i=1}^{m} (\nabla g)(X_j, X_i, X_i) = \sum_{i=1}^{m} (X_j(g(X_i, X_i)) - 2g(\nabla_{X_j} X_i, X_i)) \\ &= -2 \sum_{i=1}^{m} g(\sum_{k=1}^{m} \Gamma^k_{ji} X_k, X_i) = -2 \sum_{i=1}^{m} \Gamma^i_{ji}. \end{aligned}$$

Now, let $\{Y_1, Y_2, \ldots, Y_m\}$ be an arbitrary frame on $T_pM$. For $k \in 1, \ldots, m$ we compute

$$\nabla \omega_g(X_k, Y_1, \ldots, Y_m) := (\nabla_{X_k} \omega_g)(Y_1, \ldots, Y_m) = (\nabla_{X_k} \theta)(Y_1, \ldots, Y_m).$$

In fact, we write $Y_l = \sum_{r=1}^m a_{lr} X_r$, where $\det(\mathfrak{a}) \neq 0$ with $\mathfrak{a} = (a_{ij})$ and hence

$$\begin{aligned} \nabla \omega_g(X_k, Y_1, \ldots, Y_m) &= X_k(\omega_g(Y_1, \ldots, Y_m)) - \sum_{l=1}^{m} \omega_g(Y_1, \ldots, \nabla_{X_k} Y_l, \ldots, Y_m) \\ &= X_k(\theta(Y_1, \ldots, Y_m)) - \sum_{l=1}^{m} \theta(Y_1, \ldots, \nabla_{X_k}(\sum_{r=1}^{m} a_{lr} X_r), \ldots, Y_m). \end{aligned}$$

Also we note that: $\nabla_{X_k}(\sum_{r=1}^m a_{lr} X_r) = \sum_{r=1}^m \nabla_{X_k}(a_{lr} X_r) = \sum_{r=1}^m X_k(a_{lr}) X_r + \sum_{r=1}^m a_{lr} \nabla_{X_k} X_r$

$$= \sum_{t=1}^{m} X_k(a_{lt}) X_t + \sum_{r=1}^{m} a_{lr} \sum_{t=1}^{m} \Gamma^t_{kr} X_t = \sum_{t=1}^{m} (X_k(a_{lt}) + \sum_{r=1}^{m} a_{lr} \Gamma^t_{kr}) X_t.$$

We denote by $\widetilde{a}_{lt} = X_k(a_{lt}) + \sum_{r=1}^m a_{lr} \Gamma^t_{kr}$ and hence

$$\nabla \omega_g(X_k, Y_1, \ldots, Y_m) = X_k(\theta(Y_1, \ldots, Y_m)) - \sum_{l=1}^{m} \theta(Y_1, \ldots, \sum_{t=1}^{m} \widetilde{a}_{lt} X_t, \ldots, Y_m). \tag{3.4}$$

By definition of $\theta$ and since the determinant is multilinear

$$\theta(Y_1,\dots,\textstyle\sum_{t=1}^m \widetilde{a}_{lt}X_t,\dots,Y_m) = \begin{vmatrix} a_{11} & a_{12} & \dots & a_{1m} \\ \vdots & \vdots & \vdots & \vdots \\ \widetilde{a}_{l1} & \widetilde{a}_{l2} & \dots & \widetilde{a}_{lm} \\ \vdots & \vdots & \vdots & \vdots \\ a_{m1} & a_{m2} & \dots & a_{mm} \end{vmatrix}$$

$$= \begin{vmatrix} a_{11} & a_{12} & \dots & a_{1m} \\ \vdots & \vdots & \vdots & \vdots \\ X_k(a_{l1}) & X_k(a_{l2}) & \dots & X_k(a_{lm}) \\ \vdots & \vdots & \vdots & \vdots \\ a_{m1} & a_{m2} & \dots & a_{mm} \end{vmatrix} + \begin{vmatrix} a_{11} & a_{12} & \dots & a_{1m} \\ \vdots & \vdots & \vdots & \vdots \\ \sum_{r=1}^m a_{lr}\Gamma_{kr}^1 & \sum_{r=1}^m a_{lr}\Gamma_{kr}^2 & \dots & \sum_{r=1}^m a_{lr}\Gamma_{kr}^m \\ \vdots & \vdots & \vdots & \vdots \\ a_{m1} & a_{m2} & \dots & a_{mm} \end{vmatrix}$$

$$= \begin{vmatrix} a_{11} & a_{12} & \dots & a_{1m} \\ \vdots & \vdots & \vdots & \vdots \\ X_k(a_{l1}) & X_k(a_{l2}) & \dots & X_k(a_{lm}) \\ \vdots & \vdots & \vdots & \vdots \\ a_{m1} & a_{m2} & \dots & a_{mm} \end{vmatrix} + \sum_{r=1}^m a_{lr} \begin{vmatrix} a_{11} & a_{12} & \dots & a_{1m} \\ \vdots & \vdots & \vdots & \vdots \\ \Gamma_{kr}^1 & \Gamma_{kr}^2 & \dots & \Gamma_{kr}^m \\ \vdots & \vdots & \vdots & \vdots \\ a_{m1} & a_{m2} & \dots & a_{mm} \end{vmatrix}$$

Hence $\nabla\omega_g(X_k, Y_1,\dots,Y_m)$

$$= -\left( \sum_{r=1}^m a_{1r} \begin{vmatrix} \Gamma_{kr}^1 & \Gamma_{kr}^2 & \dots & \Gamma_{kr}^m \\ a_{21} & a_{22} & \dots & a_{2m} \\ \vdots & \vdots & \vdots & \vdots \\ a_{m1} & a_{m2} & \dots & a_{mm} \end{vmatrix} + \dots + \sum_{r=1}^m a_{mr} \begin{vmatrix} a_{11} & a_{12} & \dots & a_{1m} \\ a_{21} & a_{22} & \dots & a_{2m} \\ \vdots & \vdots & \vdots & \vdots \\ \Gamma_{kr}^1 & \Gamma_{kr}^2 & \dots & \Gamma_{kr}^m \end{vmatrix} \right). \tag{3.5}$$

By developing Equation (3.5), we obtain

$$\nabla\omega_g(X_k, Y_1,\dots,Y_m) = -\sum_{i,j=1}^m p_{ij}\Gamma_{ki}^j,$$

where

$$p_{ij} = \begin{vmatrix} a_{11} & \dots & a_{1(j-1)} & a_{1i} & a_{1(j+1)} & \dots & a_{1m} \\ a_{21} & \dots & a_{2(j-1)} & a_{2i} & a_{2(j+1)} & \dots & a_{2m} \\ \vdots & \vdots & \vdots & \vdots & \vdots & \vdots & \vdots \\ a_{m1} & \dots & a_{m(j-1)} & a_{mi} & a_{m(j+1)} & \dots & a_{mm} \end{vmatrix}.$$

Therefore for $k = 1,\dots,m$ we obtain

$$\nabla\omega_g(X_k, Y_1,\dots,Y_m) = \frac{1}{2}\det(\mathfrak{a})B_k.$$

□

LEMMA 3.3. *Let $M \subset \mathbb{R}^{m+2}$ be a locally strictly convex $m$-surface and $\xi$ a metric field. Then there exists an equiaffine plane bundle $\sigma$ defined on a neighborhood of $p \in M$.*

PROOF. Let $\mathfrak{u} = \{X_1,\dots,X_m\}$ be an orthonormal tangent frame defined on some neighborhood $U$ of $p$. Let $\overline{\sigma}$ be a transversal plane bundle defined also on $U$ and $\{\overline{\xi}_1, \overline{\xi}_2\}$ the local basis of $\overline{\sigma}$ obtained by Theorem 2.4. Now we want to construct a new equiaffine plane bundle$\sigma$ defined on $U$, with local basis $\{\xi_1, \xi_2\}$ obtained also by Theorem 2.4. By Lemma 2.6, we have

$$\xi_1 = \overline{\xi}_1 - Z_1, \qquad \xi_2 = \overline{\xi}_2 - Z_2,$$

where $Z_1$ and $Z_2$ are tangent vector fields. We denote the connection induced by $\sigma$ (resp. $\overline{\sigma}$) by $\nabla$ (resp. $\overline{\nabla}$). On the other hand, by a simple calculation we obtain

$$\overline{B}_k = B_k + 2\sum_{r=1}^m h^1(X_k, X_r)g(Z_1, X_r) + 2\sum_{r=1}^m h^2(X_k, X_r)g(Z_2, X_r).$$

Note that $\sigma$ is equiaffine if and only if $B_i = 0, \forall i = 1, \ldots, m$. By writing $Z_1 = \lambda_1 X_1 + \ldots + \lambda_m X_m$ and $Z_2 = \beta_1 X_1 + \ldots + \beta_m X_m$, this is equivalent to

$$\begin{aligned}
\overline{B}_1 &= 2h^1(X_1, X_2)\lambda_2 + \ldots + 2h^1(X_1, X_m)\lambda_m + 2\beta_1,\\
\overline{B}_2 &= 2h^1(X_2, X_1)\lambda_1 + 2h^1(X_2, X_2)\lambda_2 + \ldots + 2h^1(X_2, X_m)\lambda_m + 2\beta_2,\\
\vdots &= \vdots\\
\overline{B}_m &= 2h^1(X_m, X_1)\lambda_1 + 2h^1(X_m, X_2)\lambda_2 + \ldots + 2h^1(X_m, X_m)\lambda_m + 2\beta_m.
\end{aligned}$$

The lemma follows since the system above has a solution. For instance, set $\lambda_1 = \ldots = \lambda_m = 0$, $\beta_i = \frac{\overline{B}_i}{2}$, for $i = 1, \ldots, m$. □

Now by using the metric of the transversal vector field we introduce an affine normal plane for locally strictly convex submanifolds of codimension 2. In the case of surfaces in $\mathbb{R}^4$ the affine normal plane is symmetric equiaffine and antisymmetric equiaffine however, in general, for codimension two manifolds the affine normal plane is not equiaffine.

Let $M \subset \mathbb{R}^{m+2}$ be a locally strictly convex $m$-surface and $\xi$ a metric field. Let $\mathfrak{u} = \{X_1, \ldots, X_m\}$ be a local orthonormal tangent frame of $g = g_\xi$ and let $\sigma$ be an arbitrary transversal plane bundle. By Theorem 2.4 there exists a unique local frame $\{\xi_1, \xi_2\}$ of $\sigma$ such that $[X_1, \ldots, X_m, \xi_1, \xi_2] = 1$, $h^1(X_1, X_1) = 0$, $-\xi_1 \in [\xi]$ and $h^2(X_i, X_j) = \delta_{ij}$.

Definition 3.4. We define the affine normal plane bundle, denoted by $\mathbf{A}$, as the transversal plane bundle generated by $\{\overline{\xi}_1, \overline{\xi}_2\}$ where,

$$\begin{aligned}
\overline{\xi}_1 &= \xi_1 - \tau_1^2(X_1)X_1 - \tau_1^2(X_2)X_2 - \ldots - \tau_1^2(X_m)X_m,\\
\overline{\xi}_2 &= \xi_2 - \tau_2^2(X_1)X_1 - \tau_2^2(X_2)X_2 - \ldots - \tau_2^2(X_m)X_m.
\end{aligned}$$

Note that: if $\overline{\xi}_1 = \xi_1 - Z_1$ and $\overline{\xi}_2 = \xi_2 - Z_2$ then

$$\begin{aligned}
\overline{\tau}_1^1(X) &= \tau_1^1(X) - h^1(X, Z_1), & \overline{\tau}_2^1(X) &= \tau_2^1(X) - h^1(X, Z_2),\\
\overline{\tau}_1^2(X) &= \tau_1^2(X) - h^2(X, Z_1), & \overline{\tau}_2^2(X) &= \tau_2^2(X) - h^2(X, Z_2).
\end{aligned}$$

By a straightforward computation we obtain the torsions $\overline{\tau}_i^j$ on normal plane bundle

$$\begin{aligned}
\overline{\tau}_1^1(X_j) &= \tau_1^1(X_j) - \sum_{k=1}^m h^1(X_j, X_k)\tau_1^2(X_k),\\
\overline{\tau}_2^1(X_j) &= \tau_2^1(X_j) - \sum_{k=1}^m h^1(X_j, X_k)\tau_2^2(X_k),\\
\overline{\tau}_1^2(X_j) &= 0,\\
\overline{\tau}_2^2(X_j) &= 0.
\end{aligned}$$

Proposition 3.5. *The affine normal plane bundle* $\mathbf{A}$ *does not depend on the transversal plane bundle* $\sigma$.

Proof. Let $\hat{\sigma}$ be an other transversal plane bundle. By Theorem 2.4 there exists a frame $\{\nu_1, \nu_2\}$ on $\hat{\sigma}$ and by Lemma 2.6 there are $Z_1, Z_2$ tangent vector fields on $M$ such that $\nu_1 = \xi_1 - Z_1$ and $\nu_2 = \xi_2 - Z_2$. Now by definition we have $D_{X_i}\nu_1 = -S_{\nu_1}X_i + \hat{\tau}_1^1(X_i)\nu_1 + \hat{\tau}_1^2(X_i)\nu_2$ and $D_{X_i}\nu_2 = -S_{\nu_2}X_i + \hat{\tau}_2^1(X_i)\nu_1 + \hat{\tau}_2^2(X_i)\nu_2$. On the other hand,

$$\hat{\tau}_1^1(X_i) = \tau_1^1(X_i) - h^1(X_i, Z_1), \qquad \hat{\tau}_2^1(X_i) = \tau_2^1(X_i) - h^1(X_i, Z_2),$$
$$\hat{\tau}_1^2(X_i) = \tau_1^2(X_i) - h^2(X_i, Z_1), \qquad \hat{\tau}_2^2(X_i) = \tau_2^2(X_i) - h^2(X_i, Z_2).$$

Now we write $Z_1 = a_1X_1 + \ldots + a_mX_m$ and $Z_2 = b_1X_1 + \ldots + b_mX_m$ then

$$\begin{aligned}\overline{\nu}_1 = \nu_1 - \sum_{k=1}^m \hat{\tau}_1^2(X_k)X_k &= \xi_1 - Z_1 - \sum_{k=1}^m (\tau_1^2(X_k) - h^2(X_k, Z_1))X_k \\ &= \xi_1 - \sum_{k=1}^m \tau_1^2(X_k)X_k = \overline{\xi}_1.\end{aligned}$$

Analogously, $\overline{\nu}_2 = \overline{\xi}_2$. □

PROPOSITION 3.6. The affine normal plane bundle $\mathbf{A}$ does not depend on the local orthonormal tangent frame $\mathfrak{u}$.

PROOF. Let $\mathfrak{v} = \{Y_1, \ldots, Y_m\}$ be another orthonormal tangent frame then

$$Y_i = \sum_{j=1}^m a_{ij}X_j \tag{3.6}$$

where $\mathbf{a} = (a_{ij})$ is an orthogonal matrix and $\det(\mathbf{a}) = 1$. By Lemma 2.5 we have: $\xi_1 = \hat{\xi}_1$ and $\xi_2 = \lambda\hat{\xi}_1 + \hat{\xi}_2$ for some function $\lambda$. By a computation, $\hat{\tau}_1^2(X) = \tau_1^2(X)$ and $\hat{\tau}_2^2(X) = \tau_2^2(X) - \lambda\tau_1^2(X)$. Now by linearity of $\tau_i^j$ follows

$$\begin{aligned}\hat{\xi}_1 - \sum_{k=1}^m \hat{\tau}_1^2(Y_k)Y_k &= \xi_1 - \sum_{k=1}^m \tau_1^2(Y_k)\sum_{j=1}^m a_{kj}X_j \\ \overline{\hat{\xi}}_1 &= \xi_1 - \sum_{j=1}^m\Big(\sum_{k=1}^m \tau_1^2(Y_k)a_{kj}\Big)X_j \\ &= \xi_1 - \sum_{j=1}^m\left(\sum_{k=1}^m \tau_1^2\Big(\sum_{l=1}^m a_{kl}X_l\Big)a_{kj}\right)X_j \\ &= \xi_1 - \sum_{j=1}^m\left(\sum_{k=1}^m\left(\sum_{l=1}^m \tau_1^2(X_l)a_{kl}a_{kj}\right)\right)X_j \\ &= \xi_1 - \sum_{j=1}^m \tau_1^2(X_j)X_j = \overline{\xi}_1.\end{aligned}$$

Analogously we obtain $\overline{\xi}_2 = -\lambda\overline{\hat{\xi}}_1 + \overline{\hat{\xi}}_2$. □

Propositions 3.5 and 3.6 allow us to define the affine distance and height functions for submanifolds of codimension 2. Next we define these functions and characterize the singularities of them.

## 4. Affine distance and height functions

Let $M \subset \mathbb{R}^{m+2}$ be a locally strictly convex $m$-surface and $\xi$ a metric field. Let $\mathfrak{u} = \{X_1, \ldots, X_m\}$ be a local orthonormal tangent frame of $g = g_\xi$ and let $\sigma$ be an arbitrary transversal plane bundle. By Theorem 2.4 there exists a unique local frame $\{\xi_1, \xi_2\}$ of $\sigma_p$.

DEFINITION 4.1. We define the family of affine distance functions

$$\Delta : \mathbb{R}^{m+2} \times M \to \mathbb{R},$$

as follows: for $x \in \mathbb{R}^{m+2}$ and $p \in M$, $\Delta(x, p)$ is given by

$$p - x = z(x, p) + \Delta(x, p)\xi_2$$

where $z(x, p) \in \pi_p = T_pM \oplus \xi_p$.

LEMMA 4.2. *The family of affine distance functions is independent of the tangent frame $\mathfrak{u}$ and independent of the transversal plane bundle $\sigma$.*

PROOF. Let $\mathfrak{v} = \{Y_1, \ldots, Y_m\}$ be another orthonormal tangent frame, then write for $i = 1, \ldots, m$

$$Y_i = \sum_{j=1}^{m} a_{ij} X_j \tag{4.1}$$

where $\mathbf{a} = (a_{ij})$ is an orthogonal matrix and $\det(\mathbf{a}) = 1$.

By Lemma 2.5 we have

$$\begin{aligned} \xi_1 &= \overline{\xi}_1, \\ \xi_2 &= \lambda\overline{\xi}_1 + \overline{\xi}_2, \end{aligned}$$

for some function $\lambda$.

Now, by expanding

$$p - x = \sum_{i=1}^{m} \lambda_i Y_i + r\overline{\xi}_1 + \overline{\Delta}(x, p)\overline{\xi}_2$$

we obtain

$$\begin{aligned} p - x &= \sum_{i=1}^{m} \lambda_i \Big(\sum_{j=1}^{m} a_{ij} X_j\Big) + r\xi_1 + \overline{\Delta}(x, p)(-\lambda\xi_1 + \xi_2) \\ &= \sum_{i,j=1}^{m} \lambda_i a_{ij} X_j + (r - \overline{\Delta}(x, p)\lambda)\xi_1 + \overline{\Delta}(x, p)\xi_2. \end{aligned}$$

On the other hand, let $\overline{\sigma}$ be another transversal plane. By Lemma 2.6 there are tangent vector fields $Z_1$ and $Z_2$ such that $\overline{\xi}_1 = \xi_1 - Z_1$, and $\overline{\xi}_2 = \xi_2 - Z_2$. Then

$$\begin{aligned} p - x &= \lambda_1 X_1 + \ldots + \lambda_m X_m + r\overline{\xi}_1 + \overline{\Delta}(x, p)\overline{\xi}_2, \\ &= \lambda_1 X_1 + \ldots + \lambda_m X_m - rZ_1 - \overline{\Delta}(x, p)Z_2 + r\xi_1 + \overline{\Delta}(x, p)\xi_2. \end{aligned}$$

□

THEOREM 4.3. *The affine distance function $\Delta_x$ has a singularity if and only if $x - X$ belongs to the affine normal plane $\mathbf{A}$.*

PROOF. Since $\mathfrak{u} = \{X_1, \ldots, X_m\}$ is an orthonormal tangent frame and $\{\xi_1, \xi_2\}$ is the frame on $\sigma$ obtained by Theorem 2.4 then we can write

$$x - X = r_1 X_1 + \ldots + r_m X_m + \alpha_1 \xi_1 + \alpha_2 \xi_2.$$

We have

$$\Delta = [X_1, \ldots, X_m, \xi_1, x - X].$$

By properties of derivation and determinant we have

$$X_k(\Delta) = [D_{X_1} X_1, \ldots, X_m, \xi_1, x - X] + \ldots + \\ [X_1, \ldots, D_{X_1} X_m, \xi_1, x - X] + [X_1, \ldots, X_m, D_{X_1} \xi_1, x - X]$$

$$=(\alpha_2 \Gamma_{k1}^1 - h^2(X_k, X_1) r_1) + \ldots + (\alpha_2 \Gamma_{km}^m - h^2(X_k, X_m) r_m) + (\alpha_2 \tau_1^1(X_k) - \alpha_1 \tau_1^2(X_k))$$
$$=\alpha_2(\Gamma_{k1}^1 + \ldots \Gamma_{km}^m + \tau_1^1(X_k)) - \alpha_1 \tau_1^2(X_k) - r_k.$$

Therefore $X_k(\Delta) = 0$ if and only if

$$r_k = \alpha_2(\Gamma_{k1}^1 + \ldots \Gamma_{km}^m + \tau_1^1(X_k)) - \alpha_1 \tau_1^2(X_k).$$

By derivation of $[X_1, \ldots, X_m, \xi_1, \xi_2] = 1$ we obtain

$$\Gamma_{k1}^1 + \ldots \Gamma_{km}^m + \tau_1^1(X_k) = -\tau_2^2(X_k).$$

It follows that $X_k(\Delta) = 0$ if and only if $r_k = -\alpha_2 \tau_2^2(X_k) - \alpha_1 \tau_1^2(X_k)$.

Therefore $X_1(\Delta) = \ldots = X_m(\Delta) = 0$ if and only if

$$x - X = \alpha_1 \left( \xi_1 - \sum_{k=1}^{m} \tau_1^2(X_k) X_k \right) + \alpha_2 \left( \xi_2 - \sum_{k=1}^{m} \tau_2^2(X_k) X_k \right).$$

□

Analogously to the definition of affine distance functions we define the family of affine height functions.

DEFINITION 4.4. The family of affine height functions

$$H : \mathbb{R}^{m+2} \times M \to \mathbb{R},$$

is defined as follows: for $x \in \mathbb{R}^{m+2}$ and $p \in M$, $H(x,p)$ is given by

$$x = z(x,p) + H(x,p)\xi_2$$

where $z(x,p) \in \pi_p = T_p M \oplus \xi_p$.

LEMMA 4.5. *The family of affine height functions is independent of the tangent frame* $\mathfrak{u}$ *and independent of the transversal plane bundle* $\sigma$.

PROOF. Analogous to the proof of Lemma 4.2. □

We can write the affine distance function as $H(x,p) = [X_1, \ldots, X_m, \xi_1, x]$.

THEOREM 4.6. *The affine height function* $H_x$ *has a singularity if and only if* $x$ *is in the affine normal plane* $\mathbf{A}$.

PROOF. Analogous to the proof of Theorem 4.3. □

## 5. $m$-Surfaces in hypersurfaces in affine $(m+2)$-space

We recall the definition of the Blaschke metric of a hypersurface $N \subset \mathbb{R}^{m+2}$, for more details see [**6**]. Let $\mathfrak{u}' = \{X_1, \dots, X_{m+1}\}$ be a tangent frame defined in some neighborhood $U$ of a point $p$ in $N$. Now we consider

$$H_{\mathfrak{u}'}(Y,Z) = [X_1, \dots, X_{m+1}, D_Z Y], \quad \forall Y, Z \in T_pN.$$

Then $H_{\mathfrak{u}'}$ defines a symmetric bilinear form on $N$ that initially depends on the tangent frame $\mathfrak{u}'$. However, if we suppose that $H_{\mathfrak{u}'}$ is non-degenerate then we can normalize it and the symmetric bilinear form

$$\mathfrak{G}(Y,Z) = \frac{H_{\mathfrak{u}'}(Y,Z)}{(\det_{\mathfrak{u}'} H_{\mathfrak{u}'})^{\frac{1}{m+3}}}, \quad \forall Y, Z \in T_pN,$$

does not depend on the choice of the tangent frame $\mathfrak{u}'$. Here $\det_{\mathfrak{u}'} H_{\mathfrak{u}'}$ denotes the determinant of the matrix $(H_{\mathfrak{u}'}(X_i, X_j))$. The metric $\mathfrak{G}$ is called the *Blaschke metric* of $N$.

If $N$ is locally strictly convex, then $H_{\mathfrak{u}'}$ is always non-degenerate and positive definite, moreover the tangent space $T_pN$ is a support hyperplane with a non-degenerate contact. In particular, given any immersed $m$-surface $M \subset N$ we have $T_pM \subset T_pN \subset \mathbb{R}^{m+2}$, and hence, $M$ is also locally strictly convex. Moreover, we can consider the Blaschke metric $\mathfrak{G}$ restricted to $M$.

REMARK 5.1. We can choose a transversal vector field $\xi$ such that $g_{[\xi]}$ coincides with $\mathfrak{G}$ in $T_pM$. In fact, let $\mathfrak{u} = \{X_1, \dots, X_m\}$ be a frame in $T_pM$ and we choose a tangent vector field $-X_{m+1} \in T_pN$ such that $\mathfrak{u}' = \{X_1, \dots, X_{m+1}\}$ is a frame in $T_pN$, then

$$G_{\mathfrak{u}}(Y,Z) = H_{\mathfrak{u}'}(Y,Z), \quad \forall Y, Z \in T_pM.$$

In particular, we have that $g_{[-X_{m+1}]} = \lambda \mathfrak{G}$ where $\lambda$ is given by

$$\lambda = \frac{(\det_{\mathfrak{u}'} H_{\mathfrak{u}'})^{\frac{1}{m+3}}}{(\det_{\mathfrak{u}} G_{\mathfrak{u}})^{\frac{1}{m+2}}}.$$

Then, since

$$g_{[\beta\xi]} = \beta^{\frac{2}{m+2}} g_{[\xi]} \ \forall \beta > 0,$$

it is enough to change the transversal vector field $-X_{m+1}$ by $\xi = -\left(\frac{1}{\lambda}\right)^{\frac{m+2}{2}} X_{m+1}$, so that $g_{[\xi]} = \mathfrak{G}$.

LEMMA 5.2. *Let $M \subset N \subset \mathbb{R}^{m+2}$ be an immersed $m$-surface in a locally strictly convex hypersurface and let $\xi$ be the metric field such that $g_{[\xi]} = \mathfrak{G}$ on $M$. If $\mathfrak{u} = \{X_1, \dots, X_m\}$ is a local orthonormal tangent frame and $\{\xi_1, \xi_2\}$ given by Theorem 2.4 on $\sigma$. Then the frame $\{X_1, \dots, X_m, \xi_1 - \tau_1^2(X_1)X_1 - \dots - \tau_1^2(X_m)X_m\}$ is orthonormal relative to metric $\mathfrak{G}$ on $N$ and there are functions $r_1, \dots, r_{m+1}$ such that $\xi_2 = \mathbf{Y} + r_1X_1 + \dots + r_mX_m + r_{m+1}\xi_1$, where $\mathbf{Y}$ is the affine normal vector field on the hypersurface $N$.*

PROOF. We have that $\{X_1, \dots, X_m\}$ is orthonormal on $N$ relative to metric $\mathfrak{G}$, now by using orthonormalization, we write $\overline{\xi}_1 = \lambda\xi_1 - \lambda_1X_1 - \dots - \lambda_mX_m$ such that the frame $\{X_1, \dots, X_m, \overline{\xi}_1\}$ is a local orthonormal frame with the Blaschke metric. Since $\{X_1, \dots, X_m, \overline{\xi}_1\}$ is an orthonormal frame with the Blaschke metric $[X_1, \dots, X_m, \overline{\xi}_1, \mathbf{Y}] = 1$ it follows that $\lambda[X_1, \dots, X_m, \xi_1, \mathbf{Y}] = 1$. On the other hand

$[X_1,\ldots,X_m,\xi_1,\xi_2]=1$ and we obtain $[X_1,\ldots,X_m,\xi_1,\xi_2-\lambda\mathbf{Y}]=0$. Therefore there exist $r_i, i=1,\ldots,m+1$ such that

$$\xi_2=\lambda\mathbf{Y}+r_1X_1+\ldots+r_mX_m+r_{m+1}\xi_1.$$

On the other hand, $\mathfrak{G}(\overline{\xi}_1,X_1)=0,\ \ldots,\mathfrak{G}(\overline{\xi}_1,X_m)=0$ and $\mathfrak{G}(\overline{\xi}_1,\overline{\xi}_1)=1$, it follows that:

$$\lambda_i=\lambda\mathfrak{G}(\xi_1,X_i),\forall i=1,\ldots,m.$$
$$1+\lambda_1^2+\ldots+\lambda_m^2=\lambda^2\mathfrak{G}(\xi_1,\xi_1).$$

Now we compute the affine fundamental form $h$ on $N$. By Theorem 2.4

$$\begin{aligned}D_{X_1}X_1&=\nabla_{X_1}X_1+\xi_2=\nabla_{X_1}X_1+\lambda Y+r_1X_1+\ldots+r_mX_m+r_{m+1}\xi_1\\&=(\nabla_{X_1}X_1+r_1X_1+\ldots+r_mX_m+r_{m+1}\xi_1)+\lambda\mathbf{Y}.\end{aligned}$$

Hence $h(X_1,X_1)=\lambda$, analogously we obtain

$$h(X_i,X_j)=\lambda h^2(X_i,X_j),\qquad\qquad h(X_i,\overline{\xi}_1)=\lambda^2\tau_1^2(X_i)-\lambda_i\lambda.$$

The equations $\mathfrak{G}(X_1,\overline{\xi}_1)=0$, $h(X_1,\overline{\xi}_1)=0$ imply $\lambda_i=\lambda\tau_1^2(X_i),\forall i=1,\ldots,m$. Now by a simple computation

$$H=\det(h_{ij})=\lambda^m h(\overline{\xi}_1,\overline{\xi}_1).$$

By using the definition of the Blaschke metric

$$\begin{aligned}1=\mathfrak{G}(X_i,X_i)&=\frac{\lambda}{\lambda^{m/(m+3)}h(\overline{\xi}_1,\overline{\xi}_1)^{1/(m+3)}}=\frac{\lambda^{3/(m+3)}}{h(\overline{\xi}_1,\overline{\xi}_1)^{1/(m+3)}},\\1=\mathfrak{G}(\overline{\xi}_1,\overline{\xi}_1)&=\frac{h(\overline{\xi}_1,\overline{\xi}_1)}{\lambda^{m/(m+3)}h(\overline{\xi}_1,\overline{\xi}_1)^{1/(m+3)}}=\frac{h(\overline{\xi}_1,\overline{\xi}_1)^{m+2/m+3}}{\lambda^{m/m+3}}.\end{aligned}$$

We conclude that $\lambda=h(\overline{\xi}_1,\overline{\xi}_1)=1$ and hence

$$\xi_2=\mathbf{Y}+r_1X_1+\ldots+r_mX_m+r_{m+1}\xi_1,$$
$$\overline{\xi}_1=\xi_1-\tau_1^2(X_1)X_1-\ldots-\tau_1^2(X_m)X_m.$$

□

The next results give a relation between the family of affine distance functions on $N$ and the family of affine distance functions on $M$. First we recall the definition of affine distance on hypersurfaces, see [**3, 4**].

DEFINITION 5.3. The family of affine distance functions on $N$ is given by $\widetilde{\Delta}:\mathbb{R}^{m+2}\times N\to\mathbb{R}$ as follows: given a point $x\in\mathbb{R}^{m+2}$ and a point $p\in N$ it is defined the affine distance from $x$ to $p$ implicitly by,

$$p-x=z(x,p)+\widetilde{\Delta}(x,p)\mathbf{Y}(p)$$

where $z\in T_pN$ and $\mathbf{Y}$ is the affine normal vector field on $N$.

THEOREM 5.4. *Let $M\subset N\subset\mathbb{R}^{m+2}$ be an immersed m-surface in a locally strictly convex hypersurface and let $-\xi$ be the metric field such that $g_{-\xi}=\mathfrak{G}$ on $M$. Then the affine distance function $\widetilde{\Delta}(x,p)$ coincides with the affine distance function $\Delta(x,p)$ and the affine normal vector field $\mathbf{Y}$ belongs to the affine normal plane $\mathbf{A}$.*

PROOF. Let $\mathfrak{u} = \{X_1, \ldots, X_m\}$ be a local orthonormal tangent frame and $\sigma$ be a transversal plane bundle. By Theorem 2.4 there is a frame $\{\xi_1, \xi_2\}$ on $\sigma$ and by Lemma 5.2 $\{X_1, \ldots, X_m, \overline{\xi}_1\}$ is an orthonormal frame with the Blaschke metric and there are functions $r_1, \ldots, r_{m+1}$ such that

$$\xi_2 = \mathbf{Y} + r_1 X_1 + \ldots + r_m X_m + r_{m+1}\xi_1,$$
$$\overline{\xi}_1 = \xi_1 - \tau_1^2(X_1)X_1 - \ldots - \tau_1^2(X_m)X_m.$$

Now we compute the affine distance

$$\Delta = [X_1, \ldots, X_m, \xi_1, p - x] = \frac{[X_1, \ldots, X_m, \xi_1, p - x]}{[X_1, \ldots, X_m, \xi_1, \xi_2]} = \frac{[X_1, \ldots, X_m, \xi_1, p - x]}{[X_1, \ldots, X_m, \xi_1, \mathbf{Y}]} = \widetilde{\Delta}.$$

Now we see that $\mathbf{Y}$ belongs to the affine normal plane $\mathbf{A}$. In fact,

$$D_X\mathbf{Y} = z + (\tau_2^2(X) - r_1 h^2(X_1, X) - \ldots - r_m h^2(X_m, X) - r_{m+1}\tau_1^2(X))\mathbf{Y},$$

for some $z \in T_pN$, and since $\mathbf{Y}$ is the affine normal on $N$ it follows

$$\tau_2^2(X) - r_1 h^2(X_1, X) - \ldots - r_m h^2(X_m, X) - r_{m+1}\tau_1^2(X) = 0.$$

In particular for $i = 1, \ldots, m$ we obtain

$$\tau_2^2(X_i) = r_i + r_{m+1}\tau_1^2(X_i).$$

Finally

$$\begin{aligned}
\mathbf{Y} &= \xi_2 - r_1 X_1 - \ldots - r_m X_m - r_{m+1}\xi_1 \\
&= \xi_2 - (\tau_2^2(X_1) - r_{m+1}\tau_1^2(X_1))X_1 - \ldots - (\tau_2^2(X_m) - r_{m+1}\tau_1^2(X_m))X_m - r_{m+1}\xi_1 \\
&= (\xi_2 - \tau_2^2(X_1)X_1 - \ldots - \tau_2^2(X_m)X_m) - r_{m+1}(\xi_1 - \tau_1^2(X_1)X_1 - \ldots - \tau_1^2(X_m)X_m).
\end{aligned}$$

Therefore

$$\mathbf{Y} = \overline{\xi}_2 - r_{m+1}\overline{\xi}_1 \in \mathbf{A}.$$

□

Let $M \subset \mathbb{R}^{m+2}$ be an immersed $m$-surface in a locally strictly convex hypersurface $N$. We can consider on $M$ the extended Blaschke metric, by writing $\mathfrak{G}(e_i, Y) = 0$ for all $i = 1, \ldots, m+1$. Here $\{e_1, \ldots, e_{m+1}\}$ is a unimodular frame and $Y$ is the affine normal to $N$.

COROLLARY 5.5. With the same hypothesis as in Theorem 5.4, the affine normal plane is the orthogonal plane to the tangent space with respect to the extended Blaschke metric.

PROOF. By Theorem 5.4 we have $\mathbf{Y} = \overline{\xi}_2 - r_{m+1}\overline{\xi}_1$ where

$$\overline{\xi}_1 = \xi_1 - \tau_1^2(X_1)X_1 - \ldots - \tau_1^2(X_m)X_m,$$
$$\overline{\xi}_2 = \xi_2 - \tau_2^2(X_1)X_1 - \ldots - \tau_2^2(X_m)X_m.$$

By Lemma 5.2 the frame $\mathfrak{u}' = \{X_1, \ldots, X_m, \overline{\xi}_1\}$ is orthonormal relative to the Blaschke metric.

$$\mathfrak{G}(\overline{\xi}_2, X_i) = \mathfrak{G}(\mathbf{Y} + r_{m+1}\overline{\xi}_1, X_1) = \mathfrak{G}(\mathbf{Y}, X_1) + r_{m+1}\mathfrak{G}(\overline{\xi}_1, X_1) = 0.$$

□

## References


[1] W. Blaschke, *Differentialgeometrie der Kreise und Kugeln, Vorlesungen über Differentialgeometrie*, Grundlehren der Mathematischen Wissenschaften XXIX, vol. 3, Springer-Verlag, Berlin, 1929.

[2] C. Burstin and W. Mayer, *Über affine Geometrie XLI: Die Geometrie zweifach ausgedehnter Mannigfaltigkeiten $F^2$ im affinen $R^4$* (German), Math. Z. **26** (1927), no. 1, 373–407, DOI 10.1007/BF01475460. MR1544861

[3] Thomas E. Cecil, *Focal points and support functions in affine differential geometry*, Geom. Dedicata **50** (1994), no. 3, 291–300, DOI 10.1007/BF01267871. MR1286382

[4] Declan Davis, *Affine normal curvature of hypersurfaces from the point of view of singularity theory*, Geom. Dedicata **141** (2009), 137–145, DOI 10.1007/s10711-008-9348-8. MR2520068

[5] Ronaldo Alves Garcia, Dirce Kiyomi Hayashida Mochida, Maria Del Carmen Romero Fuster, and Maria Aparecida Soares Ruas, *Inflection points and topology of surfaces in 4-space*, Trans. Amer. Math. Soc. **352** (2000), no. 7, 3029–3043, DOI 10.1090/S0002-9947-00-02404-1. MR1638242

[6] An Min Li, Udo Simon, and Guo Song Zhao, *Global affine differential geometry of hypersurfaces*, de Gruyter Expositions in Mathematics, vol. 11, Walter de Gruyter & Co., Berlin, 1993. MR1257186

[7] Dirce Kiyomi Hayashida Mochida, Maria Del Carmen Romero Fuster, and Maria Aparecida Soares Ruas, *The geometry of surfaces in 4-space from a contact viewpoint*, Geom. Dedicata **54** (1995), no. 3, 323–332, DOI 10.1007/BF01265348. MR1326737

[8] Katsumi Nomizu and Takeshi Sasaki, *Affine differential geometry*, Cambridge Tracts in Mathematics, vol. 111, Cambridge University Press, Cambridge, 1994. Geometry of affine immersions. MR1311248

[9] Katsumi Nomizu and Luc Vrancken, *A new equiaffine theory for surfaces in* $\mathbf{R}^4$, Internat. J. Math. **4** (1993), no. 1, 127–165, DOI 10.1142/S0129167X9300008X. MR1209963

[10] Juan J. Nuño-Ballesteros, *Submanifolds with a non-degenerate parallel normal vector field in Euclidean spaces*, Singularity theory and its applications, Adv. Stud. Pure Math., vol. 43, Math. Soc. Japan, Tokyo, 2006, pp. 311–332. MR2325143

[11] J. J. Nuño-Ballesteros and M. C. Romero-Fuster, *Contact properties of codimension 2 submanifolds with flat normal bundle*, Rev. Mat. Iberoam. **26** (2010), no. 3, 799–824, DOI 10.4171/RMI/617. MR2789366

[12] J.J. Nuño-Ballesteros, L. Sánchez, *Affine metrics of locally strictly convex surface in affine 4-space*, Preprint. arXiv:1404.2609v1[math.DG].

[13] María del Carmen Romero-Fuster and Federico Sánchez-Bringas, *Umbilicity of surfaces with orthogonal asymptotic lines in* $\mathbb{R}^4$, Differential Geom. Appl. **16** (2002), no. 3, 213–224, DOI 10.1016/S0926-2245(02)00068-2. MR1900745



Universidade de São Paulo, ICMC-SMA, Caixa Postal 668, 13560-970 São Carlos (SP), Brazil

*E-mail address*: `mjsaia@icmc.usp.br`

Universidade Federal de Uberlândia, FAMAT, Rua Goiás, 2000, 38500-000. Monte Carmelo (MG), Brazil

*E-mail address*: `luis.sanchez@ufu.br`

Contemporary Mathematics
Volume **675**, 2016
http://dx.doi.org/10.1090/conm/675/13599

# Criteria for Morin singularities for maps into lower dimensions, and applications

Kentaro Saji

*Dedicated to Professor María del Carmen Romero Fuster on the occasion of her sixtieth birthday*

ABSTRACT. We give criteria for Morin singularities for germs of maps into lower dimensions. As an application, we study the bifurcation of Lefschetz singularities.

## 1. Introduction

A map-germ $f : (\boldsymbol{R}^m, 0) \to (\boldsymbol{R}^n, 0)$ $(m > n)$ is called a *$k$-Morin singularity* $(1 \leq k \leq n)$ if it is $\mathcal{A}$-equivalent to the following map-germ at the origin:

$$\begin{aligned} &h_{0,k}(x_1, \ldots, x_{n-1}, y_1, \ldots, y_{m-n}, z) \\ &\qquad = \left(x_1, \ldots, x_{n-1}, q(y_1, \ldots, y_{m-n}) + z^{k+1} + \sum_{i=1}^{k-1} x_i z^i\right) \end{aligned} \tag{1.1}$$

if $k \geq 2$, and $h_{0,1}(x_1, \ldots, x_{n-1}, y_1, \ldots, y_{m-n+1}) = (x_1, \ldots, x_{n-1}, q(y_1, \ldots, y_{m-n+1}))$ if $k = 1$, where $q$ is a non-degenerate quadratic germ of function. The 1-Morin singularity is also called the *fold*, and the 2-Morin singularity is also called the *cusp*. We say that two map-germs $f, g : (\boldsymbol{R}^m, 0) \to (\boldsymbol{R}^n, 0)$ are *$\mathcal{A}$-equivalent* if there exist germs of diffeomorphism $\varphi : (\boldsymbol{R}^m, 0) \to (\boldsymbol{R}^m, 0)$ and $\Phi : (\boldsymbol{R}^n, 0) \to (\boldsymbol{R}^n, 0)$ such that $\Phi \circ f \circ \varphi = g$. Morin singularities are stable, and conversely, all corank one and stable map-germs are Morin singularities. This means that Morin singularities are fundamental and frequently appear as singularities of maps from one manifold to another. If $\operatorname{corank} df_0 = 1$, then one can choose a coordinate system $(x, y)$ such that $f(x, y) = (x_1, \ldots, x_{n-1}, h(x, y))$, where $x = (x_1, \ldots, x_{n-1})$, $y = (y_1, \ldots, y_{m-n+1})$. We call this procedure a *normalization*. Morin [**17**] gave a characterization of those singularities in terms of transversality of the jet extensions to the Thom-Boardman singularity set, and also gave criteria for germs with respect to a normalized form $(x_1, \ldots, x_{n-1}, h(x, y))$. Morin singularities are also characterized using the intrinsic derivative due to Porteous ([**20**] see also [**1, 7**]). Criteria for singularities without

2010 *Mathematics Subject Classification.* Primary 57R45; Secondary 58K60, 58K65.

Partly supported by Japan Society for the Promotion of Science (JSPS) and Coordenadoria de Aperfeiçoamento de Pessoal de Nível Superior under the Japan-Brazil research cooperative program and Grant-in-Aid for Scientific Research (C) No. 26400087, from JSPS.

using normalization are not only more convenient but also indispensable in some cases. We refer to criteria which are independent of normalization as *general criteria*. In fact, in the case of wave front surfaces in 3-space, general criteria for cuspidal edges and swallowtails were given in [**14**], where we used them to study the local and global behavior of flat fronts in hyperbolic 3-space. Recently general criteria for other singularities and several of their applications have been given in [**11**–**13**, **19**, **26**–**28**]. In this paper, we give general criteria for Morin singularities. Using them, we give applications to bifurcation of the Lefschetz singularity which plays important roles in low-dimensional topology. See [**5**, **6**, **10**, **21**–**24**, **29**, **30**] for other investigations of Morin singularities.

## 2. Singular sets and Hesse matrix of corank one singularities

DEFINITION 2.1. Let $f : (\boldsymbol{R}^m, 0) \to (\boldsymbol{R}^n, 0)$ be a map-germ and denote by $S(f)$ the singular locus of $f$. A collection of vector fields

$$(\xi, \eta) = (\xi_1, \dots, \xi_{n-1}, \eta_1, \dots, \eta_{m-n+1}) \tag{2.1}$$

on $(\boldsymbol{R}^m, 0)$ is said to be *adapted with respect to* $f$ if $\xi_1, \dots, \xi_{n-1}, \eta_1, \dots, \eta_{m-n+1}$ generates $T_0\boldsymbol{R}^m$ at 0, and $\langle \eta_1(p), \dots, \eta_{m-n+1}(p) \rangle_{\boldsymbol{R}} = \ker df_p$ for any $p \in S(f)$ near 0.

LEMMA 2.2. *Let $f : (\boldsymbol{R}^m, 0) \to (\boldsymbol{R}^n, 0)$ be a map-germ satisfying* $\operatorname{rank} df_0 = n - 1$. *Then there exists a collection of vector fields $(\xi, \eta)$ as in (2.1) which is adapted with respect to $f$.*

PROOF. Since the result does not depend on the choice of coordinate system and $\operatorname{rank} df_0 = n - 1$, then we can take a coordinate system $(x, y) = (x_1, \dots, x_{n-1}, y_1, \dots, y_{m-n+1})$ in a neighborhood of the origin $U$ on the source space, such that

$$f(x, y) = (x, h(x, y)), \quad dh_0 = 0. \tag{2.2}$$

Then $S(f) = \{(x, y) \in U \mid h_{y_1}(x, y) = \cdots = h_{y_{m-n+1}}(x, y) = 0\}$ holds. Thus $\partial x_1, \dots, \partial x_{n-1}, \partial y_1, \dots, \partial y_{m-n+1}$ are the desired vector fields. □

Let $f : (\boldsymbol{R}^m, 0) \to (\boldsymbol{R}^n, 0)$ be a map-germ satisfying $\operatorname{rank} df_0 = n - 1$, and $(\xi, \eta) = (\xi_1, \dots, \xi_{n-1}, \eta_1, \dots, \eta_{m-n+1})$ an adapted collection of vector fields with respect to $f$. Set

$$\lambda_i = \det(\xi_1 f, \dots, \xi_{n-1} f, \eta_i f), \quad i = 1, \dots, m - n + 1$$

and

$$\Lambda = (\lambda_1, \dots, \lambda_{m-n+1}),$$

where $\zeta f$ stands for the directional derivative of $f$ along the vector field $\zeta$. Then $S(f) = \{\Lambda = 0\}$.

DEFINITION 2.3. Let 0 be a singular point of $f = (f_1, \dots, f_n) : (\boldsymbol{R}^m, 0) \to (\boldsymbol{R}^n, 0)$ satisfying $\operatorname{rank} df_0 = n - 1$. We say that 0 is *non-degenerate* if $\operatorname{rank} d\Lambda_0 = m - n + 1$.

This condition is a special case of the condition called *critical normalization.* See [**4**] for details.

LEMMA 2.4. *The non-degeneracy condition above does not depend on the choice of coordinate systems on the source space nor on the target space.*

PROOF. One can easily show that it does not depend on the coordinate system on the target. In fact, let $\Phi : (\boldsymbol{R}^n, 0) \to (\boldsymbol{R}^n, 0)$ be a germ of diffeomorphism, and we regard $d\Phi_x$ as the matrix representation of $d\Phi_x$ with respect to the standard basis at $x \in \boldsymbol{R}^n$. Set $\overline{\lambda}_i = \det\big(\xi_1(\Phi\circ f), \dots, \xi_{n-1}(\Phi\circ f), \eta_i(\Phi\circ f)\big)$, and $\overline{\Lambda} = \big(\overline{\lambda}_1, \dots, \overline{\lambda}_{m-n+1}\big)$. Then $\overline{\Lambda}(x) = \det(d\Phi_{f(x)})\Lambda(x)$ holds. Thus $\operatorname{rank} d\Lambda_0$ does not depend on the choice of the coordinate system on the target.

Secondly, we show that it does not depend on the choice of an adapted collection of vector fields. Since it does not depend on the coordinate system on the target, we may assume that $f = (f_1, \dots, f_n)$ satisfies $d(f_n)_0 = 0$. Then for any vector field $\zeta$, it holds that $\zeta\lambda_i = \det(\xi_1 f, \dots, \xi_{n-1} f, \zeta\eta_i f)(0) = \big(\Delta\zeta\eta_i f_n\big)(0)$, where $\Delta = \det(\xi_1\hat{f}, \dots, \xi_{n-1}\hat{f})$, and $\hat{f} = (f_1, \dots, f_{n-1})$. Let $(\bar{\xi}_1, \dots, \bar{\xi}_{n-1}, \bar{\eta}_1, \dots, \bar{\eta}_{m-n+1})$ be an adapted collection of vector fields satisfying

$$
\begin{aligned}
\begin{pmatrix} \bar{\xi}_1 \\ \vdots \\ \bar{\xi}_{n-1} \\ \hline \bar{\eta}_1 \\ \vdots \\ \bar{\eta}_{m-n+1} \end{pmatrix}
&= \left(\begin{array}{c|c} A^1 & A^2 \\ \hline B^1 & B^2 \end{array}\right)
\begin{pmatrix} \xi_1 \\ \vdots \\ \xi_{n-1} \\ \hline \eta_1 \\ \vdots \\ \eta_{m-n+1} \end{pmatrix}, \quad \text{where} \quad \left(\begin{array}{c|c} A^1 & A^2 \\ \hline B^1 & B^2 \end{array}\right) \\
&= \left(\begin{array}{ccc|ccc}
a^1_{1,1} & \cdots & a^1_{1,n-1} & a^2_{1,1} & \cdots & a^2_{1,m-n+1} \\
\vdots & \ddots & \vdots & \vdots & \ddots & \vdots \\
a^1_{n-1,1} & \cdots & a^1_{n-1,n-1} & a^2_{n-1,1} & \cdots & a^2_{n-1,m-n+1} \\ \hline
b^1_{1,1} & \cdots & b^1_{1,n-1} & b^2_{1,1} & \cdots & b^2_{1,m-n+1} \\
\vdots & \ddots & \vdots & \vdots & \ddots & \vdots \\
b^1_{m-n+1,1} & \cdots & b^1_{m-n+1,n-1} & b^2_{m-n+11} & \cdots & b^2_{m-n+1,m-n+1}
\end{array}\right),
\end{aligned} \tag{2.3}
$$

where $A^1, B^2$ are regular matrices at 0, and $B^1 = O$ on $S(f)$. Set

$$
\bar{\lambda}_i = \det\big(\bar{\xi}_1 f, \dots, \bar{\xi}_{n-1} f, \bar{\eta}_i f\big) \quad (i = 1, \dots, m-n+1), \quad \bar{\Lambda} = \big(\bar{\lambda}_1, \dots, \bar{\lambda}_{m-n+1}\big). \tag{2.4}
$$

Then for any vector field $\zeta$, we see that

$$
\zeta\overline{\lambda}_i(0) = \big(\det A^1\,\Delta\,\zeta\bar{\eta}_i f_n\big)(0),
$$

and $d\overline{\Lambda}_0 = \big((\det A^1\Delta)B^2 d\Lambda\big)(0)$. Thus we have the conclusion. $\square$

For a non-degenerate singularity 0, we define a matrix $\mathcal{H}(\xi,\eta)$ (abbreviated as $\mathcal{H}_\eta$) by

$$\mathcal{H}_\eta = (\eta_j\lambda_i)_{1\le i,j\le m-n+1} = \begin{pmatrix} \eta_1\lambda_1 & \cdots & \eta_1\lambda_{m-n+1} \\ \vdots & \ddots & \vdots \\ \eta_{m-n+1}\lambda_1 & \cdots & \eta_{m-n+1}\lambda_{m-n+1} \end{pmatrix}. \tag{2.5}$$

Then $\mathcal{H}_\eta$ is symmetric on $S(f)$. In fact, since $[\eta_j,\eta_i](p) \in T_p\boldsymbol{R}^m$, there exist functions $\alpha_i$ $(i=1,\dots,n-1)$ and $\beta_j$ $(j=1,\dots,m-n+1)$ such that

$$[\eta_j,\eta_i](p) = \sum_{i=1}^{n-1}\alpha_i\xi_i(p) + \sum_{j=1}^{m-n+1}\beta_j\eta_j(p). \tag{2.6}$$

If $p\in S(f)$, then by $\eta_j f(p)=0$ and (2.6) it follows that

$$\eta_j\lambda_i = \det(\xi_1 f,\dots,\xi_{n-1}f,\eta_j\eta_i f) = \det(\xi_1 f,\dots,\xi_{n-1}f,\eta_i\eta_j f) = \eta_i\lambda_j$$

on $S(f)$.

LEMMA 2.5. *Let 0 be a non-degenerate singular point of $f=(f_1,\dots,f_n):(\boldsymbol{R}^m,0)\to(\boldsymbol{R}^n,0)$. The matrix-valued function $\mathcal{H}_\eta$ on $S(f)$ does not depend on the choice of an adapted collection of vector fields with respect to $f$, nor on the coordinate systems on the target up to regular matrices multiplications. In particular,* rank $\mathcal{H}_\eta$ *on $S(f)$ does not depend on the choice of adapted collections of vector fields with respect to $f$ nor on the coordinate systems on the target.*

PROOF. Let $(\xi_1,\dots,\xi_{n-1},\eta_1,\dots,\eta_{m-n+1})$ and $(\bar\xi_1,\dots,\bar\xi_{n-1},\bar\eta_1,\dots,\bar\eta_{m-n+1})$ be adapted collections of vector fields with respect to $f$ satisfying (2.3). By the conditions, it holds that $B^1=0$, $\det A^1\neq 0$ and $\det B^2\neq 0$ on $S(f)$. Set

$$\bar\lambda_i = \det(\bar\xi_1 f,\dots,\bar\xi_{n-1}f,\bar\eta_i f),\quad \mathcal{H}_{\bar\eta} = (\bar\eta_j\bar\lambda_i)_{1\le i,j\le m-n+1}.$$

Since $\eta_i$ and $\bar\eta_i$ $(i=1,\dots,m-n+1)$ are included in $\ker df$ on $S(f)$, one can see that:

$$\begin{aligned}
\bar\eta_j\bar\lambda_i &= \bar\eta_j\det(\bar\xi_1 f,\dots,\bar\xi_{n-1}f,\bar\eta_i f)\\
&= \langle \bar\xi_1 f\times\cdots\times\bar\xi_{n-1}f, \bar\eta_j\bar\eta_i f\rangle\\
&= \det A^1\,\langle \xi_1 f\times\cdots\times\xi_{n-1}f, \bar\eta_j\bar\eta_i f\rangle\\
&= \det A^1\left\langle \xi_1 f\times\cdots\times\xi_{n-1}f, \sum_{k,l}\eta_l b^1_{i,k}\,\xi_l f + \sum_{k,l} b^2_{jl}b^2_{ik}\eta_l\eta_k f\right\rangle\\
&= \det A^1\sum_{k,l} b^2_{jl}b^2_{ik}\,\langle \xi_1 f\times\cdots\times\xi_{n-1}f, \eta_l\eta_k f\rangle\\
&= \det A^1\sum_{k,l} b^2_{jl}b^2_{ik}\,\eta_l\lambda_k.
\end{aligned}$$

Thus on $S(f)$, we have that

$$\mathcal{H}_{\bar\eta} = (\det A^1)^{m-n+1}B^2\mathcal{H}_\eta{}^tB^2, \tag{2.7}$$

where ${}^tB^2$ stands for the transportation of $B^2$.

This proves the first assertion. One can show the independence for the target coordinate systems easily by following the same method as used in the proof of Lemma 2.4. □

If 0 is a non-degenerate singularity, then $S(f)$ is a manifold. Thus we can consider $g = f|_{S(f)}$. Then we have the following lemma.

LEMMA 2.6. *Let* 0 *be a non-degenerate singular point of* $f = (f_1, \dots, f_n)$. *Then*

$$S(f|_S) = S(g) = \{p \in S(f) \mid \det \mathcal{H}_\eta(p) = 0\}$$

*near* 0. *Moreover, by the identification*

$$\mathcal{H}_\eta : \sum_{i=1}^{m-n+1} a_i \eta_i \mapsto \sum_{i=1}^{m-n+1} \left( \sum_{j=1}^{m-n+1} a_j \eta_i \lambda_j \right) \eta_i, \tag{2.8}$$

*it holds that* $\ker dg_p = \ker \mathcal{H}_\eta(p) = \ker df_p \cap T_p S(f)$.

PROOF. The assumption and results do not depend on the choice of coordinate systems, so we may assume that $f$ has the form (2.2). Let us assume that $\operatorname{rank} \operatorname{Hess}_0 h(0, y) = k$. Then by the parametrized Morse Lemma (see [**9**, p.502], [**2**, p.97]), there exist a coordinate system $\tilde{y} = (\tilde{y}_1, \dots, \tilde{y}_{m-n+1})$ and a function $\tilde{h}$ such that

$$\begin{aligned} h(x, y) &= q(\tilde{y}) + \tilde{h}(x, \tilde{y}_{k+1}, \dots, \tilde{y}_{m-n+1}), \\ q(\tilde{y}) &= \textstyle\sum_{i=1}^{k} e_i \tilde{y}_i^2, \quad \tilde{y} = (\tilde{y}_1, \dots, \tilde{y}_k), \quad e_i = \pm 1 \end{aligned}$$

holds. We rewrite the coordinate as $(y_1, \dots, y_k) = (\tilde{y}_1, \dots, \tilde{y}_k)$ and $z = (z_1, \dots, z_{k'})$ $= (\tilde{y}_{k+1}, \dots, \tilde{y}_{m-n+1})$, where $k' = m - n + 1 - k$. Then $f(x, y, z) = (x, f_n(x, y, z))$ has the form

$$f_n(x, y, z) = q(y) + \tilde{\tilde{h}}(x, z), \quad q(y) = \sum_{i=1}^{k} e_i y_i^2, \quad \operatorname{Hess} \tilde{\tilde{h}}(0, z)(0) = 0. \tag{2.9}$$

We rewrite $\tilde{\tilde{h}}(x, z) = h(x, z)$. Furthermore, by Lemma 2.5, one can take an adapted collection of vector fields

$$\xi_i = \partial x_i \ (i = 1, \dots, n-1), \ \eta_j = \partial y_j \ (j = 1, \dots, k), \ \eta_{k+j} = \partial z_j \ (j = 1, \dots, l).$$

Then we see that $\langle \eta_{k+1}, \dots, \eta_{k+l} \rangle_{\boldsymbol{R}} = \ker \mathcal{H}_\eta$ on $S(f)$. Set

$$\lambda_j = \det(\xi_1 f, \dots, \xi_{n-1} f, \eta_j f), \quad j = 1, \dots, m - n + 1, \quad \Lambda = (\lambda_1, \dots, \lambda_{m-n+1}).$$

It follows that

$$\Lambda = (2e_1y_1, \dots, 2e_ky_k, h_{z_1}(x,z), \dots, h_{z_l}(x,z))$$

and $S(f) = \{\Lambda = 0\}$. By non-degeneracy, we have $d\lambda_0 \neq 0$. The matrix which represents $d\lambda_0$ is given by

$$A = \left(\begin{array}{c|c|c} \Big((f_n)_{y_ix_j}\Big)_{\substack{i=1,\dots,k,\\ j=1,\dots,n-1}} & \Big((f_n)_{y_iy_j}\Big)_{i,j=1,\dots,k} & \Big((f_n)_{y_iz_j}\Big)_{\substack{i=1,\dots,k,\\ j=1,\dots,l}} \\ \hline \Big((f_n)_{z_ix_j}\Big)_{\substack{i=1,\dots,l,\\ j=1,\dots,n-1}} & \Big((f_n)_{z_iy_j}\Big)_{\substack{i=1,\dots,l,\\ j=1,\dots,k}} & \Big((f_n)_{z_iz_j}\Big)_{i,j=1,\dots,l} \end{array}\right)(0)$$

$$= \left(\begin{array}{c|c|c|c} * & * & \operatorname{Hess} q(y) & O \\ \hline \Big(h_{z_ix_j}\Big)_{\substack{i=1,\dots,l,\\ j=1,\dots,l}} & \Big(h_{z_ix_j}\Big)_{\substack{i=1,\dots,l,\\ j=l+1,\dots,n-1}} & O & \Big(h_{z_iz_j}\Big)_{i,j=1,\dots,l} \end{array}\right)(0)$$

$$=: \left(\begin{array}{c|c|c|c} * & * & \operatorname{Hess} q(y) & O \\ \hline M_1 & M_2 & O & M_3 \end{array}\right)(0),$$

where $O$ stands for a zero matrix. Since $M_3(0) = O$, we may assume $M_1$ is regular by a coordinate change if necessary. By the implicit function theorem, there exist functions

$$x_i(x_{l+1}, \dots, x_{n-1}, z) \quad (i = 1, \dots, l), \quad z = (z_1, \dots, z_l),$$

such that

$$h_{z_j}\Big(x_1(x_{\overrightarrow{l+1}}, z), \dots, x_l(x_{\overrightarrow{l+1}}, z), x_{\overrightarrow{l+1}}, z\Big) = 0 \quad (j = 1, \dots, l), \qquad x_{\overrightarrow{l+1}} = (x_{l+1}, \dots, x_{n-1}). \tag{2.10}$$

Then $g := f|_{S(f)}$ is expressed by

$$g(x_{l+1}, z) = f\Big(x_1(x_{\overrightarrow{l+1}}), \dots, x_l(x_{\overrightarrow{l+1}}), x_{\overrightarrow{l+1}}, 0, z\Big).$$

Hence the transportation matrix which represent $d(f|_{S(f)})$ is given by

$$B = \left(\begin{array}{ccc|c|c} (x_1)_{x_{l+1}} & \cdots & (x_l)_{x_{l+1}} & & \\ \vdots & \vdots & \vdots & I & * \\ (x_1)_{x_{n-1}} & \cdots & (x_l)_{x_{n-1}} & & \\ \hline (x_1)_{z_1} & \cdots & (x_l)_{z_1} & & \sum_{i=1}^{l} h_{x_i}(x_i)_{z_1} + h_{z_1} \\ \vdots & \vdots & \vdots & O & \vdots \\ (x_1)_{z_l} & \cdots & (x_l)_{z_l} & & \sum_{i=1}^{l} h_{x_i}(x_i)_{z_l} + h_{z_l} \end{array}\right) =: \left(\begin{array}{c|c|c} * & I & * \\ \hline N_1 & O & v \end{array}\right),$$

where $I$ stands for the identity matrix. Since $\partial z_1, \ldots, \partial z_l$ are contained in $\ker df$ on $S(f)$, the derivatives $h_{z_1}, \ldots, h_{z_l}$ vanishes on $S(f)$, and we have

$$v = N_1 \begin{pmatrix} h_{x_1} \\ \vdots \\ h_{x_l} \end{pmatrix}.$$

Hence, by elementary row operations $B$ changes to

$$\left(\begin{array}{c|c|c} * & I & * \\ \hline N_1 & O & O \end{array}\right). \tag{2.11}$$

Thus $(x, 0, z) \in S(f|_{S(f)})$ is equivalent to the determinant of $N_1(x, 0, z)$ being zero. Differentiating (2.10), we have

$$N_1{}^tM_1 = -\begin{pmatrix} h_{z_1z_1} & \cdots & h_{z_1z_l} \\ \vdots & \vdots & \vdots \\ h_{z_1z_l} & \cdots & h_{z_lz_l} \end{pmatrix}.$$

Since $M_1$ is regular, $(x, 0, z) \in S(f|_{S(f)})$ is equivalent to $\det \operatorname{Hess} h(0, z) = 0$. On the other hand, $\eta_j\lambda_i = h_{z_iz_j}$ holds on $S(f)$, and we have $\operatorname{Hess} h(0, z) = \mathcal{H}_\eta$. Since $\ker dg = \langle \partial z_1, \ldots, \partial z_l \rangle_{\boldsymbol{R}}$ by (2.11), one can easily see that the last assertion holds true. □

Set

$$H = \det \mathcal{H}_\eta.$$

DEFINITION 2.7. A non-degenerate singular point 0 is called *2-singular* if $H(0) = 0$.

This is equivalent to $\ker df_0 \cap T_0S(f) \neq \emptyset$. Set $S_2(f) = \{H = 0\}$. The 2-singularity of a non-degenerate singular point does not depend on the choice of $\eta$. By Lemma 2.6, it follows that $S_2(f) = S(g)$.

DEFINITION 2.8. A 2-singular point 0 is *2-non-degenerate* if $d(H|_{S(f)})_0 \neq 0$.

The condition is equivalent to $\ker dH_0 \not\supset T_0S(f)$. By the definition, we see that the 2-non-degeneracy condition does not depend on the choice of $\eta$, and if $p$ is 2-non-degenerate, then $S_2(f)$ is a manifold near $p$. Moreover, $\operatorname{rank} \mathcal{H}_\eta(0) = m-n$. In fact, if we assume that $\operatorname{rank} \mathcal{H}_\eta(0) < m - n$, then all the minor $m - n - 1$ determinants of $\mathcal{H}_\eta(0)$ vanish. Since $dH_0$ is expressed by these minor determinants, we have $dH_0 = 0$.

Let $p$ be a 2-singular point. Since $H(p) = 0$, dimension of $\ker \mathcal{H}_\eta(p)$ is positive. Let $\theta_p$ be a non-zero element of $\ker \mathcal{H}_\eta(p)$.

LEMMA 2.9. *If* $\operatorname{rank} \mathcal{H}_\eta = m - n$, *then there exists a vector field* $\theta$ *on* $(\boldsymbol{R}^m, 0)$ *such that* $\theta_p$ *generates* $\ker \mathcal{H}_\eta(p)$ *when* $p \in S_2(f)(= \{H = 0\})$. *Namely,* $\langle \theta_p \rangle_{\boldsymbol{R}} = \ker \mathcal{H}_\eta(p)$.

PROOF. The matrix $\mathcal{H}_\eta$ is symmetric on $S(f)$, and has only one zero-eigenvalue at 0. Thus the eigenvalue $\kappa$, that has minimum absolute value, is well-defined on a neighborhood $U$ of 0, and it takes a real value on $U$. We denote that by $\theta$ the non-zero eigenvector with respect to $\kappa$. Then $\theta$ is an eigenvector of the zero eigenvalue on $S_2(f)$ and so, one can extend $\theta$ on $(\boldsymbol{R}^m, 0)$, and get the desired vector field. □

We state a condition that $\theta$ is in the kernel of $\mathcal{H}_\eta$.

LEMMA 2.10. *For $p \in S(f)$, the condition $\theta_p \in \ker \mathcal{H}_\eta(p)$ is equivalent to $\theta\lambda_1 = \cdots = \theta\lambda_{m-n+1} = 0$ at $p$.*

PROOF. Let $\eta_1, \ldots, \eta_{m-n+1}$ be vector fields generating $\ker df$, and set $\theta = \sum_{i=1}^{m-n+1} \theta_i \eta_i$. Then by (2.8) and symmetry of $\mathcal{H}_\eta$, we see that

$$\mathcal{H}_\eta(\theta) = \sum_{i=1}^{m-n+1} \left( \sum_{j=1}^{m-n+1} \theta_j \eta_i \lambda_j \right) \eta_i = \sum_{i=1}^{m-n+1} \left( \sum_{j=1}^{m-n+1} \theta_j \eta_j \lambda_i \right) \eta_i = \sum_{i=1}^{m-n+1} (\theta\lambda_i)\eta_i.$$

Thus the assertion holds. □

If $p$ is a 2-non-degenerate singular point, then $S_2(f)$ is a manifold near $p$. Thus the condition that $\theta$ is tangent to $S_2(f)$ at a point on $S_2(f)$ is well-defined. Hence we introduce the definition below. In what follows, we denote by $'$ the directional derivative along the direction $\theta$. Namely, $H' = \theta H$. Furthermore, $H^{(i)} = (H^{(i-1)})'$ $(i = 2, 3, \ldots)$ and $H^{(1)} = H'$, $H^{(0)} = H$.

DEFINITION 2.11. A 2-non-degenerate singular point 0 is called 3-singular if $\theta(0) \in T_0 S_2(f)$.

Since the 3-singularity is determined by $\theta$ at $p$, it does not depend on the extension of $\theta$, and $S_2(f)$ does not depend on the extension of $\eta$, so the 3-singularity does not depend on the extension of $\eta$. We remark that the 3-singularity is equivalent to $H'(0) = 0$. Let us set $S_3(f) = \{q \mid \theta_q \in T_q S_2(f)\}$. Then $S_3(f)$ is determined by $\theta$ on $S_2(f)$. Thus $S_3(f)$ does not depend on the extension of $\eta, \theta$. Furthermore, we see that

$$S_3(f) = \{p \in S_2(f) \mid H'(p) = 0\} = \{p \in S(f) \mid H(p) = H'(p) = 0\}.$$

Using this terminology, 3-singularity is equivalent to $0 \in S_3(f)$. Moreover, we have:

LEMMA 2.12. *It holds that $S_3(f) = S(f|_{S_2(f)})$.*

PROOF. If $p \in S_2(f)$, it holds that $\ker d(f|_{S_2(f)})_p = \langle \theta_p \rangle_{\boldsymbol{R}}$. Thus we obtain the result. □

DEFINITION 2.13. A 3-singular point 0 is 3-*non-degenerate* if $d(H'|_{S_2(f)})_0 \neq 0$ holds.

LEMMA 2.14. *The 3-non-degeneracy condition on a 3-singular point does not depend on the extension of $\eta$, on the extension of $\theta$, nor on the coordinate system on the target.*

PROOF. Let $\tilde{\theta}$ be another extension of $\theta$. Then $\tilde{\theta} H|_{S_2(f)} = \theta H|_{S_2(f)}$ holds on $S_2(f)$, since the 3-non-degeneracy depends only on the first differential by $\theta$. Thus the 3-non-degeneracy does not depend on the extension of $\theta$. On the other hand, let $\tilde{\eta}$ be another extension of $\eta$, and set $\tilde{H} = \det \mathcal{H}_{\tilde{\eta}}$. Then we have $\tilde{H} = \alpha H + \beta$, where $\alpha|_{S(f)} \neq 0$ and $\beta|_{S(f)} = 0$. Thus it holds that $\tilde{H}' = \alpha' H + \alpha H' + \beta'$. We restrict this formula to $S_2(f)$. We see that $\beta' = 0$ on $S_2(f)$, because $H = 0$ and $p \in S_2(f)$ then $\theta_p \in T_p S(f)$ holds. Thus

$$\tilde{H}'|_{S_2(f)} = \alpha H'|_{S_2(f)}$$

holds. On the other hand, if 0 is 3-singular, then by $H'(0) = 0$, we see $d(\tilde{H}'|_{S_2(f)})_0 = \alpha d(H'|_{S_2(f)})_0$. Thus it does not depend on the extension of $\eta$. □

The 3-non-degeneracy is equivalent to $\ker d(H')_0 \not\supset T_0S_2(f)$. Thus if 0 is 3-non-degenerate, then $S_3(f)$ is a manifold.

LEMMA 2.15. *Let* 0 *be a non-degenerate singular point. Then* 0 *is* 3*-non-degenerate if and only if* $H = H' = 0$ *at* 0 *and* $\operatorname{rank} d(H,H')_0|_{T_0S(f)} = 2$.

PROOF. Since both conditions imply the 2-non-degeneracy, we assume 0 is 2-non-degenerate. Since 0 is non-degenerate, we take a coordinate systems on the source and target such that $f(x,y,z) = (x, f_n(x,y,z))$ has the form (2.9), and $(f_n)_{z_1x_1}(0) \neq 0$. Moreover $dH_0 \neq 0$, we see $l = 1$. Then we take an adapted collection of vector fields

$$\xi_1 = \partial x_1,\ \xi_i = -(f_n)_{z_1x_i}\partial x_1 + (f_n)_{z_1x_1}\partial x_i,\ (i = 2,\dots,n-1),$$
$$\eta_j = \partial y_j\ (j = 1,\dots,m-n),\ \eta_{m-n+1} = \partial z_1.$$

Then we see that

$$T_0S(f) = \langle \partial x_2,\dots,\partial x_{n-1},\partial z_1\rangle_{\boldsymbol{R}},\quad \partial z_1 \in T_0S_2(f).$$

We assume that 0 is 3-non-degenerate. Then $H = H' = 0$ at 0, and $dH'_0|_{T_0S_2(f)} \neq 0$ holds. Thus there exists a vector $\xi \in T_0S_2(f)$ such that $\xi H'(0) \neq 0$. Since $S_2(f) = \{H = 0\}$, it holds that $\xi H = 0$. By $dH_0|_{T_0S(f)} \neq 0$, it holds that $\operatorname{rank} d(H,H')_0|_{T_0S(f)} = 2$. On the other hand, $\xi H = 0$ holds for $\xi \in T_0S_2(f)$. Hence we see that $\operatorname{rank} d(H,H')_0|_{T_0S(f)} = 2$ implies $d(H')_0|_{S_2(f)} \neq 0$. □

We define $(i+1)$-singularity and $(i+1)$-non-degeneracy inductively. Let the notion of $j$-singularity, the set of $j$-singular points $S_j(f) = \{p \in (\boldsymbol{R}^m,0) \mid H(p) = \cdots = H^{(i-2)}(p) = 0\}$ as a manifold, and $j$-non-degeneracy already be defined for $f : (\boldsymbol{R}^m,0) \to (\boldsymbol{R}^n,0)$ $(j = 1,\dots,i)$. Moreover, we assume that these notions do not depend on the extensions of $\eta$ and $\theta$. Here 1-non-degenerate means non-degenerate, and 1-singular point means singular point.

DEFINITION 2.16. An $i$-non-degenerate singular point 0 is $(i+1)$*-singular* if $\theta \in T_0S_i(f)$.

We remark that, since the $(i+1)$-singularity is defined only by the condition of $\theta$ be on $S_i(f)$ and $S_i(f)$ itself, then it does not depend on the extension of $\eta$ and $\theta$. We set $S_{i+1}(f) = \{\theta_p \in T_pS_i(f)\}$. Then $S_{i+1}(f)$ also does not depend on the extension of $\eta$ and $\theta$, and we have

$$S_{i+1}(f) = \{p \in (\boldsymbol{R}^m,0) \mid H(p) = \cdots = H^{(i-1)}(p) = 0\}.$$

DEFINITION 2.17. An $(i+1)$-singular point 0 is $(i+1)$*-non-degenerate* if $d(H^{(i-1)}|_{S_i(f)})_0 \neq 0$ holds.

LEMMA 2.18. *The* $(i+1)$*-non-degeneracy does not depend on the extensions of* $\theta$ *and* $\eta$.

PROOF. We show this for the extension of $\theta$. Let $\tilde\theta$ be a vector field satisfying that $\tilde\theta|_{S_2(f)} = \delta\theta|_{S_2(f)}$ $(\delta \neq 0)$. It is enough to show that $\delta^{i-1}H^{(i-1)}|_{S_i(f)} = \tilde\theta^{i-1}H|_{S_i(f)}$. We show it by induction. We set $\tilde\theta = \delta\theta + \gamma$, where $\gamma$ is a vector field which satisfies $\gamma|_{S_2(f)} = 0$. When $i = 2$, we see the conclusion. We assume that $(H^{(i-2)} - \delta^{i-2}\tilde\theta^{i-2}H)|_{S_{i-1}(f)} = 0$. Then by

$$\tilde\theta^{i-1}H|_{S_{i-1}(f)} = \tilde\theta\,\tilde\theta^{i-2}H|_{S_{i-1}(f)} = (\delta\theta+\gamma)\tilde\theta^{i-2}H|_{S_{i-1}(f)} = \delta\theta\,\tilde\theta^{i-2}H|_{S_{i-1}(f)},$$

we see that

$$\begin{aligned}
&(\delta^{i-1}H^{(i-1)}-\tilde{\theta}^{i-1}H)|_{S_{i-1}(f)}\\
=\ &\delta\Big(\delta^{i-2}\theta H^{(i-2)}-\theta\tilde{\theta}^{i-2}H)\Big)\Big|_{S_{i-1}(f)}\\
=\ &\Big(\delta\theta\big(\delta^{i-2}H^{(i-2)}-\tilde{\theta}^{i-2}H\big)-(\theta\delta^{i-2})H^{(i-2)}\Big)\Big|_{S_{i-1}(f)}.
\end{aligned}$$

By the assumption of induction, $(\delta^{i-2}H^{(i-2)}-\tilde{\theta}^{i-2}H)|_{S_{i-1}(f)}=0$ holds. Since $\theta\in TS_{i-1}(f)$ and $\theta^{i-2}H=0$ hold on $S_i(f)$, we see that

$$\Big(\theta\big(\delta^{i-2}H^{(i-2)}-\tilde{\theta}^{i-2}H\big)\Big)\Big|_{S_{i-1}(f)}=(\theta\delta^{i-2})H^{(i-2)}\Big|_{S_{i-1}(f)}=0.$$

(2) We take another extension $\tilde{\eta}$ of $\eta$, and $\det\mathcal{H}_{\tilde{\eta}}=\tilde{H}$. Then by Lemma 2.5 we see that $\tilde{H}=\alpha H+\beta$ holds, where $\alpha|_{S(f)}\neq 0$ and $\beta|_{S(f)}=0$. Then by the same method as in the proof of Lemma 2.14, one can see $(\alpha H^{(i-1)}-\tilde{H}^{(i-1)})|_{S_i(f)}=0$, by using $S_i(f)=\{p\in S_{i-1}(f)\,|\,H^{(i-2)}(p)=0\}$, which proves the $(i+1)$-non-degeneracy does not depend on the extension of $\eta$. □

We remark that $(i+1)$-non-degeneracy is equivalent to $\ker d(H^{(i-1)})_0\not\supset T_0S_i$, and we can continue until $S_i(f)$ becomes a point, namely $i=n$. Since $T_0S_n=\{0\}$, the $(n+1)$-singularity always fails. On other words, $n$-non-degeneracy implies $(n+1)$-non-degeneracy and so on. In fact, by the definition, $n$-non-degeneracy implies $d(H^{(n-2)}|_{S_{n-1}(f)})_0\neq 0$. Since $S_{n-1}$ is one-dimensional, if $\theta_p\in T_pS_{n-1}(f)$, then $\langle\theta_p\rangle_{\boldsymbol{R}}=T_pS_{n-1}(f)$ holds, and $\theta(H^{(n-2)}|_{S_{n-1}(f)})(0)\neq 0$ follows.

LEMMA 2.19. *Let us assume that $i\leq n$, and $0$ is a non-degenerate singular point. Then the $i$-non-degeneracy is equivalent to $H(0)=H'(0)=\cdots=H^{(i-2)}(0)=0$ and $\operatorname{rank} d(H,H',\cdots,H^{(i-2)})_0|_{T_0S(f)}=i-1$.*

PROOF. By induction we assume that the conclusion is true for $1,\ldots,i-1$, and that $0$ is an $i$-non-degenerate singular point. Then we have $H(0)=H'(0)=\cdots=H^{(i-2)}(0)=0$ and $\operatorname{rank} d(H,H',\cdots,H^{(i-3)})_0=i-2$.

We take a coordinate system $(x_1,\ldots,x_{n-2},z)$ of $S$ satisfying

$$\operatorname{rank}\begin{pmatrix} H_{x_1} & \cdots & H_{x_{i-2}}\\ H'_{x_1} & \cdots & H'_{x_{i-2}}\\ \vdots & \vdots & \vdots\\ H^{(i-3)}_{x_1} & \cdots & H^{(i-3)}_{x_{i-2}}\end{pmatrix}(0)=i-2$$

and $\theta=\partial z$. The transposition of the matrix representation of $d(H,H',\cdots,H^{(i-2)})_0$ with respect to this coordinate system is

$$\left(\begin{array}{c|c} K_1 & L_1\\ \hline K_2 & L_2\end{array}\right):=\left(\begin{array}{ccc|c}
H_{x_1} & \cdots & H^{(i-3)}_{x_1} & H^{(i-2)}_{x_1}\\
\vdots & \vdots & \vdots & \vdots\\
H_{x_{i-2}} & \cdots & H^{(i-3)}_{x_{i-2}} & H^{(i-2)}_{x_{i-2}}\\ \hline
H_{x_{i-1}} & \cdots & H^{(i-3)}_{x_{i-1}} & H^{(i-2)}_{x_{i-1}}\\
\vdots & \vdots & \vdots & \vdots\\
H_{n-2} & \cdots & H^{(i-3)}_{x_{n-2}} & H^{(i-2)}_{x_{n-2}}\\
H' & \cdots & H^{(i-2)} & H^{(i-1)}
\end{array}\right).$$

By elementary matrix operations, the above matrix is deformed to

$$\left(\begin{array}{c|c} K_1 & O \\ \hline K_2 & L_2 - K_2K_1^{-1}L_1 \end{array}\right).$$

Now applying the implicit function theorem, we see that

$$X := \begin{pmatrix} (x_1)_{x_{i+1}} & \cdots & (x_i)_{x_{i+1}} \\ \vdots & \vdots & \vdots \\ (x_1)_{x_{n-2}} & \cdots & (x_i)_{x_{n-2}} \\ (x_1)' & \cdots & (x_i)' \end{pmatrix} = -K_2K_1^{-1}.$$

Since

$$\begin{pmatrix} (H^{(i-2)}|_{S_i})_{x_{i+1}} \\ \vdots \\ (H^{(i-2)}|_{S_i})_{x_{n-2}} \\ (H^{(i-2)}|_{S_i})' \end{pmatrix} = {}^t\Big(XL_1 + L_2\Big) = {}^t\Big(-K_2K_1^{-1}L_1 + L_2\Big),$$

we have the conclusion.

For the converse, if we assume that $H(0) = H'(0) = \cdots = H^{(i-2)}(0) = 0$ and $\operatorname{rank} d(H, H', \cdots, H^{(i-3)})_0 = i - 2$, we can see the conclusion just following the arguments above from the bottom up. $\square$

## 3. Criteria

THEOREM 3.1. *A map-germ $f : (\boldsymbol{R}^m, 0) \to (\boldsymbol{R}^n, 0)$ is $\mathcal{A}$-equivalent to the $k$-Morin singularity ($k = 1, \ldots, n$) if and only if $0$ is a $k$-non-degenerate singularity but not $(k+1)$-singular.*

PROOF. Since the $k$-non-degenerate conditions and $(k+1)$-singularity conditions do not depend on the coordinate systems on the source nor the target space, the sufficiency is obvious by just checking the normal form (1.1) of the Morin singularity. We now show the necessity. Let us assume that $0$ is a $k$-non-degenerate singularity but not $(k+1)$-singular. Since the assumption does not depend on the coordinate systems on the source and the target space, and since $\operatorname{rank} df_0 = n-1$, we take a coordinate system such that (2.2) holds. When $k = 1$, we see that $\operatorname{Hess} h(0, y)$ at $y = 0$ is regular. By the parametrized Morse lemma, we have the conclusion. We now assume $2 \le k \le n$. It is known that the assumption does not depend on the extension of $\theta$, and by Lemma 2.5 and (2.7), $H_f$ is multiplied by a non-zero function when changing $\eta$. Moreover, the condition also does not depend on the extension of $\eta$, hence we may take $\eta_i = \partial y_i$, and coordinate $z$ such that $\theta = \partial z$ at 0. Under this coordinate system, we rewrite $f$ as

$$f(x, y, z) = (x, h(x, y, z)), \quad x = (x_1, \ldots, x_{n-1}), \quad y = (y_1, \ldots, y_{m-n}).$$

Then Hess $h(0,y,0)$ is regular at $y=0$, by the parametrized Morse lemma, we may choose coordinates $y$ such that $f$ takes the form

$$f(x,y,z)=(x,q(y)+h(x,z)),\ q(y)=\sum_{i=1}^{m-n}\pm y_i^2,$$
$$x=(x_1,\dots,x_{n-1}),\ y=(y_1,\dots,y_{m-n}).$$

Then we see that $\theta=\partial z$. Under this coordinate system, $H(0)=\theta^2h(0)$ and $\theta H(0)=\theta^3h(0)$ hold. Moreover, the $i$-non-degeneracy implies that

$$(\theta h_{x_1},\dots,\theta h_{x_{n-1}})(0)\neq(0,\dots,0).$$

We set

$$\bar f(x,z)=f(x,0,z)=(x,g(x,z)):(\boldsymbol{R}^n,0)\to(\boldsymbol{R}^n,0).$$

Then $\bar f(x,z)$ at 0 is an $A_k$-Morin singularity in the sense of [**25**]. Because the Jacobian is $\lambda:=\det J_{\bar f}=g_z$, and $\eta:=\partial z$ generates the kernel of $d\bar f$ at the singular set. Thus by the assumption of $i$-non-degeneracy, we have $\lambda=\eta\lambda=\cdots=\eta^{k-1}\lambda=0$, $\eta^k\lambda\neq0$, $\operatorname{rank} d(\lambda,\eta\lambda,\dots,\eta^{k-1}\lambda)_0=k$. Hence, it follows by ([**25**, Theorem A1]), that $\bar f(x,z)$ is $\mathcal{A}$-equivalent to $\left(x,z^{k+1}+\sum_{i=1}^{k-1}x_iz^i\right)$. Since $\det\operatorname{Hess}q(0)\neq0$, we see the assertion. $\square$

The proof here is based on that of Morin [**17**]. By Lemma 2.19, we have the following:

THEOREM 3.2. *Let 0 be a non-degenerate singular point of $f:(\boldsymbol{R}^m,0)\to(\boldsymbol{R}^n,0)$. Then $f$ at 0 is a $k$-Morin singularity $(2\leq k\leq n)$ if and only if both conditions above holds true:*

(1) $H=H'=\cdots=H^{(k-2)}=0,\ H^{(k-1)}\neq0,$
(2) $\operatorname{rank} d\big(H,H',\dots,H^{(k-2)}\big)_0|_{T_0S(f)}=k-1.$

*Here, $\mathcal{H}_\eta$ is determined by (2.5) for an adapted collection of vector fields $(\xi,\eta)$ with respect to $f$, $H=\det\mathcal{H}_\eta$, and $\theta$ is a vector field such that it generates $\ker\mathcal{H}_\eta$ on $\{H=0\}$, and $'$ means the directional derivative along $\theta$.*

Moreover we have the following corollary.

COROLLARY 3.3. *Let 0 be a singular point of $f:(\boldsymbol{R}^m,0)\to(\boldsymbol{R}^n,0)$ satisfying $\operatorname{rank} df_0=n-1$. Then $f$ is a $k$-Morin singularity $(2\leq k\leq n)$ at 0 if and only if*

(a) $H=H'=\cdots=H^{(k-2)}=0,\ H^{(k-1)}\neq0,$
(b) $\operatorname{rank} d(\lambda_1,\dots,\lambda_{m-n+1},H,H',\dots,H^{(k-2)})_0=m-n+k,$

*where $H$ is the same as in Theorem 3.2.*

PROOF. Since the condition $(a)$ is the same, we show that $(b)$ is equivalent to non-degeneracy and the condition (2) of Theorem 3.2. The condition does not depend on the choice of an adapted collection of vector fields, so we choose

$\xi_1, \dots, \xi_{n-1}, \eta_1, \dots, \eta_{m-n}, \theta$ satisfying $\xi_2, \dots, \xi_{n-1} \in T_0S(f)$, since $\langle \eta_1, \dots, \eta_{m-n} \rangle_{\boldsymbol{R}} \cap T_0S(f) = \{0\}$. Then the transportation of the matrix representation of the differential $d(\lambda_1, \dots, \lambda_{m-n+1}, H, H', \dots, H^{(k-2)})_0$ has the form

$$
\left(\begin{array}{c|c|c|c}
\begin{matrix}\xi_1\lambda_1\\ \vdots \\ \xi_1\lambda_{m-n}\end{matrix} & \Big(\xi_j\lambda_i\Big)_{\substack{i=1,\dots,m-n,\\ j=2,\dots,n-1}} & \Big(\eta_j\lambda_i\Big)_{\substack{i=1,\dots,m-n,\\ j=1,\dots,m-n}} & \begin{matrix}\lambda_1'\\ \vdots \\ \lambda_{m-n}'\end{matrix} \\
\hline
\xi_1\lambda_{m-n+1} & \Big(\xi_j\lambda_{m-n+1}\Big)_{j=2,\dots,n-1} & \Big(\eta_j\lambda_{m-n+1}\Big)_{j=1,\dots,m-n} & \lambda_{m-n+1}' \\
\hline
\begin{matrix}\xi_1 H\\ \vdots \\ \xi_1 H^{(k-3)}\end{matrix} & \Big(\xi_j H^{(i)}\Big)_{\substack{i=0,\dots,k-3,\\ j=2,\dots,n-1}} & \Big(\eta_j H^{(i)}\Big)_{\substack{i=0,\dots,k-3,\\ j=1,\dots,m-n}} & \begin{matrix}H'\\ \vdots \\ H^{(k-2)}\end{matrix} \\
\hline
\xi_1 H^{(k-2)} & \Big(\xi_j H^{(k-2)}\Big)_{j=2,\dots,n-1} & \Big(\eta_j H^{(k-2)}\Big)_{j=1,\dots,m-n} & H^{(k-1)}
\end{array}\right)
$$

$$
= \left(\begin{array}{c|c|c|c}
\begin{matrix}*\\ \vdots \\ *\end{matrix} & O & \mathcal{H}_\eta & \begin{matrix}0\\ \vdots \\ 0\end{matrix} \\
\hline
\xi_1\lambda_{m-n+1} & 0 \quad \cdots \quad 0 & 0 \quad \cdots \quad 0 & 0 \\
\hline
\begin{matrix}*\\ \vdots \\ *\end{matrix} & \Big(\xi_j H^{(i)}\Big)_{\substack{i=0,\dots,k-3,\\ j=2,\dots,n-1}} & * & \begin{matrix}0\\ \vdots \\ 0\end{matrix} \\
\hline
* & * & * & H^{(k-1)}
\end{array}\right)
$$

at the origin. Thus we see that the condition $(b)$ is equivalent to $\xi_1\lambda_{m-n+1}(0) \neq 0$ and

$$
\operatorname{rank} \begin{pmatrix} \xi_2 H & \cdots & \xi_{n-1} H \\ \vdots & \vdots & \vdots \\ \xi_2 H^{(k-3)} & \cdots & \xi_{n-1} H^{(k-3)} \end{pmatrix} (0) = k-2.
$$

This is nothing but the non-degeneracy and the condition (2) of Theorem 3.2. $\square$

## 4. Criteria for small $k$

In this section, we remark that for small $k$, the criteria can be simplified. In what follows, for real numbers $a, b \in \boldsymbol{R}$, the notation $a \sim b$ implies $a = 0$ is equivalent to $b = 0$, and for functions $f, g$, the notation $f \sim g$ implies that $g$ is multiplication by a non-zero function $f$.

### 4.1. Criterion of the fold.

COROLLARY 4.1. *Let $f = (f_1, \dots, f_n) : (\boldsymbol{R}^m, 0) \to (\boldsymbol{R}^n, 0)$ be a map-germ satisfying $(df_n)_0 = 0$ and $\operatorname{rank} df_0 = n-1$. Then $f$ is a fold singularity at $0$ if and only if $\operatorname{rank} \operatorname{Hess}_{\hat\eta} f_n = m-n+1$, where $\hat\eta_1, \dots, \hat\eta_{m-n+1}$ are vector fields satisfying that $\langle \hat\eta_1, \dots, \hat\eta_{m-n+1} \rangle_{\boldsymbol{R}} = \ker df$ at $0$. Here, the number of minus signs in $q$ is equal to the number of negative eigenvalues of $\operatorname{Hess}_{\hat\eta} f_n$.*

PROOF. Let $\xi_1, \dots, \xi_{n-1}, \eta_1, \dots, \eta_{m-n+1}$ be an adapted collection of vector fields with respect to $f$. Then $f$ is a fold singularity at 0 if and only if non-degeneracy holds with $H(0) \neq 0$. Since $H(0) \neq 0$ implies that $\operatorname{rank} \mathcal{H}_\eta = m-n+1$, and $d\Lambda_0$ contains $\mathcal{H}_\eta$, the non-degeneracy follows from $H(0) \neq 0$. Thus $f$ is the fold if and only if $H(0) \neq 0$. On the other hand, by $(df_n)_0 = 0$, we see $\eta_j \lambda_i(0) = \delta \eta_j \eta_i f_n(0)$, where

$$\delta = \det \begin{pmatrix} \xi_1 f_1 & \cdots & \xi_{n-1} f_1 \\ \vdots & \vdots & \vdots \\ \xi_1 f_{n-1} & \cdots & \xi_{n-1} f_{n-1} \end{pmatrix}.$$

Thus $H = \det \operatorname{Hess}_\eta f_n(0)$. Furthermore, since $\eta_1, \dots, \eta_{m-n+1}$ satisfies $\eta_i f_n = 0$, we see that $\eta_i(0) = \hat{\eta}_i(0)$ implies $\det \operatorname{Hess}_{\hat{\eta}} f_n(0) = \det \operatorname{Hess}_\eta f_n(0)$. □

**4.2. Criterion of the cusp.** For a function-germ $t : (\boldsymbol{R}^n, 0) \to (\boldsymbol{R}, 0)$ which has a critical point at 0 and a subspace $V \subset T_0\boldsymbol{R}^n$, we consider the Hessian matrix $(v_j v_i t)_{(1 \le i,j \le k)}$ with respect to a basis $v_1, \dots, v_k$ of $V$, which is defined by $(\tilde{v}_j \tilde{v}_i t)_{(1 \le i,j \le k)}(0)$, where $\tilde{v}_i$ is an extension of $v_i$. We remark that since $t$ has a critical point at 0, it does not depend on the choice of extensions. Moreover, the $\ker(v_j v_i t)_{(1 \le i,j \le k)}$ depends only on $V$. We denote it by $\ker \operatorname{Hess}_V h(0)$.

COROLLARY 4.2. *Let $f = (f_1, \dots, f_n) : (\boldsymbol{R}^m, 0) \to (\boldsymbol{R}^n, 0)$ be a map-germ satisfying $(df_n)_0 = 0$ and $\operatorname{rank} df_0 = n-1$. Then $f$ is a cusp singularity at 0, if and only if for a vector field $\hat{\theta}$ satisfying*

$$\ker \operatorname{Hess}_{\ker df_0} f_n(0) = \langle \hat{\theta}_0 \rangle_{\boldsymbol{R}},$$

*and contained in the* $\ker df$ *on $S(f)$, it holds that*

(1) $\hat{\theta}^3 f_n(0) \neq 0$,
(2) $d(\hat{\theta} f_n)_0 \neq 0$.

*Here,* $\operatorname{rank} \operatorname{Hess}_{\ker df_0} f_n(0) = m-n$ *and the number of negative eigenvalues is equal to the number of minus signs in $q$.*

PROOF. The necessity is obvious, we show the sufficiency. By Theorem 3.2, we show non-degeneracy, 2-non-degeneracy and non-3-singularity. Namely, we show that the conditions (1) and (2) imply non-degeneracy, and $H(0) = 0$, $H'(0) \neq 0$ and $\operatorname{rank} dH_0|_{T_0S(f)} = 1$. If $\theta$ is a generator of the kernel of $\mathcal{H}_f$, then since 0 is 2-singular i.e., $\theta_0 \in T_0S(f)$, $\operatorname{rank} dH_0|_{T_0S(f)} = 1$ follows by $H'(0) \neq 0$. Thus it is enough to show non-degeneracy, $H(0) = 0$ and $H'(0) \neq 0$.

Before showing that, we give some calculations. Since the conditions do not depend on the choice of $\eta$, we take an adapted collection $(\xi_1, \dots, \xi_{n-1}, \eta_1, \dots, \eta_{m-n}, \theta)$ of vector fields with respect to $f$. Since $\theta$ belongs to the kernel of $\mathcal{H}_\eta$ on $S_2(f)$, and $0 \in S_2(f)$, so it holds that

$$\theta \lambda_i = 0 \quad (i = 1, \dots, m-n+1) \quad \text{at} \quad 0.$$

On the other hand, $\eta_i f = 0$ $(i = 1, \dots, m-n)$ and $\theta f = 0$ hold on $S(f)$, so it holds that

$$\theta \eta_i f = 0 \quad (i = 1, \dots, m-n) \quad \text{and} \quad \theta^2 f = 0 \quad \text{at} \quad 0.$$

Moreover, by $(df_n)_0 = 0$ and $[\theta, \eta_i] \in T\boldsymbol{R}^m$, it holds that $\eta_i \theta f_n = 0$ at 0. Thus the bottom column of

$$\eta_i \lambda_{m-n+1}(0) = \det(\xi_1 f, \dots, \xi_{n-1} f, \eta_i \theta f)(0)$$

is 0, and we see that $\eta_i\lambda_{m-n+1}(0) = 0$ $(i = 1, \dots, m-n)$. We remark that the kernel of $\mathrm{Hess}_\eta f_n(0)$ is $\theta_0$.

We translate the conditions $H = 0$ and $H' \neq 0$ using $f_n$. By the above calculation, it follows that

$$\begin{aligned}
\theta H(0) &= \theta \det \left( \begin{array}{ccc|c} \eta_1\lambda_1 & \cdots & \eta_{m-n}\lambda_1 & \theta\lambda_1 \\ \vdots & \vdots & \vdots & \vdots \\ \eta_1\lambda_{m-n} & \cdots & \eta_{m-n}\lambda_{m-n} & \theta\lambda_{m-n} \\ \hline \eta_1\lambda_{m-n+1} & \cdots & \eta_{m-n}\lambda_{m-n+1} & \theta\lambda_{m-n+1} \end{array} \right)(0) \\
&= \det \left( \begin{array}{ccc|c} \eta_1\lambda_1 & \cdots & \eta_{m-n}\lambda_1 & \theta^2\lambda_1 \\ \vdots & \vdots & \vdots & \vdots \\ \eta_1\lambda_{m-n} & \cdots & \eta_{m-n}\lambda_{m-n} & \theta^2\lambda_{m-n} \\ \hline \eta_1\lambda_{m-n+1} & \cdots & \eta_{m-n}\lambda_{m-n+1} & \theta^2\lambda_{m-n+1} \end{array} \right)(0) \\
&\sim \theta^2\lambda_{m-n+1}(0) \\
&= \theta^2 \det(\xi_1 f, \dots, \xi_{n-1} f, \theta f)(0) \\
&= \det(\xi_1 f, \dots, \xi_{n-1} f, \theta^3 f)(0) \\
&\sim \theta^3 f_n(0).
\end{aligned}$$

By the same calculation, $H(0) = \theta^2 f_n(0)$ also holds. Thus it is necessary to show that $\theta^2 f_n(0) \sim \hat\theta^2 f_n(0)$ and $\theta^3 f_n(0) \sim \hat\theta^3 f_n(0)$. We set

$$\hat\theta = \sum_{i=1}^{m-n} \alpha_i\eta_i + \beta\theta, \quad \alpha_1(0) = \cdots = \alpha_{m-n}(0) = 0, \ \beta(0) \neq 0.$$

Then we have

$$\begin{aligned}
\hat\theta f_n &= \sum_{i=1}^{m-n} \alpha_i\eta_i f_n + \beta\theta f_n, \\
\hat\theta^2 f_n &= \sum_{i,j=1}^{m-n} \underline{\alpha_j}(\eta_j\alpha_i\underline{\eta_i f_n} + \underline{\alpha_i}\eta_j\eta_i f_n) + \sum_{j=1}^{m-n} \underline{\alpha_j}(\eta_j\beta\underline{\theta f_n} + \beta\underline{\eta_j\theta f_n}) \\
&\quad + \beta\Big( \sum_{i=1}^{m-n} (\theta\alpha_i\underline{\underline{\eta_i f_n}} + \underline{\alpha_i}\theta\eta_i f_n) + \theta\beta\underline{\underline{\theta f}} + \beta\underline{\theta^2 f_n} \Big).
\end{aligned}$$

Here, we underline the terms that vanish at the origin, and we put double underlines under the terms that vanish at the origin and whose differentiation along $\theta$ vanishes at the origin. Thus we have

$$\hat\theta^3 f_n(0) = \theta\hat\theta^2 f_n(0) \sim \theta^3 f_n(0).$$

By the same reason, we have $\hat\theta^2 f_n(0) \sim \theta^2 f_n(0)$.

Now we show the non-degeneracy condition. We have

$$\begin{aligned}
d\Lambda_0 &= \left( \begin{array}{c|c|c} \Big(\xi_j\lambda_i\Big)_{\substack{i=1,\dots,m-n,\\ j=1,\dots,n-1}} & \Big(\eta_j\lambda_i\Big)_{i,j=1,\dots,m-n} & \begin{matrix} \lambda'_1 \\ \vdots \\ \lambda'_{m-n} \end{matrix} \\ \hline \Big(\xi_j\lambda_{m-n+1}\Big)_{j=1,\dots,n-1} & \Big(\eta_j\lambda_{m-n+1}\Big)_{j=1,\dots,m-n} & \lambda'_{m-n+1} \end{array} \right)(0) \\
&= \left( \begin{array}{ccc|ccc|c} * & \cdots & * & & A & & 0 \\ \hline \xi_1\lambda_{m-n+1} & \cdots & \xi_{n-1}\lambda_{m-n+1} & 0 & \cdots & 0 & 0 \end{array} \right)(0).
\end{aligned}$$

By 2-non-degeneracy, $A$ is regular, and the non-degeneracy is equivalent to

$$(\xi_1\lambda_{m-n+1},\dots,\xi_{n-1}\lambda_{m-n+1})(0)\neq(0,\dots,0).$$

Moreover, by

$$\xi_i\lambda_{m-n+1}(0)=\det(\xi_1 f,\dots,\xi_{n-1}f,\xi_i\theta f)(0)\sim\xi_i\theta f_n(0),$$

$\eta_i\theta f_n(0)=0$ $(i=1,\dots,m-n)$ and $\theta^2 f_n(0)=0$, the condition (2) is equivalent to the non-degeneracy. □

It should be remarked that by Corollary 4.2, one can easily see that $f:(\boldsymbol{R}^m,0)\to(\boldsymbol{R}^2,0)$ is a cusp singularity at 0 if and only if 0 is non-degenerate and $f|_{S(f)}$ is $\mathcal{A}$-equivalent to $t\mapsto(t^2,t^3)$. This criteria was also obtained in [**18**].

## 5. First degree bifurcation of Lefschetz singularity

The Lefschetz singularity is a map-germ $(\boldsymbol{R}^4,0)\to(\boldsymbol{R}^2,0)$ defined by

$$L(x_1,x_2,y_1,y_2)=(x_1x_2-y_1y_2,x_1y_2+x_2y_1).$$

This is obtained by considering a map-germ $\boldsymbol{R}^4=\boldsymbol{C}^2\ni(z,w)\mapsto zw\in\boldsymbol{C}=\boldsymbol{R}^2$. From the view point of low-dimensional topology there are many studies of bundles on surfaces with this kind of singular points called the Lefschetz fibrations (See [**3, 8**], for example.). The Lefschetz singularity is not a stable germ, and it is natural to consider stable perturbations of it. The wrinkling

$$L_w(x_1,x_2,y_1,y_2,s)=(x_1x_2-y_1y_2+s(x_1+x_2),x_1y_2+x_2y_1)$$

due to Lekili [**15**] is such a move and has been well studied. The Lefschetz singularity is not finitely $\mathcal{A}$-determined, and one cannot obtain a kind of bifurcation diagram. Let us consider

$$\begin{aligned}&\tilde{L}_n(x_1,x_2,y_1,y_2,a_1,a_2,b_1,b_2,c_{2000},\dots,d_{2000},\dots)\\&=(x_1x_2-y_1y_2+a_1x_1+a_2x_2,\ x_1y_2+x_2y_1+b_1x_1+b_2x_2)\\&\quad+\left(\sum_{i+j+k+l=2}^{n}c_{ijkl}x_1^ix_2^jy_1^ky_2^l,\ \sum_{i+j+k+l=2}^{n}d_{ijkl}x_1^ix_2^jy_1^ky_2^l\right).\end{aligned}$$

Then it holds that

$$\left( tL(\oplus^4\mathcal{E}_4)+\sum_{k=1}^{4} \boldsymbol{R}\frac{\partial \tilde{L}_n}{\partial a_k}\bigg|_p +\boldsymbol{R}\frac{\partial \tilde{L}_n}{\partial c_{2000}}\bigg|_p +\cdots+\boldsymbol{R}\frac{\partial \tilde{L}_n}{\partial d_{2000}}\bigg|_p +\cdots \right)+\oplus^2\mathcal{M}_4^{n+1} = \oplus^2\mathcal{E}_4,$$

where $a_3 = b_1, a_4 = b_2$ and $p = (x_1, x_2, y_1, y_2, 0, \dots, 0)$. Here, $\mathcal{E}_4$ is the set of function-germs $(\boldsymbol{R}^4, 0) \to \boldsymbol{R}$ and $\mathcal{M}_4$ is its unique maximal ideal, and $tL : \oplus^4\mathcal{E}_4 \to \oplus^2\mathcal{E}_4$ stands for the tangential map:

$$tL(v_1, v_2, v_3, v_4) = \begin{pmatrix} a_1 + x_2 & a_2 + x_1 & -y_2 & -y_1 \\ b_1 + y_2 & b_2 + y_1 & x_2 & x_1 \end{pmatrix} \begin{pmatrix} v_1 \\ v_2 \\ v_3 \\ v_4 \end{pmatrix}.$$

Thus we would like to say that $\tilde{L}_n$ is a "versal-like" unfolding of $L$ up to $n$-degrees. See [**16**] for the definition of the versal unfolding.

In [**10**], deformations of Brieskorn polynomials which include the Lefschetz singularity is considered, and an evaluation of the number of cusp appearing on it is obtained. Here, we would consider the set

$$N = \Big\{ C = (a_1, a_2, b_1, b_2) \,\Big|\, \text{there exists } q = (x_1, x_2, y_1, y_2) \in S(\tilde{L}_1) \text{ such that } \tilde{L}_1 \text{ at } q \text{ is not the fold nor the cusp.} \Big\}.$$

We call $N$ the *non-cusp locus.* Although the bifurcation diagram for $L$ cannot be drawn in any finite dimensional space, we can draw $N|_{b_2=\varepsilon} \subset \boldsymbol{R}^3$ for small $\varepsilon$, and the author believes that we might regard $N$ as a 1 degree bifurcation diagram of $L$.

We set $\tilde{L}(x_1, x_2, y_1, y_2) = \tilde{L}_1(x_1, x_2, y_1, y_2, a_1, a_2, b_1, b_2)$, regarding $a_1, a_2, b_1, b_2$ as constants. To detect $N$, we consider the following three conditions for $q = (x_1, x_2, y_1, y_2)$:

(i) $\operatorname{rank} d\tilde{L}_q = 0$,
(ii) $\operatorname{rank} d\tilde{L}_q = 1$ and $\operatorname{rank} \mathcal{H}_\eta = 0$,
(iii) $\operatorname{rank} d\tilde{L}_q = 1$, $\operatorname{rank} \mathcal{H}_\eta(q) = 1$ and $H(q) = \theta H(q) = 0$.

See Theorem 3.2 for the notations.

Let $C = (a_1, a_2, b_1, b_2) \in N$ and $q = (x_1, x_2, y_1, y_2) \in S(\tilde{L})$ satisfies the condition (iii). We assume that $a_1 + x_2 \neq 0$, then $\eta_1 = (a_2 + x_1)\partial x_1 + (a_1 + x_2)\partial x_2, \eta_2 = y_2\partial x_1 + (a_1 + x_2)\partial y_1, \eta_3 = y_1\partial x_1 + (a_1 + x_2)\partial y_2$ form a basis of the kernel of $df$ at $p = (x_1, x_2, y_1, y_2) \in S(\tilde{L})$. Moreover, $\partial x_1$ together with $\eta_1, \eta_2, \eta_3$ forms a basis of $T_p\boldsymbol{R}^4$. Then we define $\lambda_i = \det(\tilde{L}_{x_1}, \eta_i\tilde{L})$, $\mathcal{H}_{(\eta_1,\eta_2,\eta_3)} = (\eta_i\lambda_j)_{i,j=1,2,3}$ and $H = \det \mathcal{H}_{(\eta_1,\eta_2,\eta_3)}$. Then we see

$$\begin{aligned}
\lambda_1 &= -x_1y_2 + x_2y_1 - b_1x_1 + b_2x_2 + a_1y_1 - a_2y_2 + a_1b_2 - a_2b_1,\\
\lambda_2 &= x_2^2 + y_2^2 + a_1x_2 + b_1y_2,\\
\lambda_3 &= x_1x_2 + y_1y_2 + a_1x_1 + b_1y_1.
\end{aligned}$$

If $(b_1 + 2y_2, (a_1 + 2x_2)y_1) \neq (0, 0)$, then we set

$$\theta = \det\begin{pmatrix}\eta_2\lambda_1 & \eta_2\lambda_2\\ \eta_3\lambda_1 & \eta_3\lambda_2\end{pmatrix}\eta_1 + \det\begin{pmatrix}\eta_3\lambda_1 & \eta_3\lambda_2\\ \eta_1\lambda_1 & \eta_1\lambda_2\end{pmatrix}\eta_2 + \det\begin{pmatrix}\eta_1\lambda_1 & \eta_1\lambda_2\\ \eta_2\lambda_1 & \eta_2\lambda_2\end{pmatrix}\eta_3.$$

Then $\theta$ forms a basis of $\ker \mathcal{H}$ on $S_2(\tilde{L})$. We assume $(x_1, y_1) \neq (0, 0)$. If $y_1 = 0$, then $x_2 = -a_1$ since $\lambda_3 = 0$. Thus we may assume $y_1 \neq 0$. Moreover, we may assume $b_1 + y_2 \neq 0$, because if $b_1 + y_2 = 0$ and $x_1 \neq 0$, then we see $a_1 + x_2 = 0$ by $\lambda_3 = 0$. If $b_1 + y_2 = 0$ and $x_1 = 0$, then we see $x_2 = 0$ by $\lambda_2 = 0$. Then we see $a_1 \neq 0$. Then we have $b_2 + y_1 = 0$ since $\lambda_1 = 0$. In this case, $H = 0$ and $\theta H = 0$ can be calculated as

$$y_1 = -(a_2/a_1)y_2, \quad a_1^2a_2y_2(a_1^2 + y_2^2) = 0.$$

Thus we have $y_1 = 0$ which is a contradiction. By the above discussion, if $(x_1, y_1) \neq (0, 0)$, then $\lambda_1 = \lambda_2 = \lambda_3 = 0$ can be modified to

$$(5.1)\quad x_2 = -\frac{x_1(a_1x_1 + b_1y_1)}{x_1^2 + y_1^2},\ y_2 = -\frac{y_1(a_1x_1 + b_1y_1)}{x_1^2 + y_1^2},\ a_2x_1 + y_1b_2 + x_1^2 + y_1^2 = 0.$$

Substituting (5.1) into $H = 0$, we have

$$(b_1x_1 - a_1y_1)\big(a_1y_1(-3x_1^2 + y_1^2 - 2a_2x_1) + b_1(x_1^3 - 3x_1y_1^2 + a_2x_1^2 - a_2y_1^2)\big) = 0.$$

If we assume $b_1x_1 - a_1y_1 = 0$, then we obtain $a_1 + x_2 = 0$ by (5.1). Thus we may assume $b_1x_1 - a_1y_1 \neq 0$. Moreover, if we assume $\big(-3x_1^2 + y_1^2 - 2a_2x_1, x_1^3 - 3x_1y_1^2 + a_2x_1^2 - a_2y_1^2\big) = (0, 0)$, then we obtain $x_1 = y_1 = 0$. Thus we may assume $\big(-3x_1^2 + y_1^2 - 2a_2x_1, x_1^3 - 3x_1y_1^2 + a_2x_1^2 - a_2y_1^2\big) \neq (0, 0)$. Then we see that $\theta H = 0$ on $H = 0$ is equivalent to $(a_2 + x_1)(a_2 + 2x_1) = 0$. Hence we have a part of $N$:

$$N_1 = \{(a_1, a_2, b_1, b_2) \mid a_1(a_2^2 - b_2^2) - 2a_2b_1b_2 = 0 \text{ or } a_2(a_1^2 + b_1^2) - 2b_2(a_1 + b_1) = 0\}.$$

If $x_1 = y_1 = 0$, then $\lambda_3 = 0$ holds, and $\lambda_1 = \lambda_2 = 0$ can be modified to

$$(5.2)\qquad a_2(b_1 + y_2) - b_2(a_1 + x_2) = 0, \quad x_2(a_1 + x_2) + y_2(b_1 + y_2) = 0.$$

If $(a_2, b_2) \neq (0,0)$, since $a_1 + x_2 \neq 0$, we obtain

$$x_2 = \frac{b_2(a_2b_1 - a_1b_2)}{a_2^2 + b_2^2}, \qquad y_2 = \frac{a_2(-a_2b_1 + a_1b_2)}{a_2^2 + b_2^2}.$$

Since $a_1 + x_2 \neq 0$ and $b_1 + 2y_2 \neq 0$, we obtain a part of $N$:

$$N_2 = \{(a_1, a_2, b_1, b_2) \mid a_1a_2 + b_1b_2 = 0\}.$$

Next we assume $(b_1 + 2y_2, (a_1 + 2x_2)y_1) = (0,0)$. If $(b_1 + 2y_2, a_1 + 2x_2) = (0,0)$, then we have $x_2 = y_2 = 0$ by $\lambda_2 = 0$. Thus we see $a_1 = b_1 = 0$. In this case, $C = (0,0,a_2,b_2) \in N_2$. We assume $(b_1 + 2y_2, y_1) = (0,0)$, and $a_1 + 2x_2 \neq 0$. Then we have $x_1 = 0$ by $\lambda_3 = 0$, and $a_1x_2 + x_2^2 - y_2^2 = 0$ by $\lambda_2 = 0$. Then $\theta = a_2\eta_2 + (a_1 + x_2)\eta_3$ is a generator of kernel of $\mathcal{H}_{(\eta_1,\eta_2,\eta_3)}$. Then $\theta H$ is a non-zero multiplication of $a_2(3a_1^2 + 7a_1x_2 + 4x_2^2 + 2y_2^2)$. Substituting $a_1x_2 + x_2^2 - y_2^2 = 0$ into this formula, we have $3a_2(a_1 + x_2)(a_1 + 2x_2) = 0$, which implies $a_2 = 0$. Then we have $b_2 = 0$ by $\lambda_1 = 0$. In this case, $C \in N_2$. On the other hand, we also see that if $C = (a_1, a_2, b_1, b_2) \in N$ and $q = (x_1, x_2, y_1, y_2) \in S(\tilde{L})$ satisfies the condition (ii), then $C \in N_1 \cup N_2$. Summarizing the above arguments, if $a_1 + x_2 \neq 0$, then we have a part of the non-cusp locus $N_1 \cup N_2$. By symmetry, we may interchange the subscript 1 with 2. Thus we obtain another part of $N$ in the case of $a_2 + x_1 \neq 0$:

$$N_3 = \{(a_1, a_2, b_1, b_2) \mid a_2(a_1^2 - b_1^2) - 2a_1b_1b_2 = 0 \text{ or } a_1(a_2^2 + b_2^2) - 2b_1(a_2 + b_2) = 0\}.$$

Next, we assume $a_1 + x_2 = a_2 + x_1 = 0$ and $(b_1 + y_2, b_2 + y_1) \neq (0,0)$. Then by the same method, we see $C \in N_1 \cup N_2 \cup N_3$. Also if $a_1 + x_2 = a_2 + x_1 = b_1 + y_2 = b_2 + y_1 = 0$, then we see $C \in N_1 \cup N_2 \cup N_3$. On the other hand, if $C \in N$ and $q \in S(\tilde{L})$ satisfies the condition (i), then we also see $C \in N_1 \cup N_2 \cup N_3$. Summarizing all these arguments, we have

$$\begin{aligned} N =& N_1 \cup N_2 \cup N_3 \\ =& \{C = (a_1, a_2, b_1, b_2) \mid a_1(a_2^2 - b_2^2) - 2a_2b_1b_2 = 0\} \\ & \cup \{C \mid a_2(a_1^2 + b_1^2) - 2b_2(a_1 + b_1) = 0\} \cup \{C \mid a_1a_2 + b_1b_2 = 0\} \\ & \cup \{C \mid a_2(a_1^2 - b_1^2) - 2a_1b_1b_2 = 0\} \cup \{C \mid a_1(a_2^2 + b_2^2) - 2b_1(a_2 + b_2) = 0\}. \end{aligned}$$

We draw the pictures of $N|_{b_2=-\varepsilon}$, $N|_{b_2=0}$ and $N|_{b_2=\varepsilon}$ in the $(a_1, a_2, b_1)$-space for small $\varepsilon$ in Figure 1. Here, the thick line in $N|_{b_2=0}$ stands for the wrinkling.

ACKNOWLEDGEMENTS. The author thanks María del Carmen Romero Fuster for fruitful advice and constant encouragement. He also thanks Kenta Hayano, Toru Ohmoto and Kazuto Takao for valuable comments and encouragements. The author is grateful to the referee for careful reading and valuable comments.

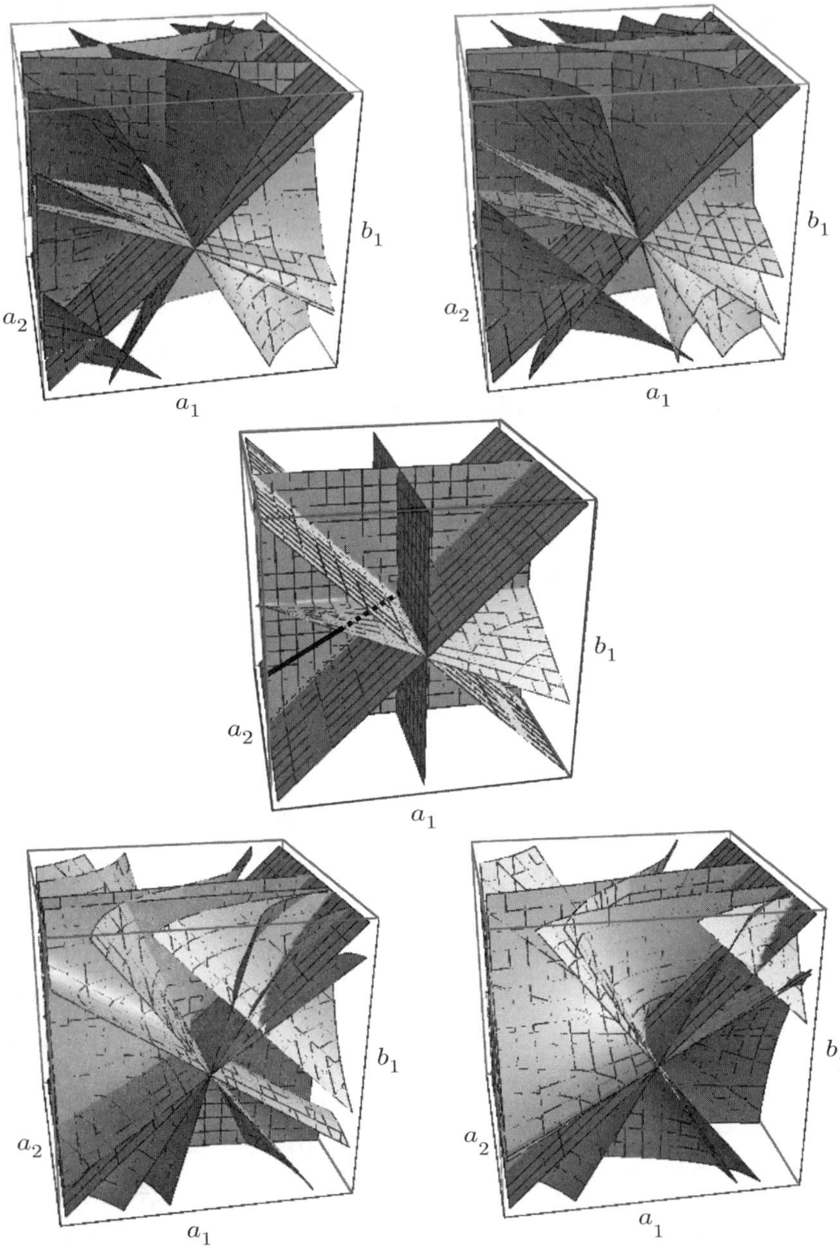


FIGURE 1. Non-cusp locus. From upper left to lower right, $N|_{b_2=-1/2}$, $N|_{b_2=-1/4}$, $N|_{b_2=0}$, $N|_{b_2=1/4}$, $N|_{b_2=1/2}$

## References


[1] Yoshifumi Ando, *On local structures of the singularities $A_k$ $D_k$ and $E_k$ of smooth maps*, Trans. Amer. Math. Soc. **331** (1992), no. 2, 639–651, DOI 10.2307/2154132. MR1055564

[2] J. W. Bruce and P. J. Giblin, *Curves and singularities*, 2nd ed., Cambridge University Press, Cambridge, 1992. A geometrical introduction to singularity theory. MR1206472

[3] Simon K. Donaldson, *Lefschetz pencils on symplectic manifolds*, J. Differential Geom. **53** (1999), no. 2, 205–236. MR1802722

[4] Andrew du Plessis, Terence Gaffney, and Leslie C. Wilson, *Map-germs determined by their discriminants*, Stratifications, singularities and differential equations, I (Marseille, 1990; Honolulu, HI, 1990), Travaux en Cours, vol. 54, Hermann, Paris, 1997, pp. 1–40. MR1473230

[5] Nicolas Dutertre and Toshizumi Fukui, *On the topology of stable maps*, J. Math. Soc. Japan **66** (2014), no. 1, 161–203, DOI 10.2969/jmsj/06610161. MR3161397

[6] Takuo Fukuda, *Topology of folds, cusps and Morin singularities*, A fête of topology, Academic Press, Boston, MA, 1988, pp. 331–353. MR928406

[7] M. Golubitsky and V. Guillemin, *Stable mappings and their singularities*, Springer-Verlag, New York-Heidelberg, 1973. Graduate Texts in Mathematics, Vol. 14. MR0341518

[8] Robert E. Gompf and András I. Stipsicz, *4-manifolds and Kirby calculus*, Graduate Studies in Mathematics, vol. 20, American Mathematical Society, Providence, RI, 1999. MR1707327

[9] Lars Hörmander, *The analysis of linear partial differential operators. III*, Grundlehren der Mathematischen Wissenschaften [Fundamental Principles of Mathematical Sciences], vol. 274, Springer-Verlag, Berlin, 1994. Pseudo-differential operators; Corrected reprint of the 1985 original. MR1313500

[10] Kazumasa Inaba, Masaharu Ishikawa, Masayuki Kawashima and Tat Thang Nguyen, *On linear deformations of Brieskorn singularities of two variables into generic maps*, arXiv:1412.0310.

[11] Go-O Ishikawa and Yoshinori Machida, *Singularities of improper affine spheres and surfaces of constant Gaussian curvature*, Internat. J. Math. **17** (2006), no. 3, 269–293, DOI 10.1142/S0129167X06003485. MR2215151

[12] Shyuichi Izumiya, Kentaro Saji, and Masatomo Takahashi, *Horospherical flat surfaces in hyperbolic 3-space*, J. Math. Soc. Japan **62** (2010), no. 3, 789–849. MR2648063

[13] Masatoshi Kokubu, Wayne Rossman, Masaaki Umehara, and Kotaro Yamada, *Flat fronts in hyperbolic 3-space and their caustics*, J. Math. Soc. Japan **59** (2007), no. 1, 265–299. MR2302672

[14] Masatoshi Kokubu, Wayne Rossman, Kentaro Saji, Masaaki Umehara, and Kotaro Yamada, *Singularities of flat fronts in hyperbolic space*, Pacific J. Math. **221** (2005), no. 2, 303–351, DOI 10.2140/pjm.2005.221.303. MR2196639

[15] Yanki Lekili, *Wrinkled fibrations on near-symplectic manifolds*, Geom. Topol. **13** (2009), no. 1, 277–318, DOI 10.2140/gt.2009.13.277. Appendix B by R. İnanç Baykur. MR2469519

[16] Jean Martinet, *Déploiements versels des applications différentiables et classification des applications stables* (French), Singularités d'applications différentiables (Sém., Plans-sur-Bex, 1975), Springer, Berlin, 1976, pp. 1–44. Lecture Notes in Math., Vol. 535. MR0649264

[17] Bernard Morin, *Formes canoniques des singularités d'une application différentiable* (French), C. R. Acad. Sci. Paris **260** (1965), 5662–5665. MR0180982

[18] Ana Claudia Nabarro, *Projections of hypersurfaces in* $\mathbf{R}^4$ *to planes*, Real and complex singularities, Lecture Notes in Pure and Appl. Math., vol. 232, Dekker, New York, 2003, pp. 283–299. MR2075073

[19] Takashi Nishimura, *Whitney umbrellas and swallowtails*, Pacific J. Math. **252** (2011), no. 2, 459–471, DOI 10.2140/pjm.2011.252.459. MR2860434

[20] I. R. Porteous, *Simple singularities of maps*, Proceedings of Liverpool Singularities Symposium, I (1969/70), Springer, Berlin, 1971, pp. 286–307. Lecture Notes in Math., Vol. 192. MR0293646

[21] Osamu Saeki, *Studying the topology of Morin singularities from a global viewpoint*, Math. Proc. Cambridge Philos. Soc. **117** (1995), no. 2, 223–235, DOI 10.1017/S0305004100073072. MR1307077

[22] Osamu Saeki and Kazuhiro Sakuma, *Maps with only Morin singularities and the Hopf invariant one problem*, Math. Proc. Cambridge Philos. Soc. **124** (1998), no. 3, 501–511, DOI 10.1017/S0305004197002478. MR1636580

[23] K. Saji, *Criteria for Morin singularities into higher dimensions*, RIMS Kôkyûroku Bessatsu B55 (2016) 205-224.

[24] Kentaro Saji, *Isotopy of Morin singularities*, Houston J. Math., 42 (2016), 499-519.

[25] Kentaro Saji, Masaaki Umehara, and Kotaro Yamada, $A_k$ *singularities of wave fronts*, Math. Proc. Cambridge Philos. Soc. **146** (2009), no. 3, 731–746, DOI 10.1017/S0305004108001977. MR2496355

[26] Kentaro Saji, Masaaki Umehara, and Kotaro Yamada, *Coherent tangent bundles and Gauss-Bonnet formulas for wave fronts*, J. Geom. Anal. **22** (2012), no. 2, 383–409, DOI 10.1007/s12220-010-9193-5. MR2891731

[27] K. Saji, M. Umehara and K. Yamada, *An index formula for hypersurfaces which admit only generic corank one singularities*, to appear in J. Math. Soc. Japan, arXiv:1202.3854.

[28] Takeshi Sasaki, Kotaro Yamada, and Masaaki Yoshida, *The hyperbolic Schwarz map for the hypergeometric differential equation*, Experiment. Math. **17** (2008), no. 3, 269–282. MR2455700

[29] Endre Szabó, András Szűcs, and Tamás Terpai, *On bordism and cobordism groups of Morin maps*, J. Singul. **1** (2010), 134–145. MR2783151

[30] András Szücs, *On the cobordism groups of cooriented, codimension one Morin maps*, J. Singul. **4** (2012), 196–205. MR3044496

Department of Mathematics, Graduate School of Science, Kobe University, Rokko, Nada, Kobe 657-8501, Japan

*E-mail address*: `saji@math.kobe-u.ac.jp`

Contemporary Mathematics
Volume **675**, 2016
http://dx.doi.org/10.1090/conm/675/13600

# Legendre curves in the unit spherical bundle over the unit sphere and evolutes

Masatomo Takahashi

*Dedicated to Professor Maria del Carmen Romero Fuster on the occasion of her 60th birthday*

ABSTRACT. In order to consider singular curves in the unit sphere, we consider Legendre curves in the unit spherical bundle over the unit sphere. By using a moving frame, we define the curvature of Legendre curves in the unit spherical bundle. As applications, we give a relationship among Legendre curves in the unit spherical bundle, Legendre curves in the unit tangent bundle and framed curves in the Euclidean space, respectively. Moreover, we define not only an evolute of a front, but also an evolute of a frontal in the unit sphere under certain conditions. Since the evolute of a front is also a front, we can take evolute again. On the other hand, the evolute of a frontal if exists, is also a frontal. We give an existence and uniqueness conditions of the evolute of a frontal.

## 1. Introduction

For regular curves in the unit sphere, the Frenet Serret formula and the geodesic curvature are important to investigate geometric properties of the regular curves. On the other hand, for singular curves in the unit sphere, we can not construct the Frenet Serret formula and the geodesic curvature at singular points of the curve. For singular curves, V. I. Arnold established the spherical geometry by using Legendre singularity theory [**2**]. It studied fronts in the unit sphere and gave properties of fronts. Some results in this paper have already considered in [**2, 16, 17, 20, 21**]. However, we clarify the notations and calculations by using the curvature of Legendre curves in the unit spherical bundle over the unit sphere. By using the curvature of the Legendre curves, we give existence and uniqueness theorems of Legendre curves in the unit spherical bundle in §2. We also give relationships among Legendre curves in the unit spherical bundle, Legendre curves in the unit tangent bundle and framed curves in the Euclidean space, respectively in §3. Moreover, we define not only an evolute of a front in §4, but also an evolute of a frontal in the unit sphere under certain conditions in §5. Since the evolute of a front is also a front, we can take evolute again. We give $k$-th evolute of the front and its curvature inductively. On the other hand, the evolute of a frontal if exists, is also a frontal. We give an existence and uniqueness conditions of the evolute of a

2010 *Mathematics Subject Classification.* Primary 58K05; Secondary 53A40, 53D35.
*Key words and phrases.* Spherical Legendre curve, frontal, front, curvature, evolute.
Supported by JSPS KAKENHI Grant Number No.26400078.

frontal. It is a quit different property between the evolute of a frontal in the sphere and in the Euclidean plane (cf. [**8**]). We also give examples of evolutes of a front and a frontal in §6.

All maps and manifolds considered here are differentiable of class $C^\infty$.

The author would like to thank the referee for helpful comments to improve the original manuscript.

## 2. Legendre curves in the unit spherical bundle

Let $\mathbb{R}^3$ be the 3-dimensional Euclidean space. The inner product on $\mathbb{R}^3$ is given by $\boldsymbol{a}\cdot\boldsymbol{b} = a_1b_1 + a_2b_2 + a_3b_3$ and the vector product of $\boldsymbol{a}$ and $\boldsymbol{b}$ on $\mathbb{R}^3$ is given by

$$\boldsymbol{a}\times\boldsymbol{b} = \begin{vmatrix} \boldsymbol{e}_1 & \boldsymbol{e}_2 & \boldsymbol{e}_3 \\ a_1 & a_2 & a_3 \\ b_1 & b_2 & b_3 \end{vmatrix},$$

where $\boldsymbol{e}_1, \boldsymbol{e}_2, \boldsymbol{e}_3$ is the canonical basis on $\mathbb{R}^3$, $\boldsymbol{a} = (a_1, a_2, a_3)$ and $\boldsymbol{b} = (b_1, b_2, b_3)$. We denote the unit sphere $S^2 = \{\boldsymbol{x}\in\mathbb{R}^3 | \boldsymbol{x}\cdot\boldsymbol{x} = 1\}$.

Let $\gamma : I \to S^2$ be a regular curve. We define the unit tangent vector $\boldsymbol{t}(t) = \dot\gamma(t)/|\dot\gamma(t)|$ and the unit normal vector $\boldsymbol{n}(t) = \gamma(t)\times\dot\gamma(t)/|\dot\gamma(t)|$, where $|\dot\gamma(t)| = \sqrt{\dot\gamma(t)\cdot\dot\gamma(t)}$ and $\dot\gamma(t) = (d\gamma/dt)(t)$. Then $\{\gamma(t), \boldsymbol{t}(t), \boldsymbol{n}(t)\}$ is a moving frame along $\gamma(t)$ and the Frenet Serret formula is given by

$$\begin{pmatrix} \dot\gamma(t) \\ \dot{\boldsymbol{t}}(t) \\ \dot{\boldsymbol{n}}(t) \end{pmatrix} = \begin{pmatrix} 0 & |\dot\gamma(t)| & 0 \\ -|\dot\gamma(t)| & 0 & |\dot\gamma(t)|\kappa_g(t) \\ 0 & -|\dot\gamma(t)|\kappa_g(t) & 0 \end{pmatrix} \begin{pmatrix} \gamma(t) \\ \boldsymbol{t}(t) \\ \boldsymbol{n}(t) \end{pmatrix},$$

where the geodesic curvature $\kappa_g$ is given by

$$\kappa_g(t) = \frac{\dot{\boldsymbol{t}}(t)\cdot\boldsymbol{n}(t)}{|\dot\gamma(t)|} = \frac{\det(\gamma(t), \dot\gamma(t), \ddot\gamma(t))}{|\dot\gamma(t)|^3}.$$

The evolute $Ev(\gamma) : I \to S^2$ of a regular curve $\gamma : I \to S^2$ is given by

$$Ev(\gamma)(t) = \pm\frac{\kappa_g(t)}{\sqrt{\kappa_g^2(t)+1}}\gamma(t) \pm \frac{1}{\sqrt{\kappa_g^2(t)+1}}\boldsymbol{n}(t). \tag{2.1}$$

By definition, we can not construct the Frenet Serret formula at singular points of $\gamma : I \to S^2$. In this paper, we would like to consider singular curves in the unit sphere.

We denote $\Delta = \{(\boldsymbol{a}, \boldsymbol{b}) \in S^2\times S^2 \mid \boldsymbol{a}\cdot\boldsymbol{b} = 0\}$ and is a 3-dimensional manifold .

Definition 2.1. We say that $(\gamma, \nu) : I \to \Delta \subset S^2\times S^2$ is a *Legendre curve* (or, *spherical Legendre curve*) if $\dot\gamma(t)\cdot\nu(t) = 0$ for all $t \in I$. We call $\gamma$ a *frontal* and $\nu$ a *dual* of $\gamma$. Moreover, if $(\gamma, \nu)$ is an immersion, we call $\gamma$ a *front*.

We consider the canonical contact structure on the unit spherical bundle $T_1S^2$ over $S^2$. If $(\gamma, \nu)$ is a Legendre curve, then $(\gamma, \nu)$ is an integral curve with respect to the contact structure (cf. [**2**]).

We define $\boldsymbol{\mu}(t) = \gamma(t)\times\nu(t)$. Then $\boldsymbol{\mu}(t) \in S^2$, $\gamma(t)\cdot\boldsymbol{\mu}(t) = 0$ and $\nu(t)\cdot\boldsymbol{\mu}(t) = 0$. It follows that $\{\gamma(t), \nu(t), \boldsymbol{\mu}(t)\}$ is a moving frame along the frontal $\gamma(t)$. By the standard arguments, we have the Frenet Serret type formula as follows:

PROPOSITION 2.2. *Let $(\gamma,\nu): I \to \Delta$ be a Legendre curve. Then we have*

$$\begin{pmatrix} \dot{\gamma}(t) \\ \dot{\nu}(t) \\ \dot{\boldsymbol{\mu}}(t) \end{pmatrix} = \begin{pmatrix} 0 & 0 & m(t) \\ 0 & 0 & n(t) \\ -m(t) & -n(t) & 0 \end{pmatrix} \begin{pmatrix} \gamma(t) \\ \nu(t) \\ \boldsymbol{\mu}(t) \end{pmatrix},$$

*where $m(t) = \dot{\gamma}(t)\cdot\boldsymbol{\mu}(t)$ and $n(t) = \dot{\nu}(t)\cdot\boldsymbol{\mu}(t)$.*

We say that the pair of the functions $(m,n)$ is the *curvature* of the Legendre curve $(\gamma,\nu): I \to \Delta \subset S^2 \times S^2$.

Note that $t_0$ is a singular point of $\gamma$ (respectively, $\nu$) if and only if $m(t_0) = 0$ (respectively, $n(t_0)$=0).

REMARK 2.3. If $(\gamma,\nu): I \to \Delta \subset S^2 \times S^2$ is a Legendre curve with the curvature $(m,n)$, then $(\gamma,-\nu)$ is a Legendre curve with the curvature $(-m,n)$. Also $(-\gamma,\nu)$ is a Legendre curve with the curvature $(m,-n)$. Moreover, $(\nu,\gamma)$ is a Legendre curve with the curvature $(-n,-m)$.

DEFINITION 2.4. Let $(\gamma,\nu),(\widetilde{\gamma},\widetilde{\nu}): I \to \Delta \subset S^2 \times S^2$ be Legendre curves. We say that $(\gamma,\nu)$ and $(\widetilde{\gamma},\widetilde{\nu})$ are *congruent as Legendre curves* if there exists a special orthogonal matrix $A \in SO(3)$ such that

$$\widetilde{\gamma}(t) = A(\gamma(t)),\ \widetilde{\nu}(t) = A(\nu(t)),$$

for all $t \in I$.

Then we have the following existence and uniqueness theorems in terms of the curvature of the Legendre curve.

THEOREM 2.5 (The Existence Theorem). *Let $(m,n): I \to \mathbb{R}\times\mathbb{R}$ be a smooth mapping. There exists a Legendre curve $(\gamma,\nu): I \to \Delta \subset S^2 \times S^2$ whose associated curvature is $(m,n)$.*

THEOREM 2.6 (The Uniqueness Theorem). *Let $(\gamma,\nu)$ and $(\widetilde{\gamma},\widetilde{\nu}): I \to \Delta \subset S^2 \times S^2$ be Legendre curves whose curvatures $(m,n)$ and $(\widetilde{m},\widetilde{n})$ coincide. Then $(\gamma,\nu)$ and $(\widetilde{\gamma},\widetilde{\nu})$ are congruent as Legendre curves.*

By using the theorems of the existence and uniqueness of the solution of a system of linear ordinary differential equations, these proofs are similar to the cases of regular space curves ([**11**]), Legendre curves in the unit tangent bundle ([**6**]) and framed curves ([**12**]), we omit it.

EXAMPLE 2.7. Let $\gamma: I \to S^2$ be a regular curve. We consider a Legendre immersion $(\gamma,\boldsymbol{n}): I \to \Delta \subset S^2 \times S^2$. Then the relationship between the geodesic curvature $\kappa_g$ of $\gamma$ and the curvature $(m,n)$ of $(\gamma,\boldsymbol{n})$ is given by $\kappa_g(t) = n(t)/|m(t)|$.

EXAMPLE 2.8. Let $n, m$ and $k$ be natural numbers with $m = k+n$. We give a mapping $(\gamma,\nu): \mathbb{R} \to \Delta \subset S^2 \times S^2$ by

$$\gamma(t) = \frac{1}{\sqrt{1+t^{2n}+t^{2m}}}(1,t^n,t^m),\ \nu(t) = \frac{1}{\sqrt{n^2+m^2t^{2k}+k^2t^{2m}}}(kt^m,-mt^k,n).$$

Then $(\gamma,\nu)$ is a Legendre curve. By definition, we have

$$\boldsymbol{\mu}(t) = \frac{(nt^n+mt^{m+k},-n+kt^{2m},-mt^k-kt^{m+n})}{\sqrt{(1+t^{2n}+t^{2m})(n^2+m^2t^{2k}+k^2t^{2m})}}$$

and the curvature

$$m(t) = \frac{-t^{n-1}\sqrt{n^2+m^2t^{2k}+k^2t^{2m}}}{1+t^{2n}+t^{2m}}, \ n(t) = \frac{knmt^{k-1}\sqrt{1+t^{2n}+t^{2m}}}{n^2+m^2t^{2k}+k^2t^{2m}}.$$

Let $\gamma : (I, t_0) \to S^2$ be a smooth curve germ and denote $\gamma(t) = (x(t), y(t), z(t))$. It can be shown that, if either $x(t), y(t)$ or $z(t)$ does not belong to $\mathfrak{m}_1^\infty$, then $\gamma$ is a frontal. Here $\mathfrak{m}_1^\infty$ is the ideal of infinitely flat function germs (cf. [**9**]).

Without loss of generality, we suppose that $x(t)$ does not belong to $\mathfrak{m}_1^\infty$ such that

$$\text{order } x(t) \le \text{order } y(t) \le \text{order } z(t).$$

Assume that $x(t_0) > 0$. By the assumptions and $\gamma(t) \in S^2$, there exist smooth function germs $a(t), b(t), c(t)$ around $t_0$ such that $y(t) = a(t)x(t)$, $z(t) = b(t)x(t)$ and $\dot{b}(t) = c(t)\dot{a}(t)$. It follows that $\gamma$ is given by

$$\gamma(t) = \frac{1}{\sqrt{1+a^2(t)+b^2(t)}}(1, a(t), b(t)).$$

If we take

$$\nu(t) = \frac{1}{\sqrt{(a(t)c(t)-b(t))^2+c^2(t)+1}}(a(t)c(t)-b(t), -c(t), 1),$$

then $(\gamma, \nu)$ is a Legendre curve.

On the other hand, constant maps in $S^2$ are also frontal, which do not satisfy the above sufficient condition. In particular an analytic curve germ is always frontal, because if it is infinitely flat, then it is constant.

Let $I$ and $\widetilde{I}$ be intervals. A smooth function $u : \widetilde{I} \to I$ is a *(positive) change of parameter* when $u$ is surjective and has a positive derivative at every point. It follows that $u$ is a diffeomorphism.

Let $(\gamma, \nu) : I \to \Delta$ and $(\widetilde{\gamma}, \widetilde{\nu}) : \widetilde{I} \to \Delta$ be Legendre curves whose curvatures are $(m, n)$ and $(\widetilde{m}, \widetilde{n})$ respectively. Suppose that $(\gamma, \nu)$ and $(\widetilde{\gamma}, \widetilde{\nu})$ are parametrically equivalent via the change of parameter $t : \widetilde{I} \to I$, that is, $(\widetilde{\gamma}(u), \widetilde{\nu}(u)) = (\gamma(t(u)), \nu(t(u)))$ for all $u \in \widetilde{I}$. By differentiation, we have

$$\widetilde{m}(u) = m(t(u))\dot{t}(u), \ \widetilde{n}(u) = n(t(u))\dot{t}(u). \tag{2.2}$$

Hence the curvature is dependent of the parametrization.

Let $(\gamma, \nu) : I \to \Delta \subset S^2 \times S^2$ be a Legendre curve with the curvature $(m, n)$. We define a parallel curve $\gamma_\theta : I \to S^2$ by

$$\gamma_\theta(t) = \cos\theta\gamma(t) + \sin\theta\nu(t),$$

where $\theta \in [0, 2\pi)$. Then $\gamma_\theta$ is a frontal. More precisely, we have the following. We denote $\nu_\theta : I \to S^2$ by $\nu_\theta(t) = -\sin\theta\gamma(t) + \cos\theta\nu(t)$.

PROPOSITION 2.9. *Under the above notations, $(\gamma_\theta, \nu_\theta) : I \to \Delta \subset S^2 \times S^2$ is a Legendre curve with the curvature*

$$(m(t)\cos\theta + n(t)\sin\theta, -m(t)\sin\theta + n(t)\cos\theta). \tag{2.3}$$

PROOF. By definition, $\gamma_\theta(t)\cdot\nu_\theta(t)=0$. Since $\dot\gamma_\theta(t)=(m(t)\cos\theta+n(t)\sin\theta)\boldsymbol{\mu}(t)$, then $\dot\gamma_\theta(t)\cdot\nu_\theta(t)=0$. It follows that $(\gamma_\theta,\nu_\theta)$ is a Legendre curve. Moreover, we have $\boldsymbol{\mu}_\theta(t)=\gamma_\theta(t)\times\nu_\theta(t)=\boldsymbol{\mu}(t)$ and $\dot\nu_\theta(t)=(-m(t)\sin\theta+n(t)\cos\theta)\boldsymbol{\mu}(t)$. The curvature of the Legendre curve is given by $(m(t)\cos\theta+n(t)\sin\theta,-m(t)\sin\theta+n(t)\cos\theta)$. $\square$

We say that $(\gamma_\theta,\nu_\theta)$ is a *parallel Legendre curve* of the Legendre curve $(\gamma,\nu)$. Note that if $(\gamma,\nu)$ is a Legendre immersion, then $(\gamma_\theta,\nu_\theta)$ is also a Legendre immersion.

## 3. Relationships among spherical Legendre curves, Legendre curves and framed curves

First, we give a relationship between spherical Legendre curves and Legendre curves in the unit tangent bundle over $\mathbb{R}^2$.

We review on the Legendre curves in the unit tangent bundle over $\mathbb{R}^2$, for more detail see [**6**]. We say that $(\gamma,\nu):I\to\mathbb{R}^2\times S^1$ is *a Legendre curve* if $(\gamma(t),\nu(t))^*\theta=0$ for all $t\in I$, where $\theta$ is a canonical contact 1-form on the unit tangent bundle $T_1\mathbb{R}^2=\mathbb{R}^2\times S^1$ (cf. [**1, 2**]). This condition is equivalent to $\dot\gamma(t)\cdot\nu(t)=0$ for all $t\in I$. We say that $\gamma:I\to\mathbb{R}^2$ is *a frontal* if there exists a smooth mapping $\nu:I\to S^1$ such that $(\gamma,\nu)$ is a Legendre curve.

Let $(\gamma,\nu):I\to\mathbb{R}^2\times S^1$ be a Legendre curve. Then we have the Frenet formula of the frontal $\gamma$ as follows. We put $\boldsymbol{\mu}(t)=J(\nu(t))$, where $J$ is the anti-clockwise rotation by $\pi/2$ on $\mathbb{R}^2$. We call the pair $\{\nu(t),\boldsymbol{\mu}(t)\}$ a *moving frame along the frontal* $\gamma(t)$ in $\mathbb{R}^2$ and the Frenet formula of the frontal (or, the Legendre curve) which is given by

$$\begin{pmatrix}\dot\nu(t)\\ \dot{\boldsymbol{\mu}}(t)\end{pmatrix}=\begin{pmatrix}0 & \ell(t)\\ -\ell(t) & 0\end{pmatrix}\begin{pmatrix}\nu(t)\\ \boldsymbol{\mu}(t)\end{pmatrix},$$

where $\ell(t)=\dot\nu(t)\cdot\boldsymbol{\mu}(t)$. Moreover, there exists a smooth function $\beta(t)$ such that

$$\dot\gamma(t)=\beta(t)\boldsymbol{\mu}(t).$$

We say that the pair of functions $(\ell,\beta)$ is *the curvature of the Legendre curve* $(\gamma,\nu):I\to\mathbb{R}^2\times S^1$.

Now we consider the central and the canonical projections to give relationships between Legendre curves in $T_1S^2$ and $T_1\mathbb{R}^2$. We denote a hemisphere $S^+=\{(x,y,z)\in S^2|z>0\}$.

We consider the central projection $\Phi:S^+\to\mathbb{R}^2$ by

$$\Phi(x,y,z)=\left(\frac{x}{z},\frac{y}{z}\right).$$

The central projection is useful to analyze the pedal curves (cf. [**14, 15**]).

PROPOSITION 3.1. *Let $(\gamma,\nu):I\to\Delta\subset S^2\times S^2$ be a spherical Legendre curve with the curvature $(m,n)$ and $\gamma(I)\subset S^+$. We denote $\gamma(t)=(x(t),y(t),z(t))$ and $\nu(t)=(a(t),b(t),c(t))$. Then $\widetilde\gamma=\Phi\circ\gamma$ is a frontal in $\mathbb{R}^2$. More preciously, $(\widetilde\gamma,\widetilde\nu):I\to\mathbb{R}^2\times S^1$ is a Legendre curve, where*

$$\widetilde\gamma(t)=\left(\frac{x(t)}{z(t)},\frac{y(t)}{z(t)}\right),\ \widetilde\nu(t)=\frac{1}{\sqrt{a^2(t)+b^2(t)}}(a(t),b(t))$$

*with the curvature*

$$\ell(t) = \frac{n(t)z(t)}{a^2(t)+b^2(t)},\ \beta(t) = \frac{m(t)z^2(t)+(x(t)b(t)-y(t)a(t))\dot{z}(t)}{z^2(t)\sqrt{a^2(t)+b^2(t)}}.$$

PROOF. Since $(\gamma,\nu)$ is a spherical Legendre curve, we have

$$x(t)a(t)+y(t)b(t)+z(t)c(t)=0,\ \dot{x}(t)a(t)+\dot{y}(t)b(t)+\dot{z}(t)c(t)=0.$$

It follows that $x(t)\dot{a}(t)+y(t)\dot{b}(t)+z(t)\dot{c}(t)=0$. By definition, we have

$$\boldsymbol{\mu}(t)=\gamma(t)\times\nu(t)=(y(t)c(t)-z(t)b(t),z(t)a(t)-x(t)c(t),x(t)b(t)-y(t)a(t)).$$

By a direct calculation, we have

$$\begin{aligned} m(t) &= \dot{\gamma}(t)\cdot\boldsymbol{\mu}(t) = \frac{-\dot{x}(t)b(t)+\dot{y}(t)a(t)}{z(t)},\\ n(t) &= \dot{\nu}(t)\cdot\boldsymbol{\mu}(t) = \frac{-\dot{a}(t)b(t)+a(t)\dot{b}(t)}{z(t)}. \end{aligned}$$

By the assumption $\gamma(t)\in S^+$, $c(t)\neq\pm1$ and hence $a^2(t)+b^2(t)\neq 0$. It follows that $\widetilde{\nu}: I\to S^1, \widetilde{\nu}(t)=(a(t),b(t))/\sqrt{a^2(t)+b^2(t)}$ is a smooth mapping. Moreover, we have

$$\dot{\widetilde{\gamma}}(t)=\left(\frac{\dot{x}(t)z(t)-x(t)\dot{z}(t)}{z^2(t)},\frac{\dot{y}(t)z(t)-y(t)\dot{z}(t)}{z^2(t)}\right)$$

and $\dot{\widetilde{\gamma}}(t)\cdot\widetilde{\nu}(t)=0$. Therefore $(\widetilde{\gamma},\widetilde{\nu}): I\to\mathbb{R}^2\times S^1$ is a Legendre curve.

By definition, we have $\widetilde{\boldsymbol{\mu}}(t)=J(\widetilde{\nu}(t))=(-b(t),a(t))/\sqrt{a^2(t)+b^2(t)}$ and the curvature

$$\begin{aligned} \ell(t) &= \dot{\widetilde{\nu}}(t)\cdot\widetilde{\boldsymbol{\mu}}(t)=\frac{a(t)\dot{b}(t)-\dot{a}(t)b(t)}{a^2(t)+b^2(t)}=\frac{n(t)z(t)}{a^2(t)+b^2(t)},\\ \beta(t) &= \dot{\widetilde{\gamma}}(t)\cdot\widetilde{\boldsymbol{\mu}}(t)=\frac{(-\dot{x}(t)b(t)+\dot{y}(t)a(t))z(t)+(x(t)b(t)-y(t)a(t))\dot{z}(t)}{z^2(t)\sqrt{a^2(t)+b^2(t)}}\\ &= \frac{m(t)z^2(t)+(x(t)b(t)-y(t)a(t))\dot{z}(t)}{z^2(t)\sqrt{a^2(t)+b^2(t)}}. \end{aligned}$$

□

Also, we consider the canonical projection $\pi: S^+\to D^2\subset\mathbb{R}^2$ by $\pi(x,y,z)=(x,y)$, where $D^2=\{(x,y)\in\mathbb{R}^2 | x^2+y^2<1\}$.

PROPOSITION 3.2. *Let $(\gamma,\nu): I\to\Delta\subset S^2\times S^2$ be a spherical Legendre curve with the curvature $(m,n)$ and $\gamma(I)\subset S^+$. We denote $\gamma(t)=(x(t),y(t),z(t))$ and $\nu(t)=(a(t),b(t),c(t))$. Then $\widetilde{\gamma}=\pi\circ\gamma$ is a frontal in $D^2\subset\mathbb{R}^2$. More preciously, $(\widetilde{\gamma},\widetilde{\nu}): I\to D^2\times S^1$ is a Legendre curve, where*

$$\widetilde{\gamma}(t)=(x(t),y(t)),\ \widetilde{\nu}(t)=\frac{(z(t)a(t)-x(t)c(t),z(t)b(t)-y(t)c(t))}{\sqrt{(z(t)a(t)-x(t)c(t))^2+(z(t)b(t)-y(t)c(t))^2}}$$

*with the curvature*

$$\begin{aligned} \ell(t) &= \frac{n(t)z(t)+x(t)\dot{y}(t)-\dot{x}(t)y(t)}{(z(t)a(t)-x(t)c(t))^2+(z(t)b(t)-y(t)c(t))^2},\\ \beta(t) &= \frac{m(t)-(x(t)b(t)-y(t)a(t))\dot{z}(t)}{\sqrt{(z(t)a(t)-x(t)c(t))^2+(z(t)b(t)-y(t)c(t))^2}}. \end{aligned}$$

PROOF. If $z(t)a(t) - x(t)c(t) = 0$ and $z(t)b(t) - y(t)c(t) = 0$, then $a(t) = x(t)c(t)/z(t)$ and $b(t) = y(t)c(t)/z(t)$. Since $\nu(t) \in S^2$, we have $c^2(t) = z^2(t)$ and hence $c(t) = \pm z(t)$. It follows that $a(t) = \pm x(t)$ and $b(t) = \pm y(t)$. It is contradict the fact that $\gamma(t) \cdot \nu(t) = 0$. Hence $\widetilde{\nu}$ is a smooth mapping. By $\dot{x}(t)a(t) + \dot{y}(t)b(t) + \dot{z}(t)c(t) = 0$ and $\dot{x}(t)x(t) + \dot{y}(t)y(t) + \dot{z}(t)z(t) = 0$, we have $\dot{\widetilde{\gamma}}(t) \cdot \widetilde{\nu}(t) = 0$. Therefore $(\widetilde{\gamma}, \widetilde{\nu}) : I \to D^2 \times S^1$ is a Legendre curve. By a similar calculation as in Proposition 3.1, we have the curvature $(\ell, \beta)$ of the Legendre curve $(\widetilde{\gamma}, \widetilde{\nu})$. □

REMARK 3.3. As a projection from the sphere to the plane, how about the stereographic projection. The properties of the stereographic projection see [**11**, **18**, **19**], for example. Does it hold the similar results of Propositions 3.1 and 3.2 or not?

Conversely, for a Legendre curve in the unit tangent bundle, we have a spherical Legendre curve as follows.

PROPOSITION 3.4. *Let $(\widetilde{\gamma}, \widetilde{\nu}) : I \to \mathbb{R}^2 \times S^1$ be a Legendre curve with the curvature $(\ell, \beta)$. We denote $\widetilde{\gamma}(t) = (x(t), y(t))$ and $\widetilde{\nu}(t) = (a(t), b(t))$. Then $\gamma = \Phi^{-1} \circ \widetilde{\gamma}$ is a frontal in $S^+$. More preciously, $(\gamma, \nu) : I \to \Delta \subset S^+ \times S^2$ is a spherical Legendre curve, where*

$$\gamma(t) = \frac{(x(t), y(t), 1)}{\sqrt{1 + x^2(t) + y^2(t)}}, \quad \nu(t) = \frac{(a(t), b(t), -(x(t)a(t) + y(t)b(t)))}{\sqrt{1 + (x(t)a(t) + y(t)b(t))^2}}$$

*with the curvature*

$$\begin{aligned} m(t) &= \frac{\beta(t) + (x(t)\dot{y}(t) - \dot{x}(t)y(t))(x(t)a(t) + y(t)b(t))}{(1 + x^2(t) + y^2(t))\sqrt{1 + (x(t)a(t) + y(t)b(t))^2}}, \\ n(t) &= \frac{\ell(t)\sqrt{1 + x^2(t) + y^2(t)}}{1 + (x(t)a(t) + y(t)b(t))^2}. \end{aligned}$$

PROOF. Since $(\widetilde{\gamma}, \widetilde{\nu}) : I \to \mathbb{R}^2 \times S^1$ is a Legendre curve, we have $\dot{x}(t)a(t) + \dot{y}(t)b(t) = 0$. By definition, we have $\widetilde{\boldsymbol{\mu}}(t) = J(\widetilde{\nu}(t)) = (-b(t), a(t))$. It follows that

$$\ell(t) = -\dot{a}(t)b(t) + a(t)\dot{b}(t), \ \beta(t) = -\dot{x}(t)b(t) + \dot{y}(t)a(t).$$

By a direct calculation, we have

$$\begin{aligned} \dot{\gamma}(t) &= \frac{1}{1 + x^2(t) + y^2(t)}((1 + y^2(t))\dot{x}(t) - x(t)y(t)\dot{y}(t), \\ &\qquad (1 + x^2(t))\dot{y}(t) - x(t)\dot{x}(t)y(t), -(x(t)\dot{x}(t) + y(t)\dot{y}(t))). \end{aligned}$$

Then $\gamma(t) \cdot \nu(t) = 0$ and $\dot{\gamma}(t) \cdot \nu(t) = 0$ for all $t \in I$. Therefore $(\gamma, \nu) : I \to \Delta \subset S^+ \times S^2$ is a spherical Legendre curve. By definition, $\boldsymbol{\mu}(t)$ is given by

$$\frac{(-x(t)y(t)a(t) - (1 + y^2(t))b(t), (1 + x^2(t))a(t) + x(t)y(t)b(t), x(t)b(t) - y(t)a(t))}{\sqrt{(1 + x^2(t) + y^2(t))(1 + (x(t)a(t) + y(t)b(t))^2)}}$$

and the curvature

$$\begin{aligned} m(t) &= \frac{(x(t)\dot{y}(t)-\dot{x}(t)y(t))(x(t)a(t)+y(t)b(t))-\dot{x}(t)b(t)+\dot{y}(t)a(t)}{(1+x^2(t)+y^2(t))\sqrt{1+(x(t)a(t)+y(t)b(t))^2}} \\ &= \frac{\beta(t)+(x(t)\dot{y}(t)-\dot{x}(t)y(t))(x(t)a(t)+y(t)b(t))}{(1+x^2(t)+y^2(t))\sqrt{1+(x(t)a(t)+y(t)b(t))^2}}, \\ n(t) &= \frac{(a(t)\dot{b}(t)-\dot{a}(t)b(t))\sqrt{1+x^2(t)+y^2(t)}}{1+(x(t)a(t)+y(t)b(t))^2} \\ &= \frac{\ell(t)\sqrt{1+x^2(t)+y^2(t)}}{1+(x(t)a(t)+y(t)b(t))^2}. \end{aligned}$$

□

PROPOSITION 3.5. *Let $(\widetilde{\gamma},\widetilde{\nu}) : I \to D^2 \times S^1$ be a Legendre curve with the curvature $(\ell,\beta)$. We denote $\widetilde{\gamma}(t) = (x(t),y(t))$ and $\widetilde{\nu}(t) = (a(t),b(t))$. Then $\gamma = \pi^{-1}\circ\widetilde{\gamma}$ is a frontal in $S^+$. More preciously, $(\gamma,\nu) : I \to \Delta \subset S^+ \times S^2$ is a spherical Legendre curve, where*

$$\begin{aligned} \gamma(t) &= (x(t),y(t),z(t)), \\ \nu(t) &= \frac{1}{\sqrt{1-(x(t)a(t)+y(t)b(t))^2}}(a(t)-x(t)(x(t)a(t)-y(t)b(t)), \\ &\qquad b(t)-y(t)(x(t)a(t)-y(t)b(t)),-z(t)(x(t)a(t)+y(t)b(t))) \end{aligned}$$

*with the curvature*

$$\begin{aligned} m(t) &= \frac{\beta(t)+(\dot{x}(t)y(t)-x(t)\dot{y}(t))(x(t)a(t)-y(t)b(t))}{z(t)\sqrt{1-(x(t)a(t)+y(t)b(t))^2}}, \\ n(t) &= \frac{1}{z(t)(1-(x(t)a(t)+y(t)b(t))^2)}(\ell(t)z^2(t)-\beta(t)(x(t)a(t)+y(t)b(t)) \\ &\qquad +(x(t)\dot{y}(t)-\dot{x}(t)y(t))(x(t)a(t)+y(t)b(t))^2). \end{aligned}$$

*Here we put $z(t) = \sqrt{1-x^2(t)-y^2(t)}$.*

PROOF. Since $\widetilde{\gamma}(t)\cdot\widetilde{\gamma}(t) < 1$ and $\widetilde{\nu}(t)\cdot\widetilde{\nu}(t) = 1$, we have $x(t)a(t)+y(t)b(t) < 1$ for all $t \in I$. Therefore $\nu : I \to S^2$ is a smooth mapping. By the same argument as in Proposition 3.4, we have

$$\ell(t) = -\dot{a}(t)b(t)+a(t)\dot{b}(t),\ \beta(t) = -\dot{x}(t)b(t)+\dot{y}(t)a(t).$$

Since

$$\dot{\gamma}(t) = \left(\dot{x}(t),\dot{y}(t),-\frac{x(t)\dot{x}(t)-y(t)\dot{y}(t)}{z(t)}\right)$$

and $x^2(t)+y^2(t)+z^2(t) = 1$, we have $\gamma(t)\cdot\nu(t) = 0$ and $\dot{\gamma}(t)\cdot\nu(t) = 0$ for all $t \in I$. Therefore $(\gamma,\nu) : I \to \Delta \subset S^+ \times S^2$ is a spherical Legendre curve. By definition and a direct calculation, we have

$$\boldsymbol{\mu}(t) = \frac{1}{\sqrt{1-(x(t)a(t)+y(t)b(t))^2}}(-z(t)b(t),z(t)a(t),x(t)b(t)-y(t)a(t))$$

and the curvature $(m,n)$ of the spherical Legendre curve $(\gamma,\nu)$. □

Second, we discuss relationships between framed curves in the Euclidean space and spherical Legendre curves.

We review on the framed curves, for more detail see [**12**]. We say that $(\gamma, \nu_1, \nu_2) : I \to \mathbb{R}^3 \times S^2 \times S^2$ is *a framed curve* if

$$\dot{\gamma}(t) \cdot \nu_1(t) = 0, \ \dot{\gamma}(t) \cdot \nu_2(t) = 0, \ \nu_1(t) \cdot \nu_2(t) = 0$$

for all $t \in I$. Then $(\nu_1, \nu_2) \in \Delta$.

Let $(\gamma, \nu_1, \nu_2) : I \to \mathbb{R}^3 \times \Delta$ be a framed curve and denote $\boldsymbol{\mu}(t) = \nu_1(t) \times \nu_2(t)$. The Frenet Serret type formula is given by

$$\begin{pmatrix} \dot{\nu_1}(t) \\ \dot{\nu_2}(t) \\ \dot{\boldsymbol{\mu}}(t) \end{pmatrix} = \begin{pmatrix} 0 & \ell(t) & m(t) \\ -\ell(t) & 0 & n(t) \\ -m(t) & -n(t) & 0 \end{pmatrix} \begin{pmatrix} \nu_1(t) \\ \nu_2(t) \\ \boldsymbol{\mu}(t) \end{pmatrix},$$

where $\ell(t) = \dot{\nu_1}(t) \cdot \nu_2(t), m(t) = \dot{\nu_1}(t) \cdot \boldsymbol{\mu}(t)$ and $n(t) = \dot{\nu_2}(t) \cdot \boldsymbol{\mu}(t)$. Moreover, there exists a smooth mapping $\alpha : I \to \mathbb{R}$ such that

$$\dot{\gamma}(t) = \alpha(t)\boldsymbol{\mu}(t).$$

We say that the pair of the functions $(\ell, m, n, \alpha)$ is the curvature of the framed curve $(\gamma, \nu_1, \nu_2) : I \to \mathbb{R}^3 \times \Delta$.

Let $(\gamma, \nu_1, \nu_2) : I \to \mathbb{R}^3 \times \Delta$ be a framed curve with the curvature of the framed curve $(\ell, m, n, \alpha)$. For the normal plane of $\gamma(t)$, spanned by $\nu_1(t)$ and $\nu_2(t)$, there is some ambient of framed curves similarly to the case of the Bishop frame of a regular space curve (cf. [**4**]). We define $(\overline{\nu}_1(t), \overline{\nu}_2(t)) \in \Delta$ by

$$\begin{pmatrix} \overline{\nu}_1(t) \\ \overline{\nu}_2(t) \end{pmatrix} = \begin{pmatrix} \cos\theta(t) & -\sin\theta(t) \\ \sin\theta(t) & \cos\theta(t) \end{pmatrix} \begin{pmatrix} \nu_1(t) \\ \nu_2(t) \end{pmatrix},$$

where $\theta(t)$ is a smooth function. Then $(\gamma, \overline{\nu}_1, \overline{\nu}_2) : I \to \mathbb{R}^3 \times \Delta$ is also a framed curve and

$$\begin{aligned} \overline{\boldsymbol{\mu}}(t) &= \overline{\nu}_1(t) \times \overline{\nu}_2(t) \\ &= (\cos\theta(t)\nu_1(t) - \sin\theta(t)\nu_2(t)) \times (\sin\theta(t)\nu_1(t) + \cos\theta(t)\nu_2(t)) \\ &= \nu_1(t) \times \nu_2(t) = \boldsymbol{\mu}(t). \end{aligned}$$

By a direct calculation, we have

$$\begin{aligned} \dot{\overline{\nu}}_1(t) &= (\ell(t) - \dot{\theta}(t))\sin\theta(t)\nu_1(t) + (\ell(t) - \dot{\theta}(t))\cos\theta(t)\nu_2(t) \\ &\quad +(m(t)\cos\theta(t) - n(t)\sin\theta(t))\boldsymbol{\mu}(t), \\ \dot{\overline{\nu}}_2(t) &= -(\ell(t) - \dot{\theta}(t))\cos\theta(t)\nu_1(t) + (\ell(t) - \dot{\theta}(t))\sin\theta(t)\nu_2(t) \\ &\quad +(m(t)\sin\theta(t) + n(t)\cos\theta(t))\boldsymbol{\mu}(t). \end{aligned}$$

If we take a smooth function $\theta : I \to \mathbb{R}$ which satisfies $\dot{\theta}(t) = \ell(t)$, then we call the frame $\{\overline{\nu}_1(t), \overline{\nu}_2(t), \boldsymbol{\mu}(t)\}$ an *adapted frame* along the framed base curve $\gamma(t)$. It follows that the Frenet Serret type formula is given by

$$\begin{pmatrix} \dot{\overline{\nu}}_1(t) \\ \dot{\overline{\nu}}_2(t) \\ \dot{\boldsymbol{\mu}}(t) \end{pmatrix} = \begin{pmatrix} 0 & 0 & \overline{m}(t) \\ 0 & 0 & \overline{n}(t) \\ -\overline{m}(t) & -\overline{n}(t) & 0 \end{pmatrix} \begin{pmatrix} \overline{\nu}_1(t) \\ \overline{\nu}_2(t) \\ \boldsymbol{\mu}(t) \end{pmatrix},$$

where $\overline{m}(t)$ and $\overline{n}(t)$ are given by

$$\begin{pmatrix} \overline{m}(t) \\ \overline{n}(t) \end{pmatrix} = \begin{pmatrix} \cos\theta(t) & -\sin\theta(t) \\ \sin\theta(t) & \cos\theta(t) \end{pmatrix} \begin{pmatrix} m(t) \\ n(t) \end{pmatrix}.$$

PROPOSITION 3.6. *Let $(\gamma, \nu_1, \nu_2) : I \to \mathbb{R}^3 \times \Delta$ be a framed curve with the curvature $(\ell, m, n, \alpha)$.*

*(1) Suppose that $\{\overline{\nu}_1(t), \overline{\nu}_2(t), \boldsymbol{\mu}(t)\}$ is an adapted frame of $\gamma(t)$. Then $(\overline{\nu}_1, \overline{\nu}_2) : I \to \Delta \subset S^2 \times S^2$ is a spherical Legendre curve with the curvature $(\overline{m}(t), \overline{n}(t))$.*

*(2) Let $\gamma(t)$ be non-zero. We denote $\widetilde{\gamma}(t) = \gamma(t)/|\gamma(t)|$ and $\widetilde{\gamma}(t) = a(t)\nu_1(t) + b(t)\nu_2(t) + c(t)\boldsymbol{\mu}(t)$ with $a^2(t) + b^2(t) + c^2(t) = 1$. Suppose that $a^2(t) + b^2(t) \neq 0$. Then $\widetilde{\gamma}(t)$ is a frontal in $S^2$. More preciously, $(\widetilde{\gamma}, \widetilde{\nu}) : I \to \Delta \subset S^2 \times S^2$ is a spherical Legendre curve, where*

$$\widetilde{\nu}(t) = \frac{\widetilde{\gamma}(t) \times \boldsymbol{\mu}(t)}{|\widetilde{\gamma}(t) \times \boldsymbol{\mu}(t)|},$$

*with the curvature*

$$\begin{aligned}
\widetilde{m}(t) &= -\frac{a(t)m(t) + b(t)n(t) + \dot{c}(t)}{\sqrt{a^2(t) + b^2(t)}}, \\
\widetilde{n}(t) &= \frac{(a^2(t) + b^2(t))(a(t)n(t) - b(t)m(t) + c(t)\ell(t)) + (a(t)\dot{b}(t) - \dot{a}(t)b(t))c(t)}{\sqrt{a^2(t) + b^2(t)}}.
\end{aligned}$$

PROOF. (1) By definition, $(\overline{\nu}_1, \overline{\nu}_2)$ is a spherical Legendre curve with the curvature $(\overline{m}(t), \overline{n}(t))$.

(2) Since $\widetilde{\nu}(t) = (b(t)\nu_1(t) - a(t)\nu_2(t))/\sqrt{a^2(t) + b^2(t)}$, we have

$$\widetilde{\boldsymbol{\mu}}(t) = \frac{1}{\sqrt{a^2(t) + b^2(t)}}(a(t)c(t)\nu_1(t) + b(t)c(t)\nu_2(t) - (a^2(t) + b^2(t))\boldsymbol{\mu}(t)).$$

By using the Frenet Serret type formula, we have

$$\begin{aligned}
\dot{\widetilde{\gamma}}(t) &= (\dot{a}(t) - b(t)\ell(t) - c(t)m(t))\nu_1(t) + (\dot{b}(t) + a(t)\ell(t) - c(t)n(t))\nu_2(t) \\
&\quad + (\dot{c}(t) + a(t)m(t) + b(t)n(t))\boldsymbol{\mu}(t), \\
\dot{\widetilde{\nu}}(t) &= \frac{1}{(a^2(t) + b^2(t))^{\frac{3}{2}}}\Big((\dot{b}(t)a^2(t) + a(t)(a^2(t) + b^2(t))\ell(t) - \dot{a}(t)a(t)b(t))\nu_1(t) \\
&\quad + (-\dot{a}(t)b^2(t) + b(t)(a^2(t) + b^2(t))\ell(t) + \dot{b}(t)a(t)b(t))\nu_2(t) \\
&\quad + (a^2(t) + b^2(t))(-a(t)n(t) + b(t)m(t))\boldsymbol{\mu}(t)\Big).
\end{aligned}$$

By a direct calculation, we have

$$\begin{aligned}
\widetilde{m}(t) &= \dot{\widetilde{\gamma}}(t) \cdot \widetilde{\boldsymbol{\mu}}(t) = -\frac{a(t)m(t) + b(t)n(t) + \dot{c}(t)}{\sqrt{a^2(t) + b^2(t)}}, \\
\widetilde{n}(t) &= \dot{\widetilde{\nu}}(t) \cdot \widetilde{\boldsymbol{\mu}}(t) \\
&= \frac{(a^2(t) + b^2(t))(a(t)n(t) - b(t)m(t) + c(t)\ell(t)) + (a(t)\dot{b}(t) - \dot{a}(t)b(t))c(t)}{\sqrt{a^2(t) + b^2(t)}}.
\end{aligned}$$

□

Conversely, for a spherical Legendre curve, we have a framed curve as follows.

PROPOSITION 3.7. *Let $(\gamma, \nu) : I \to \Delta \subset S^2 \times S^2$ be a spherical Legendre curve with the curvature $(m, n)$. Then $(\gamma, \gamma, \nu) : I \to S^2 \times \Delta \subset \mathbb{R}^3 \times \Delta$ is a framed curve with the curvature $(\ell, m, n, \alpha) = (0, m, n, m)$.*

PROOF. Since $(\gamma, \nu) : I \to \Delta \subset S^2 \times S^2$ is a spherical Legendre curve, $(\gamma, \gamma, \nu) : I \to S^2 \times \Delta \subset \mathbb{R}^3 \times \Delta$ is a framed curve with the curvature $(0, m, n, m)$. □

## 4. Evolutes of fronts in the sphere

In this section, we assume that $(\gamma,\nu): I \to \Delta \subset S^2 \times S^2$ is a Legendre immersion. It follows that $(m(t),n(t)) \neq (0,0)$ for all $t \in I$. We define an evolute of the front and give properties of the evolute in the sphere. For the evolutes of curves in the Euclidean plane see [**5**, **7**, **8**, **10**].

DEFINITION 4.1. We define an evolute $\mathcal{E}v(\gamma): I \to S^2$ of the front $\gamma$ by

$$\mathcal{E}v(\gamma)(t) = \pm \frac{n(t)}{\sqrt{m^2(t)+n^2(t)}}\gamma(t) \mp \frac{m(t)}{\sqrt{m^2(t)+n^2(t)}}\nu(t). \tag{4.1}$$

REMARK 4.2. If $(\gamma,\nu)$ is a Legendre immersion with the curvature $(m,n)$, then $(\gamma,-\nu)$ (respectively, $(-\gamma,\nu)$) is a Legendre immersion with the curvature $(-m,n)$ (respectively, $(m,-n)$) by Remark 2.3. It is easy to see that the evolute $\mathcal{E}v(\gamma)$ does not change. For the case $(\nu,\gamma)$, see below Corollary 4.6.

PROPOSITION 4.3. *Let $\gamma : I \to S^2$ be a regular curve. Then the evolute of the regular curve and the evolute of the front are coincide.*

PROOF. We consider a Legendre immersion $(\gamma,\boldsymbol{n}): I \to \Delta \subset S^2 \times S^2$ with the curvature $(m,n)$, see Example 2.7. Since $\boldsymbol{n}(t) = \nu(t)$ and $\boldsymbol{t}(t) = -\boldsymbol{\mu}(t)$, we have $m(t) < 0$. The geodesic curvature of the regular curve is given by $\kappa_g(t) = n(t)/|m(t)| = -n(t)/m(t)$. By the definition of the evolute of the regular curve (2.1), we have

$$\begin{aligned} Ev(\gamma)(t) &= \pm\frac{\kappa_g(t)}{\sqrt{\kappa_g^2(t)+1}}\gamma(t) \pm \frac{1}{\sqrt{\kappa_g^2(t)+1}}\boldsymbol{n}(t) \\ &= \pm\frac{n(t)}{\sqrt{m^2(t)+n^2(t)}}\gamma(t) \mp \frac{m(t)}{\sqrt{m^2(t)+n^2(t)}}\nu(t) = \mathcal{E}v(\gamma)(t). \end{aligned}$$

□

PROPOSITION 4.4. *Suppose that $(\gamma,\nu): I \to \Delta \subset S^2 \times S^2$ and $(\widetilde{\gamma},\widetilde{\nu}): \widetilde{I} \to \Delta \subset S^2 \times S^2$ are parametrically equivalent via the change of parameter $t : \widetilde{I} \to I$. Then $\mathcal{E}v(\widetilde{\gamma})(u) = \mathcal{E}v(\gamma)(t(u))$.*

PROOF. We denote $(m(t),n(t))$ and $(\widetilde{m}(u),\widetilde{n}(u))$ the curvature of $(\gamma(t),\nu(t))$ and $(\widetilde{\gamma}(u),\widetilde{\nu}(u))$ respectively. By the assumption, $(\widetilde{\gamma}(u),\widetilde{\nu}(u)) = (\gamma(t(u)),\nu(t(u)))$. By using the relationship between the curvature (2.2), we have the result. □

PROPOSITION 4.5. *Let $\theta \in [0,2\pi)$ and $(\gamma_\theta,\nu_\theta): I \to \Delta \subset S^2 \times S^2$ be a parallel Legendre immersion of $(\gamma,\nu)$. Then the evolute of the parallel curve and the evolute of the front are coincide.*

PROOF. By Proposition 2.9, $(m_\theta(t),n_\theta(t)) = (m(t)\cos\theta + n(t)\sin\theta, -m(t)\cos\theta + n(t)\cos\theta)$ is the curvature of $(\gamma_\theta,\nu_\theta)$. Then we have $m_\theta^2(t)+n_\theta^2(t) = m^2(t)+n^2(t)$.

It follows that

$$\begin{aligned}\mathcal{E}v(\gamma_\theta)(t) &= \pm\frac{n_\theta(t)}{\sqrt{m_\theta^2(t)+n_\theta^2(t)}}\gamma_\theta(t) \mp \frac{m_\theta(t)}{\sqrt{m_\theta^2(t)+n_\theta^2(t)}}\nu_\theta(t) \\ &= \pm\frac{(-m(t)\sin\theta+n(t)\cos\theta)}{\sqrt{m^2(t)+n^2(t)}}(\cos\theta\gamma(t)+\sin\theta\nu(t)) \\ &\quad \mp\frac{(m(t)\cos\theta+n(t)\sin\theta)}{\sqrt{m^2(t)+n^2(t)}}(-\sin\theta\gamma(t)+\cos\theta\nu(t)) \\ &= \pm\frac{n(t)}{\sqrt{m^2(t)+n^2(t)}}\gamma(t) \mp \frac{m(t)}{\sqrt{m^2(t)+n^2(t)}}\nu(t) = \mathcal{E}v(\gamma)(t).\end{aligned}$$

□

If we take $\theta = \pi/2$, then $(\gamma_\theta, \nu_\theta) = (\nu, -\gamma)$. By Proposition 4.5 and Remark 4.2, we have the following Corollary.

COROLLARY 4.6. *For a Legendre immersion* $(\gamma, \nu) : I \to \Delta \subset S^2 \times S^2$, $(\nu, \gamma)$ *is also a Legendre immersion. Then the evolute of the front and the evolute of the dual curve are coincide, that is,* $\mathcal{E}v(\gamma)(t) = \mathcal{E}v(\nu)(t)$.

We define a family of functions $H : I \times S^2 \to \mathbb{R}$ by

$$H(t, \boldsymbol{v}) = \boldsymbol{\mu}(t) \cdot \boldsymbol{v}.$$

PROPOSITION 4.7. *Let* $(\gamma, \nu) : I \to \Delta \subset S^2 \times S^2$ *be a Legendre immersion with the curvature* $(m, n)$. *We have the following.*

(1) $H(t, \boldsymbol{v}) = 0$ *if and only if* $\boldsymbol{v} = a\gamma(t) + b\nu(t)$ *for some* $a, b \in \mathbb{R}$ *with* $a^2+b^2 = 1$.

(2) $H(t, \boldsymbol{v}) = \frac{\partial H}{\partial t}(t, \boldsymbol{v}) = 0$ *if and only if*

$$\boldsymbol{v} = \pm\frac{n(t)}{\sqrt{m^2(t)+n^2(t)}}\gamma(t) \mp \frac{m(t)}{\sqrt{m^2(t)+n^2(t)}}\nu(t).$$

PROOF. (1) Since $\{\gamma(t), \nu(t), \boldsymbol{\mu}(t)\}$ is an orthogonal base on $\mathbb{R}^3$, we have $\boldsymbol{\mu}(t) \cdot \boldsymbol{v} = 0$ if and only if there exist real numbers $a, b \in \mathbb{R}$ such that $\boldsymbol{v} = a\gamma(t) + b\nu(t) \in S^2$.

(2) Since $(d/dt)H(t, \boldsymbol{v}) = (-m(t)\gamma(t) - n(t)\nu(t)) \cdot \boldsymbol{v}$, we have $a^2 + b^2 = 1$ and $am(t) + bn(t) = 0$. It follows that

$$a = \pm\frac{n(t)}{\sqrt{m^2(t)+n^2(t)}}, \quad b = \mp\frac{m(t)}{\sqrt{m^2(t)+n^2(t)}}.$$

By a direct calculation, the converse holds. □

We can show that $H$ is a *Morse family*, in the sense of Legendrian singularity theory (cf. [**1**, **3**, **13**, **22**]), namely, $(H, \partial H/\partial t) : I \times S^2 \to \mathbb{R} \times \mathbb{R}$ is a submersion at $(t, \boldsymbol{v}) \in D(H)$, where

$$D(H) = \{(t, \boldsymbol{v}) \mid H(t, \boldsymbol{v}) = (\partial H/\partial t)(t, \boldsymbol{v}) = 0\}.$$

It follows that the evolute of the front $\mathcal{E}v(\gamma)$ is a (wave) front of a Legendre immersion and is given by the envelope of the parallel of the front. In fact, we have the following.

PROPOSITION 4.8. *Let $(\gamma,\nu): I \to \Delta \subset S^2 \times S^2$ be a Legendre immersion with the curvature $(m,n)$. Then $\mathcal{E}v(\gamma)$ is a front. More precisely, $(\mathcal{E}v(\gamma),\boldsymbol{\mu}): I \to \Delta \subset S^2 \times S^2$ is a Legendre immersion with the curvature*

$$m_{\mathcal{E}v}(t) = \frac{\dot{m}(t)n(t) - m(t)\dot{n}(t)}{m^2(t)+n^2(t)}, \ n_{\mathcal{E}v}(t) = \pm\sqrt{m^2(t)+n^2(t)}.$$

PROOF. Here we denote $(\gamma_{\mathcal{E}v},\nu_{\mathcal{E}v}) = (\mathcal{E}v(\gamma),\boldsymbol{\mu})$. By definition of the evolute of the front $\mathcal{E}v(t)$, $\gamma_{\mathcal{E}v}(t)\cdot\nu_{\mathcal{E}v}(t) = 0$ for all $t \in I$. Moreover, since

$$\dot{\gamma}_{\mathcal{E}v}(t) = \pm\frac{d}{dt}\left(\frac{n(t)}{\sqrt{m^2(t)+n^2(t)}}\right)\gamma(t) \mp \frac{d}{dt}\left(\frac{m(t)}{\sqrt{m^2(t)+n^2(t)}}\right)\nu(t),$$

we have $\dot{\gamma}_{\mathcal{E}v}(t)\cdot\nu_{\mathcal{E}v}(t) = 0$ for all $t \in I$. Hence $(\gamma_{\mathcal{E}v},\nu_{\mathcal{E}v}): I \to \Delta \subset S^2 \times S^2$ is a Legendre curve. By a direct calculation, we have $\dot{\nu}_{\mathcal{E}v}(t) = \dot{\boldsymbol{\mu}}(t) = -m(t)\gamma(t) - n(t)\nu(t)$ and

$$\boldsymbol{\mu}_{\mathcal{E}v}(t) = \gamma_{\mathcal{E}v}(t)\times\nu_{\mathcal{E}v}(t) = \mp\frac{m(t)}{\sqrt{m^2(t)+n^2(t)}}\gamma(t) \mp \frac{n(t)}{\sqrt{m^2(t)+n^2(t)}}\nu(t).$$

Then the curvature is given by

$$\begin{aligned} m_{\mathcal{E}v}(t) &= \dot{\gamma}_{\mathcal{E}v}(t)\cdot\boldsymbol{\mu}_{\mathcal{E}v}(t) \\ &= -\frac{m(t)}{\sqrt{m^2(t)+n^2(t)}}\frac{d}{dt}\left(\frac{n(t)}{\sqrt{m^2(t)+n^2(t)}}\right) \\ &\quad +\frac{n(t)}{\sqrt{m^2(t)+n^2(t)}}\frac{d}{dt}\left(\frac{m(t)}{\sqrt{m^2(t)+n^2(t)}}\right) \\ &= \frac{\dot{m}(t)n(t)-m(t)\dot{n}(t)}{m^2(t)+n^2(t)}, \\ n_{\mathcal{E}v}(t) &= \dot{\nu}_{\mathcal{E}v}(t)\cdot\boldsymbol{\mu}_{\mathcal{E}v}(t) \\ &= \pm\frac{m^2(t)}{\sqrt{m^2(t)+n^2(t)}} \pm \frac{n^2(t)}{\sqrt{m^2(t)+n^2(t)}} = \pm\sqrt{m^2(t)+n^2(t)}. \end{aligned}$$

It follows from $n_{\mathcal{E}v}(t) \neq 0$ for all $t \in I$ that $(\gamma_{\mathcal{E}v},\nu_{\mathcal{E}v})$ is a Legendre immersion. □

REMARK 4.9. The evolute of $\gamma$ and of $\nu$ are coincide by Corollary 4.6. It follows that the evolute of $\boldsymbol{\mu}$ is given by the second evolute of $\gamma$, see Theorem 4.11 below, that is, $\mathcal{E}v(\boldsymbol{\mu}) = \mathcal{E}v(\mathcal{E}v(\gamma))$ by Proposition 4.8

We denote a plane by $P(\boldsymbol{v},a) = \{\boldsymbol{x} \in \mathbb{R}^3 | \boldsymbol{x}\cdot\boldsymbol{v} = a\}$, where $\boldsymbol{v} \in S^2$ is a constant vector and $a \in \mathbb{R}$ is a constant.

PROPOSITION 4.10. *Let $(\gamma,\nu): I \to \Delta \subset S^2 \times S^2$ be a Legendre immersion with the curvature $(m,n)$. Then $\mathcal{E}v(\gamma)$ is constant if and only if there exist a vector $\boldsymbol{v} \in S^2$ and $a,b \in \mathbb{R}$ with $a^2+b^2 = 1$ such that $\gamma(t) \in P(\boldsymbol{v},a)\cap S^2$ and $\nu(t) \in P(\boldsymbol{v},b)\cap S^2$ for all $t \in I$.*

PROOF. By Proposition 4.8 and $\dot{\mathcal{E}v}(\gamma)(t) = 0$, we have $\dot{m}(t)n(t) - m(t)\dot{n}(t) = 0$ and $m^2(t)+n^2(t) \neq 0$ for all $t \in I$. Then $m$ and $n$ are linearly dependent, that is, there exist $a,b \in \mathbb{R}$ with $a^2+b^2 = 1$ such that $am(t)+bn(t) = 0$ for all $t \in I$. By the Frenet Serret formula, we have $a\dot{\gamma}(t)+b\dot{\nu}(t) = 0$ for all $t \in I$. There exists a constant vector $\boldsymbol{v} \in S^2$ such that $a\gamma(t)+b\nu(t) = \boldsymbol{v}$. Therefore, we have $\gamma(t) \in P(\boldsymbol{v},a)\cap S^2$ and $\nu(t) \in P(\boldsymbol{v},b)\cap S^2$ for all $t \in I$.

Conversely, if $\gamma(t)\cdot \boldsymbol{v}=a$ and $\nu(t)\cdot\boldsymbol{v}=b$ for all $t\in I$, then $m(t)\boldsymbol{\mu}(t)\cdot\boldsymbol{v}=0$ and $n(t)\boldsymbol{\mu}(t)\cdot\boldsymbol{v}=0$. It follows that $\boldsymbol{\mu}(t)\cdot\boldsymbol{v}=0$. Since $\{\gamma(t),\nu(t),\boldsymbol{\mu}(t)\}$ is an orthogonal basis on $\mathbb{R}^3$, we can denote $\boldsymbol{v}=a\gamma(t)+b\nu(t)$. By differentiate, we have $am(t)+bn(t)=0$ and $a\dot{m}(t)+b\dot{n}(t)=0$. Since $a^2+b^2=1$, we have $\dot{m}(t)n(t)-m(t)\dot{n}(t)=0$ for all $t\in I$. It follows that $\mathcal{E}v(\gamma)$ is constant. $\square$

Let $(\gamma,\nu):I\to\Delta\subset S^2\times S^2$ be a Legendre immersion with the curvature $(m,n)$. We give the form of the $k$-th evolute of the front, where $k$ is a natural number. We denote

$$\mathcal{E}v^0(\gamma)(t)=\gamma(t),\ \nu_0(t)=\nu(t),\ \boldsymbol{\mu}_0(t)=\boldsymbol{\mu}(t),\ m_0=m(t),\ n_0=n(t),$$

for convenience. We define

$$\mathcal{E}v^k(\gamma)(t)=\mathcal{E}v(\mathcal{E}v^{k-1}(\gamma))(t),\ \nu_k(t)=\boldsymbol{\mu}_{k-1}(t),\ \boldsymbol{\mu}_k(t)=\mathcal{E}v^k(\gamma)(t)\times\nu_k(t),$$

$$m_k(t)=\frac{\dot{m}_{k-1}(t)n_{k-1}(t)-m_{k-1}(t)\dot{n}_{k-1}(t)}{m_{k-1}^2(t)+n_{k-1}^2(t)},\ n_k(t)=\pm\sqrt{m_{k-1}^2(t)+n_{k-1}^2(t)},$$

inductively. Then we have the following theorem.

THEOREM 4.11. *Let $(\gamma,\nu):I\to\Delta\subset S^2\times S^2$ be a Legendre immersion with the curvature $(m,n)$. Then $\mathcal{E}v^k(\gamma)$ is a front. More precisely, $(\mathcal{E}v^k(\gamma),\nu_k):I\to\Delta\subset S^2\times S^2$ is a Legendre immersion with the curvature $(m_k,n_k)$, where*

$$\mathcal{E}v^k(\gamma)(t)=\pm\frac{n_{k-1}(t)}{\sqrt{m_{k-1}^2(t)+n_{k-1}^2(t)}}\mathcal{E}v^{k-1}(\gamma)(t)\mp\frac{m_{k-1}(t)}{\sqrt{m_{k-1}^2(t)+n_{k-1}^2(t)}}\nu_{k-1}(t).$$

PROOF. By Proposition 4.8, the case of $k=1$ holds.

Suppose that the case of $k$ holds. We consider $\mathcal{E}v(\mathcal{E}v^k(\gamma))$. By the assumption, $(\mathcal{E}v^k(\gamma),\nu_k)$ is a Legendre immersion with the curvature $(m_k,n_k)$. By Proposition 4.8, the $(k+1)$-th evolute of the front is given by

$$\mathcal{E}v^{k+1}(\gamma)(t)=\pm\frac{n_k(t)}{\sqrt{m_k^2(t)+n_k^2(t)}}\mathcal{E}v^k(\gamma)(t)\mp\frac{m_k(t)}{\sqrt{m_k^2(t)+n_k^2(t)}}\nu_k(t).$$

Since $(d/dt)\mathcal{E}v^{k+1}(\gamma)(t)$

$$\begin{aligned}
&= \pm\frac{d}{dt}\left(\frac{n_k(t)}{\sqrt{m_k^2(t)+n_k^2(t)}}\right)\mathcal{E}v^k(\gamma)(t)\pm\frac{n_k(t)}{\sqrt{m_k^2(t)+n_k^2(t)}}m_k(t)\boldsymbol{\mu}_k(t)\\
&\quad\mp\frac{d}{dt}\left(\frac{m_k(t)}{\sqrt{m_k^2(t)+n_k^2(t)}}\right)\nu_k(t)\mp\frac{m_k(t)}{\sqrt{m_k^2(t)+n_k^2(t)}}n_k(t)\boldsymbol{\mu}_k(t)\\
&= \pm\frac{d}{dt}\left(\frac{n_k(t)}{\sqrt{m_k^2(t)+n_k^2(t)}}\right)\mathcal{E}v^k(\gamma)(t)\mp\frac{d}{dt}\left(\frac{m_k(t)}{\sqrt{m_k^2(t)+n_k^2(t)}}\right)\nu_k(t),
\end{aligned}$$

and $\nu_{k+1}(t)=\boldsymbol{\mu}_k(t)=\mathcal{E}v^k(\gamma)(t)\times\nu_k(t)$, we have $\mathcal{E}v^{k+1}(\gamma)(t)\cdot\nu_{k+1}(t)=0$ and $\dot{\mathcal{E}v}^{k+1}(\gamma)(t)\cdot\nu_{k+1}(t)=0$. It follows that $(\mathcal{E}v^{k+1}(\gamma),\nu_{k+1}):I\to\Delta\subset S^2\times S^2$ is a

Legendre curve. By definition, we have $\boldsymbol{\mu}_{k+1}(t) = \mathcal{E}v^{k+1}(\gamma)(t) \times \nu_{k+1}(t)$ and

$$\begin{aligned} m_{k+1}(t) &= \frac{d}{dt}\mathcal{E}v^{k+1}(\gamma)(t) \cdot \boldsymbol{\mu}_{k+1}(t) \\ &= -\frac{d}{dt}\left(\frac{n_k(t)}{\sqrt{m_k^2(t)+n_k^2(t)}}\right)\frac{m_k(t)}{\sqrt{m_k^2(t)+n_k^2(t)}}\mathcal{E}v^k(\gamma)(t)\cdot(\nu_k(t)\times\boldsymbol{\mu}_k(t)) \\ &\quad -\frac{d}{dt}\left(\frac{m_k(t)}{\sqrt{m_k^2(t)+n_k^2(t)}}\right)\frac{n_k(t)}{\sqrt{m_k^2(t)+n_k^2(t)}}\nu_k(t)\cdot(\mathcal{E}v^k(\gamma)(t)\times\boldsymbol{\mu}_k(t)) \\ &= \frac{\dot{m}_k(t)n_k(t)-m_k(t)\dot{n}_k(t)}{m_k^2(t)+n_k^2(t)}. \end{aligned}$$

Moreover, since $\dot{\nu}_{k+1}(t) = \dot{\boldsymbol{\mu}}_k(t) = -m_k(t)\mathcal{E}v^k(\gamma)(t) - n_k(t)\nu_k(t)$, we have

$$\begin{aligned} n_{k+1}(t) &= \dot{\nu}_{k+1}(t)\cdot\boldsymbol{\mu}_{k+1}(t) \\ &= \pm\frac{m_k^2(t)}{\sqrt{m_k^2(t)+n_k^2(t)}}\mathcal{E}v^k(\gamma)(t)\cdot(\nu_k(t)\times\boldsymbol{\mu}_k(t)) \\ &\quad \mp\frac{n_k^2(t)}{\sqrt{m_k^2(t)+n_k^2(t)}}\nu_k(t)\cdot(\mathcal{E}v^k(\gamma)(t)\times\boldsymbol{\mu}_k(t)) \\ &= \pm\sqrt{m_k^2(t)+n_k^2(t)}. \end{aligned}$$

It follows from $n_{k+1}(t) \neq 0$ for all $t \in I$ that $(\mathcal{E}v^{k+1}(\gamma), \nu_{k+1})$ is also a Legendre immersion with the curvature $(m_{k+1}, n_{k+1})$. This completes the proof of Theorem. □

## 5. Evolutes of frontals in the sphere

Let $(\gamma, \nu) : I \to \Delta \subset S^2 \times S^2$ be a Legendre curve with the curvature $(m, n)$. We define an evolute of the frontal as follows.

DEFINITION 5.1. The *evolute* $\mathcal{E}v(\gamma) : I \to S^2$ of the frontal $\gamma$ is given by

$$\mathcal{E}v(\gamma)(t) = \pm p(t)\gamma(t) \pm q(t)\nu(t),$$

if there exists a smooth mapping $(p, q) : I \to S^1$ such that

$$m(t)p(t) + n(t)q(t) = 0 \tag{5.1}$$

for all $t \in I$. In this case, we say that *the evolute* $\mathcal{E}v(\gamma)$ exists.

REMARK 5.2. If $m(t) = n(t) = 0$ for all $t \in I$, that is, $\gamma(t)$ and $\nu(t)$ are constant vectors in $S^2$, then for any smooth mapping $(p, q) : I \to S^1$ satisfies the condition $m(t)p(t) + n(t)q(t) = 0$. Then the evolute exists but does not unique.

The uniqueness condition is well-known as a topological condition.

LEMMA 5.3. *Suppose that there exists a continuous mapping $(p, q) : I \to S^1$ such that $p(t) = n(t)/\sqrt{m^2(t)+n^2(t)}$ and $q(t) = -m(t)/\sqrt{m^2(t)+n^2(t)}$ on $X = \{t \in I \mid m^2(t)+n^2(t) \neq 0\}$. Then the mapping $(p, q)$ is a unique if and only if $X$ is a dense subset of $I$.*

Let $(\gamma, \nu) : I \to \Delta \subset S^2 \times S^2$ be a Legendre curve with the curvature $(m, n)$. In this section, we assume that $X = \{t \in I \mid m^2(t) + n^2(t) \neq 0\}$ is a dense subset of $I$, that is, the set of regular points of the Legendre curve $(\gamma, \nu)$ is a dense subset of

$I$. This condition follows that if such a smooth mapping $(p,q): I \to S^2$ exists, then the uniqueness condition is satisfied by Lemma 5.3. Note that if the singular points $(\gamma,\nu)$ are isolated, then the condition that $X$ is a dense subset of $I$ is satisfied.

The existence condition of the evolute of a frontal is as follows. It is a quit different property between the evolute of a frontal in the sphere and in the Euclidean plane (cf. [**8**]).

PROPOSITION 5.4. *Let $(\gamma,\nu): I \to \Delta \subset S^2 \times S^2$ be a Legendre curve with the curvature $(m,n)$. If $m(t)$ or $n(t)$ does not belong to $\mathfrak{m}_1^\infty$ around $t_0$, then the evolute of the frontal $\mathcal{E}v(\gamma)$ exists around $t_0$.*

PROOF. Suppose that $m(t) \notin \mathfrak{m}_1^\infty$ around $t_0$. There exists a smooth function $\lambda: (I,t_0) \to \mathbb{R}$ such that $n(t) = \lambda(t)m(t)$ around $t_0$. We put

$$p(t) = \frac{\lambda(t)}{\sqrt{\lambda(t)^2+1}}, \quad q(t) = -\frac{1}{\sqrt{\lambda(t)^2+1}}.$$

Then the condition $m(t)p(t)+n(t)q(t)=0$ holds around $t_0$. Therefore the evolute of the frontal $\mathcal{E}v(\gamma)$ exists around $t_0$. By the similar arguments, we can prove the case of $n(t) \notin \mathfrak{m}_1^\infty$ around $t_0$. □

PROPOSITION 5.5. *Let $(\gamma,\nu): I \to \Delta \subset S^2 \times S^2$ be a Legendre curve with the curvature $(m,n)$. If the evolute $\mathcal{E}v(\gamma)$ of the frontal exists with $(p,q): I \to S^1$ satisfies* (5.1), *then the evolute $\mathcal{E}v(\gamma)$ is also a frontal. More precisely, $(\mathcal{E}v(\gamma),\boldsymbol{\mu}):$ $I \to \Delta \subset S^2 \times S^2$ is a Legendre curve with the curvature*

$$m_{\mathcal{E}v}(t) = \dot{p}(t)q(t) - p(t)\dot{q}(t), \quad n_{\mathcal{E}v}(t) = \mp m(t)q(t) \pm n(t)p(t).$$

PROOF. By the Frenet Serret type formula (Proposition 2.2), we have

$$\begin{aligned}
\dot{\mathcal{E}v}(\gamma)(t) &= \pm\dot{p}(t)\gamma(t) \pm p(t)\dot{\gamma}(t) \pm \dot{q}(t)\nu(t) \pm q(t)\dot{\nu}(t) \\
&= \pm\dot{p}(t)\gamma(t) \pm \dot{q}(t)\nu(t) \pm (m(t)p(t)+n(t)q(t))\boldsymbol{\mu}(t) \\
&= \pm\dot{p}(t)\gamma(t) \pm \dot{q}(t)\nu(t).
\end{aligned}$$

By definition, $\boldsymbol{\mu}(t) = \gamma(t) \times \nu(t)$. Then $(\mathcal{E}v(\gamma),\boldsymbol{\mu})$ is a Legendre curve. We denote $\boldsymbol{\mu}_{\mathcal{E}v}(t) = \mathcal{E}v(\gamma)(t) \times \boldsymbol{\mu}(t) = \pm q(t)\gamma(t) \mp p(t)\nu(t)$. Thus, the curvature is given by

$$\begin{aligned}
m_{\mathcal{E}v}(t) &= \dot{\mathcal{E}v}(\gamma)(t) \cdot \boldsymbol{\mu}_{\mathcal{E}v}(t) = \dot{p}(t)q(t) - p(t)\dot{q}(t), \\
n_{\mathcal{E}v}(t) &= \dot{\boldsymbol{\mu}}(t) \cdot \boldsymbol{\mu}_{\mathcal{E}v}(t) = \mp m(t)q(t) \pm n(t)p(t).
\end{aligned}$$

□

REMARK 5.6. By Proposition 5.5, if $n_{\mathcal{E}v}(t) = 0$, then we have $m(t) = n(t) = 0$. Hence if the set of regular points of the Legendre curve $(\gamma,\nu)$ is a dense subset of $I$, then the set of regular points of $(\mathcal{E}v(\gamma),\boldsymbol{\mu})$ is also a dense subset of $I$. By Proposition 5.4, if $m_{\mathcal{E}v}(t)$ or $n_{\mathcal{E}v}(t)$ dose not belong to $\mathfrak{m}_1^\infty$, then there exists unique the second evolute $\mathcal{E}v^2(\gamma)$ of the Legendre curve $(\gamma,\nu)$ at least locally.

## 6. Examples

We give examples of the evolutes of fronts and frontals.

EXAMPLE 6.1. (Spherical nephroid) Let $(\gamma,\nu):[0,2\pi)\to\Delta\subset S^2\times S^2$ be

$$\begin{aligned}\gamma(t) &= \left(\frac{3}{4}\cos t-\frac{1}{4}\cos 3t, \frac{3}{4}\sin t-\frac{1}{4}\sin 3t, \frac{\sqrt{3}}{2}\cos t\right),\\ \nu(t) &= \left(\frac{3}{4}\sin t-\frac{1}{4}\sin 3t, -\frac{3}{4}\cos t-\frac{1}{4}\cos 3t, -\frac{\sqrt{3}}{2}\sin t\right).\end{aligned}$$

Since

$$\dot\gamma(t)=\left(-\frac{3}{4}\sin t+\frac{3}{4}\sin 3t, \frac{3}{4}\cos t-\frac{3}{4}\cos 3t, -\frac{\sqrt{3}}{2}\sin t\right),$$

we have $\gamma(t)\cdot\nu(t)=0$ and $\dot\gamma(t)\cdot\nu(t)=0$. Hence $(\gamma,\nu):[0,2\pi)\to\Delta\subset S^2\times S^2$ is a Legendre curve. By definition, we have

$$\boldsymbol{\mu}(t)=\left(\frac{\sqrt{3}}{2}\cos 2t, \frac{\sqrt{3}}{2}\sin 2t, -\frac{1}{2}\right),$$

and the curvature $(m(t),n(t))=(\sqrt{3}\sin t,\sqrt{3}\cos t)$. It follows that $(\gamma,\nu)$ is a Legendre immersion. The evolute of the front is given by

$$\begin{aligned}\mathcal{E}v(\gamma)(t) &= \pm\frac{n(t)}{\sqrt{m^2(t)+n^2(t)}}\gamma(t)\mp\frac{m(t)}{\sqrt{m^2(t)+n^2(t)}}\nu(t)\\ &= \pm\cos t\gamma(t)\mp\sin t\nu(t)\\ &= \pm\left(\frac{1}{2}\cos 2t, \frac{1}{2}\sin 2t, \frac{\sqrt{3}}{2}\right)\end{aligned}$$

and the curvature of $(\mathcal{E}v(\gamma),\boldsymbol{\mu})$ is $(m_{\mathcal{E}v}(t),n_{\mathcal{E}v}(t))=(1,\pm\sqrt{3})$ by Proposition 4.8. Then the second evolute of the front is given by

$$\mathcal{E}v^2(\gamma)(t)=\pm\frac{n_{\mathcal{E}v}(t)}{\sqrt{m_{\mathcal{E}v}^2(t)+n_{\mathcal{E}v}^2(t)}}\mathcal{E}v(t)\mp\frac{m_{\mathcal{E}v}(t)}{\sqrt{m_{\mathcal{E}v}^2(t)+n_{\mathcal{E}v}^2(t)}}\boldsymbol{\mu}(t)=\pm(0,0,1).$$

EXAMPLE 6.2. Let $n,m$ and $k$ be natural numbers with $m=k+n$. Consider a Legendre curve $(\gamma,\nu):\mathbb{R}\to\Delta\subset S^2\times S^2$ by

$$\gamma(t)=\frac{1}{\sqrt{1+t^{2n}+t^{2m}}}(1,t^n,t^m),\ \nu(t)=\frac{1}{\sqrt{n^2+m^2t^{2k}+k^2t^{2m}}}(kt^m,-mt^k,n),$$

see Example 2.8. Then the curvature is given by

$$m(t)=\frac{-t^{n-1}\sqrt{n^2+m^2t^{2k}+k^2t^{2m}}}{1+t^{2n}+t^{2m}},\ n(t)=\frac{knmt^{k-1}\sqrt{1+t^{2n}+t^{2m}}}{n^2+m^2t^{2k}+k^2t^{2m}}.$$

Note that $(\gamma,\nu)$ is a Legendre immersion when $k=1$ or $n=1$. We put $1<n\le k=n+r$ for some natural number $r$. Then the evolute of the frontal is given by

$$\mathcal{E}v(\gamma)(t)=\pm p(t)\gamma(t)\pm q(t)\nu(t),$$

where

$$p(t) = \frac{(1+t^{2n}+t^{2m})^{\frac{3}{2}} knmt^r}{\sqrt{(1+t^{2n}+t^{2m})^3(knm)^2t^{2r}+(n^2+m^2t^{2k}+k^2t^{2m})^3}},$$

$$q(t) = \frac{(n^2+m^2t^{2k}+k^2t^{2m})^{\frac{3}{2}}}{\sqrt{(1+t^{2n}+t^{2m})^3(knm)^2t^{2r}+(n^2+m^2t^{2k}+k^2t^{2m})^3}}.$$

For example, when $n=2, m=5, k=3$ and $r=1$, then we have

$$\gamma(t) = \frac{1}{\sqrt{1+t^4+t^{10}}}(1,t^2,t^5),\ \nu(t) = \frac{1}{\sqrt{4+25t^6+9t^{10}}}(3t^5,-5t^3,2)$$

and

$$p(t) = \frac{30t(1+t^4+t^{10})^{\frac{3}{2}}}{\sqrt{30^2t^2(1+t^4+t^{10})^3+(4+25t^6+9t^{10})^3}},$$

$$q(t) = \frac{(4+25t^6+9t^{10})^{\frac{3}{2}}}{\sqrt{30^2t^2(1+t^4+t^{10})^3+(4+25t^6+9t^{10})^3}}.$$

Then the evolute of the frontal is given by

$$\mathcal{E}v(\gamma)(t) = \pm\frac{30t(1+t^4+t^{10})(1,t^2,t^5)+(4+25t^6+9t^{10})(3t^5-5t^3,2)}{\sqrt{30^2t^2(1+t^4+t^{10})^3+(4+25t^6+9t^{10})^3}}.$$

For a smooth curve $\gamma$ on $S^2$, we say that $t$ is an ordinary rhamphoid cusp if $\dot\gamma(t)=0, \ddot\gamma(t)\neq 0, \gamma^{(3)}(t)=3\lambda\ddot\gamma(t)$ for some $\lambda\in\mathbb{R}$, and $\gamma^{(5)}(t)$ is linearly independent of $\ddot\gamma(t)$ and $\gamma^{(4)}(t)$ (cf. [**16**, **17**]). Therefore, this is an example that 0 is an ordinary rhamphoid cusp of $\gamma$, but 0 is not of $\nu$ (cf. Proposition 2 in [**16**] page 219).

## References


[1] V. I. Arnol'd, *Singularities of Caustics and Wave Fronts*, Mathematics and Its Applications **62** Kluwer Academic Publishers, 1990.

[2] V. I. Arnol'd, *The geometry of spherical curves and quaternion algebra*, Russian Math. Surveys, **50** (1995), 1–68

[3] V. I. Arnol'd, S. M. Gusein-Zade and A. N. Varchenko, *Singularities of Differentiable Maps vol.* I, Birkhäuser, 1986.

[4] R. L. Bishop, *There is more than one way to frame a curve*. Amer. Math. Monthly, **82** (1975), 246–251.

[5] J. W. Bruce and P. J. Giblin, *Curves and singularities. A geometrical introduction to singularity theory. Second edition*, Cambridge University Press, Cambridge, 1992.

[6] T. Fukunaga and M. Takahashi, *Existence and uniqueness for Legendre curves*, J. Geometry. **104** (2013), 297–307.

[7] T. Fukunaga and M. Takahashi, *Evolutes of fronts in the Euclidean plane*, J. Singularity. **10** (2014), 92–107.

[8] T. Fukunaga and M. Takahashi, *Evolutes and involutes of frontals in the Euclidean plane*, Demonstratio Math. **48** (2015), 147–166.

[9] C. G. Gibson, *Singular points of smooth mappings*, Research Notes in Mathematics, 25, Pitman (Advanced Publishing Program), Boston, Mass.-London, 1979.

[10] C. G. Gibson, *Elementary geometry of differentiable curves. An undergraduate introduction*, Cambridge University Press, Cambridge, 2001.

[11] A. Gray, E. Abbena, and S. Salamon, *Modern differential geometry of curves and surfaces with Mathematica. Third edition*, Studies in Advanced Mathematics. Chapman and Hall/CRC, Boca Raton, FL, 2006.

[12] S. Honda and M. Takahashi, *Framed curves in the Euclidean space*, to appear in Advances geometry (2015).

[13] T. Nagai, *The Gauss map of a hypersurface in Euclidean sphere and the spherical Legendrian duality.* Topology Appl. **159** (2012), 545–554.
[14] T. Nishimura, *Normal forms for singularities of pedal curves produced by non-singular dual curve germs in $S^n$.* Geom. Dedicata. **133** (2008), 59–66.
[15] T. Nishimura, *Singularities of pedal curves produced by singular dual curve germs in $S^n$.* Demonstratio Math. **43** (2010), 447–459.
[16] I. R. Porteous, *Some remarks on duality in $S^3$*, Geometry and topology of caustics-CAUSTICS '98, Banach Center Publ., **50** (1999), 217–226.
[17] I. R. Porteous, *Geometric differentiation. For the intelligence of curves and surfaces. Second edition.* Cambridge University Press, Cambridge, 2001.
[18] M. C. Romero Fuster, *Stereographic projections and geometric singularities.* Workshop on Real and Complex Singularities (SÃo Carlos, 1996). Mat. Contemp. **12** (1997), 167–182.
[19] R. Uribe-Vargas, *On polar duality, Lagrange and Legendre singularities and stereographic projection to quadrics.* Proc. London Math. Soc. **87** (2003), 701–724.
[20] R. Uribe-Vargas, *Rigid body motions and Arnol'd's theory of fronts on $S^2 \subset \mathbb{R}^3$*, J. Geom. Phys. **45** (2003), 91–104.
[21] R. Uribe-Vargas, *Theory of fronts on the 2-sphere and the theory of space curves*, J. Math. Sci. (N.Y.) **126** (2005), 1344–1353.
[22] V. M. Zakalyukin, *Reconstructions of fronts and caustics depending on a parameter and versality of mappings.* J. Soviet Math. **27** (1983), 2713–2735.

MURORAN INSTITUTE OF TECHNOLOGY, MURORAN 050-8585, JAPAN
*E-mail address*: masatomo@mmm.muroran-it.ac.jp